Antioxidant Activity of Honey Bee Products

Antioxidant Activity of Honey Bee Products

Guest Editors

Josipa Vlainić
Ivana Tlak Gajger

MDPI

Basel • Beijing • Wuhan • Barcelona • Belgrade • Novi Sad • Cluj • Manchester

Guest Editors
Josipa Vlainić
Department of Molecular Medicine
Institute Ruer Bošković
Zagreb
Croatia

Ivana Tlak Gajger
Department for Biology and Pathology of Fish and Bees
Faculty of Veterinary Medicine
University of Zagreb
Zagreb
Croatia

Editorial Office
MDPI AG
Grosspeteranlage 5
4052 Basel, Switzerland

This is a reprint of the Special Issue, published open access by the journal *Antioxidants* (ISSN 2076-3921), freely accessible at: www.mdpi.com/journal/antioxidants/special_issues/Honey_Bee_Products.

For citation purposes, cite each article independently as indicated on the article page online and using the guide below:

Lastname, A.A.; Lastname, B.B. Article Title. *Journal Name* **Year**, *Volume Number*, Page Range.

ISBN 978-3-7258-3208-8 (Hbk)
ISBN 978-3-7258-3207-1 (PDF)
https://doi.org/10.3390/books978-3-7258-3207-1

Contents

About the Editors

Josipa Vlainić

Josipa Vlainić graduated from the Faculty of Veterinary Medicine, University of Zagreb in 2002. That same year, she was employed at the Laboratory for Neuropharmacology, Rudjer Boskovic Institute as assistant in a project: deciphering molecular mechanisms undelying tolerance and dependence on GABAergic drugs. She obtained her PhD degree in 2009 from the Faculty of Veterinary Medicine University of Zagreb. Within IRB, she moved to the Laboratory of Advance Genomics and, recently, to the Laboratory of Oxidative Stress. Her research includes the antimicrobial effectiveness-testing of naturally occuring products as well as their antioxidant activity. She is the author of several scientific publications in international journals.

Ivana Tlak Gajger

Ivana Tlak Gajger is an accomplished professor at the Faculty of Veterinary Medicine, University of Zagreb. She graduated in 2005, and in 2010, she completed her Ph.D. dissertation with the title: Establishing system of transposon mutagenesis for bacteria *Paenibacillus larvae*. Since 2006, she has been employed at the Department for Biology and Pathology of Fish and Bees, Faculty of Veterinary Medicine, where she is still working as a Full Professor. Besides activities in scientific and research work, she is active as a lecturer in the field of biology and pathology of beneficial insects. She was the supervisor of 47 graduate and 7 doctoral students. She is the Head of the postgraduate master study program Honeybee Health Protection and the establisher of the Educational-archive station for beekeeping. Also, she is Head of accredited (according to HRN EN ISO/IEC 17025), official and Croatian National Reference Laboratory for Honeybee Diseases – APISlab.

She works in close collaboration with beekeepers and other stakeholders through various science-based activities to identify opportunities that can be made to simultaneously promote beekeeping, healthy food production, and natural biodiversity and improve the implementation of honeybee health protection measures. Her goal is to find smart solutions for real beekeepers' problems that ultimately inform policymakers, promote education and good apiary–veterinary–environmental practices, and help primary producers.

As the greatest recognition of her work, she received the Award of the Croatian Academy of Sciences and Arts for the highest scientific achievements in the Republic of Croatia in the field of medical sciences. In 2023, she was elected as a full member of the Collegium of Veterinary Sciences of the Croatian Academy of Medical Sciences. According to research by Stanford University in the USA, she is on the list of the 2% most influential scientists in 2022 and 2023.

antioxidants

Editorial

Antioxidant Activity of Honey Bee Products

Ivana Tlak Gajger [1,*] and Josipa Vlainić [2,*]

1 NRL for Honeybee Diseases, Faculty of Veterinary Medicine, University of Zagreb, Heinzelova 55, 10000 Zagreb, Croatia
2 Institute Ruđer Bošković, Bijenička Cesta 54, 10000 Zagreb, Croatia
* Correspondence: itlak@vef.unizg.hr (I.T.G.); josipa.vlainic@irb.hr (J.V.); Tel.: +385-91-2390-041 (I.T.G.)

Antioxidants have gained significant importance in modern nutrition. They can be sourced from various natural products, including those from bees [1]. Honey bee products are royal jelly, venom, beeswax, honey, propolis, pollen, and fermented pollen (bee bread). These substances are known for their antimicrobial, anti-inflammatory, antitumor, antioxidant properties, and different biological activities [2]. Throughout history, they have been used for various apitherapy purposes and are highly valued for their medicinal, cosmetic, and nutritional benefits [3,4]. These products are rich in antioxidants, which play a crucial role in combating oxidative stress linked to chronic diseases like cancer and cardiovascular conditions [5]. This Special Issue, entitled "Antioxidant Activity of Honey Bee Products", focuses on the antioxidant properties of these products and their associated health benefits.

Honey is especially noteworthy for its high levels of phenolic compounds and flavonoids [6–8], all of which exhibit significant antioxidant activity. Propolis is similarly rich in flavonoids and phenolic acids, which contribute to its powerful antioxidant and anti-inflammatory properties [9,10]. Royal jelly, recognized for its protein-rich composition, contains several phenolic compounds that also demonstrate antioxidant effects [11]. While beeswax is not as commonly researched, it does show some mild antioxidant activity as well [12]. Also, pollen and bee venom are promising natural sources of antioxidants that may help prevent diseases related to oxidative stress, although more research is needed to explore their mechanisms and potential therapeutic applications [13–15]. The beneficial effects of honey bee products have been widely recognized in traditional medicine, and they show potential for contemporary therapeutic uses. Additionally, these products are often linked to health benefits and are considered valuable natural foods.

The primary focus of this Special Issue was to summarize previous findings, present new research, and align this information with current global health needs as remedies and agents that promote health, reducing the risk of various diseases. In particular, oxidative stress is a key factor in many modern diseases, and honey bee products may provide supportive benefits due to their antioxidant capacity. We aimed to discuss the bioaccessibility and bioavailability of these complex natural products, which can vary based on their composition and geographical origin. To enable accurate comparisons, it is essential to standardize their quality parameters and ensure that the same methods and measurement units are used across different matrices of bee products [16,17]. This Special Issue brings together three reviews and twelve original articles that explore recent research focused on enhancing the potential functionalities of various bioactive compounds, such as phytochemicals, found in bee products. These studies provide insights into the effectiveness of these compounds as natural alternatives for managing health conditions related to oxidative stress.

check for updates

Received: 30 December 2024
Accepted: 7 January 2025
Published: 8 January 2025

Citation: Tlak Gajger, I.; Vlainić, J. Antioxidant Activity of Honey Bee Products. *Antioxidants* **2025**, *14*, 64. https://doi.org/10.3390/antiox14010064

Bee pollen is increasingly recognized as a functional food because of its high-quality composition and therapeutic potential in medical and culinary applications. A review by El Ghouizi et al. (Contribution 1) highlights important details about the composition of bee pollen, particularly its phenolic compounds, as well as its biological properties and molecular pathways. Although its diverse composition provides various pharmacological benefits, the significant variability in its content poses a challenge to its use in phytomedicine. On this basis, more attention should be paid to standardization, including phenolic composition and nutritional value of different plant-origin pollen; to controls in the frame of production beekeeping practices; to more pharmacological and biochemical examinations; to enhance the bioavailability of bee pollen bioactive compounds; to the bio- and techno-functional value of bee pollen as food, such as novel bee pollen-enriched food products or dietary supplements; and to more clinical trials to investigate the beneficial effect of pollen as functional food on human and animal health.

Martiniakova et al. (Contribution 2) summarized the current animal and human studies regarding using honey as a potential therapeutic agent for osteoporosis and breast cancer which are posing significant socioeconomic challenges. Preclinical in vitro studies indicate that honey has a beneficial impact on bone health and breast tissue health where honey has a significant impact on the microstructure and strength of bones, as well as on oxidative stress. It also influences breast cancer by affecting cell proliferation, apoptosis, tumor growth rate, and volume. Clinical studies focused on breast cancer have shown that honey can effectively increase blood cell counts, interleukin-3 levels, and overall quality of patient life. These findings suggest that honey may serve as a promising therapeutic supplement for promoting the health of bone and breast tissue.

Honeybee products offer known pharmacological and health benefits, particularly concerning periodontal disorders, which are caused by dental biofilm and an inflammatory response to bacterial overgrowth resulting from dysbiosis in the oral microbiome. Choudhary et al. (Contribution 3) reviewed the potential role of these bee products in preventing oral diseases, highlighting their diverse biologically active compounds, including flavonoids, phenolic acids, and terpenoids. These findings suggest that bee products could serve as a therapeutic option for individuals suffering from various oral disorders.

Bae et al. (Contribution 4) propose that properly purified Korean-origin propolis could be used as a cosmetic ingredient that helps mitigate human skin toxicity caused by air pollutants. This effect is attributed to propolis hydrophilic and lipophilic fractions and several key bioactive components that influence the viability and oxidative stress of HaCaT epidermal keratinocytes exposed to particulates with a diameter smaller than 10 μm.

Honeybee products were assessed for their potential to protect against nephrotoxicity induced by doxorubicin. The combined treatment of honey, royal jelly, and propolis significantly improved biochemical, histological, and immunohistochemical parameters in rat renal tissue, leading to a notable enhancement in the expression of the PARP-1 and Bcl-2 genes (Contribution 5).

Moskwa et al. (Contribution 6) compared the chemical composition, total phenolic content, and concentration of toxic elements in Polish propolis extracts and New Zealand Manuka propolis extracts and evaluated their anticancer potential against diffuse astrocytoma derived from patient cells and glioblastoma cell lines (T98G, LN-18). Both propolis extracts showed antioxidant capacity and exhibited similar activities, demonstrating promising anti-glioma potential for in vitro experimental conditions. To support the authenticity of Polish honeybee varieties, phenolic acids were analyzed as indicators. Specifically, syringic acid, vanillic acid, and caffeic acid are characteristic of linden honey, while p-coumaric acid and 4-hydroxybenzoic acid are associated with buckwheat honey and vanillic acid is notable in honeydew honey. Of these, buckwheat honey has the highest median total phenolic content, indicating a rich concentration of phenolic compounds; thus, it is recommended for enriching human diets with its antioxidant ingredients (Contribution 7).

Kelulut honey has excellent antioxidative and anti-inflammatory properties and unique physicochemical characteristics. It is being studied for its isolated and combined effects with metformin or clomiphene in addressing oxidative stress and reproductive and metabolic abnormalities associated with polycystic ovary syndrome. The results indicate that this combination can improve oxidative stress, hormonal profiles, and the estrous cycle in rats, suggesting a potential complementary treatment option for women with this condition (Contribution 8).

Montaser et al. (Contribution 9) conducted chromatography on the HC2 fraction and found that the secondary metabolites in citrus honey and marjoram honey, specifically hesperetin, linalool, and caffeic acid, are responsible for increasing antioxidant activities in comparison with clover honey.

The color indexes of 39 propolis samples from various locations in Turkiye were determined for the first time using the Lovibond Tintometer. This study also examined the relationship between the color index, total phenolic content, and the cytotoxic and antioxidant activities of the propolis samples, including two commercial ones. The research highlighted how these samples can be characterized by their color indices, chemical contents, and potential activities, such as antioxidant, antiviral, and cytotoxic properties. These findings suggest that propolis could be useful in various fields, ranging from medicine to cosmetics (Contribution 10).

Zakaria et al. (Contribution 11) investigated the therapeutic effects of stingless bee (*Heterotrigona itama*) bee bread (fermented pollen) from Malaysia on obesity-related disorders in hepatic lipid metabolism. They focused on its Keap1/Nrf2 pathway regulation and proposed that it could serve as a natural supplement for treating obesity-related fatty liver disease.

Dimitriu et al. (Contribution 12) reported that honey enriched with polyphenols from raspberry extracts maintains the characteristic properties of honey while enhancing the synergistic antioxidative activity between honey and the polyphenols. Although a honey-biomimetic natural deep eutectic solvent, which has similar properties to honey, demonstrated comparable antioxidant activity when mixed with polyphenols, honey appears to possess additional qualities that further enhance synergism and reduce antagonism.

The total phenolic content, antioxidant activity, and phenolic compounds of honey from Western Australia were measured, and the results were ranked as follows: *Calothamnus* spp. (Red Bell) had the highest levels, followed by *Eucalyptus marginata* (Jarrah), *Agonis flexuosa* (Coastal Peppermint), and *Corymbia calophylla* (Marri). Similar trends were observed in their respective Ferric Reducing Antioxidant Power (FRAP) and 2,2-diphenyl-1-picrylhydrazyl (DPPH) antioxidant activities (Contribution 13).

Cerumen is a bee product made exclusively by stingless bees, composited as a mixture of beeswax and plant resins. Ferreira et al. (Contribution 14) investigated the chemical composition and antioxidant activity of cerumen produced by the Geotrigona sp. and Tetragonisca fiebrigi stingless bees. They conducted both in vitro and in vivo analyses using HPLC, GC, and ICP OES techniques. The results showed promising effects against oxidative stress and related diseases.

Various interactions within complex chemical systems influence the antioxidant and prebiotic properties of mixtures that combine honey with polyphenol-rich extracts. Typically, the modulation of antioxidant activities varies between the two extracts, which can be attributed to differences in the composition of polyphenols found in the tested plant extracts. Notably, the honeysuckle flower extract demonstrated higher prebiotic activity than the raspberry extract, and the effect on lactic acid production resulted in a hormetic response (Contribution 15).

The published scientific papers highlight the significance of utilizing the antioxidative properties found in various compounds of natural bee products. By exploring a diverse range of bioactive substances present in propolis, honey, pollen, cerumen, and other products, the authors are paving the way for sustainable beekeeping and apitherapeutic practices. Additionally, incorporating natural antioxidants into functional foods presents an innovative and promising opportunity to enhance humans' and animals' health and well-being.

Funding: This research received no external funding.

Acknowledgments: We thank all the authors for their valuable contributions to the Special Issue on the antioxidant activity of honey bee products. We are also grateful to the reviewers for their work and constructive recommendations, which significantly improved the quality of the articles. Also, we would like to acknowledge the COST Action CA22105-Beekeeping products valorization and biomonitoring for the SAFEty of BEEs and HONEY (BeSafeBeeHoney).

Conflicts of Interest: The authors declare no conflicts of interest.

List of Contributions

1. El Ghouizi, A.; Bakour, M.; Laaroussi, H.; Ousaaid, D.; El Menyiy, N.; Hano, C.; Lyoussi, B. Bee Pollen as Functional Food: Insights into Its Composition and Therapeutic Properties. *Antioxidants* **2023**, *12*, 557. https://doi.org/10.3390/antiox12030557.
2. Martiniakova, M.; Kovacova, V.; Mondockova, V.; Zemanova, N.; Babikova, M.; Biro, R.; Ciernikova, S.; Omelka, R. Honey: A Promising Therapeutic Supplement for the Prevention and Management of Osteoporosis and Breast Cancer. *Antioxidants* **2023**, *12*, 567. https://doi.org/10.3390/antiox12030567.
3. Choudhary, P.; Tushir, S.; Bala, M.; Sharma, S.; Sangha, M.K.; Rani, H.; Yewle, N.R.; Kumar, P.; Singla, D.; Chandran, D.; et al. Exploring the Potential of Bee-Derived Antioxidants for Maintaining Oral Hygiene and Dental Health: A Comprehensive Review. *Antioxidants* **2023**, *12*, 1452. https://doi.org/10.3390/antiox12071452.
4. Bae, I.A.; Ha, J.W.; Choi, J.Y.; Boo, Y.C. Antioxidant Effects of Korean Propolis in HaCaT Keratinocytes Exposed to Particulate Matter 10. *Antioxidants* **2022**, *11*, 781. https://doi.org/10.3390/antiox11040781.
5. Mohamed, H.K.; Mobasher, M.A.; Ebiya, R.A.; Hassen, M.T.; Hagag, H.M.; El-Sayed, R.; Abdel-Ghany, S.; Said, M.M.; Awad, N.S. Anti-Inflammatory, Anti-Apoptotic, and Antioxidant Roles of Honey, Royal Jelly, and Propolis in Suppressing Nephrotoxicity Induced by Doxorubicin in Male Albino Rats. *Antioxidants* **2022**, *11*, 1029. https://doi.org/10.3390/antiox11051029.
6. Moskwa, J.; Naliwajko, S.K.; Markiewicz-Żukowska, R.; Gromkowska-Kępka, K.J.; Soroczyńska, J.; Puścion-Jakubik, A.; Borawska, M.H.; Isidorov, V.; Socha, K. Polish and New Zealand Propolis as Sources of Antioxidant Compounds Inhibit Glioblastoma (T98G, LN-18) Cell Lines and Astrocytoma Cells Derived from Patient. *Antioxidants* **2022**, *11*, 1305. https://doi.org/10.3390/antiox11071305.
7. Puścion-Jakubik, A.; Karpińska, E.; Moskwa, J.; Socha, K. Content of Phenolic Acids as a Marker of Polish Honey Varieties and Relationship with Selected Honey-Quality-Influencing Variables. *Antioxidants* **2022**, *11*, 1312. https://doi.org/10.3390/antiox11071312.
8. Kamal, D.A.M.; Ibrahim, S.F.; Ugusman, A.; Mokhtar, M.H. Kelulut Honey Ameliorates Oestrus Cycle, Hormonal Profiles, and Oxidative Stress in Letrozole-Induced Polycystic Ovary Syndrome Rats. *Antioxidants* **2022**, *11*, 1879. https://doi.org/10.3390/antiox11101879.
9. Montaser, M.; Ali, A.T.; Sayed, A.M.; Abdelmohsen, U.R.; Zidan, E.W.; Orfali, R.; Rateb, M.E.; Zaki, M.A.; Hassan, H.M.; Mohammed, R.; et al. ^{1}H-NMR Metabolic Profiling, Antioxidant Activity, and Docking Study of Common Medicinal Plant-Derived Honey. *Antioxidants* **2022**, *11*, 1880. https://doi.org/10.3390/antiox11101880.

10. Boke Sarikahya, N.; Varol, E.; Sumer Okkali, G.; Yucel, B.; Margaoan, R.; Nalbantsoy, A. Comparative Study of Antiviral, Cytotoxic, Antioxidant Activities, Total Phenolic Profile and Chemical Content of Propolis Samples in Different Colors from Turkiye. *Antioxidants* **2022**, *11*, 2075. https://doi.org/10.3390/antiox11102075.
11. Zakaria, Z.; Othman, Z.A.; Suleiman, J.B.; Mustaffa, K.M.F.; Jalil, N.A.C.; Ghazali, W.S.W.; Zulkipli, N.N.; Mohamed, M.; Kamaruzaman, K.A. Therapeutic Effects of Heterotrigona itama (Stingless Bee) Bee Bread in Improving Hepatic Lipid Metabolism through the Activation of the Keap1/Nrf2 Signaling Pathway in an Obese Rat Model. *Antioxidants* **2022**, *11*, 2190. https://doi.org/10.3390/antiox11112190.
12. Dimitriu, L.; Constantinescu-Aruxandei, D.; Preda, D.; Nichițean, A.-L.; Nicolae, C.-A.; Faraon, V.A.; Ghiurea, M.; Ganciarov, M.; Băbeanu, N.E.; Oancea, F. Honey and Its Biomimetic Deep Eutectic Solvent Modulate the Antioxidant Activity of Polyphenols. *Antioxidants* **2022**, *11*, 2194. https://doi.org/10.3390/antiox11112194.
13. Lawag, I.L.; Islam, M.K.; Sostaric, T.; Lim, L.Y.; Hammer, K.; Locher, C. Antioxidant Activity and Phenolic Compound Identification and Quantification in Western Australian Honeys. *Antioxidants* **2023**, *12*, 189. https://doi.org/10.3390/antiox12010189.
14. Ferreira, I.C.; Côrrea, R.C.D.; Orué, S.L.; Leite, D.F.; da Rocha, P.d.S.; Cardoso, C.A.L.; Mussury, R.M.; Vit, P.; de Picoli Souza, K.; dos Santos, E.L.; et al. Chemical Components and Antioxidant Activity of *Geotrigona* sp. and *Tetragonisca fiebrigi* Stingless Bee Cerumen Reduce Juglone-Induced Oxidative Stress in *Caenorhabditis elegans*. *Antioxidants* **2023**, *12*, 1276. https://doi.org/10.3390/antiox12061276.
15. Dimitriu, L.; Constantinescu-Aruxandei, D.; Preda, D.; Moraru, I.; Băbeanu, N.E.; Oancea, F. The Antioxidant and Prebiotic Activities of Mixtures Honey/Biomimetic NaDES and Polyphenols Show Differences between Honeysuckle and Raspberry Extracts. *Antioxidants* **2023**, *12*, 1678. https://doi.org/10.3390/antiox12091678.

References

1. Giampieri, F.; Quiles, J.L.; Cianciosi, D.; Forbes-Hernández, T.Y.; Orantes-Bermejo, F.J.; Alvarez-Suarez, J.M.; Battino, M. Bee Products: An Emblematic Example of Underutilized Sources of Bioactive Compounds. *J. Agric. Food Chem.* **2022**, *15*, 6833–6848. [CrossRef] [PubMed]
2. Sawicki, T.; Starowicz, M.; Kłębukowska, L.; Hanus, P. The Profile of Polyphenolic Compounds, Contents of Total Phenolics and Flavonoids, and Antioxidant and Antimicrobial Properties of Bee Products. *Molecules* **2022**, *27*, 1301. [CrossRef] [PubMed]
3. Bartkiene, E.; Lele, V.; Sakiene, V.; Zavistanaviciute, P.; Zokaityte, E.; Dauksiene, A.; Jagminas, P.; Klupsaite, D.; Bliznikas, S.; Ruzauskas, M. Variations of the antimicrobial, antioxidant, sensory attributes and biogenic amines content in Lithuania-derived bee products. *LWT Food Sci. Technol.* **2020**, *118*, 108793. [CrossRef]
4. Manyi-Loh, C.M.; Clarke, A.; Nidp, R.N. An overview of honey: Therapeutic properties and contribution in nutrition and human health. *Afr. J. Microbiol. Res.* **2011**, *5*, 844–852. [CrossRef]
5. Cornara, L.; Biagi, M.; Xiao, J.; Burlando, B. Therapeutic properties of bioactive compounds from different honeybee products. *Front. Pharmacol.* **2017**, *8*, 412. [CrossRef] [PubMed]
6. Pena Júnior, D.S.; Almeida, C.A.; Santos, M.C.F.; Fonseca, P.H.V.; Menezes, E.V.; de Melo Junior, A.F.; Brandão, M.M.; de Oliveira, D.A.; de Souza, L.F.; Silva, J.C.; et al. Antioxidant activities of some monofloral honey types produced across Minas Gerais (Brazil). *PLoS ONE* **2022**, *17*, e0262038. [CrossRef] [PubMed]
7. Bertoncelj, J.; Doberšek, U.; Jamnik, M.; Golob, T. Evaluation of the phenolic content, antioxidant activity and colour of Slovenian honey. *Food Chem.* **2007**, *105*, 822–828. [CrossRef]
8. Da Silva, P.M.; Gauche, C.; Gonzaga, L.V.; Costa, A.C.O.; Fett, R. Honey: Chemical composition, stability and authenticity. *Food Chem.* **2016**, *196*, 309–323. [CrossRef]
9. Šuran, J.; Cepanec, I.; Mašek, T.; Radić, B.; Radić, S.; Tlak Gajger, I.; Vlainić, J. Propolis Extract and Its Bioactive Compounds—From Traditional to Modern Extraction Technologies. *Molecules* **2021**, *26*, 2930. [CrossRef] [PubMed]
10. Tlak Gajger, I.; Pavlović, I.; Bojić, M.; Kosalec, I.; Srečec, S.; Vlainić, T.; Vlainić, J. The components responsible for the antimicrobial activity of propolis from continental and Mediterranean regions in Croatia. *Czech J. Food Sci.* **2017**, *35*, 376–385. [CrossRef]
11. Ramanathan, A.N.K.G.; Nair, A.J.; Sugunan, V.S. A review on royal jelly proteins and peptides. *J. Funct. Foods* **2018**, *44*, 255–264. [CrossRef]

12. Giampieri, F.; Quiles, J.L.; Orantes-Bermejo, F.J.; Gasparrini, M.; Forbes-Hernandez, T.Y.; Sánchez-González, C.; Llopis, J.; Rivas-García, L.; Afrin, S.; Varela-López, A.; et al. Are by-products from beeswax recycling process a new promising source of bioactive compounds with biomedical properties? *Food Chem. Toxicol.* **2018**, *112*, 126–133. [CrossRef] [PubMed]
13. Denisow, B.; Denisow-Pietrzyk, M. Biological and therapeutic properties of bee pollen: A review. *J. Sci. Food Agric.* **2016**, *96*, 4303–4309. [CrossRef] [PubMed]
14. Anjum, S.I.; Ullah, A.; Gohar, F.; Raza, G.; Khan, M.I.; Hameed, M.; Ali, A.; Chen, C.-C.; Tlak Gajger, I. Bee pollen as a food and feed supplement and a therapeutic remedy: Recent trends in nanotechnology. *Front. Nutr.* **2024**, *11*, 1371672. [CrossRef] [PubMed]
15. Ullah, A.; Aldakheel, F.M.; Anjum, S.I.; Raza, G.; Khan, S.A.; Tlak Gajger, I. Pharmacological properties and therapeutic potential of honeybee venom. *Saudi Pharm. J.* **2023**, *31*, 96–109. [CrossRef] [PubMed]
16. Shakoori, Z.; Salaseh, E.; Mehrabian, A.R.; Tehrani, D.M.; Dardashti, N.F.; Salmanpour, F. The amount of antioxidants in honey has a strong relationship with the plants selected by honey bees. *Sci. Rep.* **2024**, *14*, 351. [CrossRef] [PubMed]
17. Martinello, M.; Mutinelli, F. Antioxidant Activity in Bee Products: A Review. *Antioxidants* **2021**, *10*, 71. [CrossRef] [PubMed]

antioxidants

Article

Anti-Inflammatory, Anti-Apoptotic, and Antioxidant Roles of Honey, Royal Jelly, and Propolis in Suppressing Nephrotoxicity Induced by Doxorubicin in Male Albino Rats

Hanaa K. Mohamed [1], Maysa A. Mobasher [2,*], Rasha A. Ebiya [1], Marwa T. Hassen [1], Howaida M. Hagag [3,4], Radwa El-Sayed [1], Shaimaa Abdel-Ghany [5], Manal M. Said [5] and Nabil S. Awad [5,6]

1 Department of Zoology, Faculty of Women for Arts, Science and Education, Ain Shams University, Cairo 11757, Egypt; hanaa.khairy@women.asu.edu.eg (H.K.M.); rasha.ebiya@women.asu.edu.eg (R.A.E.); marwa.hassen@women.asu.edu.eg (M.T.H.); radwa.ahmed@women.asu.edu.eg (R.E.-S.)
2 Department of Pathology, Biochemistry Division, College of Medicine, Jouf University, Sakaka 41412, Saudi Arabia
3 Department of Pathology, Faculty of Medicine, Al-Azhar University, Nasr City, Cairo 11884, Egypt; howaida.m@tu.edu.sa
4 Department of Clinical Laboratory Sciences, College of Applied Medical Sciences, Taif University, Taif 21944, Saudi Arabia
5 College of Biotechnology, Misr University for Science and Technology, Giza 12563, Egypt; shaimaa.ibraheem@must.edu.eg (S.A.-G.); manalfaris@yahoo.com (M.M.S.); nabil.faris@must.edu.eg (N.S.A.)
6 Department of Genetics, Faculty of Agriculture and Natural Resources, Aswan University, Aswan 81528, Egypt
* Correspondence: mmobasher@ju.edu.sa

check for updates

Citation: Mohamed, H.K.; Mobasher, M.A.; Ebiya, R.A.; Hassen, M.T.; Hagag, H.M.; El-Sayed, R.; Abdel-Ghany, S.; Said, M.M.; Awad, N.S. Anti-Inflammatory, Anti-Apoptotic, and Antioxidant Roles of Honey, Royal Jelly, and Propolis in Suppressing Nephrotoxicity Induced by Doxorubicin in Male Albino Rats. *Antioxidants* **2022**, *11*, 1029. https://doi.org/10.3390/antiox11051029

Academic Editors: Josipa Vlainić and Ivana Tlak Gajger

Received: 27 April 2022
Accepted: 19 May 2022
Published: 23 May 2022

Abstract: Nephrotoxicity is one of the limiting factors for using doxorubicin (DOX). Honey, propolis, and royal jelly were evaluated for their ability to protect against nephrotoxicity caused by DOX. Forty-two adult albino rats were divided into control groups. The DOX group was injected i.p. with a weekly dose of 3 mg/kg of DOX for six weeks. The DOX plus honey treated group was injected with DOX and on the next day, received 500 mg/kg/day of honey orally for 21 days. The DOX plus royal jelly treated group was injected with DOX and on the following day, received 100 mg/kg/day of royal jelly orally for 21 days. The DOX plus propolis treated group received DOX and on the following day, was treated orally with 50 mg/kg/day of propolis for 21 days. The DOX plus combined treatment group received DOX and on the following day, was treated with a mix of honey, royal jelly, and propolis orally for 21 days. Results confirmed that DOX raised creatinine, urea, MDA, and TNF-α while decreasing GPX and SOD. Damages and elevated caspase-3 expression were discovered during renal tissue's histopathological and immunohistochemical studies. Combined treatment with honey, royal jelly, and propolis improved biochemical, histological, and immunohistochemical studies in the renal tissue. qRT-PCR revealed increased expression of poly (ADP-Ribose) polymerase-1 (PARP-1) and a decline of Bcl-2 in the DOX group. However, combined treatment induced a significant decrease in the PARP-1 gene and increased Bcl-2 expression levels. In addition, the combined treatment led to significant improvement in the expression of both PARP-1 and Bcl-2 genes. In conclusion, the combined treatment effectively inhibited nephrotoxicity induced by DOX.

Keywords: doxorubicin; nephrotoxicity; PARP-1; Bcl-2; TNF-α; honey; royal jelly; propolis

1. Introduction

Doxorubicin (DOX), an anthracycline antibiotic, has been applied as an effective anti-cancer therapy since 1969 [1,2]. Despite its potent anti-cancer properties, the clinical usefulness of DOX in cancer chemotherapy is limited by its severe consequences on non-targeted organs, including the kidneys, liver, and brain [3]. DOX was reported to raise the permeability of glomerular capillaries and atrophy of the glomeruli in rat kidneys [4]. After

DOX accumulation, severe glomerular damage was produced due to oxidative stress [5]. Lipid peroxidation and diminished antioxidant enzyme activity are common mediators promoting nephrotic syndrome [6]. Inflammation plays an influential role in renal damage exhibited by DOX through cytokines and other cytotoxic factors [7]. Protein excretion due to renal failure is primarily associated with damage to the filtration barrier [7]. This filtration barrier deterioration is caused by DOX [8]. Several researchers reported that DOX caused alterations in the kidney function parameters. Manal et al. (2019) published that DOX caused significant increases in creatinine and urea levels [9]. DOX causes oxidative stress in kidney tissues characterized by low antioxidant enzyme activity and enhanced malondialdehyde activity [10]. DOX-induced nephrotoxicity causes increased vascular porousness and glomerular contraction. It is marked by increased kidney functions and raised lactate dehydrogenase activity, along with decreased renal Ca2+- ATPase, Mg2+-ATPase, and Na+, K+-ATPase rates [11,12]. Natural antioxidants reduce harmful side effects and improve the antitumor activity of anti-cancer medications [13–15]. Different studies have demonstrated that a diet high in honey and bee products (propolis and royal jelly) provides significant health benefits against various diseases due to their antioxidant properties [16,17].

Honey is a naturally occurring food that is produced by bees. It is known worldwide for its great nutrient ingredients beneficial to humans. Major sugars and vitamins, phenolic acids, minerals, and phytochemicals are found in honey. Egyptians, Greeks, Romans, and Chinese have all utilized honey to treat wounds and intestinal disorders, including gastric ulcers. It has also been used to treat earache, coughs, and sore throats [18,19]. Honey has lately been used for its anti-inflammatory, antimicrobial, and antioxidant activities and for enhancing the immune response [20]. Honey's biological activity can be related to its polyphenolic content, linked to its antioxidant and anti-inflammatory properties, and its cardiovascular, anti-proliferative, and antibacterial benefit [21,22].

Royal jelly is made from the mandibular glands and hypopharyngeal of worker honeybees as a milky secretion [23]. It contains royalactin proteins, monosaccharides, lipids, fatty acids, minerals, free amino acids, and vitamins. Antitumor, anti-inflammatory, antioxidant, hypoglycemic, and hypo-cholesterolemic actions are considered biological benefits of royal jelly [24,25].

Propolis, another defense product from the bee, is collected from different plant sources. Additionally, propolis has several biological properties, including antioxidant and free radical scavenger action, antimicrobial activity against a broad spectrum of pathogens, anti-malignant action, anti-inflammatory activities, and immune-boosting properties by promoting numerous pro-inflammatory cytokines [19]. Propolis was reported to enhance the protective effect on kidney failure induced by paracetamol, carbon tetrachloride, and Doxorubicin [26,27]. The current study aims to assess the preventive roles of natural honey, royal jelly, and propolis administration, both alone and combined, on Dox-induced renal toxicity. The biochemical, histopathological, immunohistochemical, and gene expression modifications in albino rats were examined.

2. Materials and Methods

2.1. Chemicals and Natural Products

Doxorubicin (DOX) hydrochloride, honey, royal jelly, and propolis were obtained from Sigma chemical company (St. Louis, MO, USA).

2.2. Experimental Animals

The protocol of the Institutes of Health (NIH) and the Commission for Control and Supervision of Experiments on Animals (CPCSEA) (registration number: 13/165) were performed in our experimental animal study. Our research was conducted in the biology lab of Ain Shams University's Department of Zoology, Faculty of Women for Arts, Science, and Education. The rats used in this investigation were 42 adult male albino rats weighing (150–160 g). Rats were kept in a properly ventilated room and exposed to natural light

(12:12 h light-dark cycle). In the laboratory, the animals were housed in metabolic labeled cages at a constant temperature of 25 °C. They had unrestricted access to conventional dry pellet food and water. They were acclimated for a week before the experiment started.

2.3. Induction of Nephrotoxicity

Doxorubicin (DOX) hydrochloride 10 mg vial (Pharmacia Italia, Milano, Italy) was injected intraperitoneally in six equal doses (i.p., 3 mg/kg b.w.) for six weeks (one dosage each week for a total dose was18 mg/kg b.w.) as previously described [28]. All other chemicals and reagents used were of analytical grade.

2.4. Experimental Design

Forty-two male albino rats were divided into six groups of seven animals each. The control group (NG) was given saline; the Doxorubicin group (DOX-G) was administrated with a single dose of 3 mg/kg/week i.p. of DOX for six weeks for kidney toxicity induction; the Dox plus honey treated group (DOX-H) was introduced with DOX (single dosage of 3 mg/kg/week i.p. for six weeks) and on the next day receiving 500 mg/kg/day of honey orally for 21 days; the Dox plus royal jelly treated group (DOX-R) treated with DOX (only one dose 3 mg/kg/week i.p. for six weeks) and on the next day treated orally with 100 mg/kg/day royal jelly for 21 days; the Dox plus propolis treated group (DOXP) received DOX (single dose 3 mg/kg/week i.p.) and on the next day treated orally with 50 mg/kg/day propolis for 21 days; the Dox plus honey, royal jelly, and propolis combined group (DOXHRP) received DOX (15 mg/kg, i.p.) and on the subsequent day treated with 500 mg/kg/day of honey, 100 mg/kg/day Royal jelly, and 50 mg/kg/day propolis orally for 21 days.

At the end of the experiment, non-heparinized capillary tubes were used to collect blood samples from the retro-orbital plexus. To analyze kidney functions, serum specimens were produced after centrifuging for 20 min and collecting the supernatants. The kidneys were promptly divided and cleaned with saline after decapitating the rats. One portion of the kidney was homogenized in phosphate-buffered saline and centrifuged to prepare 25% *w/v* tissue homogenates. After that, the supernatants were collected and kept at −80 °C until analysis. For histopathology and immunohistochemistry examination, another kidney section was rinsed with saline and then put in 10% formal saline. The third part was kept at −80 °C in trizol for gene expression analyses.

2.5. Biochemical Determinations

Renal Biomarkers: Biodiagnostic Co. Egypt kit was used as a colorimetric method to assay urea, as previously described [29]. Creatinine was measured using a Kit obtained from Diamond Diagnostics Co., Cairo, Egypt, according to [30].

Antioxidants and oxidative markers: Biodiagnostic Co., Egypt kits were used in colorimetric methods to determine renal malondialdehyde (MDA), superoxidase dismutase (SOD), and glutathione peroxidase (GPX) according to [31–34], respectively.

Tumor necrosis factor-α (TNF-α) and B-Cell Leukemia/Lymphoma 2 (Bcl2): Tumor necrosis factor-α (TNF-α) content was evaluated by ELISA technique using a TNF-α assay kit acquired from Assay Pro., Co., Charles city, IA, USA, following the procedure described by [35]. B-Cell Leukemia/Lymphoma 2 (Bcl2) was measured using the ELISA technique following the manufacturer's instructions for ELISA KIT of rat Bcl2 procured from Cloud-Clone Corp. (CCC, Houston, TX, USA).

2.6. Cell Examination of Kidney Tissues

Dissected renal tissue samples were rinsed in normal saline and fixed in 10% saline for 72 h. After that, the specimens were cut and dried in alcohol, then cleared in xylene, filtered in wax, and finally blocked out into Paraplast tissue embedding media. A rotatory microtome was used to cut 5 μm thick sections of each sample. According to the previously reported methodology, the sections were stained with hematoxylin and eosin (H&E) stain [36]. Slides were examined under a microscope at 400× magnification.

2.7. Immuno-Histochemical Study (Caspase-3)

Caspase 3 antibodies were stained immunohistochemically at dilution (1:50) and on 4-µm, paraffin-embedded sections. Antigen retrieval in all samples was accomplished by heating the plates for 30 min in a solution of EDTA (1-mmol/L, pH 8.0), followed by endogenous biotin inhibition. Staining with an automated immune Stainer (DAKO) was performed, then detected with a streptavidin-biotin detection system. Additionally, positive and negative control sections were applied.

2.8. Gene Expression Analysis

The changes in mRNA levels of PARP1 and Bcl2 genes were assessed using quantitative real-time PCR [37,38]. An internal control GAPDH was used as a standard for the RT- PCR analysis. Trizol reagent (Invitrogen Life Technologies, Carlsbad, CA, USA) was used to extract the total RNA of kidney tissue following the supported manufacture. The spectrophotometer was used to investigate the purity and concentration of total RNA. The cDNA synthesis kit (Takara, Kyoto, Japan) was used for cDNA synthesizing according to the manufacturer's instructions. Quantitative RT-PCR was conducted using the SYBR Green mix kit (Applied Biosystems, Foster City, CA, USA) applying Mini TM thermocycler (Bio-Rad Laboratories Inc., New York, CA, USA). In a final volume of 25 µL, 2 µL cDNA was added to 12.5 µL 2× SYBR Green mix, 1 µL of each forward and reverse primers, and 8.5 µL of deionized distal water. The primers sequences were designed using primer design software Primer 3 version 4.1.0 online at https://primer3.ut.ee/ (accessed on 31 March 2022): Bcl-2 (F) 5′-TTTGATTTCTCCTGGCTGTCT-3′ and (R) 5′-CTGATTTGACCATTTGCCTG-3′; PARP-1 (F) 5′-TCTCCAATCGCT TCTACACCCT-3′ and (R) 5′-TACTGCTGTCATCAGACCCACC-3′; GAPDH (F) 5-GCAAGTTC GCAAGTTCAACGGCACAGTCAAG-3 and (R) 5-GTACTCAGCACCAGCAT CACC-3(The PCR reactions followed programs of first 95 °C for 10 min, then 40 cycles consisting of 94 °C for 15 s, and 60 °C for 1 min, finally melting curve analysis was used to specify the amplification. gene accession number Bcl_2, PARP-1, NM_013063.2 and NM_016993.2) The 2ΔΔCT method was used to analyze the obtained data. The mRNA levels results were generalized against the amount of housekeeping gene GAPDH Results were presented as fold change relative to the negative control (RFC) [39].

2.9. Statistical Analysis

The SPSS program version 25 (IBM Corp., Armonk, NY, USA) was used to analyze the obtained data statistically. The mean and standard error were used to summarize our data. For comparisons between groups, variance analysis (ANOVA) was used with a multiple comparisons post hoc test for every two groups.

3. Results

3.1. Renal Biomarkers

The findings in Figure 1A,B show that serum creatinine and urea levels in the DOX group were significantly higher ($p \leq 0.05$) than in the control group. However, compared to the DOX group, treatment of H, R, or P alone or a mixture of H, R, and P led to a substantial reduction ($p \leq 0.05$) in the serum urea and creatinine levels.

3.2. Antioxidants Status

The effects of H, R, and P, alone or combined of all H, R, and P on MDA and enzymatic antioxidants in DOX-treated rats are shown in Figure 2A,B. The DOX group had a significant rise in renal MDA content ($p \leq 0.05$) in contrast to the control group. Treatment of the DOX group with H, R, and P or all of H, R, and P together, on the other hand, reversed this rise as evidenced by a significant decline ($p \leq 0.05$) in MDA concentration when compared to the DOX group. Compared to controls, the DOX group had significantly lower SOD and GPX. As the DOX group was treated with H, R, and P, alone or combined of all H, R, and P together, the level of SOD and GPX in the kidney homogenate increased ($p \leq 0.05$) relative to the DOX group.

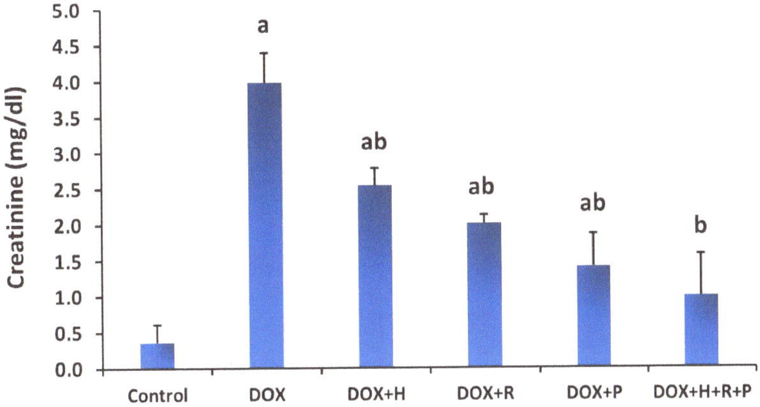

(A)

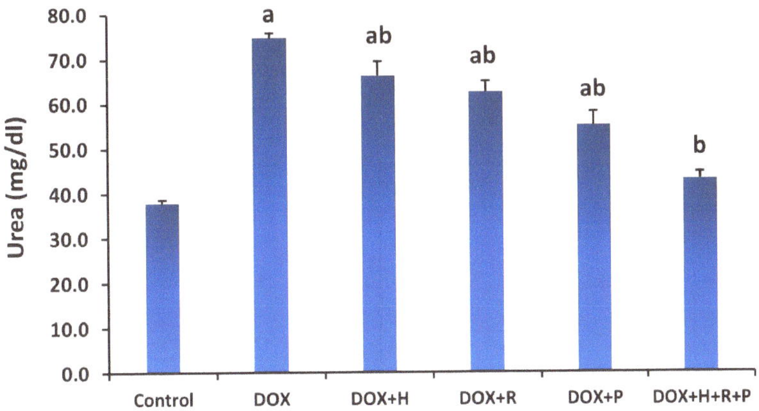

(B)

Figure 1. Effect of honey (H), royal jelly (R), and propolis (P) on the renal functions in rats treated with Doxorubicin, (**A**) creatinine, and (**B**) urea. Values are shown as mean ± SE. Different lowercase letters indicate significant differences compared to the corresponding value in the control group at $p < 0.05$.

3.3. *Inflammatory Markers*

Results in Figure 3 demonstrate the treatment with H, R, and P, alone or combined of H, R, and P together on renal TNF-α in the DOX group. The DOX group showed a significant increase ($p \leq 0.05$) in TNF-α levels compared to the control group. In contrast, when the DOX group was treated with H, R, and P, alone or combined with H, R, and P, the renal levels of TNF-α were markedly decreased ($p \leq 0.05$) compared to the untreated DOX group.

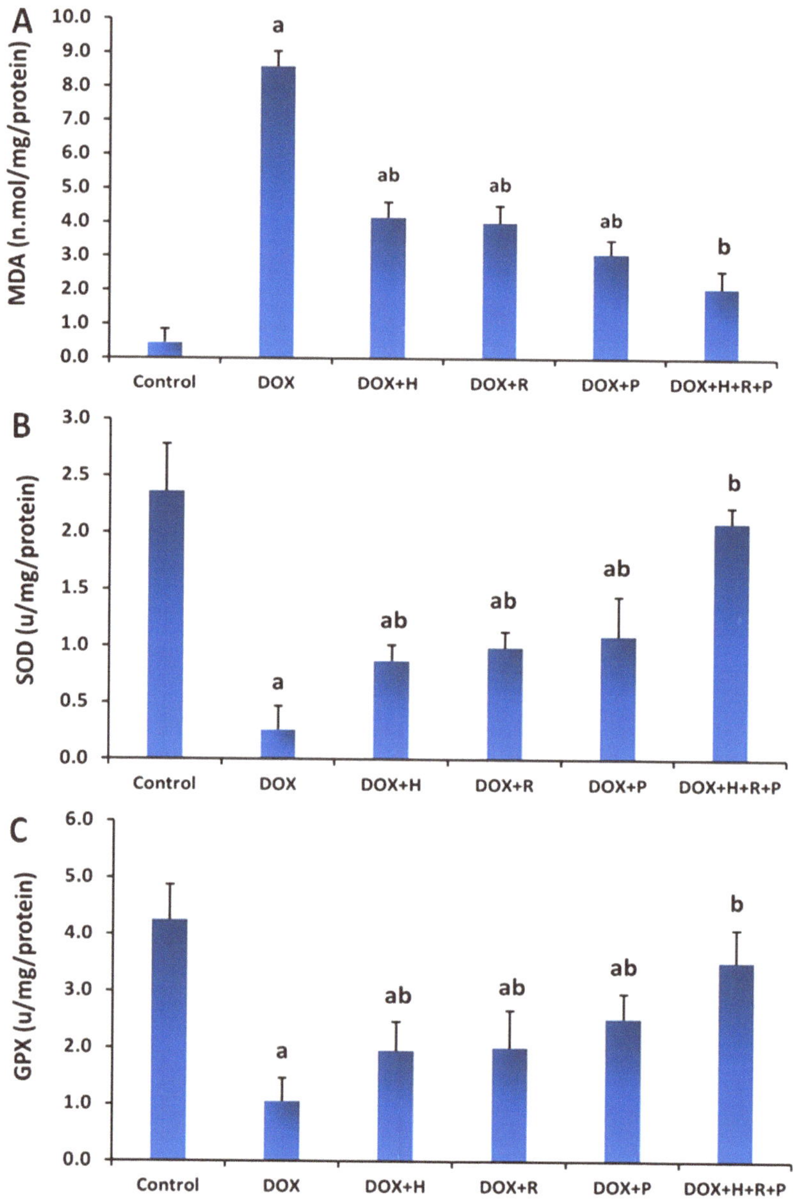

Figure 2. Effect of honey, royal jelly, and propolis on the oxidative marker and antioxidant enzyme activity in renal tissues of rats treated with Doxorubicin, (**A**) MDA, (**B**) SOD, and (**C**) GPX. Values are shown as mean ± SE. Statistically, Different lowercase letters indicate significant differences compared to the corresponding value in the control group at $p < 0.05$.

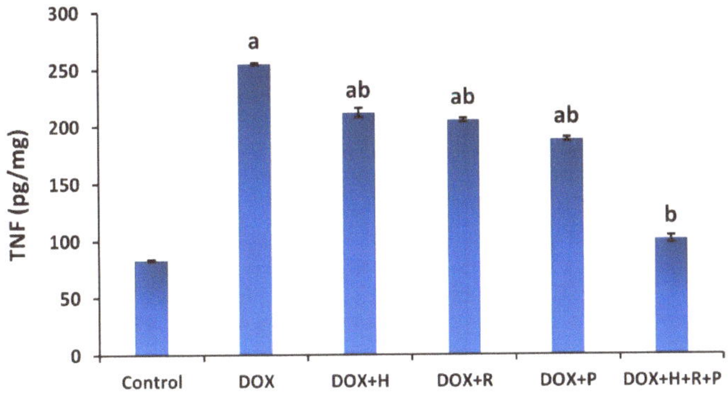

Figure 3. Effect of honey, royal jelly, and propolis on TNF-α of rats treated with Doxorubicin. Values are shown as mean ± SE. Statistically. Different lowercase letters indicate significant differences compared to the corresponding value in the control group at $p < 0.05$.

3.4. Histopathological Analysis: Haematoxylin and Eosin

The control group's kidney sections (Figure 4A) revealed an average renal capsule, average glomeruli with average Bowman's spaces, average proximal tubules with preserved brush borders, average distal tubules, and an average renal medulla with average collecting tubules, average epithelial lining, and average interstitium. In the DOX group, the kidney showed average renal capsule, atrophied glomeruli with enlarged Bowman's gaps, proximal tubules with apoptotic epithelial lining, partial loss of brush borders, and intra-tubular debris notably dilated congested interstitial blood vessels with areas of hemorrhage. The renal medulla demonstrated collecting tubules with apoptotic epithelial lining and congested peri-tubular capillaries (Figure 4B). Regarding the renal medulla of the Dox + H group (Figure 4C), no histopathological changes were detected except for a few sections that exhibited marked congestion. Similarly, the renal medulla of the Dox + R group (Figure 4D) was apparently normal except for some sections that showed renal tubular degeneration and necrosis. Figure 4E (Dox + P) and Figure 4F (DOX + H + R + P) show an apparently normal renal cortex with congestion in some instances while exhibiting an apparently normal renal medulla.

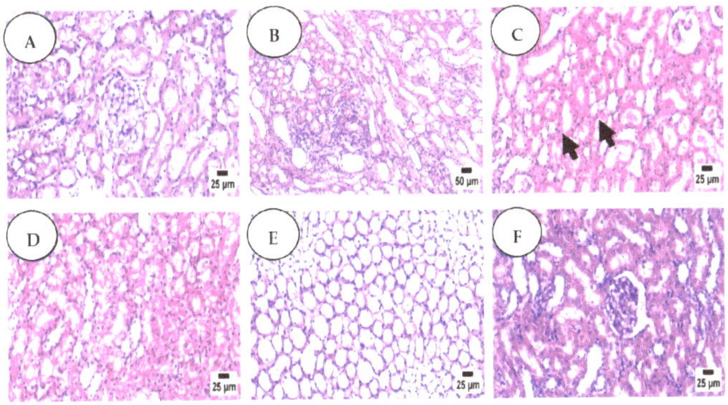

Figure 4. Control group (**A**) high power view shows average glomeruli with average Bowman's spaces, average proximal tubules with preserved brush borders (black arrow), average distal tubules, and average interstitium (H&E); Dox group (**B**) showing focal cystic dilation of renal tubules with focal interstitial nephritis (H&E); Dox + H group (**C**) showing apparently normal renal cortex with few tubules containing renal cast (arrow) (H&E); (**D**) (Dox + R), showing some degenerating renal tubules (H&E); Dox + P group (**E**) showing apparently normal renal medulla (H&E); Dox + H + R + P group (**F**) showing apparently normal renal cortex (H&E).

3.5. Immuno-Histochemical Studies: Caspase 3

In the control group, the kidneys showed negative reactivity (0) for caspase-3 in glomeruli, negative cytoplasmic reactivity (0) in proximal tubules, and negative reactivity (0) in collecting tubules (Figure 5A); however, in the DOX Group, the kidneys showed an average renal capsule, atrophied glomeruli with widened Bowman's spaces, proximal tubules with apoptotic epithelial lining with high expression of caspase 3% when compared with the control group at $p \leq 0.05$ (Figure 5G), partial loss of brush borders and intra-tubular debris, and markedly dilated congested interstitial blood vessels with areas of hemorrhage. The renal medulla showed collecting tubules with apoptotic epithelial lining and congested peri-tubular capillaries (Figure 5B). In particular Dox + H and Dox + R groups, the kidneys showed moderate cytoplasmic reactivity (++) for caspase-3 in glomeruli, moderate (++) in proximal tubules, and moderate reactivity (++) in collecting tubules (Figure 5C,D) with a significant decrease in optical density % of Caspase 3 in comparison with Dox group at $p \leq 0.05$ (Figure 5G). On the other hand, kidneys in both Dox + P and Dox + H + R + P groups showed improvement in weak cytoplasmic reactivity (+) for caspase-3 (Figure 5E,F) with significant inhibition in optical density % of Caspase 3 in comparison with Dox group as well as Dox + H and Dox + R groups at $p \leq 0.05$ (Figure 5G).

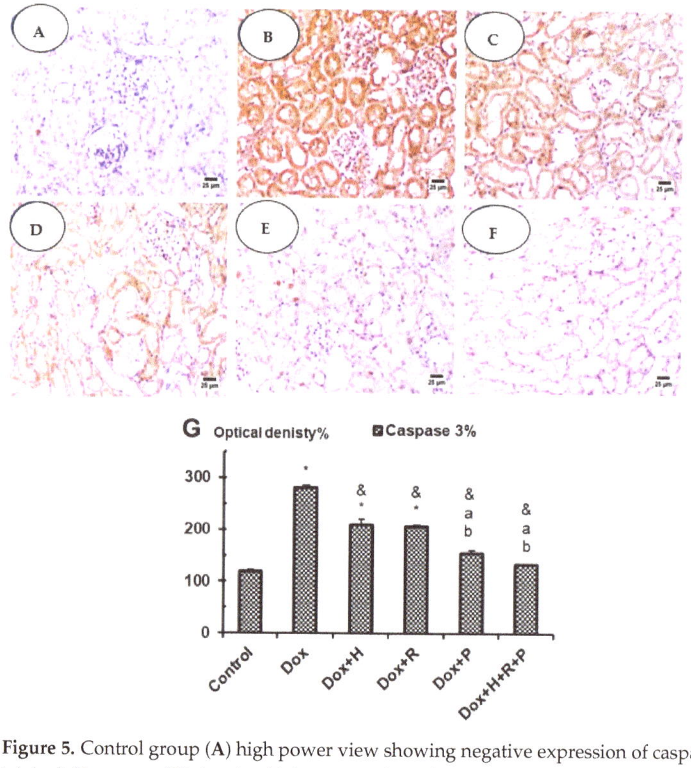

Figure 5. Control group (**A**) high power view showing negative expression of caspase 3 (immunostaining); Dox group (**B**) showing higher expression of caspase 3 (immunostaining); Dox + H group (**C**) and Dox + R group (**D**) showing moderate expression of caspase 3 (immunostaining); Dox + P group (**E**) and Dox + H + R + P group (**F**) show weak expression of caspase 3 (immunostaining). (**G**) Optical density% for Caspase 3 in experimental groups. Results are expressed as mean ± S.E.M. and analyzed using one-way ANOVA followed by Bonferroni's test for multiple comparisons. * $p \leq 0.05$ versus control group. $^{\&}$ $p \leq 0.05$ versus Dox group. ^{a} $p \leq 0.05$ versus Dox + H group. ^{b} $p \leq 0.05$ versus Dox + R group. $n = 5$.

3.6. Gene Expression Analysis

The present study detects changes in the mRNA expression levels of two genes as molecular biomarkers using real-time PCR. The inflammatory impact and anti-apoptotic effect of DOX and treatment with honey, royal jelly, and propolis were detected in kidney tissues. A statistically significant ($p \leq 0.05$) increase in PARP-1 gene expression was determined after DOX injection compared to the control group (5.5, 1.06, respectively). The protective effect of honey, propolis, and royal jelly treatment was detected in the significant downregulation ($p \leq 0.05$) of the PARP-1 gene expression level, whereas highly significant ($p \leq 0.01$) downregulation was demonstrated after combined treatment of honey, royal jelly, and propolis as compared to the DOX group (4.3, 4.7, 3.9, 2.6, 5.5, respectively). The apoptotic molecular biomarker Bcl2 showed significant downregulation of Bcl2 gene expression after treatment with DOX (1.3) compared to the control group (5.9). Treatment with honey, propolis, and royal jelly after nephrotoxicity induction with DOX resulted in a significant ($p \leq 0.05$) increase in Bcl2 mRNA expression and was highly significant with a combined honey, royal jelly, and propolis treatment as compared with the DOX group (1.7, 1.8, 1.9, 2.4, 1.3, respectively) (Figure 6).

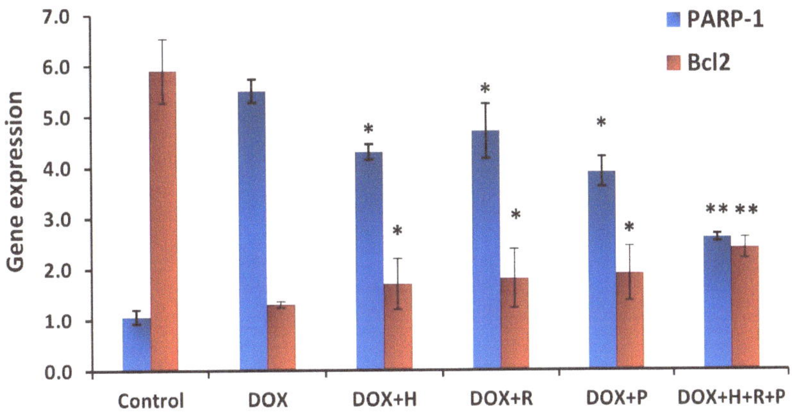

Figure 6. Histogram for poly (ADP-ribose) polymerase 1 (PARP-1) and Bcl2 genes expression in response to treatment with honey (H), propolis (P), and royal jelly (R) mix (H + P + R) on rats treated with Doxorubicin. The groups are negative control; Dox treated group (Dox); the Dox +hony treated group (Dox + H); the Dox + royal jelly treated group (Dox + R); the Dox +propolis treated group (Dox + P); the Dox + mix of honey, royal jelly, and propolis treated group (Dox + H + P + R). * mean significant p-value less than 0.05. ** mean significant p-value less than 0.01.

4. Discussion

Our study revealed a decline in the glomerular filtration level and a considerable increase in blood creatinine and urea after DOX administration. These findings were consistent with data previously published by [8,40]. They stated that chemotherapy causes acute renal failure with severe renal tubular impairments. The mechanism of DOX-inducing renal injury is through inflammation, which stimulates ROS production and apoptosis, with a decrease in antioxidant enzymes in the kidneys [41]. The most sensitive markers of nephrotoxicity are serum urea and creatinine [42]. The elevated creatinine level in the DOX group is related to DOX toxicity disrupting kidney function, which profoundly affects total body metabolism (Figure 7). Our investigation parallels the earlier analyses [43].

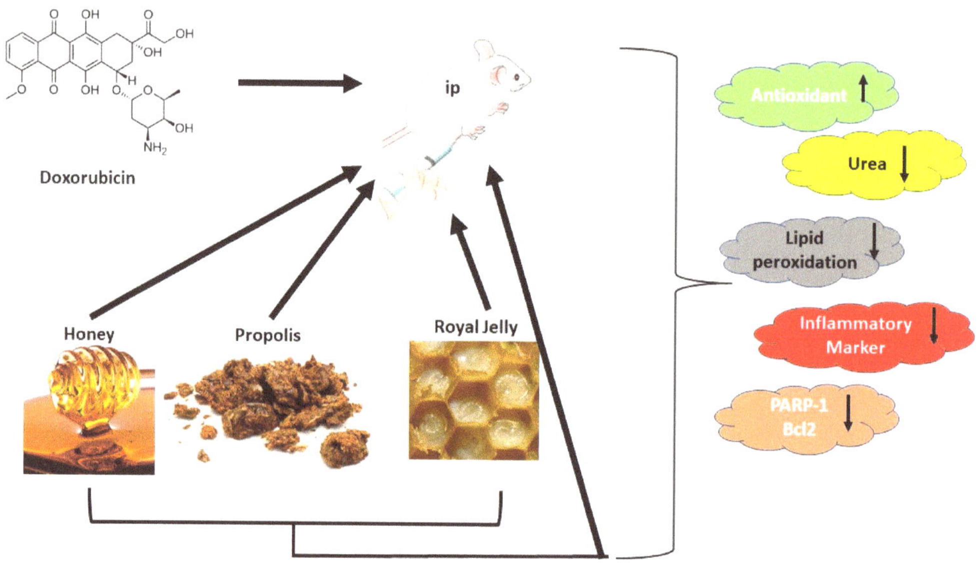

Figure 7. Schematic diagram showing the experimental design and relevant results. ↑ Upregulation of the parameter. ↓ Downlegulation of the parameter.

In contrast, the treatment with H, R, and P, or all of them as a mixture, improved abnormalities in renal parameters (serum creatinine and urea) caused by DOX. Previously, the utilization of honey was revealed to protect against cisplatin-induced kidney toxicity via the suppression of inflammation [44]. These findings established the protecting role of honey against DOX-induced kidney toxicity in rats. These results agree with the results found by Omotayo et al. (2012), Waykar et al. (2018), and Alhumaydhi (2020), who indicated that royal jelly and honey have preventive properties on renal dysfunctions [22,45,46]. Honey and royal jelly are both beneficial foods with high antioxidant capacity. They have hepato-protective, hypoglycemic, reproductive, and antihypertensive benefits. Several investigations declared the protective effect of honey in kidney functions against many drugs [47]. Another study reported the capability of honey to preclude hepato-nephrotoxicity-induced rats treated with cadmium. Additionally, the nephroprotective effect of propolis was evaluated by Baykara et al. by improving renal oxidation and decreasing serum creatinine urea levels which are similar to our data [48].

Promsan et al. studied pretreatment with one of the main constituents of propolis flavonoids, pinocembrin (5,7-dihydroxyflavone), which enhanced renal function and diminished apoptotic and oxidative stress markers [49]. These conclusions determined the protective effect of pinocembrin against nephrotoxicity because of its antioxidant and anti-apoptotic roles. It regulates the antioxidant enzymes and attenuates the rise in oxidative stress through Nrf2/HO-1 and NQO1 pathways [50]. An increase of MDA, an indicator of lipid peroxidation, is directly associated with free radical impairment to the glomerular basement membrane. The dismutation of O_2 to H_2O_2 and molecular oxygen is catalyzed by the SOD enzyme, while GPX catalyzes the degradation of H_2O_2 to O_2 and H_2O. The reduction of SOD and GPX activities and increment of MDA content were revealed after DOX injection, resulting in diminished kidney ability to scavenge toxic H_2O_2 and lipid peroxides. These conclusions agree with El-Sheikh et al., who discussed the mechanism by which DOX-induced nephrotoxicity and cytotoxicity, evidenced by the breakdown of cell membranes and cellular components, is accelerated by oxidative stress generated by excess reactive oxygen species (ROS) [51]. ROS activity changes specific intracellular components, including proteins, lipids, and nuclear DNA [52].

The treatment with H, R, and P, separated or mixed, was proven to lower MDA levels and alter SOD and GPX levels. These results are consistent with earlier research that indicated an increase in antioxidant levels in honey use; this effect might be accompanied by the honey composition, such as many nutrients and antioxidants [53,54]. Additionally, these findings suggest a potent protective effect against oxidative stress, resulting in honey administration. Moreover, royal jelly's high antioxidant capacity facilitates scavenging free radicals, lowering the nitric oxide level and subsequently reducing lipid oxidation and inhibiting protein oxidation, as reflected through the decline in renal function parameters. Additionally, honey and royal jelly serve an effective role in developing normal cellular immunity [55]. Propolis, a strong antioxidant rich in flavonoids, can scavenge free radicals and therefore protect the cell membrane against lipid peroxidation. Caffeic acid phenethyl ester (CAPE) is one of the main components of propolis, which can block ROS production in several systems [56]. Additionally, propolis induced upregulation of Nrf2 expression, the main intracellular transcription factor. It is released under oxidative stress from its repressor (Keap1) and thus restores antioxidant enzyme function. The released Nrf2 binds to the antioxidant response element (ARE) in the gene promoter of cytoprotective genes, stimulating their expression. Subsequently, to remove the effect of cytotoxic oxidants, the expression of free radical-scavenging enzymes occurred [57,58]. DOX administration produced a significant increase in TNF-α levels, a pro-inflammatory cytokine created by glomerular and tubular cells and outside injected inflammatory cells, and acts via mitogen-activated protein kinases (MAPKs) and nuclear factor kappa B (NF-κB) signaling pathways [59]. The initiation of these paths upregulates the expression of some inflammatory cytokines, such as TNF-α [60]. Al-Saedi et al. DOX-produced superoxide anion was found responsible for TNF-induced nuclear factor (NF) stimulation [61] and TNF upregulation [62].

After the DOX group was treated with H, R, and P, or a mixture of them, there was a noticeable increase in TNF-α levels. These findings were in accordance with Thi Lan Nguyen et al., who found that the anti-inflammatory activity of honey is due to its phenolic mixes and other minor constituents [63]. Ahmad et al. and Kassim et al. detected that quercetin, chrysin, ellagic acid, and ferulic acid hesperetin in honey are protective supplements for different inflammatory diseases [64,65]. Royal jelly treatment controlled the alterations of measured pro-inflammatory cytokine. Several reports documented the beneficial impact of royal jelly and its ingredients on anti-inflammatory activity in different experimental models. Moreover, one of the major lipid constituents in royal jelly is 10-hydroxy-2-decenoic acid, which was said to exert anti-inflammatory consequences in colon cancer cells passing through inhibiting NF-κB, which further inhibited the release of TNF-α. CAPE, the main constituent in propolis, may be responsible for propolis's anti-inflammatory effects by lowering the inflammatory cytokines in the inflammatory cells [66].

The histological examination in the present study revealed glomerular congestion, tubular degeneration, vacuolization, necrosis, hyaline cast, brush border loss of proximal cells, and epithelial cell detachment in the DOX group. These alterations were linked to the failure of renal functions, such as elevated creatinine and urea levels. These results were attributed to DOX, which was absorbed by the kidney's tubular cells, especially in proximal tubules. Additionally, this result could be associated with the high concentration of free radicals that cause lipid peroxidation due to the induction of DOX. Similarly, Köse et al. reported that DOX administration resulted in renal cell degeneration with detectable apoptotic bodies due to ROS production [67]. Administration of H, R, and P provided renal histological treatment as normal tubules, glomeruli, and interstitial nephritis were detected. These facts are in harmony with [67,68]. The current findings point to the antioxidant involvement of H, R, and P in scavenging ROS and furthermore their anti-apoptotic and anti-inflammatory properties.

Apoptosis recreates a causative function in developing DOX toxicity in different tissues [69]. In the current work, DOX produced a significant elevation in the immunological reactivity of caspase-3 in the cytoplasm of renal cells. These results were attributed to the reactive oxidative stress-producing oxidative stress in addition to inflammation, leading to the apoptosis of tubular cells. These statements agree with Rashid et al., who attributed the activation of apoptotic DOX, which resulted in DNA injury and mitochondrial DNA damage [70]. The present work also explained the sufficient suppression of apoptosis by propolis extract in kidneys exposed to DOX. This anti-apoptotic effect of honey, royal jelly, and propolis was reported by other investigators [71]. To clarify the protective mechanisms of H, P, and R, and (H + P + R) combined at the molecular level, the expression levels of Bcl-2 and PARP-1 were evaluated. Poly (ADP-ribose) polymerase (PARP) is known to stimulate a specific response wherein the ADP-ribose moiety of NAD+ is transmitted to an amino acid receptor, producing poly (ADP-ribose) polymers. The PARP family includes 17 enzymes participating in a conserved catalytic domain [72].

Additionally, PARP-1 is mainly depicted as a key enzyme for detecting and repairing DNA damage; however, excessive activation of PARP-1 causes necrotic cell death by depleting intracellular ATP [73]. Our results indicated that treatment with DOX induces kidney injury and increases PARP-1 gene expression level two-fold. These effects were modulated by treatment with the H, P, and R, and (H + P + R) mixture. These findings were in line with other studies that point out PARP-1 inhibition of expression of adhesion molecules, infiltration in the inflammatory cells, and secondary oxidative injury in the kidney [74–76]. Mixed administration of royal jelly and honey diminished the cisplatin-induced alterations in diagnostic markers of both kidney and liver functions, under the effect of the capability of honey and royal jelly to scavenge free radicals, lipid peroxidation inhibition, and its anti-inflammatory roles [46,77]. Bcl-2, an anti-apoptotic gene, and its product of the Bcl-2 protein, inhibits the progression of apoptosis by the variability of oxidative stress and through interaction with mitochondrial superoxide dismutase SOD [78]. Obtained results showed that Bcl-2 gene expression was lessened in the DOX treatment group and increased significantly due to treatment with H, P, and R, and (H + P + R) combined. This effect might be due to the antioxidant potentiality of H, P, and R. Honey, propolis, and royal jelly are shown to have antioxidant properties [79,80].

5. Conclusions

We conclude that the administration of honey, propolis, and royal jelly significantly enhanced and resolved the renal injuries and toxicity that took place after the Doxorubicin injection. The mitigation of renal damage was in the significant improvement of kidney function assays, renal histopathology and immunohistochemistry, and mRNA of Bcl-2 and PARP-1 expression levels. Furthermore, the combined treatment significantly inhibited nephrotoxicity induced by DOX compared to the honey, propolis, and royal jelly alone. These natural products induce antioxidant effects, prevent oxidative stress, and enhance the expression level of anti-apoptotic genes.

Author Contributions: Conceptualization, H.K.M., M.A.M., R.A.E., H.M.H. and N.S.A.; methodology, H.K.M., M.A.M., R.A.E., M.T.H., R.E.-S., S.A.-G., M.M.S. and N.S.A.; software, H.K.M., M.A.M., R.A.E., M.T.H., H.M.H., S.A.-G. and M.M.S.; validation, M.T.H., H.M.H., R.E.-S. and N.S.A.; formal analysis, H.K.M., M.A.M., R.A.E., M.T.H., R.E.-S., S.A.-G., M.M.S. and N.S.A.; investigation, M.T.H., H.M.H., R.E.-S. and N.S.A.; resources, R.E.-S., S.A.-G. and M.M.S.; data curation, H.K.M., M.A.M., R.A.E., H.M.H., S.A.-G. and M.M.S.; writing—original draft preparation, H.K.M., M.A.M., R.A.E., M.T.H., S.A.-G., M.M.S. and N.S.A.; writing—review and editing, H.K.M., M.A.M., R.A.E., M.T.H., H.M.H., R.E.-S., S.A.-G., M.M.S. and N.S.A.; visualization, R.A.E., M.T.H., H.M.H., R.E.-S., S.A.-G., M.M.S. and N.S.A.; supervision, M.T.H., S.A.-G., M.M.S. and N.S.A.; project administration, H.K.M., M.A.M., R.A.E., M.T.H., H.M.H. and N.S.A.; funding acquisition, R.E.-S. All authors have read and agreed to the published version of the manuscript.

Funding: This work was funded by the Deanship of Scientific Research at Jouf University under grant No (DSR-2021-01-03146).

Institutional Review Board Statement: Not applicable.

Informed Consent Statement: Not applicable.

Data Availability Statement: Data is contained within the article.

Conflicts of Interest: The authors declare no conflict of interest.

References

1. Refaie, M.M.; Amin, E.F.; El-Tahawy, N.F.; Abdelrahman, A.M. Possible protective effect of diacerein on doxorubicin-induced nephrotoxicity in rats. *J. Toxicol.* **2016**, *2016*, 9507563. [CrossRef] [PubMed]
2. Hekmat, A.S.; Chenari, A.; Alipanah, H.; Javanmardi, K. Protective effect of alamandine on doxorubicin-induced nephrotoxicity in rats. *BMC Pharm. Toxicol.* **2021**, *22*, 31.
3. Özgermen, B.; Bulut, G.; Alpaslan Pinarli, F.; Gültekin, S.; Özen, D.; Yavuz, O.; Haydardedeoglu, A.E. Investigation of the effects of fetal rat kidney-derived mesenchymal stem cells implementation on doxorubicin-induced nephropathy in male Sprague–Dawley rats. *Ank. Univ. Vet. Fak. Derg.* **2022**, *69*, 201–209. [CrossRef]
4. Saad, S.Y.; Najjar, T.A.; Al-Rýkabý, A.C. The preventive role of deferoxamine against acute doxorubicin-induced cardiac, renal and hepatic toxicity in rats. *Pharmacol. Res.* **2001**, *43*, 211–218. [CrossRef] [PubMed]
5. Khodeer, D.M.; Mehanna, E.T.; Abushouk, A.I.; Abdel-Daim, M.M. Protective Effects of Evening Primrose Oil against Cyclophosphamide-Induced Biochemical, Histopathological, and Genotoxic Alterations in Mice. *Pathogens* **2020**, *9*, 98. [CrossRef]
6. Rafiee, Z.; Moaiedi, M.Z.; Gorji, A.V.; Mansouri, E. p-Coumaric Acid Mitigates Doxorubicin-Induced Nephrotoxicity Through Suppression of Oxidative Stress, Inflammation and Apoptosis. *Arch. Med. Res.* **2020**, *51*, 32–40. [CrossRef]
7. Saad, Z.A.; Khodeer, D.M.; Zaitone, S.A.; Ahmed, A.A.M.; Moustafa, Y.M. Exenatide Ameliorates Experimental Non-Alcoholic Fatty Liver in Rats via Suppression of Toll-like Receptor 4/NFκB Signaling: Comparison to Metformin. *Life Sci.* **2020**, *253*, 117725. [CrossRef]
8. Afsar, T.; Razak, S.; Almajwal, A.; Al-Disi, D. Doxorubicin-induced alterations in kidney functioning, oxidative stress, DNA damage, and renal tissue morphology; Improvement by *Acacia hydaspica* tannin-rich ethyl acetate fraction. *Saudi J. Biol. Sci.* **2020**, *27*, 2251–2260. [CrossRef]
9. Sami, M.M.; Ali, E.A.; Galhom, R.A.; Youssef, A.M.; Mohammad, H.M. Boswellic acids ameliorate doxorubicin-induced nephrotoxicity in mice: A focus on antioxidant and anti-apoptotic effects. *Egypt. J. Basic Appl. Sci.* **2019**, *6*, 10–24. [CrossRef]
10. Deaval, D.G.; Martin, E.A.; Horner, J.M. Drug-induced oxidative stress and toxicity. *J. Toxicol.* **2012**, *21*, 645460. [CrossRef]
11. Crook, M.A. *Clinical Biochemistry and Metabolic Medicine*, 8th ed.; Hodder and Stoughton Ltd.: London, UK, 2012.
12. El-Rahman, G.I.A.; Behairy, A.; Elseddawy, N.M.; Batiha, G.E.-S.; Hozzein, W.N.; Khodeer, D.M.; Abd-Elhakim, Y.M. Saussurea Lappa Ethanolic Extract Attenuates Triamcinolone Acetonide-Induced Pulmonary and Splenic Tissue Damage in Rats via Modulation of Oxidative Stress, Inflammation, and Apoptosis. *Antioxidants* **2020**, *9*, 396. [CrossRef] [PubMed]
13. Montaser, M.; Ebiya, R.A.; Afifi, M.; Saddick, S.; Allogmani, A.S.; Almaghrabi, O.A. Effect of natural and synthetic food colorants on spermatogenesis and the expression of its controlling genes. *Slov. Vet. Res.* **2018**, *55*, 187–199. [CrossRef]
14. Madbouly, S.; Ebiya, R. Histopathological and molecular assessment of moringa and sesame oil as protective potential in gentamicin induced nephrotoxicity. *Cienc. Tec. Vitivinic.* **2018**, *33*, 248–267.
15. Zhang, X.; Yeung, E.D.; Wang, J.; Panzhinskiy, E.E.; Tong, C.; Li, W.; Li, J. Isoliquiritigenin, a natural antioxidant, selectively inhibits the proliferation of prostate cancer cells. *Clin. Exp. Pharmacol. Physiol.* **2010**, *4*, 841–847. [CrossRef] [PubMed]
16. Atrooz, O.M.; Al-Sabayleh, M.A.; Al-Abbadi, S.Y. Studies on physical and chemical analysis of various honey samples and their antioxidant activities. *J. Biol. Sci.* **2008**, *8*, 1338–1342. [CrossRef]
17. Ahmed, S.; Othman, N.H. Honey as a potential natural anti-cancer agent: A review of its mechanisms. *Evid.-Based Complement. Altern. Med.* **2013**, *2013*, 829070. [CrossRef] [PubMed]
18. Subramanian, A.P.; John, A.A.; Vellayappan, M.V.; Balaji, A.; Jaganathan, S.K.; Mandal, M.; Supriyanto, E. Honey and its phytochemicals: Plausible agents in combating colon cancer through its diversified actions. *J. Food Biochem.* **2016**, *40*, 613–629. [CrossRef]
19. Pasupuleti, V.; Sammugam, L.; Ramesh, N.; Gan, S.H. Honey, Propolis, and royal jelly: A comprehensive review of their biological actions and health benefits. *Oxid. Med. Cell Longev.* **2017**, *2017*, 21. [CrossRef]
20. Abeshu, M.A.; Geleta, B. Medicinal uses of honey. *Biol. Med.* **2016**, *8*, 1–7.
21. Alvarez-Suarez, J.M.; Giampieri, F.; Battino, M. Honey as a source of dietary antioxidants: Structures, bioavailability, and evidence of protective effects against human chronic diseases. *Curr. Med. Chem.* **2013**, *20*, 621–638. [CrossRef]
22. Alhumaydhi, F.A. Biochemical studies on the protective effect of honey against doxorubicin-induced toxicity in BALB/C mice. *Int. J. Health Sci.* **2020**, *14*, 31.
23. Ibrahim, R.S.; El-Banna, A.A. Royal jelly fatty acids bioprofiling using TLC-MS and digital image analysis coupled with chemometrics and non-parametric regression for discovering efficient biomarkers against melanoma. *RSC Adv.* **2021**, *11*, 18717. [CrossRef] [PubMed]
24. Pan, Y. Royal Jelly Reduces Cholesterol Levels, Ameliorates Abeta Pathology and Enhances Neuronal Metabolic Activities in a Rabbit Model of Alzheimer's Disease. *Front. Aging Neurosci.* **2018**, *10*, 50. [CrossRef] [PubMed]
25. Kunugi, H.; Ali, M.A. Royal Jelly and Its Components Promote Healthy Aging and Longevity: From Animal Models to Humans. *J. Mol. Sci.* **2019**, *20*, 4662. [CrossRef] [PubMed]
26. Bhadauria, M. Propolis prevents hepatorenal injury induced by chronic exposure to carbon tetrachloride. *Evid.-Based Complement. Altern. Med.* **2012**, *2012*, 235358. [CrossRef] [PubMed]

27. Mohamed, E.K.; Osman, A.A.; Moghazy, A.M.; Rahman, A.; AS, A. Propolis Protective Effects Against Doxorubicin-Induced Multi-Organ Toxicity via Suppression of Oxidative Stress, Inflammation, Apoptosis, and Histopathological Alterations in Female Albino Rats. *Biointerface Res. Appl. Chem.* **2021**, *12*, 1762–1777.
28. Zhao, X.; Zhang, J.; Tong, N.; Chen, Y.; Luo, Y. Protective effects of berberine on doxorubicin-induced hepatotoxicity in mice. *Biol. Pharm. Bull.* **2012**, *35*, 796–800. [CrossRef]
29. Fawcett, J.K.; Scott, J.E. A rapid and precise method for the determination of urea. *J. Clin. Pathol.* **1960**, *13*, 156–159. [CrossRef]
30. Schirmeister, J.; Willmann, H.; Kiefer, H. Plasma Creatinine as Rough Indicator of Renal Function. *Dtsch. Med. Wochenschr.* **1964**, *89*, 1940.
31. Satoh, K. Serum lipid peroxide in cerebrovascular disorders determined by a new colorimetric method. *Clin. Chim. Acta* **1978**, *90*, 37.
32. Ohkawa, H.; Ohishi, W.; Yagi, K. Anal Assay for lipid peroxides in animal tissues by thiobarbituric acid reaction. *Anal. Biochem.* **1979**, *95*, 351. [CrossRef]
33. Nishikimi, M.; Roa, N.A.; Yogi, K. The occurrence of superoxide anion in the reaction of reduced phenazine methosulfate and molecular oxygen. *Biochem. Biophys. Res. Commun.* **1972**, *46*, 849–854. [CrossRef]
34. Paglia, D.E.; Valentine, W.N. Studies on the quantitative and qualitative characterization of erythrocyte glutathione peroxidase. *J. Lab. Clin. Med.* **1967**, *70*, 158–169. [PubMed]
35. Taylor, P.C. Anti-TNF therapy for rheumatoid arthritis and other inflammatory diseases. *Mol. Biotechnol.* **2001**, *19*, 153–168. [CrossRef]
36. Bancroft, J.; Gamble, M. *Theory and Practice of Histological Techniques*; Churchill Livingstone Elsevier: London, UK, 2002.
37. Wang, L.; Liu, X.; Chen, H.; Chen, Z.; Weng, X.; Qiu, T.; Liu, L. Effect of picroside II on apoptosis induced by renal ischemia/reperfusion injury in rats. *Exp. Ther. Med.* **2015**, *9*, 817–822. [CrossRef]
38. Ebiya, R.A.; Montaser, M.M.; Darwish, S.M. Downregulated StAR gene and male reproductive dysfunction caused by nifedipine and ethosuximide. *J. Basic Appl. Zool.* **2016**, *76*, 42–51. [CrossRef]
39. Montaser, M.; Mahmoud, S.; Mahdi, E.; El-Hallous, E.; Ebiya, R.A. Molecular and histopathological assessment of Arabic Coffee effect on fertility in Male Albino Rats. *Res. J. Pharm. Biol. Chem. Sci.* **2017**, *8*, 666–677.
40. Ruggiero, A.; Ferrara, P.; Attinà, G.; Rizzo, D.; Riccardi, R. Renal toxicity and chemotherapy in children with cancer. *Brit. J. Clin. Pharmacol.* **2017**, *83*, 2605–2614. [CrossRef]
41. Abdelmeguid, N.E.; Chmaisse, H.N.; Zeinab, N.A. Protective effect of silymarin on cisplatin-induced nephrotoxicity in rats. *Pak. J. Nutr.* **2010**, *9*, 624–636. [CrossRef]
42. Khan, N.; Sultana, S. Abrogation of potassium bromate induced renal oxidative stress and subsequent cell proliferation response by soy isoflavones in Wistar rats. *Toxicology* **2004**, *201*, 173–184. [CrossRef]
43. Razaa, Z.; Naureen, Z. Melatonin ameliorates the drug induced nephrotoxicity: Molecular insights. *Nefrologia* **2020**, *40*, 12–25. [CrossRef] [PubMed]
44. Hamad, R.; Jayakumar, C.; Ranganathan, P.; Mohamed, R.; El-Hamamy, M.M.; Dessouki, A.A.; Ibrahim, A.; Ramesh, G. Honey feeding protects kidney against cisplatin nephrotoxicity through suppression of inflammation. *Clin. Exp. Pharm. Physiol.* **2015**, *42*, 843–848. [CrossRef] [PubMed]
45. Oktem, G.; Ayla, S.; Seckin, I.; Tanriverdi, G.; Cengiz, M.; Eser, M. Doxorubicin induced nephrotoxicity: Protective effect of nicotinamide. *Int. J. Cell Biol.* **2011**, *2011*, 390238.
46. Bhalchandra, W.; Alqadhi, Y.A. Administration of Honey and Royal Jelly Ameliorate Cisplatin-Induced Changes in Liver and Kidney Function in Rat. *Biomed. Pharmacol. J.* **2018**, *11*, 2191–2199.
47. Al-Yahya, M.; Mothana, R.; Al-Said, M.; Al-Dosari, M.; Al-Musayeib, N.; Al-Sohaibani, M.; Parvez, M.K.; Rafatullah, S. Attenuation of CCl4-Induced Oxidative Stress and Hepatonephrotoxicity by Saudi Sidr Honeyin Rats. *Evid.-Based Complement. Altern. Med.* **2013**, *10*, 569037.
48. Baykara, M.; Silici, S.; Özçelik, M.; Güler, O.; Erdoğan, N.; Bilgen, M. In vivo nephroprotective efficacy of Propolis against contrast induced nephropathy. *Diagn. Interv. Radiol.* **2015**, *21*, 317. [CrossRef]
49. Promsan, S.; Jaikumkao, K.; Pongchaidecha, A.; Chattipakorn, N.; Chatsudthipong, V.; Arjinajarn, P.; Lungkaphin, A. Pinocembrin attenuates gentamicin-induced nephrotoxicity in rats. *Can. J. Physiol. Pharmacol.* **2016**, *94*, 808–818. [CrossRef]
50. Cauich-Kumul, R.; Campos, M.R. Bee Propolis: Properties, Chemical Composition, Applications, and Potential Health Effects. In *Bioactive Compounds*; Woodhead Publishing: Sawston, UK, 2019; pp. 227–243.
51. El-Sheikh, A.A.; Morsy, M.A.; Mahmoud, M.M.; Rifaai, R.A.; Abdelrahman, A.M. Effect of coenzyme-Q10 on doxorubicin-induced nephrotoxicity in rats. *Adv. Pharm. Sci.* **2012**, *2012*, 981461. [CrossRef]
52. Balli, E.; Mete, U.O.; Tuli, A. Effect of melatonin on the cardiotoxicity of Doxorubicin. *Histol. Histopathol.* **2004**, *19*, 1101–1108.
53. Al-Waili, N.S. Identification of nitric oxide metabolites in various honeys: Effects of intravenous honey on plasma and urinary nitric oxide metabolites concentrations. *J. Med. Food* **2003**, *6*, 359–364. [CrossRef]
54. Al-Seeni, M.N.; El Rabey, H.A.; Al-Solamy, S.M. The protective role of bee honey against the toxic effect of melamine in the male rat kidney. *Toxicol. Ind. Health* **2015**, *31*, 485–493. [CrossRef] [PubMed]
55. Singh, K.P.; Zaidi, A.I.A.; Raisuddin, S.; Saxena, A.K.; Murthy, R.C.; Ray, P.K. Effect of zink on immune functions and host resistance against infectionand tumor challence. *Immunopharmacol. Immunotoxicol.* **1992**, *14*, 813–840. [CrossRef] [PubMed]

56. Geiger, S.; Lange, V.; Suhl, P.; Heinemann, V.; Stemmler, H.J. Anti-cancer therapy induced cardiotoxicity: Review of thl literature. *Anti-Cancer Drugs* **2010**, *21*, 578–590. [CrossRef] [PubMed]
57. Erejuwa, O.O.; Sulaiman, S.A.; Ab Wahab, M.S. Honey: A novel antioxidant. *Molecules* **2012**, *17*, 4400–4423. [CrossRef]
58. Yassin, S.A.; Abdel-Moneim, A.; Abdel-Reheim, E.S. Protective Effect of Propolis on Doxorubicin Induced Cardio- and Nephrotoxicity. *Ann. Cardiol. Vasc. Med.* **2020**, *3*, 1028.
59. Neale, T.J.; Rüger, B.; Macaulay, H.; Dunbar, P.R.; Hasan, Q.; Bourke, A. Tumor necrosis factor-alpha is expressed by glomerular visceral epithelial cells in human membranous nephropathy. *Am. J. Pathol.* **1995**, *146*, 1444–1454.
60. Hosseinian, S.; Khajavi, R.A.; Ebrahimzadeh, A.; Soukhtanloo, M.; Sadeghnia, H.R.; Shafei, M.N. Thymoquinone ameliorates renal damage in unilateral ureteral obstruction in rats. *Pharmacol. Rep.* **2017**, *69*, 648–657. [CrossRef]
61. Al-Saedi, H.F.; Al-Zubaidy, A.A.; Khattab, Y.I. Thepossible effects of montelukast against doxorubicin-induced nephrotoxicity in rabbits. *Int. J. Adv. Res.* **2014**, *2*, 723–729.
62. Chien, S.; Wu, Y.; Chen, Z.; Yang, W. Naturally occurring anthraquinones: Chemistry and therapeutic potential in autoimmune diabetes. *Evid.-Based Complement. Altern. Med.* **2015**, *2015*, 357357. [CrossRef]
63. Thi Lan Nguyen, H.; Panyoyai, N.; Kasapis, S.; Pang, E.; Mantri, N. Honey and Its Role in Relieving Multiple Facets of Atherosclerosis. *Nutrients* **2019**, *11*, 167. [CrossRef]
64. Ahmad, A.; Khan, R.A.; Mesaik, M.A. Anti-inflammatory effect of natural honey on bovine thrombin-induced oxidative burst in phagocytes. *Phytother. Res.* **2009**, *23*, 801–808. [CrossRef] [PubMed]
65. Kassim, M.; Yusoff, K.M.; Ong, G.; Sekaran, S.; Yusof, M.Y.B.M.; Mansor, M. Gelam honey inhibits lipopolysaccharide-induced endotoxemia in rats through the induction of heme oxygenase-1 and the inhibition of cytokines, nitric oxide, and high-mobility group protein B1. *Fitoterapia* **2012**, *83*, 1054–1059. [CrossRef] [PubMed]
66. Rizk, S.M.; Zaki, H.F.; Mina, M.A.M. Propolis Attenuates Doxorubicin-Induced Testicular Toxicity in Rats. *Food Chem. Toxicology* **2014**, *67*, 176–186. [CrossRef] [PubMed]
67. Köse, E.; Oğuz, F.; Vardi, N.; Sarihan, M.E.; Beytur, A.; Yücel, A.; Polat, A.; Ekïncï, N. Therapeutic and protective effects of montelukast against doxorubicininduced acute kidney damage in rats. *Iran. J. Basic Med. Sci.* **2019**, *22*, 407–411. [CrossRef] [PubMed]
68. Weng, M.S.; Ho, Y.S.; Lin, J.K. Chrysin induces G1 phase cell cycle arrest in C6 glioma cells through inducing p21Waf1/Cip1 expression: Involvement of p38 mitogen-activated protein kinase. *Biochem. Pharmacol.* **2005**, *69*, 1815–1827. [CrossRef] [PubMed]
69. Nakamura, T.; Katsuda, S.; Takahashi, H.; Koh, E. Fas-mediated apoptosis in adriamycin-induced cardiomyopathy in rats: In vivo study. *Circulation* **2000**, *102*, 572–578. [CrossRef]
70. Rashid, S.; Ali, N.; Nafees, S.; Ahmad, S.T.; Arjumand, W.; Hasan, S.K.; Sultana, S. Alleviation of doxorubicin-induced nephrotoxicity and hepatotoxicity by chrysin in Wistar rats. *Toxicol. Mech. Methods* **2013**, *23*, 337–345. [CrossRef]
71. Szliszka, E.; Krol, W. Polyphenols isolated from Propolis augment TRAILinduced apoptosis in cancer cells. *Evid.-Based Complement. Altern. Med.* **2013**, *2013*, 731940. [CrossRef]
72. Krishnakumar, R.; Kraus, W.L. The PARP side of the nucleus: Molecular actions, physiological outcomes, and clinical targets. *Mol. Cell* **2010**, *39*, 8–24. [CrossRef]
73. Ha, H.C.; Snyder, S.H. Poly (ADP-ribose) polymerase is a mediator of necrotic cell death by ATP depletion. *Proc. Natl. Acad. Sci. USA* **1999**, *96*, 13978–13982. [CrossRef]
74. Szabó, C.; Dawson, V.L. Role of poly (ADP-ribose) synthetase in inflammation and ischaemia-reperfusion. *Trends Pharm. Sci.* **1998**, *19*, 287–298. [CrossRef]
75. Jagtap, P.; Szabó, C. Poly(ADP-ribose) polymerase and the therapeutic effects of its inhibitors. *Nat. Rev. Drug Discov.* **2005**, *4*, 421–440. [CrossRef] [PubMed]
76. Mamdouh, A.M.; Khodeer, D.M.; Tantawy, M.A.; Moustafa, Y.M. In-Vitro and in-Vivo Investigation of Amygdalin, Metformin, and Combination of Both against Doxorubicin on Hepatocellular Carcinoma. *Life Sci.* **2021**, *285*, 119961. [CrossRef] [PubMed]
77. Hassen, M.T.; Mohamed, H.K.; Montaser, M.M.; El-Sharnouby, M.E.; Awad, N.; Ebiya, R.A. Molecular, Immunomodulatory, and Histopathological Role of Mesenchymal Stem Cells and Beetroot Extract on Cisplatin Induced Testicular Damage in Albino Rats. *Animals* **2021**, *11*, 1142. [CrossRef]
78. Elkazzaz, S.K.; Khodeer, D.M.; El Fayoumi, H.M.; Moustafa, Y.M. Role of Sodium Glucose Cotransporter Type 2 Inhibitors Dapagliflozin on Diabetic Nephropathy in Rats; Inflammation, Angiogenesis and Apoptosis. *Life Sci.* **2021**, *280*, 119018. [CrossRef]
79. Erejuwa, O.O.; Sulaiman, S.A.; Wahab, M.S.; Sirajudeen, K.N.S.; Salleh, M.M.; Gurtu, S. Antioxidant Protection of Malaysian Tualang Honey in Pancreas of Normal and Streptozotocininduced Diabetic Rats. In *Annales d'Endocrinologie*; Elsevier: Amsterdam, The Netherlands, 2010; pp. 291–296.
80. Kocot, J.; Kiełczykowska, M.; Luchowska-Kocot, D.; Kurzepa, J.; Musik, I. Antioxidant Potential of Propolis, Bee Pollen, and Royal Jelly: Possible Medical Application. *Oxid. Med. Cell. Longev.* **2018**, *2*, 7074209. [CrossRef]

antioxidants

MDPI

Review

Exploring the Potential of Bee-Derived Antioxidants for Maintaining Oral Hygiene and Dental Health: A Comprehensive Review

Poonam Choudhary [1], Surya Tushir [1,*], Manju Bala [2], Sanjula Sharma [3], Manjeet Kaur Sangha [4], Heena Rani [5], Nileshwari Raju Yewle [6], Parminder Kumar [7], Diksha Singla [5], Deepak Chandran [8], Manoj Kumar [9,*] and Mohamed Mekhemar [10,*]

[1] Department of Agricultural Structures and Environment Control, ICAR-Central Institute of Post-Harvest Engineering and Technology, Ludhiana 141004, India; poonam@icar.gov.in
[2] Department of Food Grain and Oilseed Processing, ICAR-Central Institute of Post-Harvest Engineering and Technology, Ludhiana 141004, India; manju.bala1@icar.gov.in
[3] Oilseeds Section, Department of Plant Breeding and Genetics, Punjab Agricultural University, Ludhiana 141004, India; drsanjula@pau.edu
[4] Department of Biochemistry, College of Basic Sciences and Humanities, Punjab Agricultural University, Ludhiana 141004, India; manjeet_kaursangha@pau.edu
[5] Department of Biochemistry, Punjab Agricultural University, Ludhiana 141004, India; heena-1860003@pau.edu (H.R.); singladiksha3@gmail.com (D.S.)
[6] Department of Entomology, Purdue University, West Lafayette, IN 47907, USA; nyewle@purdue.edu
[7] Department of Soil Science, Punjab Agricultural University, Ludhiana 141004, India; pk2704@pau.edu
[8] Department of Animal Husbandry, Government of Kerala, Palakkad 679335, India; drdeepakchandran24@gmail.com
[9] Chemical and Biochemical Processing Division, ICAR-Central Institute for Research on Cotton Technology, Mumbai 400019, India
[10] Clinic for Conservative Dentistry and Periodontology, School of Dental Medicine, Christian-Albrecht's University, 24105 Kiel, Germany
* Correspondence: surya@icar.gov.in (S.T.); manoj.kumar13@icar.gov.in (M.K.); mekhemar@konspar.uni-kiel.de (M.M.)

Citation: Choudhary, P.; Tushir, S.; Bala, M.; Sharma, S.; Sangha, M.K.; Rani, H.; Yewle, N.R.; Kumar, P.; Singla, D.; Chandran, D.; et al. Exploring the Potential of Bee-Derived Antioxidants for Maintaining Oral Hygiene and Dental Health: A Comprehensive Review. *Antioxidants* **2023**, *12*, 1452. https://doi.org/10.3390/antiox12071452

Academic Editor: Zdeňka Ďuračková

Received: 9 May 2023
Revised: 12 July 2023
Accepted: 14 July 2023
Published: 19 July 2023

Abstract: Honey bee products comprise various compounds, including honey, propolis, royal jelly, bee pollen, bee wax and bee venom, which have long been recognized for their pharmacological and health-promoting benefits. Scientists have discovered that periodontal disorders stem from dental biofilm, an inflammatory response to bacterial overgrowth produced by dysbiosis in the oral microbiome. The bee products have been investigated for their role in prevention of oral diseases, which are attributed to a myriad of biologically active compounds including flavonoids (pinocembrin, catechin, caffeic acid phenethyl ester (CAPE) and galangin), phenolic acids (hydroxybenzoic acid, hydroxycinnamic acid, p-coumaric, ellagic, caffeic and ferulic acids) and terpenoids. This review aims to update the current understanding of role of selected bee products, namely, honey, propolis and royal jelly, in preventing oral diseases as well as their potential biological activities and mechanism of action in relation to oral health have been discussed. Furthermore, the safety of incorporation of bee products is also critically discussed. To summarize, bee products could potentially serve as a therapy option for people suffering from a variety of oral disorders.

Keywords: bee products; honey; propolis; royal jelly; oral care; bioactivities; oral pathology

1. Introduction

Honey bees are members of the genus Apis, which means "bee" in Latin. The prefix "api" is frequently used in beekeeping terms such as apiarist or a beekeeper. Apitherapy is defined as the use of beehive produces such as honey, pollen, royal jelly, propolis, bee venom and wax for treating and healing of ailments as well as in boosting the human immune system [1,2]. The origin of apitherapy can be pinpointed to more than 6000 years back in medicine of ancient Egypt [3]. The bee products are rich in natural antioxidants, and these are commonly used as natural remedies for health maintenance in traditional medicine in many countries. Honey bee products are popular among people of all ages, and are used across cultural and ethnic boundaries, and all religious and cultural beliefs promote and embrace the usage of honey bee products [4]. Among the various bee products, honey, propolis, and royal jelly have been discussed for improving oral health in the present review.

Natural honey is the world's oldest sweetener, with records indicating that it was used around the planet several million years ago [5]. Honey is produced by the honey bees that collect nectar from flowers and process it through repeated digestion and regurgitation. It comprises fructose (38.5%), glucose (31%), other sugars (12.9%), water (17.1%), protein (0.5%) and minerals such as Ca, Co, Fe, Mg, Mn, Zn, Na, F and K [6,7]. In addition, it comprises valuable bioactive compounds, such as phenolic acids, carotenoids, flavonoids, organic acids, ascorbic acid, enzymes and other proteins [8–11].

Propolis, also known as "bee glue", is the third most important bee product after honey and wax. The term "propolis" comes from the Greek words "pro" for defense and "polis" for city or community [12]. Its complex composition includes phenolic compounds (58%), beeswax (24%), lipids and wax (8%), flavonoids (6%), terpenes (0.5%), bio-elements (0.5%) and other substances (3%) [13,14]. It provides thermal insulation and protects the hive from invaders and is used to strengthen the hive by filling cracks and holes [13,15,16].

Royal jelly is a white, viscous jelly-like material secreted by worker bees' hypopharyngeal and mandibular glands [17]. Royal jelly consists of water (50–60%), proteins (18%), carbohydrates (15%), lipids (3–6%), mineral salts (1.5%) and vitamins [18–20]. With around 185 organic compounds, royalactin is one of the major royal jelly proteins (MRJPs) [18–20]. Besides polyphenolic and flavonoid compounds, royal jelly is rich in essential amino acids, small peptides, fatty acids and vitamins. Other prominent carboxylic acids discovered in royal jelly include sebacic acid (SA) and 10-hydroxy-2-decenoic acid (10H2DA), which has not been recorded in any other natural raw material or even in any other apiculture product [21].

According to a report, oral diseases are becoming a matter of concern globally [22]. Oral health, in turn, affects the systemic health. An increased risk of cardiovascular diseases, digestive problems, diabetes and bacterial pneumonia has been reported in patients with poor oral health [23]. Various studies have explained the unique bioactive components of honey bee products that confer therapeutic and medicinal properties and improve the overall health. The bioactive compounds found in honey, propolis and royal jelly are known to possess antioxidant, anti-microbial, anti-inflammatory, anti-cancer, anti-ulcer, immunomodulatory, neuromodulatory and metabolic syndrome preventing activities [24–36]. Due to these properties, these products are being used/or show wide applications in preventing oral diseases like stomatitis, mouth ulcers, dental caries, plaque, gingivitis, periodontitis, cavity infection, dentin hypersensitivity, oral cancer, malodor and mucositis [15,37–40].

In today's health-conscious society, honey bee products are gaining attention in traditional and modern medicine [41]. Application of bee products alleviates the consequences of oral disorders in a cost-effective manner. Thus, it is a challenge for researchers all over the world to make their effective use and realize their maximum medicinal benefits. There are no critical miscellanea on key facts about the role of bee products in treatment of oral diseases. As a result, the focus of this review will be on the major bioactive components, possible pharmacological characteristics and mechanism of action of bee products in the treatment of oral diseases. Figure 1 depicts the various components as discussed in this review.

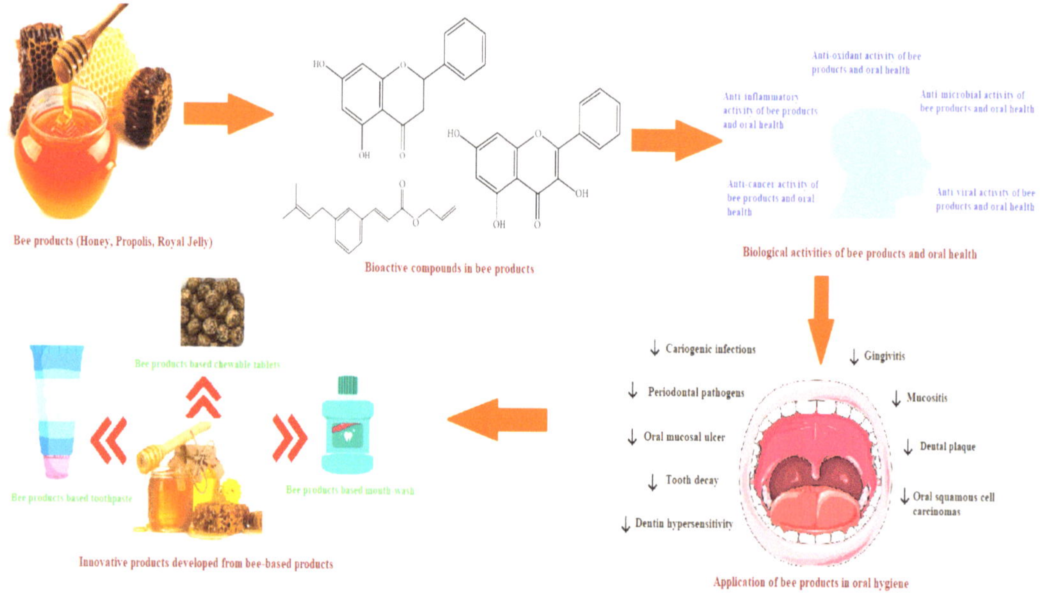

Figure 1. Various components discussed in current review.

2. Bioactive Compounds of Bee Products in Relation to Oral Health

Natural bee products (honey, propolis and royal jelly) are generally considered as high-quality sources of bioactive compounds, having impressive biomedical potential. The therapeutic effect of these products is associated with their bioactive composition and plant source from which they are derived [42].

Bioactive Compounds

Honey is rich in bioactive compounds such as phenolic acids and flavonoids, which contribute to its beneficial properties [8–10,43]. The composition of these compounds in honey can vary based on factors such as environmental conditions, geographical location, production process, and the specific flora from which the honey bees collect nectar [10,15,43,44]. Dark-colored honey is believed to contain higher levels of bioactive chemicals compared to light-colored honey. For example, manuka honey has been found to contain compounds such as pinocembrin, chrysin, pinobanksin, kaempferol, luteolin, isorhamnetin, galangin, sakuranetin, quercetin and magniferolic acid [45]. Similarly, honey produced from different plant sources, such as *Echium plantagineum* L., may exhibit variations in their phenolic acid and flavonoid content [46]. Flavonoids are another significant group of bioactive components found in honey. Their content can vary among different honey types, with values ranging from 1.93 to 21.16 mg of quercetin equivalents per 100 g in samples such as sunflower and Heather honey [47]. Flavonoids possess antioxidant properties and contribute to protecting cell membranes by reducing lipid peroxidation and scavenging free radicals [48,49]. The composition of flavonoids in honey is influenced by floral origin and geographical location. For example, chrysin and apigenin are commonly found flavonoids in honey from Spain and New Zealand, while orange blossom honey has a significant amount of quercetin [50]. Propolis, another bee-derived product, contains bioactive compounds such as flavonoids (apigenin, acacetin, chrysin, catechin, naringenin, luteolin, galangin, kaempferol, rutin, pinocembrin, myricetin and quercetin) and phenolic acids (cinnamic acid and caffeic acid) [51–53]. The composition of propolis varies depending on its geographical and botanical origins [54]. Cinnamic acid derivatives and flavonoids, collectively known as citrin and vitamin P, are considered prime bioactive compounds in propolis [4,55]. The specific composition of propolis differs among regions, with caffeic acid phenethyl ester (CAPE) being the most common compound in temperate propolis, geranyl flavanones prevalent in Pacific and African propolis, and prenylated phenyl propanoids, acetophenones, terpenoids and aromatic acid derivatives found in tropical regions [4,55].

Royal jelly is composed of a small amount of minor bioactive compounds, including flavonoids such as flavanones (hesperetin, naringenin, isosakuranetin), flavones (chrysin, acacetin, luteolin, epigenin and its glycoside), flavonols (kaempferol and isorhamnetin glycosides) and isoflavonoids (coumestrol, genistein, formononetin) [18,56,57]. Phenolic acids such as octanoic acid, dodecanoic acid, pinobanksin and 1,2-benzenedicarboxylic acid, along with their esters, are also present in royal jelly [18,56]. The concentration of these phenolic acids and flavonoids in royal jelly is influenced by several factors, including the type of plants used by bees, seasonal variations and environmental factors [58,59]. In a study by Nagai and Inoue [60], the total phenolic compound content in royal jelly powder was reported as 21.2 μg/mg in water and 22.8 μg/mg in an alkaline extract. Different bioactive compounds in bee products and their actions are represented in Table 1.

Table 1. Bioactive compounds with their biological activities in bee products.

Source of Bee Product	Group	Bioactive Compounds	Biological Activity	References
Honey	Flavonoids	flavonoles ([a,b,c,g] quercetin, [b,g] galangin, [b,f,g] fisetin, [b] myricetin) flavanones ([a,b] pinocembrin, [c] pinobanksin, [e] naringenin, [c,e] hesperetin) flavones ([c,f] apigenin, [b,g] acacetin, [b,c,f] chrysin, [b,c,e] Luteolin, [c] genistein, [b] wagonin) and [b] caffeic acid phenethyl ester	[a] anti-microbial [b] anti-cancer [c] anti-inflammatory [d] anti-fungal [e] antioxidant [f] anti-bacterial [g] anti-allergic [h] anti-genotoxic [i] neuroprotective [j] anti-anxiolytic [k] chemoprotective [l] anti-proliferative	[61–65]
	Phenolic acid	[h,i] p-coumaric acid, [j] gallic acid, [e,k,l] ellagic acid, [c,e,i] ferulic acid, [b,e] syringic acid		[64]
Propolis	Flavonoids	flavonoles ([a,b,c,g] quercetin, [b,g] galangin, [b,f,g] fisetin) flavanones([a,b] pinocembrin) flavones ([c,f] Apigenin, [b,g] acacetin, [b,c,f] chrysin) and [b] caffeic acid phenethyl ester (CAPE)		[61–63]
	Phenolic acid	[a] 2,2-dimethyl-8-phenylchromene, [a,b,c] 4-hydroxy-3,5-diprenyl cinnamic acid (artepillin C), [a] 3-prenyl cinnamic acid allyl ester, [b] kaempferide, [d] benzofuran,		[61]
	Terpenoid	[d] isocupressic acid, [b] symphyoreticulic acid, [a,b,e] procrim a and b, [a,b,e] lupeol, [d] farnesol		[61,66]
Royal jelly	Flavonoids	flavonoles (e.g., [a,b,c,g] quercetin, [b] kaempherol, [b,g] galangin and [b,f,g] fisetin) flavanones (e.g., [a,b] pinocembrin, [c] naringin, [c,e] hesperidin and isosakuranetin) flavones (e.g., [c,f] apigenin, [b,g] acacetin, [b,c,f] chrysin and [b,c,e] luteolin)		[67]
	Phenolic acid	2,2-dimethyl-8-prenylchromene, 3-prenyl cinnamic acid allyl ester, artepillin C		[63]
	Terpenoid	isocupressic acid, labdane diterpenoid		[63]

The bioactive compounds and their associated biological activities are presented in subscripted small letters.

3. Biological Activities of Bee Products in Relation to Oral Health

3.1. Antioxidant Activity of Bee Products and Its Effects on Diseases of the Oral Cavity

Honey has been reported to show antioxidant activity since the phenolic level is related to radical scavenging activity of honey and other bee products [64–66]. Polyphenols are regarded to be the key components responsible for honey and bee products' antioxidant properties. These components have the ability to counteract the effects of oxidative stress, which is a key factor in the development of many diseases. [67]. In the area of dental medicine, the awareness of the potential benefits of bee products as antioxidants is rising. During normal metabolism, reactive oxygen species (ROS) are generated, which include free radicals (superoxide anion radical, hydroxyl radical, etc.) and non-radical molecules (hydrogen peroxide, singlet oxygen, lipoperoxides, etc.). These ROS interact with lipids, protein components in the cell membranes, enzymes and DNA of cell and may cause damage to them, which, in turn, may lead to various diseases. The antioxidants are compounds that seize free radicals before they can cause damage to the cell as they donate hydrogen atoms from their hydroxyl groups to free radicals [68]. According to Zhang et al. [32], polyphenols have the capability to control the nuclear factor kappa B (NF-κB), which is associated with modulating cell signaling pathways involved in cancer and inflammation. Antioxidants are generally given as oral ingestion, diet or vitamin supplements and in the form of nutraceuticals. Natural antioxidants that can be applied topically such as mouth rinse, gel, paste, gum, or lozenge compositions are nowadays gaining importance. These topical antioxidants may help to reduce ROS, which may act as inflammatory factors in the development of gingival and periodontal problems [69].

Honey has a high level of antioxidant activity and can lessen the effects of oxidative reactions that produce free radicals. Buckwheat and Heather honey are typically dark brown or black in color and are the richest sources of antioxidants and can ameliorate oxidative stress [70,71] while manuka honey has been reported to be rich source of flavonoids and also possesses higher antioxidant properties in comparison to other types of honey. Kishore et al. [72] compared antioxidant activity of Tualang, Indian forest, pineapple and Gelam honey samples and reported highest antioxidants in Tualang honey collected from Tulang honey bees in forests of Malaysia. Rosa et al. [73] studied TPC and antioxidant activities in Asphodel, Acacia, Eucalyptus, Heather, Citrus, Honeydew and Strawberry tree honey samples and found that Strawberry tree honey expressed the highest 2,2-diphenylpicrylhydrazyl (DPPH) and ferric reducing ability of plasma (FRAP) activities. They further reported 2,5-dihydroxyphenylacetic acid and homogentisic acid (HGA), major phenolic compound and a chemical marker for strawberry tree honey, were the potent antioxidants. Alvarez–Suarez et al. [74] reported significant radical scavenging activity in the ether-soluble phenolic extracts of Cuban honeys. Yuslianti et al. [75] used Rambutan honey to investigate free radical scavenging capacity (in vitro) and lipid peroxidation inhibition of oral mucosal wounds (in vivo) in male Wistar rats. They reported that 1 mg/mL honey showed 45.3% DPPH inhibition. Significant ($p = 0.028$) reduction in lipid peroxidation in oral mucosa wound tissue was observed. Toczewska et al. [76] reported that the antioxidant capacity in gingival crevicular fluid and saliva of periodontitis patients is diminished as a result of the chronic inflammatory process, which further makes proteins, lipids and DNA prone to oxidative damage and may result in the progressive devastation of the periodontal attachment apparatus. The beneficial effect of honey rich in polyphenolic compounds in enhancing the antioxidant capacity of oral fluids has been reviewed [77]. According to Ding et al. [78], the antioxidant properties of polyphenols appear to give significant protection against oral malignancies. Phenolic chemicals aid in the treatment of periodontal disease, the strengthening of teeth and the prevention of tooth decay [79–81]. Patients with recurrent aphthous stomatitis (RAS) exhibited considerably lower salivary antioxidant levels than healthy controls, according to Babaee et al. [82]. They correlated that ulcerations created ROS, which resulted in a reduction in antioxidant compounds in the mouth. Honey, when applied topically, has antioxidant components that can help minimize the effects of ROS activity during the development of ulcers.

Propolis possesses high concentration of phenolics [83] and is well-known for its antioxidant and radical-scavenging properties [61,81]. Bioactive substances such as pinocembrin, chrysin and pinobanksin have high antioxidant and anti-radical properties [82]. In DPPH and ORAC tests, pinobanksin-3-acetate was found to be the most powerful antioxidant component [84]. Fabris et al. [85] studied ethanol extracts of Russian and Italian propolis and found similarity in their total phenol content and antioxidant activity while low levels of phenolics and antioxidants were observed in ethanolic extract of Brazilian propolis. Zhang et al. [32] reported that antioxidant activity in Brazilian green propolis was due to 4,5-dicaffeoylquinic acid, 3,4,5-tricaffeoylquinic acid, artepillin C compounds and 3,5-dicaffeoylquinic acid. Total polyphenol and total flavonoid levels in poplar propolis contribute to antioxidant action [86]. In vitro investigations by Kumari et al. [87] and

Bonamigo et al. [88] reported that propolis extracts were discovered to have antioxidant properties similar to the synthetic antioxidants, butylated hydroxytoluene or ascorbic acid. The total phenolic content (TPC), antioxidant activity and free radical scavenging activities (FRSA) of 70 Turkish samples of various honey bee products (honey, pollen, royal jelly and propolis) and their mixtures were examined by Ozkök and Silici [89]. Among the studied samples, honey bee propolis exhibited highest antioxidant and FRSA activity. Moreover, TPC was also found to have a positive relationship with antioxidant activity and FRSA. Kocot et al. [90] evaluated the total phenolic content (30 to 200 mg gallic acid equivalents/g DW), DPPH free radical-scavenging activity (20 to 190 g/mL) and flavonoid content (30 to 70 mg quercetin equivalents/g) of several propolis extracts. Aghel et al. [91] studied the useful result of propolis antioxidants on saliva in the test animals. El-Sharkawy [92] used propolis as a dietary supplement and found that when compared to the control group, the propolis-treated group had a substantial decrease in pocket depth (PD) and a rise in clinical attachment level (CAL), which could be due to antioxidant activity and other propolis properties. Giammarinaro et al. [93] compared the propolis efficacy with chlorhexidine in 40 patients of gingivitis and found that propolis-treated patients had improved results in terms of oxidative stress markers in their saliva, as well as significant improvements in their periodontal health. CAPE, extracted from propolis, is a strong antioxidant compound that is potentially used in oral squamous cell carcinoma (OSCC) patients for adjuvant therapy [38].

Royal jelly has been found to have antioxidant action owing to its total phenolic content, fatty acids and proteins. The small peptides contained in royal jelly, which are made up of 2–4 amino acid residues, have a high antioxidant activity. Tyrosine residues at the C-terminus of most active peptides in royal jelly allow them to scavenge hydroxyl radicals and H_2O_2 [94]. Nagai and Inoue [60] investigated the antioxidative activity and scavenging ability of protein fractions of royal jelly against free radicals such as the DPPH (1,1-diphenyl-2-picrylhydrazyl) radical, hydroxyl radical and superoxide anion radical. The antioxidant effect of royal jelly was studied by Silici et al. [95], who found a decrease in malondialdehyde levels and an increase in catalase superoxide and glutathione peroxidase dismutase activities. Park et al. [96] studied the assays of purified recombinant *Apis mellifera* major royal jelly proteins (AmMRJPs) and found that these proteins displayed radical-scavenging activity with 1,1-diphenyl-2-picrylhydrazyl and protection against oxidative DNA damage. Effect of the harvest time and the initial larval age has significant effect on the antioxidant potential in Royal jelly. Royal jelly collected 24 h after the larval transfer had the highest antioxidant activity in terms of DPPH radicals, prevention of linoleic acid peroxidation and reducing power [97]. Anatolian royal jelly samples were examined by Kolayli et al. [21] for their chemical composition and antioxidant capabilities. The total phenolic content ranged from 91.0 to 301.0 mg gallic acid equivalents/kg fresh weight. Anatolian royal jelly samples were found to be similar with other royal jelly samples around the world. Balkanska et al. [98] evaluated the antioxidant activity of royal jelly (RJ) collected from different areas of Bulgaria. The FRAP and total polyphenols showed variability and values ranged from 0.44–8.49 mM Fe^{2+}/g and from 11.66–36.73 (mgGAE/g) for FRAP and total polyphenols, respectively. Uçar and Barlak [99] studied the antioxidant activity of water, dimethyl sulfoxide (DMSO) and methanol extracts of royal jelly from Bursa province in Turkey and found all the extracts showed total phenol content (TPC) and free radical scavenging capacity. The Table 2 represents the different biological activities of bee products and their relation in improving the oral health.

Table 2. Biological activities of bee products in relation to oral health.

Type of Bee Product	Type of Extract	Bioactive Compounds	Type of Study	Biological Activity	Key Findings	References
Honey	Aqueous extract	Mixture of phenols and flavonoids	In vitro and in vivo	Antioxidant	In male Wistar rats, application of 1 mg/mL honey showed 45.3% DPPH inhibition. Significant ($p = 0.028$) reduction in lipid per oxidation in oral mucosa wound tissue was observed.	[75]
	Aqueous extract	Flavonoids	In vitro and in vivo	Anti-bacterial	Honey mouth rinse showed antibacterial characteristics and was found effective against oral infections (in vitro). Plaque development was also inhibited (in vivo).	[100]
	Aqueous extract	Phenolic acids, Flavonoids	In vitro	Anti-fungal	The Algerian honeys with different concentrations (undiluted, 10%, 30%, 50% and 70% *w/v*) were tested against *Candida albicans* and *Rhodotorula* sp. Both species had MICs of 70.09–93.48% and 4.90–99.70% *v/v*, respectively.	[101]
	Aqueous extract	Flavonoids	Clinical trial	Antiviral	In children with primary herpetic gingivostomatitis, combining honey with oral acyclovir can yield better results than acyclovir alone.	[102]
	Aqueous extract	Flavonoids	In vitro	Anti-inflammatory	Honey flavonoid extract (HFE) considerably reduce release of pro-inflammatory cytokines such as tumor necrosis factor alpha (TNF-α) and interleukin -1 (IL-1), according to the findings. The formation of reactive oxygen intermediates and the expression of inducible nitric oxide synthase (iNOS) were also dramatically reduced.	[103]
	Aqueous extract	Phenolic acids	In vitro	Anti-cancer	Oral squamous cell carcinoma (OSCC) and osteosarcoma (HOS) cell lines showed maximum suppression of cell growth of 80% at a concentration of 15%.	[104]
Propolis	Aqueous extract	Flavonoids	Clinical trial	Antioxidant and anti-inflammatory	Propolis reduced and delayed radiation-induced mucositis in rats by being able to prevent reduction in salivary antioxidant levels.	[91]
	3% ethanolic extract (EEP)	Flavonoids:—kaempferol and quercetin and cinnamic acid derivatives	Clinical trial	Anti-bacterial	At 50 mg/L, EEP had a time-dependent microbiological effect with anti-bacterial efficacy against Gram-positive bacteria. Hygienic treatments containing 3% EEP effectively assist plaque clearance and enhance the condition of the marginal periodontium.	[105]
	70% ethanol	Esters of phenolic acids (caffeates and ferulates) and flavonoids	In vitro	Anti-bacterial, Anti-fungal and Antiviral	All the studied samples exhibited significant activity against fungal and Gram-positive bacterial test strains, with the majority also showing antiviral activity.	[106]
	Aqueous and ethanol extracts	Galangin and chrysin	In vitro	Antiviral	In viral suspension experiments, both propolis extracts were found to have significant antiviral efficacy against HSV-1, with plaque formation reduced by >98%.	[107]
	Aqueous extracts	Artepillin C	Clinical trial	Anti-inflammatory	Rinse products containing Brazilian green propolis high in artepillin C reduced gingivitis to the same extent as a NaF/cetylpyridinium chloride rinse or a chlorhexidine solution in randomized, double-blind, placebo-controlled studies.	[108]
	Ethanol extracts	Caffeic acid phenethyl ester	In vitro	Anti-cancer	In TW2.6 human oral squamous cell carcinoma (OSCC) cells, propolis extracted caffeic acid phenethyl ester CAPE treatment reduced cell proliferation and colony formation in a dose-dependent manner. CAPE treatment reduced the number of G1 phase cells, increased the number of G2/M phase cells, and caused death in TW2.6 cells.	[109]

Table 2. *Cont.*

Type of Bee Product	Type of Extract	Bioactive Compounds	Type of Study	Biological Activity	Key Findings	References
Royal Jelly	Aqueous extracts	I-IV jellein peptides	In vitro	Anti-bacterial	Four peptides were isolated from honey bee and Royal Jelly presented exclusively antimicrobial activities against Gram-positive and Gram-negative bacteria.	[29]
	Aqueous extracts	Phenolic compounds	In vitro	Anti-fungal	The MIC, MIC50 and MFC of Royal Jelly on Candida albicans were 80, 103 and 160 mg/mL, respectively, while the MIC, MIC50 and MFC of Iranian Propolis alcoholic extract were 0.030, 0.0618 and 0.0833 mg/mL, respectively.	[110]
	Aqueous extracts	Major royal jelly protein 3 (MRJP3)	Clinical trial	Anti-inflammatory	When RJ suspension was given to a culture of mouse peritoneal macrophages activated with lipopolysaccharide and IFN-gamma, the production of proinflammatory cytokines such as TNF-α, IL-6 and IL-1 was efficiently reduced in a dose-dependent manner without causing macrophage cytotoxicity.	[111]
	Aqueous extracts	10-Hydroxy-2-decenoic acid	In vitro	Anti-cancer	The 10-HDA at 20 M or higher significantly suppressed proliferation and migration of cancerous cells.	[112]

3.2. *Anti-Microbial Activity of Bee-Based Products and Its Effects on Diseases of the Oral Cavity*

The microbial flora associated with oral disease is diverse, including aerobic and anaerobic bacteria, viruses, parasites and other pathogens [113,114]. Oral microbial infections have been treated with a variety of agents, like amine fluorides, chlorhexidine, cetylpyridinium chloride and ethanol, which are commonly present in mouthwashes. But these have been proven to discolor teeth and may be toxic and cause oral malignancies, as well as have an undesirable taste [115]. Furthermore, the interest in such natural product as a possible source of novel antimicrobials has grown due to a concurrent decline in the number of effective antibiotics available and the ever-increasing prevalence of antibiotic resistance among pathogenic bacteria [116,117]. Natural substances discovered in bee-based product such as honey, propolis and royal jelly have been used for its anti-bacterial, anti-fungal and antiviral activities as well as anti-microbial benefits [118]. The antimicrobial properties of propolis against oral pathogens are attributed to the flavonone pinocembrin, amyrins, flavonol galangin and the caffeic acid phenethyl ester by inhibiting the bacterial RNA polymerase [119].

3.2.1. Anti-Bacterial Activity of Bee-Based Products and Its Effects on Diseases of the Oral Cavity

Numerous bacterial microbiotas inhabit in the mouth cavity with *Streptococcus mutans, Porphyromonas gingivalis, Staphylococcus* and *Lactobacillus* being the common oral bacteria [120]. *Lactobacillus* is a bacterium that produces lactic acid by fermenting sugar, which can easily cause caries [121]. *P. gingivalis* is a periodontal pathogen that is a Gram-negative anaerobic bacterium of non-glycolytic nature. *P. gingivalis,* if left untreated, could cause the teeth to fall off from gums. Further, anti-bacterial resistance has become a major problem globally that has prompted scientist community to study active ingredients used before the antibiotic era. Van Ketel, a Dutch scientist, was the first to identify honey's bactericidal properties in 1892 [122]. This anti-bacterial activity is attributed to its properties which include: (i) its hygroscopic nature that can draw moisture out of the environment and dehydrate bacteria, (ii) its high sugar content and acidity (low pH) that prevents the microbes from growth [123], (iii) presence of hydrogen peroxide and phenolic acids [124], flavonoids [125] and lysozyme [126] as oxidizing agents restrict the bacterial responses to proliferative signals due to which bacterial growth remains arrested [127], (iv) phytochemical components like methylglyoxal (MGO) (non-peroxide) trigger modifications in the shape of bacterial flagella and fimbriae that impede bacterial adhesion and motility [128,129] and (v) an anti-microbial peptide, bee defensin-1 [130]. Honey's pH (3.2–4.5) is low enough to suppress specific bacterial infections, such as *Salmonella* spp. (4.0), *E. coli* (4.3), *P. aeruginosa* (4.4) and *S. pyogenes* (4.5) [131], and enhances the healing process of wound through epithelialization [132]. Moreover, its anti-bacterial action was not shown to be affected by loss of its acidity after dilution [133]. Shiga et al. [134] reported the antibacterial action of honey was derived from methylglyoxal component, which help in oral disorders like halitosis (a bad breath condition). Honey is also used to prevent dental plaque, gingivitis, mouth ulcers and periodontitis. Honey may decrease dental plaque production and aid in reducing gingivitis associated with orthodontic operations, according to a study conducted by Patel et al. [135] on bacterial isolates collected from patients receiving orthodontic treatment. Honey inhibits anaerobic bacteria, which helps to prevent periodontal disease caused by *Porphyromonas gingivalis* [136]. In a trial of 150 dyspeptic individuals, honey consumption at least once a week dramatically reduced the chance of *Helicobacter pylori* infection. [137]. Honey can be used to replace glucose in oral rehydration, and its anti-bacterial qualities helped to shorten the duration of bacterial diarrhea [130].

The anti-bacterial properties of propolis have also been reported due to presence of caffeic acids, benzophenone derivatives, ferulic acid, prenylated coumaric acid and diterpenic acids [12,138–141]. The anti-bacterial properties of propolis have been shown to prevent the formation of bacterial plaques [142]. When compared to chlorhexidine, propolis solutions had a lesser cytotoxic effect on human gum fibroblasts. Mouthwashes containing propolis were discovered to have anti-bacterial action towards *S. mutans* and has been used as an alternative treatment for dental caries prevention [143]. Propolis fluoride was administered to teeth of children with dental caries surface to investigate the efficacy of propolis in combating dental caries. In another study, researchers treated individuals with a 10% hydroalcoholic solution of propolis extract, who were suffering from chronic periodontitis, and discovered a 95% reduction in chronic periodontitis [144]. It played an important role in the repair of tooth pulp [145]. The combined use of mouthwash and toothpaste with ethanolic extract of propolis improved the prevention of microbial infection and the treatment of gum inflammation [15,146].

The protein components of royal jelly like RJ proteins, royalisin, jellenies and enzymes, such as glucose oxidase, have significant anti-microbial potential [147]. The fatty acid 10-HDA, which also has immunomodulatory effects, is responsible for royal jelly's anti-bacterial capabilities [148]. Terada et al. [149] found that the fatty acids contained 32% (10-HDA), 24% gluconic acid, 22% 10-hydroxydecanoic acid (HDAA), 5% dicarboxylic acids and several other acids, which effectively suppressed oral pathogens *S. viridans* and *S. mutans* as well as *S. aureus* and *S. epidermis*. Overall, royal jelly is a natural product having minimal side effects that works in synergy with other anti-plaque agents. Therefore, it can be used as an alternative to synthetic antibiotics [150].

3.2.2. Anti-Fungal Activity of Bee-Based Products and Its Effects on Diseases of the Oral Cavity

In human mouth, around 85 different fungi could be found and the most significant one is *Candida* [151]. When the oral microbiota is balanced, Candida sp. remains neutral; nevertheless, if the equilibrium is disrupted, this fungus will seek a chance to harm oral health. The *Candida* species, particularly *Candida albicans*, are major human fungal pathogens that can cause superficial oral mucosal infections often known as Candidiasis. According to reports, honey, propolis and royal jelly have anti-fungal properties against *Candida albicans*. Honey is also effective against dermatophytes (*Microsporum ferrugineum*, *Trichophyton mentagrophyte*, *Trichophyton longfeuseus*, *Trichophyton semmie*, *Trichophyton tonsurance*), parasitic fungi (*Allescheria boydii*), saprophytes (*Mucor mucaralis*) and *Aspergillus* species, according to Sheikh et al. [152]. Honey's anti-fungal action has been associated with: (i) H_2O_2, which is formed in honey after dilution as a result of glucose oxidation [153,154], (ii) MGO, bee defensin-1 and other bee substances (e.g., phenolic compounds and flavonoids of floral origin and lysozyme), and (iii) honey's osmotic impact. The honey's mode of action, which inhibits biofilm development or accelerates the rupture of mature biofilms, includes the breakdown of cell membrane integrity and suppression of extracellular polysaccharide matrix creation [155]. Honey could be used to treat oral candidiasis so as to prevent infections from becoming more severe. In pilot research conducted by English et al. [156], patients were given chewable 'honey leather' and showed a substantial reduction in mean plaque scores and bleeding sites; the same strategy could be used to treat oral candidiasis. The flavonoid content of propolis confers its anti-fungal properties and also prevents division of fungal cell that further disrupts fungal cell wall and cytoplasm and which is comparable to several antibiotics [157]. Geopropolis produced by bees such as *M. fasciculata* possesses anti-fungal characteristics, according to Feres et al. [158]. In recent in vivo studies, geopropolis produced by *M. fasciculata* was reported to lower salivary *S. mutans* populations. This anti-fungal effect is attributable to caffeic acid derivatives (benzyl ester and pterostilbene) and flavonoids (sakuranetin, pinobanksin and pinocembrin) found in bee propolis. Propolis extract applied topically to oral *Candida albicans* lesions achieved remission in three weeks, with treatment efficacy comparable to nystatin, the most commonly used anti-fungal medication [159]. Royalisin is an insect defensin extracted from the royal jelly (RJ) and its inhibitory effect against a large spectrum of fungi was observed [160–162]. Royalisin has also been shown to have anti-fungal effect against *Botrytis cinerea* [163]. According to Melliou and Chinou [164], royal jelly carboxylic acids such as sebacic acid have high anti-fungal action against *Candida albicans*, *Candida tropicalis* and *Candida glabrata*.

3.2.3. Antiviral Activity of Bee-Based Products and Its Effects on Diseases of the Oral Cavity

The oral microbiome also contains viruses, primarily phages [165]. When human body is acquainted with certain diseases, other certain viruses may also arise in the mouth, for example, mumps virus [166] and HIV [167] are the most frequent. The DNA and RNA based viruses are responsible for oral cavity infection in the mucosal epithelium, which may further lead to ulceration or blistering [168]. Human herpes virus (HHV) members including varicella-zoster virus (VZV), HSV-1 and 2, HHV-6, HHV-7, and HHV-8, cytomegalovirus (CMV), human papillomaviruses (HPV) and Epstein-Barr virus (EBV) cause primary oral infections and diseases such as herpes ulcers, tumors, herpes zoster, precancerous lesions, periodontitis and herpes chicken pox etc. The oral mucosa may also be affected by secondary pathological processes [169]. Honey's efficacy in the treatment of recurrent HSV lesions and oral mucositis in cancer patients has been well documented [30,170,171]. Antiviral effects of propolis have been established against a wide spectrum of viruses. Pagani [172] studied the antiviral activity of propolis flavonoids such as acacetin, kaempferol, chrysin, quercetin and galangin against diverse types of adenoviruses, influenza viruses, rotavirus, herpesvirus and coronavirus. Serkedjieva et al. [173] found that propolis phenolics, particularly isopentyl ferulate, have potent antiviral properties against the H3N2 influenza A virus. It is often applied in the treatment of diseases involving the oral cavity and gums. Antiviral effects of propolis against herpes viruses are also promising. In vitro studies have also revealed that the flavonoids present in propolis inhibited herpes virus strains from replicating intracellularly [174]. Using the plaque assay approach, in an extra-somatic environment, Hashemipour et al. [175] studied the antiviral properties of honey, royal jelly and acyclovir on herpes simplex virus-1. Honey, royal jelly and acyclovir at concentrations of 500, 250 and 100 g/mL, respectively, had the strongest inhibitory effects on HSV-1.

3.2.4. Anti-Inflammatory Activity of Honey Bee-Derived Products and Its Effects on Diseases of the Oral Cavity

Inflammation is the immune system's natural innate response to infections, resulting in the development of diverse cellular and humoral immunological responses [176]. Parallelly, oxidative stress occurs when the equilibrium favors free radical formation over antioxidant components. Inflammation and oxidative stress are linked through a number of signaling mechanisms [27]. Reactive oxygen species (ROS) produced by mitochondria activate a number of transcription factors involved in the production of pro-inflammatory cytokines and mediators (NF-κB, extracellular signal-regulated kinase/ERK, janus kinase/JNK, mitogen-activated protein kinase/MAPK). Similarly, a few cytokines, such as TNF-α and IL-1, can cause mitochondrial ROS generation [177]. Metabolic and cellular alterations are caused by the interaction of ROS with pro-inflammatory cytokines. In other words, the initiation of an uncontrolled inflammatory process in the presence of oxidative stress is a key element in the pathophysiology of chronic diseases like mental, cardiovascular, traumatic, metabolic and autoimmune diseases. Stomatitis, or inflammation of the mouth's mucous membranes, can cause redness and swelling of the oral tissues as well as prominent and painful ulcers.

Honey has been proposed as an immunologic modulator with two functions: (1) anti-inflammatory activities by downregulating inflammatory transcription factors (NF-κB and MAPK) and/or suppressing the production of pro-inflammatory cytokines, and (2) stimulation of inflammatory mediators such as prostaglandin E2 (PGE2) and cyclooxygenase-2 (COX-2) [178]. Honey is superior to dextromethorphan and diphenhydramine for the treatment of cough-induced upper respiratory infections. Honey is useful against stomatitis because it penetrates fast into the tissues [179,180].

Many studies have found that propolis has anti-inflammatory activities, which may be because of phenolic acids, notably CAPE. Anti-inflammatory properties of propolis are commonly used in mouthwashes. CAPE, in particular, has been proven to have antigingivitis properties [181]. Furthermore, in randomized, double-blind, placebo-controlled studies, rinse solutions containing Brazilian green propolis with higher artepillin C concentration decreased gingivitis to same extent as a NaF/cetylpyridinium chloride rinse or a chlorhexidine solution [169]. Royal jelly has been shown to have anti-inflammatory properties [182]. The key anti-inflammatory and immunosuppressive component of RJ that promotes anti-allergic response has been identified as major royal jelly protein 3 [183].

3.2.5. Anti-Cancer Activity of Bee Products in Relation to Oral Health

Oral cancer refers to malignant tumors that develop in mouth, and the majority includes squamous cell carcinoma, also known as mucosal variation. In clinical practice, gingival, jaw, tongue, oropharyngeal, soft and hard sputum, salivary gland, lip, maxillary sinus and face mucosal are all examples of oral cancer [184]. There are several strategies that can combat cancer, but many of them are associated with detrimental side effects on the health of patients As a result, when chemotherapy and radiotherapy are employed systemically or over a large area of tissue to kill malignant cells, they often affect healthy cells as well, resulting in undesirable side effects. Natural anti-proliferative compounds are an excellent alternative. Several recent studies have found that several natural bee products suppress tumor cell growth and spread and induce cancer cell apoptosis [185], implying that these natural chemicals (or their active components) could be used as part of an alternative medical treatment for human tumors.

Honey bee polyphenols have been demonstrated to have improved anti-cancer effects, helping to prevent oral cancer initiation, proliferation and progression. Apoptosis, cell cycle arrest, oxidative stress regulation, inflammation relief, increased mitochondrial outer membrane permeabilization and angiogenesis suppression are all involved in mechanism. Tualang honey inhibited cell proliferation and promoted apoptosis in oral squamous cell cancer under in vitro conditions [186]. The anti-cancer action of crude honey extracts in oral malignancies is most likely due to caspase 3 activation, which induces apoptosis [187]. Purified polyphenols, rather than crude honey, are currently being studied in cancer research. Caffeic acid, phenyl esters, galangin, kaempferol, acacetin, chrysin, quercetin, pinobanksin, apigenin and pinocembrin are simple polyphenols found in honey bees that are prospective pharmacological agents in cancer treatment [187], but cancer of colon, gastric tract, skin, fibrosarcoma and glioma cell have all been shown to be resistant to these phenolic chemicals.

Propolis is an effective antioxidant in the treatment of oral cancer due to presence of high concentration of phenolics and another antioxidant. When glutathione synthesis is low, tumor cells are more vulnerable to radiation impacts [188]. Propolis produces glutathione in hematopoietic tissue, which has anti-cancer properties [189]. Propolis is used as a supplement to help people avoid chronic degenerative disorders like mouth cancer [190]. Propolis contains caffeic acid phenyl ester (CAPE), which possesses anti-mitogenic and anti-cancer characteristics [191]. CAPE could be used as an adjuvant treatment for people with oral squamous cell cancer (OSCC). Because of its high oral absorption and long-term safety profile, propolis is an ideal adjuvant medication for future immunomodulatory or anti-cancer regimens [190]. Flavonoids in propolis stop oral cancer as well as esophagus, stomach, colorectal, prostate and skin cancer [67].

In experimental animals, RJ has shown pharmacological actions, including anti-tumor activity [192]. RJ's anti-tumor effects were studied using transplantable mice tumors, including advanced leukemia strains and solid tumors [185]. In murine tumor models, effects of RJ formation of tumor and metastasis were investigated [193]. A spontaneous mammary carcinoma (MCa) and a methylcholanthrene-induced fibrosarcoma (FS) of the CBA mouse were employed as transplantable murine tumors. When injected intra-peritoneally or subcutaneously, RJ had no effect on metastases while intravenous administration of tumor cells and RJ dramatically reduced the metastases growth. Royal jelly components like 10-HDA and 4-hydroperoxy-2-decenoic acid ethyl ester possess high anti-proliferative effect.

4. Applications of Bee-Based Products in Managing oral Diseases

In recent years, increased implementation of honey bee products like honey, propolis and RJ as an alternative medicine has been witnessed due to their health-promoting activities such as antioxidant, anti-microbial, anti-bacterial, anti-inflammatory, anti-cancer, etc. [25–28,194]. Honey is employed in treatment of multiple oral ailments such as dental caries, gingivitis, oral cancer, plaque, malodor, radiation-induced oral mucositis and xerostomia [195–197]. Similarly, role of propolis has been shown to help with dental caries prevention, oral mucositis reduction, oral cancer prevention and gingival and periodontal disease prevention [37–39]. It also reduces dental hypersensitivity and dentin permeability as well as obstruction of dentinal tubules. It works as a transport medium to improve the periodontal ligament cell viability of avulsed teeth as well as for direct pulp capping and analgesia [198,199]. In another study, a positive effect of royal jelly (with anti-microbial activity) on periodontium has been found and that makes it a valuable agent in dentistry [200]. Figure 2 presents the mechanism of action of polyphenols in maintaining oral health.

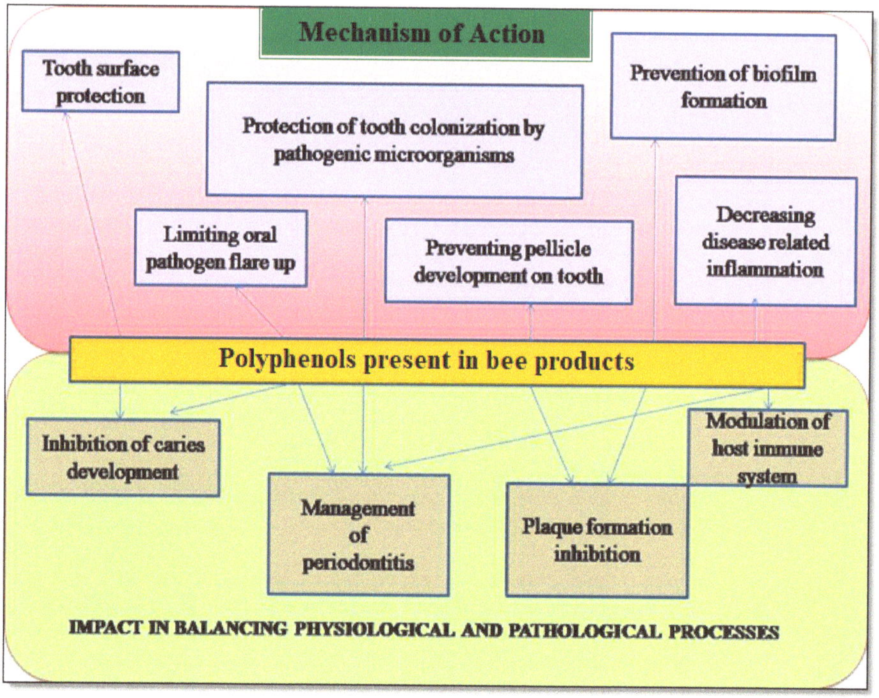

Figure 2. Importance of polyphenols in bee products for oral health.

4.1. Gingivitis

Gingivitis arises due to bacterial presence in plaque biofilm resulting in inflammation of gingival tissues. Gingivitis can be treated with good oral hygiene practice, but failing to do so can develop periodontal (gum) disorders leading to loss of alveolar bone and periodontal ligaments attachment [201]. In terms of reducing gingival scores, mouthwash containing manuka honey has been found to be equally efficient as chlorhexidine [202]. The principal infective bacterium responsible for persistent periodontal inflammation is called *Porphyromonas gingivalis*. Levels of inflammatory cytokines, chemokines and CD54 increase in inflammatory sites in periodontal lesions in response to *P. gingivalis* lipopolysaccharides. It has been observed that consumption of royal jelly can inhibit the development of periodontal infections [147].

4.2. Dental Caries

The effectiveness of honey has been shown against 60 different species of bacteria, including *S. mutans*, a heavily involved pathogen in dental caries [203]. Presence of bacteria and fermentable carbohydrates along with a susceptible tooth surface are all etiological factors for dental caries. Catechins cause irreversible damage to the microbial cytoplasmic membrane, regulate biofilm production, decrease several cariogenic virulence factors and increase dental caries prevention, according to Xu et al. [204]. Other polyphenols such as flavonols, myricetin and proanthocyanins also work in a similar way by disrupting biofilm development, which inhibit attachment of oral harmful bacteria and reduce acid production by *S. mutans* [205]. Numerous studies have demonstrated anti-bacterial activity of propolis against caries-causing bacteria with no side effects in animal/human models. Fatty acids in propolis act as an anti-caries agent by limiting acid generation and reducing the microorganism's tolerance to low pH [206]. Hayacibara et al. [207] investigated the impact of propolis on *S. mutans* vulnerability, caries development and glycosyl transferase activity in rats, concluding propolis extract as an anti-caries agent. Nam et al. [208] reported that propolis prevents dental caries by inhibiting the cell division and enzyme activity of bacteria. Propolis has also been incorporated into certain products to prevent caries.

Current research has shown that apigenin and trans-trans farnesol may have biological activity against dental caries by suppressing several virulence associated genes (GtfB and C in *S. mutans*) by apigenin [209]. Topical treatments of 1 mM apigenin, 5 mM tt-farnesol and 13 mM fluoride combination twice a day inhibited the production of *S. mutans* biofilms in the experimental organisms, according to Koo et al. [205]. Likewise, in an in vitro tooth avulsion model, royal jelly solution was observed to be more effective compared to milk and Hank's balanced salt solution for preservation and transportation [39,210].

4.3. Oral Cancer

Oral cancer is among the sixth common type of cancer reported globally [209]. Pathologically, oral squamous cell carcinoma (OSCC) is most widespread contributing remarkably for 84–97% of all cases [211,212]. Tualang honey has been found to exhibit chemopreventive activity due to its phenolic acids and flavonoids in an animal model against 4-nitroquinoline 1-oxide-induced oral cancer [167,213]. Mahmood et al. [214] revealed that *Trigona itama* honey may be an effective chemotherapeutic adjunct in managing HSC-2 cells derived from OSCC. The anti-cancer activity of crude honey extraction in oral cancer appears to be associated with polyphenolic chemical-induced apoptosis via caspase 3 activation. Chrysin, caffeic acid, acacetin, galangin, kaempferol, phenyl esters, pinobanksin, quercetin, pinocembrin and apigenin are the most promising pharmacological compounds identified in honey for cancer treatment. Because of their easy of application in the mouth and established anti-cancer influence on other malignancies, polyphenolic chemicals derived from bee products have a promising future in the creation of a natural, non-toxic, effective alternative in oral cancer treatment [215].

Similar to honey, propolis is found to be one of the natural agents against oral cancer. The anti-cancer properties of propolis are attributed to its flavonoid compounds. Dornelas et al. [216] discovered that pharmacological substances extracted from propolis (artepilin C, p-coumaric acid, CAPE, quercetin, chrysin, caffeic acid and naringenin) had lethal effects in cultured human tumor cells and shrink tumors in animals. CAPE was discovered to suppress oral cancer cell metastasis by modulating matrix metalloproteinase-2 and MAP kinase pathway in another investigation, suggesting that it could be utilized as a chemotherapeutic to prevent oral cancer metastasis [217]. Yanagita et al. [218] reported that royal jelly reduces the production of interleukin 6 and CXC chemokine ligand 10 formation from MPDL22 cells. CD54, a cell adhesion protein involved in the proliferation of leukocytes in periodontal disorders, was also suppressed by royal jelly in MPDL22 cells. These findings suggest that royal jelly has anti-inflammatory and osteoinductive effects, and it may have a role in periodontal disease prevention.

4.4. Oral Malodor (Halitosis)

Nowadays, oral malodor (halitosis) draws more attention as it is one of the causes of interpersonal interaction issues. Halitosis can be treated by removing coated tongue with a brushing tooth, tongue scraper, oral prophylaxis and anti-microbial mouth rinse. Patients with oral squamous cell cancer who used manuka honey reported less halitosis, likely due to the honey's anti-bacterial activity and its ability to divert microbes' nourishment away from the production of malodor sulfur compounds and toward the production of lactic acid [219,220]. Mouth rinse containing propolis had shown similar effects to those containing essential oils in reducing oral malodor [221].

4.5. Oral Mucositis

Honey had been reported to alleviate the severity and duration of radiation-induced oral mucositis (OM), an epithelial injury to the oral, laryngeal or pharyngeal mucosa produced by ionizing radiation during the second and third weeks of radiotherapy [37,222]. Raeessi et al. [223] observed that honey plus coffee regimen is the effective modality for treating oral mucositis. Abdulrhman et al. [37] suggested the honey usage help in faster healing of patients with chemotherapy-induced oral mucositis. RAS Noronha et al. [224] reported that propolis gel with mucoadhesive properties could be used as a potential topical treatment for preventing radiation-induced oral mucositis. Use of propolis assists in replantation of avulsed permanent teeth as well as their healing after oral surgery by reducing inflammation, accelerating the formation of granulation tissue and also imparting analgesic effect [39]. According to Erdem and Güngörmüs [225], royal jelly extract demonstrated a significant reduction in the signs, symptoms and healing time of oral mucositis. Similarly, Yamauchi et al. [226] reported that RJ-treated patients had significantly less mucositis than the non-treated group.

4.6. Xerostomia

Xerostomia is defined as a considerable reduction and/or thickening of saliva owing to reduced salivary flow during radiation therapy for neck and head cancer [227]. Charalambous et al. [228] observed that thyme honey is an effective treatment for patients with radiation-induced xerostomia. Thyme honey also enhanced overall quality of life and relieved severe pain and dysphagia. Polyphenolic extracts of honey have been shown to have synergistic effect with antibiotics (e.g., amoxicillin) and are now recommended as a viable alternative to synthetic medications in the prevention and treatment of oral illnesses [229].

4.7. Dentin Hypersensitivity

Dentin hypersensitivity (DH) occurs when dentin is exposed to any kind of stimuli such as thermal, osmotic, evaporative, tactile or chemical stimuli. This exposure activates the odontoblast process, resulting in an acute pain [230]. Recently, Tavares et al. [231] reported propolis as an effective, safe and low-cost alternative of reducing DH. Number of researchers [232–235] have shown evidences that propolis is a promising agent for reducing DH. Current application of bee products and their role in the alleviation of oral pathologies are presented in Table 3.

Table 3. Current applications of bee-based products in the improvement of oral health.

Type of Bee Product	Disease Targeted	Whole Bee Product/Extracts and Dose	Key Findings	Reference
Honey	Radiotherapy-induced xerostomia	Individuals suffering from neck and head cancer were randomly allocated to the control group (oral rinses with saline) and the intervention group (oral rinses with 20 mL of thyme honey diluted in 100 mL of purified water). Patients were required to perform oral rinses just before, immediately after and 6 h after the radiotherapy session. Radiation-induced xerostomia was assessed starting from the 4th week of radiotherapy, one and six months after the completion of radiotherapy	Thyme honey was found useful in lowering or stabilizing the degree of xerostomia over time, with progressive improvement. The better management of xerostomia showed significant effects on overall quality of life.	[228]
	Chemotherapy-induced mucositis	Patients randomly assigned to one of three treatment groups: Group 1—Empirical dosage of 0.5 g honey/kg (max.15 g) was applied topically three times daily to the afflicted oral mucosa for 10 days, or until healing. Group 2—An empirical dose of 0.25 g/kg (max. 5 g) of a 4:2:1 mixture of honey, olive oil-propolis extract, and beeswax (HOPE) administered topically three times daily to the diseased oral mucosa for 10 days, or until healing. Group 3—Served as the control group where benzocaine 7.5% gel was topically applied three times daily to the afflicted oral mucosa.	Honey (Group 1) resulted in faster healing than HOPE (Group 2) or control (Group 3) in both grades of mucositis ($p > 0.05$).	[37]
	Dental Plaque	At the commencement of the trial, each patient underwent a professional prophylaxis to completely remove plaque and calculus from the teeth. The patients were randomly assigned to below groups: Group 1—Manuka honey was applied gently to gingival sulcus of the teeth, and the procedure was repeated twice after 5 min. The honey was applied twice a day after meals. Group 2—Rinsing with 0.2% chlorhexidine (10 mL) twice a day for 60 s followed by its expectoration. Group 3—Chewing the xylitol chewing gum for 5 min, thrice a day after meals. Following the experimental time, the plaque scores were determined.	The mean plaque scores for Groups 1, 2 and 3 were correspondingly 1.37, 1.35 and 1.57. The study indicated that manuka honey and chlorhexidine mouthwash significantly reduced plaque development compared to xylitol chewing gum.	[236]
	Dental Plaque	Dental plaque score was recorded in individuals before and after gargling with Tongra original honey 5% solution for six days.	Gargling with Tongra original honey 5% solution effectively decreased the dental plaque score.	[237]
	Oral squamous cell carcinoma (OSCC) and human osteosarcoma (HOS)	Different Tualang honey (1–20%) concentrations were administered to OSCC and HOS cell lines at different intervals of 3, 6, 12, 24, 48 and 72 h.	Tualang honey had an anti-proliferative effect on both cell lines. Maximum inhibition (≥80%) of cell growth recorded at a dose of 15%.	[238]
	Oral mucosal ulcers	For the oral mucosal ulcer model, excisional wounds were conducted on 30 Wistar albino rats (240–30 g) and they were separated into 3 groups: Group 1—Apitherapeutic agent or honey treatment (0.1 ml, 2 × 1). Group 2—Glyceroloxytriester (TGO) (0.1 ml, 2 × 1) was used to treat locally. Control Group 3—On the 7th day, biopsy samples were collected from the right buccal mucosa, and on the 14th day, samples were taken from the left buccal mucosa.	Only on the 7th day a significant difference documented between groups 1 and 3, whereas on day 14, no significant difference was noted among the groups. Honey was found to be efficient in the therapy of oral mucosal ulcers and showed a greater therapeutic benefit than glyceroloxytriester (TGO).	[239]

Table 3. *Cont.*

Type of Bee Product	Disease Targeted	Whole Bee Product/Extracts and Dose	Key Findings	Reference
Propolis	Gingivitis	One of the twins got 2% pure propolis for rinsing during the gingivitis induction period while the other received a color-matched 0.05% sodium fluoride + 0.05% cetylpyridinium chloride for rinsing (positive control). For 21 days, patients were advised to rinse 20 mL of respective rinses twice a day for 30 s each time.	During a 3-week no-hygiene period, a 2% typified propolis rinse performed similar to positive control rinse.	[240]
	Radiotherapy-induced mucositis	Patients were divided into two groups at random: The Case group (n = 10) received 15 mL of water-based propolis mouthwash three times daily, while the Control group (n = 10) received 15 mL of placebo mouthwash.	Propolis water extract effectively prevented and cured radiotherapy-induced mucositis.	[241]
	Denture stomatitis	The patients were randomized to one of two therapy groups at random: Miconazole oral gel, 20 mg/g, was given to the control group (MIC) for 14 days. For 14 days, the PROP group was given a mucoadhesive formulation containing a standardized extract of 2% (20 mg/g) propolis (EPP-AF ®®). On days 1, 7 and 14, patients were assessed.	EPP-AF showed effect at par with miconazole.	[242]
	Cariogenic infections in a caries-active patient	Patients were randomly assigned to one of three experimental groups after their cavitated lesions were restored: (1) PROP-alcohol-free 2% propolis rinse (n = 20); (2) CHX- 0.12% chlorhexidine rinse; (3) PL-placebo mouth rinse. Patients were asked to rinse with 15 mL of respective rinses twice day for 60 s for 28 days. Salivary levels of Mutans *Septococci* (MS) and *Lactobacilli* (LACT) were evaluated at baseline, 7-day, 14-day and 28-day visits (experimental effects) and 45-day visits (residual effects).	Among all the treatments evaluated, propolis rinse was found to be the most efficient at suppressing cariogenic infections in caries-active participants.	[243]
	Dental Plaque and Gingivitis	A phase II clinical trial involved patients with a minimum of 20 healthy natural teeth, a mean plaque index of at least 1.5 P I, and a mean gingival index of at least 1.0 GI. Patients were advised to rinse with 10 mL of mouthwash test for one minute after brushing their teeth in the morning and at night.	Plaque and gingival index considerably reduced after 45 and 90 days of using mouthwash.	[244]
	Dentin hypersensitivity (DH)	96 patients with DH in one or more teeth were studied in this study. The teeth were allocated in one of four treatment groups at random: Group 1—10% ethanolic extract of propolis. Group 2—30% ethanolic extract of propolis. Group 3—Single Bond Universal dentin bonding agent. Group 4—Distilled water as a placebo. The degree of DH was assessed using a visual analog scale based on the patients' response to tactile and air blast stimuli.	Propolis extracts and dentin bonding agent were found to be equally effective in relieving DH. Propolis application was advised for patients experiencing mild to moderate discomfort whereas dentin bonding agent could be a preferable alternative for rapid relief.	[234]
	Oral malodor	Patients were allocated into three groups: I received a placebo (P), II received an ethanolic extract of propolis type-3 at a concentration of 3% (EEP), and III received chlorhexidine (CHX) at a concentration of 0.12%. Participants were instructed to rinse with respective rinses twice a day for 5 days. Each trial period was followed by a washout period of 21 days. Microbiological samples were obtained from the tongue dorsum at baseline and at the end of the rinse period to measure the concentration of volatile sulfur compounds (VSC) in the morning mouth breath.	Both EEP and CHX treatment resulted in a considerable reduction in the mean counts of bacterial pathogens, including certain VSC producers. The study suggested the use of mouth rinse containing 3% propolis type-3 prevents malodorous morning breath.	[245]

Table 3. *Cont.*

Type of Bee Product	Disease Targeted	Whole Bee Product/Extracts and Dose	Key Findings	Reference
Royal Jelly	Periodontal pathogens	Subgingival plaque samples were obtained and analyzed from 15 chronic periodontitis patients. RJ and chlorhexidine were tested in vitro for their ability to inhibit the growth of aerobic and anaerobic bacteria.	Chlorhexidine at low dose (50 μg/100 μL) was shown to be more sensitive in suppressing aerobic and anaerobic periodontopathic bacteria in subgingival plaque area whereas a larger dosage of RJ was required to have an inhibitory effect.	[150]
	Radiotherapy-induced mucositis	Patients who needed chemoradiation for head and neck cancer were randomly allocated to one of two groups. During chemoradiation treatment, seven patients in the experimental group received RJ three times a day while six patients in the control group did not. The development of mucositis in both groups of patients was monitored twice a week.	The use of RJs reduced radiation-induced mucositis in patients with head and neck cancer.	[226]
	Radiotherapy and chemotherapy-induced mucositis	Patients were allocated into two categories. Both groups used benzydamine hydrochloride and nystatin mouthwashes. The experimental group also received 1 g RJ twice daily in addition to mouthwash. Both experimental groups administered mouthwash and/or RJ to patients until the mucositis resolved.	The RJ group resolved oral mucositis in a fraction of the time it took the control group.	[225]

5. Bee Products-Based Innovative Products for Oral Hygiene

Oral and dental and hygiene is practice of maintaining clean and healthy mouth and teeth in order to avoid dental problems such as cavities, gingivitis, periodontal disease and bad breath. Periodontal disorders are caused by poor oral hygiene, which can occur at any age [246,247]. As a result, suitable and effective plaque reduction methods must be implemented on a regular basis [247]. Mechanical approaches, however, may not be practicable or sufficient. As a result, chemical preparations such as anti-bacterial mouthwashes have been proposed as a complement to or replacement for mechanical plaque control [248]. Chlorhexidine (CHX) is most commonly used mouthwash and considered as golden standard against dental plaque [248,249]. However, long-term application of CHX is associated with various adverse effects including changed taste sensitivity, staining of teeth and burning sensation [249,250]. As a result, pharmaceutical companies are working to develop natural-derived oral care products such as toothpaste, mouthwash and chewable tablets or chewing gums based on bee products.

5.1. Bee Product-Based Toothpaste

Herbal toothpaste containing clove fruit, neem leaves, honey and Acacia powder for maintaining oral hygiene was developed and it was found that the developed herbal toothpaste was equally effective as per Bureau of Indian standards [251]. The propolis containing toothpaste showed positive biological activity with respect to spectra of oral microbiota without causing adverse effect and can be used as natural alternative to chemical mouthwashes [252]. Similarly, the toothpaste developed using tea tree oil and propolis extract was found to be effective in maintaining oral hygiene by quantitative reduction in oral microbiota due to their anti-microbial and anti-fungal properties [253]. Propolis based toothpaste is intended to prevent the formation of bacterial plaque and pathogenic microflora, which can lead to tooth decay. It has been reported that propolis-based toothpaste was more effective than calcium hydroxide-based toothpaste and showed good anti-bacterial activity [254]. A patent application (CN102283795A) for the toothpaste preparation method with propolis is under consideration. This invention claims effective control and treatment of oral diseases like dental caries by reducing plaque bacteria [255]. Another patent application (CN110755355A) for toothpaste containing propolis extract claims to inhibit pathogenic bacteria in an oral cavity improves bleeding gum and maintain oral health [256].

5.2. Bee Product-Based Mouthwash

Mouth rinse has been quite popular due to its ease of use; however, chemical compounds found in mouth rinses such as cetylpyridinium chloride, chlorhexidine and zinc chloride may have side effects if used for long term [257,258]. Propolis-based mouth rinse or mouthwashes have been studied by several researchers and observed great potential in reducing dental plaque and gingival inflammation [259–263]. A patent (CN104739738A) was granted for a mouthwash made with propolis to improve oral health and to treat oral diseases [264].

5.3. Bee Product-Based Chewable Tablets or Chewing Gums

Chewable tablets are tablets that are chewed in the oral cavity before being swallowed. These tablets have several benefits such as oral drug delivery without any requirement of water, palatability and stability. These tablets are suitable for children, the elderly and patients suffering from dysphagia. Commercially available chewing gums containing propolis was superior to the one with xylitol gum in reducing bacterial count [265]. A patent (CN107198189B) was granted for a chewable tablet made with propolis, royal jelly, honey, beeswax, sodium citrate, mannitol, maltodextrin and xanthan gum [266]. Another patent application (CN102326723A) is under consideration for development of chewable propolis tablet [267]. Chewing gum are confectionary products that are chewed for various reasons. A worldwide patent (WO2020101601A2) was assigned for producing three types of propolis chewing gums *viz.* sugar, sweetener and sugar-free [268]. Various innovative products developed from the bee products are presented in Table 4.

Table 4. Various innovative products developed from bee products.

Type of Product	Product	Patents	Bee Product Used	Intended Use	Reference
Toothpaste	Propolis toothpaste	CN102283795A	Propolis	To prevent and treat oral diseases	[255]
	Propolis toothpaste	CN110755355A	Propolis	To improve bleeding gum and maintain oral health by inhibiting pathogenic bacteria	[256]
	Brazilian green propolis toothpaste	CN107412138B	Propolis	To prevent on gingivitis, periodontitis and halitosis	[269]
	Manuka honey toothpaste	CN105287328A	Manuka honey	To prevent prevents gingivitis, periodontitis, tooth decay and oral ulcer during pregnancy	[270]
Mouthwash	Propolis mouthwash	CN104739738A	Propolis	Improved anti-bacterial and anti-inflammatory properties, as well as the prevention and treatment of oral disorders such oral cancer	[264]
	Propolis mouthwash	KR20060041348A	Propolis	To prevent tooth decay and treat oral diseases	[271]
Chewing gum	Hive honey chewing gum	CN107751533A	Honey	To improve health of oral cavity and remove bad breath	[272]
Chewable tablet	Tablet	CN107198189B	Propolis, royal jelly, honey, and beeswax	To improve immunity and highly suitable for patients suffering from pharyngitis	[266]

6. Safety Aspects of Honey and Bee Products

Although, consumption of bee products like honey, propolis and royal jelly is safe and allergies and sensitivities to them are uncommon. However, patients who are taking them extensively should be advised about the possibility of adverse reactions and allergies. Toxic chemicals in honey have been reported including polycyclic diterpene grayanotoxins in honey from rhododendron plants such as *R. luteum* and *R. ponticum*. This honey is recognized as "mad honey" because of its severe neural intoxication and even death, particularly in Turkey's eastern Black Sea region. In spite of its toxicity, it is used as a traditional remedy for sexual dysfunction, hypertension and other diseases, probably because its activity [273,274]. Plants in the Boraginaceae, Asteraceae and Fabaceae families possess pyrrolizidine alkaloids, which are not hazardous but are transformed into injurious pyrrolic metabolites by the liver after ingesting honey. These alkaloids may provide a health danger to honey consumers because they are found in common honey botanical sources [275]. Honey poisoning has been associated to incidences of convulsions, delirium and poor memory due to contamination by the neurotoxic sesquiterpene. Honey bees gather dew produced by passionvine hoppers (*Scolypopa australis*), which feed on sap of the poisonous shrub tutu (*Coriaria spp.*) and swallow these oxygenated sesquiterpene picrotoxanes, which further target GABAergic and glycinergic receptors [276,277]. Hyoscyamine, saponins, strychnine, gelsemine, hyoscine, oleandrin and oleandrigenin are some of the other plant secondary metabolites identified in honey that may be toxic to humans. Honey can be polluted by environmental pollutants such as heavy metals, pesticides and antibiotics in addition to phytochemicals. Furthermore, honey that has been stored or heated for an extended period of time may produce Maillard reaction products [278]. Clinical and in vivo animal studies have shown that propolis is safe and non-toxic. At higher doses (>15 g/day) it shows toxic effects in humans [12]. However, incidences of propolis allergy and contact dermatitis have long been observed, primarily among beekeepers. Furthermore, it has the potential to irritate the skin, causing eczema, lesions, psoriasis and mouth sores [279]. The main sensitizers in propolis, according to Burdock [138] and Walgrave et al. [280], are 3-methyl-2-butenyl caffeate, phenylethyl caffeate, benzyl salicylate, benzyl cinnamate and 1,1-dimethylallylcaffeic acid. Similar to honey, harmful chemicals and pollutants are also present in royal jelly. Pesticides from the organochlorine, organophosphorus and carbamate families are the most common. Eczema, asthma and hypersensitivity have all been linked to royal jelly consumption, and the royal jelly proteins MRJP-1 and MRJP-2 have been recognized as key allergens [281].

7. Conclusions

Honey bee products have remarkably high biological and therapeutic properties since they are imparted with a wide variety of bioactive such as phenolic compounds, flavonoids and terpenoids. Studies with in vitro, in vivo and clinical trials showed on antimicrobial activities of bee products against various pathogenic bacteria such as *S. mutans* and *Porphyromonas gingivalis*, antioxidant, anti-inflammatory and anti-cancer against various oral pathologies periodontitis, dental caries, mucositis and dentin hypersensitivity. Many human clinical trials revealed bee products are safe and helpful in the treatment of various oral diseases. It was reported that novel products based on bee products, such as chewing gums, toothpaste, and mouthwash, could be sources of cost-effective and consumer-friendly nutritional components for improving human oral health. However, there is a lot of potential in using these qualities of bee products to cure a variety of diseases. In vitro and intervention study results have been inconsistent in identifying the various functional characteristics of each bioactive compound in bee products, as well as the method to improve their bio-accessibility. Though, honey bee products also cause some side effects like allergies, but they appear only when they are used in high concentrations. The molecular mechanism of action of bee products must be explored in order to fully comprehend their mode of action. In addition, to confirm the influence of bee products on human health, in vivo and human clinical investigations should be conducted. Based on these findings, a policy for using bee products in commercial products must be developed.

Author Contributions: Conceptualization, P.C., S.T., M.K. and M.M.; Writing—original draft preparation, M.B., S.S., M.K.S., H.R., D.C., P.C., S.T., N.R.Y., P.K. and D.S.; writing—review and editing, M.K., M.M. and P.C. All authors have read and agreed to the published version of the manuscript.

Funding: This research received no external funding.

Acknowledgments: The authors would like to thank the University of Kiel and Schleswig-Holstein for the support through the OA program.

Conflicts of Interest: The authors declare no conflict of interest.

References

1. Apimondia. *Rodusele Stupului—Hrană, Sănătate, Frumuseţe*; Apimondia: Bucureşti, Romania, 1989.
2. Apimondia. *Un Preţios Produs al Apiculturii. Propolisul*; Apimondia: Bucureşti, Romania, 1990.
3. Hellner, M.; Winter, D.; von Georgi, R.; Münstedt, K. Apitherapy: Usage and experience in German beekeepers. *Evid.-Based Complement. Altern. Med.* **2008**, *5*, 475–479. [CrossRef] [PubMed]
4. Khurshid, Z.; Naseem, M.; Zafar, M.S.; Najeeb, S.; Zohaib, S. Propolis: A natural biomaterial for dental and oral healthcare. *J. Dent. Res. Dent. Clin. Dent. Prospect.* **2017**, *11*, 265–274. [CrossRef]
5. Crane, E. *History of Honey in Honey: A Comprehensive Survey*; Crane, E., Ed.; Northern Bee Books: West Yorkshire, UK, 1975.
6. Rossi, M.; Marrazzo, P. The potential of honeybee products for biomaterial applications. *Biomimetics* **2021**, *6*, 6. [CrossRef] [PubMed]
7. Tappi, S.; Laghi, L.; Dettori, A.; Piana, L.; Ragni, L.; Rocculi, P. Investigation of water state during induced crystallization of honey. *Food Chem.* **2019**, *294*, 260–266. [CrossRef] [PubMed]
8. Nguyen, H.; Panyoyai, N.; Kasapis, S.; Pang, E.; Mantri, N. Honey and its role in relieving multiple facets of atherosclerosis. *Nutrients* **2019**, *11*, 167. [CrossRef] [PubMed]
9. Pashte, V.V.; Pashte, S.V.; Said, P.P. Nutraceutical properties of natural honey to fight health issues: A comprehensive review. *J. Pharmacogn. Phytochem.* **2020**, *9*, 234–242. [CrossRef]
10. Cianciosi, D.; Forbes-Hernández, T.Y.; Afrin, S.; Gasparrini, M.; Reboredo-Rodriguez, P.; Manna, P.P.; Zhang, J.; Lamas, L.B.; Flórez, S.M.; Toyos, P.A. Phenolic compounds in honey and their associated health benefits: A review. *Molecules* **2018**, *23*, 2322. [CrossRef]
11. Escuredo, O.; Seijo, M.C.; Salvador, J.; González-Martín, M.I. Near infrared spectroscopy for prediction of antioxidant compounds in the honey. *Food Chem.* **2013**, *141*, 3409–3414. [CrossRef]
12. Castaldo, S.; Capasso, F. Propolis, an old remedy used in modern medicine. *Fitoterapia* **2002**, *73*, S1–S6. [CrossRef]
13. Kasote, D.M. Propolis: A neglected product of value in the Indian beekeeping sector. *Bee World* **2017**, *94*, 80–83. [CrossRef]
14. Kurek-Górecka, A.; Górecki, M.; Rzepecka-Stojko, A.; Balwierz, R.; Stojko, J. Bee products in dermatology and skin care. *Molecules* **2020**, *25*, 556. [CrossRef] [PubMed]

15. Pasupuleti, V.R.; Sammugam, L.; Ramesh, N.; Gan, S.H. Honey, Propolis, and Royal Jelly: A Comprehensive Review of Their Biological Actions and Health Benefits. *Oxidative Med. Cell. Longev.* **2017**, *2017*, 1259510. [CrossRef] [PubMed]
16. Ebadi, P.; Fazeli, M. Evaluation of the potential in vitro effects of propolis and honey on wound healing in human dermal fibroblast cells. *S. Afr. J. Bot.* **2021**, *137*, 414–422. [CrossRef]
17. Buttstedt, A.; Mureşan, C.I.; Lilie, H.; Hause, G.; Ihling, C.H.; Schulze, S.-H.; Pietzsch, M.; Moritz, R.F. How honeybees defy gravity with royal jelly to raise queens. *Curr. Biol.* **2018**, *28*, 1095–1100.e1093. [CrossRef]
18. Kunugi, H.; Ali, A.M. Royal jelly and its components promote healthy aging and longevity: From animal models to humans. *Int. J. Mol. Sci.* **2019**, *20*, 4662. [CrossRef] [PubMed]
19. Kamakura, M. Royalactin induces queen differentiation in honeybees. *Nature* **2011**, *473*, 478–483. [CrossRef]
20. Detienne, G.; De Haes, W.; Ernst, U.R.; Schoofs, L.; Temmerman, L. Royalactin extends lifespan of Caenorhabditis elegans through epidermal growth factor signaling. *Exp. Gerontol.* **2014**, *60*, 129–135. [CrossRef]
21. Kolayli, S.; Sahin, H.; Can, Z.; Yildiz, O.; Malkoc, M.; Asadov, A. A member of complementary medicinal food: Anatolian royal jellies, their chemical compositions, and antioxidant properties. *J. Evid.-Based Complement. Altern. Med.* **2016**, *21*, NP43–NP48. [CrossRef]
22. WHO. Oral Health. EXECUTIVE BOARD 148th Session Provisional Agenda Item 6. Available online: https://apps.who.int/gb/ebwha/pdf_files/EB148/B148_8-en.pdf (accessed on 30 January 2023).
23. Yu, X.; Chen, Y.; Li, Y.; Hong, J.; Hua, F. A bibliometric mapping study of the literature on oral health-related quality of life. *J. Evid.-Based Dent. Pract.* **2020**, *23*, 101780. [CrossRef]
24. Almuhayawi, M.S. Propolis as a novel antibacterial agent. *Saudi J. Biol. Sci.* **2020**, *27*, 3079–3086. [CrossRef]
25. Biluca, F.C.; da Silva, B.; Caon, T.; Mohr, E.T.B.; Vieira, G.N.; Gonzaga, L.V.; Vitali, L.; Micke, G.; Fett, R.; Dalmarco, E.M. Investigation of phenolic compounds, antioxidant and anti-inflammatory activities in stingless bee honey (Meliponinae). *Food Res. Int.* **2020**, *129*, 108756. [CrossRef] [PubMed]
26. Didaras, N.A.; Karatasou, K.; Dimitriou, T.G.; Amoutzias, G.D.; Mossialos, D. Antimicrobial activity of bee-collected pollen and beebread: State of the art and future perspectives. *Antibiotics* **2020**, *9*, 811. [CrossRef] [PubMed]
27. Ranneh, Y.; Akim, A.M.; Hamid, H.A.; Khazaai, H.; Fadel, A.; Zakaria, Z.A.; Albujja, M.; Bakar, M.F.A. Honey and its nutritional and anti-inflammatory value. *BMC Complement. Med. Ther.* **2021**, *21*, 30. [CrossRef] [PubMed]
28. Almasaudi, S. The antibacterial activities of honey. *Saudi J. Biol. Sci.* **2021**, *28*, 2188. [CrossRef] [PubMed]
29. Fontana, R.; Mendes, M.A.; De Souza, B.M.; Konno, K.; César, L.l.M.M.; Malaspina, O.; Palma, M.S. Jelleines: A family of antimicrobial peptides from the Royal Jelly of honeybees (*Apis mellifera*). *Peptides* **2004**, *25*, 919–928. [CrossRef]
30. Yang, Y.-C.; Chou, W.-M.; Widowati, D.A.; Lin, I.-P.; Peng, C.-C. 10-hydroxy-2-decenoic acid of royal jelly exhibits bactericide and anti-inflammatory activity in human colon cancer cells. *BMC Complement. Altern. Med.* **2018**, *18*, 202. [CrossRef]
31. Filipič, B.; Gradišnik, L.; Rihar, K.; Šooš, E.; Pereyra, A.; Potokar, J. The influence of royal jelly and human interferon-alpha (HuIFN-αN3) on proliferation, glutathione level and lipid peroxidation in human colorectal adenocarcinoma cells in vitro. *Arhiv za Higijenu Rada i Toksikologiju* **2015**, *66*, 269–274. [CrossRef]
32. Zhang, S.; Shao, Q.; Geng, H.; Su, S. The effect of royal jelly on the growth of breast cancer in mice. *Oncol. Lett.* **2017**, *14*, 7615–7621. [CrossRef]
33. Guendouz, M.; Haddi, A.; Grar, H.; Kheroua, O.; Saidi, D.; Kaddouri, H. Preventive effects of royal jelly against anaphylactic response in a murine model of cow's milk allergy. *Pharm. Biol.* **2017**, *55*, 2145–2152. [CrossRef]
34. Taniguchi, Y.; Kohno, K.; Inoue, S.-I.; Koya-Miyata, S.; Okamoto, I.; Arai, N.; Iwaki, K.; Ikeda, M.; Kurimoto, M. Oral administration of royal jelly inhibits the development of atopic dermatitis-like skin lesions in NC/Nga mice. *Int. Immunopharmacol.* **2003**, *3*, 1313–1324. [CrossRef]
35. Kai, H.; Motomura, Y.; Saito, S.; Hashimoto, K.; Tatefuji, T.; Takamune, N.; Misumi, S. Royal jelly enhances antigen-specific mucosal IgA response. *Food Sci. Nutr.* **2013**, *1*, 222–227. [CrossRef]
36. Ramadan, M.F.; Al-Ghamdi, A. Bioactive compounds and health-promoting properties of royal jelly: A review. *J. Funct. Foods* **2012**, *4*, 39–52. [CrossRef]
37. Abdulrhman, M.; Elbarbary, N.S.; Amin, D.A.; Ebrahim, R.S. Honey and a mixture of honey, beeswax, and olive oil–propolis extract in treatment of chemotherapy-induced oral mucositis: A randomized controlled pilot study. *Pediatr. Hematol. Oncol.* **2012**, *29*, 285–292. [CrossRef]
38. Kuo, Y.-Y.; Lin, H.-P.; Huo, C.; Su, L.-C.; Yang, J.; Hsiao, P.-H.; Chiang, H.-C.; Chung, C.-J.; Wang, H.-D.; Chang, J.-Y. Caffeic acid phenethyl ester suppresses proliferation and survival of TW2. 6 human oral cancer cells via inhibition of Akt signaling. *Int. J. Mol. Sci.* **2013**, *14*, 8801–8817. [CrossRef]
39. Więckiewicz, W.; Miernik, M.; Więckiewicz, M.; Morawiec, T. Does propolis help to maintain oral health? *Evid.-Based Complement. Altern. Med.* **2013**, *2013*, 351062. [CrossRef]
40. Ahmad, S.; Campos, M.G.; Fratini, F.; Altaye, S.Z.; Li, J. New insights into the biological and pharmaceutical properties of royal jelly. *Int. J. Mol. Sci.* **2020**, *21*, 382. [CrossRef]
41. Meo, S.A.; Al-Asiri, S.A.; Mahesar, A.L.; Ansari, M.J. Role of honey in modern medicine. *Saudi J. Biol. Sci.* **2017**, *24*, 975–978. [CrossRef]
42. Zhang, H.; Tsao, R. Dietary polyphenols, oxidative stress and antioxidant and anti-inflammatory effects. *Curr. Opin. Food Sci.* **2016**, *8*, 33–42. [CrossRef]

43. Liu, J.-R.; Ye, Y.-L.; Lin, T.-Y.; Wang, Y.-W.; Peng, C.-C. Effect of floral sources on the antioxidant, antimicrobial, and anti-inflammatory activities of honeys in Taiwan. *Food Chem.* **2013**, *139*, 938–943. [CrossRef]
44. Abbas, M.; Saeed, F.; Anjum, F.M.; Afzaal, M.; Tufail, T.; Bashir, M.S.; Ishtiaq, A.; Hussain, S.; Suleria, H.A.R. Natural polyphenols: An overview. *Int. J. Food Prop.* **2017**, *20*, 1689–1699. [CrossRef]
45. Ahmed, S.; Othman, N.H. Review of the medicinal effects of tualang honey and a comparison with manuka honey. *Malays. J. Med. Sci. MJMS* **2013**, *20*, 6. [PubMed]
46. Olas, B. Honey and its phenolic compounds as an effective natural medicine for cardiovascular diseases in humans? *Nutrients* **2020**, *12*, 283. [CrossRef] [PubMed]
47. Socha, R.; Juszczak, L.; Pietrzyk, S.; Gałkowska, D.; Fortuna, T.; Witczak, T. Phenolic profile and antioxidant properties of Polish honeys. *Int. J. Food Sci. Technol.* **2011**, *46*, 528–534. [CrossRef]
48. Hung, H.-C.; Joshipura, K.J.; Jiang, R.; Hu, F.B.; Hunter, D.; Smith-Warner, S.A.; Colditz, G.A.; Rosner, B.; Spiegelman, D.; Willett, W.C. Fruit and vegetable intake and risk of major chronic disease. *J. Natl. Cancer Inst.* **2004**, *96*, 1577–1584. [CrossRef] [PubMed]
49. Chaudhuri, S.; Banerjee, A.; Basu, K.; Sengupta, B.; Sengupta, P.K. Interaction of flavonoids with red blood cell membrane lipids and proteins: Antioxidant and antihemolytic effects. *Int. J. Biol. Macromol.* **2007**, *41*, 42–48. [CrossRef] [PubMed]
50. Cheung, Y.; Meenu, M.; Yu, X.; Xu, B. Phenolic acids and flavonoids profiles of commercial honey from different floral sources and geographic sources. *Int. J. Food Prop.* **2019**, *22*, 290–308. [CrossRef]
51. Curti, V.; Zaccaria, V.; Sokeng, A.J.T.; Dacrema, M.; Masiello, I.; Mascaro, A.; D'Antona, G.; Daglia, M. Bioavailability and In Vivo Antioxidant Activity of a Standardized Polyphenol Mixture Extracted from Brown Propolis. *Int. J. Mol. Sci.* **2019**, *20*, 1250. [CrossRef]
52. Anjum, S.I.; Ullah, A.; Khan, K.A.; Attaullah, M.; Khan, H.; Ali, H.; Bashir, M.A.; Tahir, M.; Ansari, M.J.; Ghramh, H.A.; et al. Composition and functional properties of propolis (bee glue): A review. *Saudi J. Biol. Sci.* **2019**, *26*, 1695–1703. [CrossRef]
53. Dezmirean, D.S.; Paşca, C.; Moise, A.R.; Bobiş, O. Plant Sources Responsible for the Chemical Composition and Main Bioactive Properties of Poplar-Type Propolis. *Plants* **2021**, *10*, 22. [CrossRef]
54. Almaraz-Abarca, N.; da Graça Campos, M.; Avila-Reyes, J.A.; Naranjo-Jimenez, N.; Corral, J.H.; Gonzalez-Valdez, L.S. Antioxidant activity of polyphenolic extract of monofloral honeybee-collected pollen from mesquite (*Prosopis juliflora*, Leguminosae). *J. Food Compos. Anal.* **2007**, *20*, 119–124. [CrossRef]
55. Ahuja, V.; Ahuja, A. Apitherapy-A sweet approach to dental diseases. Part II: Propolis. *J. Adv. Oral Res.* **2011**, *2*, 1–8. [CrossRef]
56. López-Gutiérrez, N.; del Mar Aguilera-Luiz, M.; Romero-González, R.; Vidal, J.L.M.; Frenich, A.G. Fast analysis of polyphenols in royal jelly products using automated TurboFlow™-liquid chromatography–Orbitrap high resolution mass spectrometry. *J. Chromatogr. B* **2014**, *973*, 17–28. [CrossRef] [PubMed]
57. Shi, Z.; Enayatullah, H.; Lv, Z.; Dai, H.; Wei, Q.; Shen, L.; Karwand, B.; Shi, F. Freeze-dried royal jelly proteins enhanced the testicular development and spermatogenesis in pubescent male mice. *Animals* **2019**, *9*, 977. [CrossRef] [PubMed]
58. Gómez-Caravaca, A.; Gómez-Romero, M.; Arráez-Román, D.; Segura-Carretero, A.; Fernández-Gutiérrez, A. Advances in the analysis of phenolic compounds in products derived from bees. *J. Pharm. Biomed. Anal.* **2006**, *41*, 1220–1234. [CrossRef]
59. Küçük, M.; Kolaylı, S.; Karaoğlu, Ş.; Ulusoy, E.; Baltacı, C.; Candan, F. Biological activities and chemical composition of three honeys of different types from Anatolia. *Food Chem.* **2007**, *100*, 526–534. [CrossRef]
60. Nagai, T.; Inoue, R. Preparation and the functional properties of water extract and alkaline extract of royal jelly. *Food Chem.* **2004**, *84*, 181–186. [CrossRef]
61. Viuda-Martos, M.; Ruiz-Navajas, Y.; Fernandez-Lopez, J.; Perez-Alvarez, J.A. Functional properties of honey, propolis, and royal jelly. *J. Food Sci.* **2008**, *73*, R117–R124. [CrossRef]
62. Khalil, M.; Sulaiman, S.A. The potential role of honey and its polyphenols in preventing heart disease: A review. *Afr. J. Tradit. Complement. Altern. Med.* **2010**, *7*, 315–321. [CrossRef]
63. Ratajczak, M.; Kaminska, D.; Matuszewska, E.; Hołderna-Kedzia, E.; Rogacki, J.; Matysiak, J. Promising antimicrobial properties of bioactive compounds from different honeybee products. *Molecules* **2021**, *26*, 4007. [CrossRef]
64. Pérez-Pérez, E.; Vit, P.; Huq, F. Flavonoids and polyphenols in studies of honey antioxidant activity. *Int. J. Med. Plant Altern. Med.* **2013**, *1*, 63–72.
65. Gheldof, N.; Wang, X.-H.; Engeseth, N.J. Buckwheat honey increases serum antioxidant capacity in humans. *J. Agric. Food Chem.* **2003**, *51*, 1500–1505. [CrossRef] [PubMed]
66. Hunter, M.; Kellett, J.; D'Cunha, N.M.; Toohey, K.; McKune, A.; Naumovski, N. The effect of honey as a treatment for oral ulcerative lesions: A systematic review. *Explor. Res. Hypothesis Med.* **2020**, *5*, 27–37. [CrossRef]
67. Martinello, M.; Mutinelli, F. Antioxidant activity in bee products: A review. *Antioxidants* **2021**, *10*, 71. [CrossRef] [PubMed]
68. Martinotti, S.; Ranzato, E. Propolis: A new frontier for wound healing? *Burn. Trauma* **2015**, *3*, 9. [CrossRef]
69. Miguel, S.M.S.; Opperman, L.A.; Allen, E.P.; Svoboda, K.K. Use of antioxidants in oral healthcare. *Compend. Contin. Educ. Dent.* **2011**, *32*, E156–E159.
70. Kuś, P.M.; Congiu, F.; Teper, D.; Sroka, Z.; Jerković, I.; Tuberoso, C.I.G. Antioxidant activity, color characteristics, total phenol content and general HPLC fingerprints of six Polish unifloral honey types. *LWT-Food Sci. Technol.* **2014**, *55*, 124–130. [CrossRef]
71. Dżugan, M.; Grabek-Lejko, D.; Swacha, S.; Tomczyk, M.; Bednarska, S.; Kapusta, I. Physicochemical quality parameters, antibacterial properties and cellular antioxidant activity of Polish buckwheat honey. *Food Biosci.* **2020**, *34*, 100538. [CrossRef]

72. Kishore, R.K.; Halim, A.S.; Syazana, M.N.; Sirajudeen, K. Tualang honey has higher phenolic content and greater radical scavenging activity compared with other honey sources. *Nutr. Res.* **2011**, *31*, 322–325. [CrossRef]
73. Rosa, A.; Tuberoso, C.I.G.; Atzeri, A.; Melis, M.P.; Bifulco, E.; Dessì, M.A. Antioxidant profile of strawberry tree honey and its marker homogentisic acid in several models of oxidative stress. *Food Chem.* **2011**, *129*, 1045–1053. [CrossRef]
74. Alvarez-Suarez, J.M.; Giampieri, F.; González-Paramás, A.M.; Damiani, E.; Astolfi, P.; Martinez-Sanchez, G.; Bompadre, S.; Quiles, J.L.; Santos-Buelga, C.; Battino, M. Phenolics from monofloral honeys protect human erythrocyte membranes against oxidative damage. *Food Chem. Toxicol.* **2012**, *50*, 1508–1516. [CrossRef]
75. Yuslianti, E.R.; Bachtiar, B.M.; Sastradipura, D.F.S.; Sutjiatmo, A.B. Antioxidant Activity of Rambutan Honey: The free radical-scavenging activity in vitro and lipid peroxidation inhibition of oral mucosa wound tissue in vivo. *Res. J. Med. Plant* **2015**, *9*, 284–292. [CrossRef]
76. Toczewska, J.; Maciejczyk, M.; Konopka, T.; Zalewska, A. Total oxidant and antioxidant capacity of gingival crevicular fluid and saliva in patients with periodontitis: Review and clinical study. *Antioxidants* **2020**, *9*, 450. [CrossRef] [PubMed]
77. Petti, S.; Scully, C. Polyphenols, oral health and disease: A review. *J. Dent.* **2009**, *37*, 413–423. [CrossRef] [PubMed]
78. Ding, Y.; Yao, H.; Yao, Y.; Fai, L.Y.; Zhang, Z. Protection of dietary polyphenols against oral cancer. *Nutrients* **2013**, *5*, 2173–2191. [CrossRef] [PubMed]
79. Chatterjee, A.; Saluja, M.; Agarwal, G.; Alam, M. Green tea: A boon for periodontal and general health. *J. Indian Soc. Periodontol.* **2012**, *16*, 161. [PubMed]
80. Khan, H.; Sureda, A.; Belwal, T.; Çetinkaya, S.; Süntar, İ.; Tejada, S.; Devkota, H.P.; Ullah, H.; Aschner, M. Polyphenols in the treatment of autoimmune diseases. *Autoimmun. Rev.* **2019**, *18*, 647–657. [CrossRef]
81. Abbasi, A.J.; Mohammadi, F.; Bayat, M.; Gema, S.M.; Ghadirian, H.; Seifi, H.; Bayat, H.; Bahrami, N. Applications of Propolis in Dentistry: A Review. *Ethiop. J. Health Sci.* **2018**, *28*, 505–512. [CrossRef]
82. Babaee, N.; Hosseinkazemi, H.; Pouramir, M.; Baboli, O.K.; Salehi, M.; Khadir, F.; Bijani, A.; Mehryari, M. Salivary oxidant/antioxidant status and hematological parameters in patients with recurrent aphthous stomatitis. *Casp. J. Intern. Med.* **2016**, *7*, 13.
83. Samarghandian, S.; Farkhondeh, T.; Samini, F. Honey and Health: A Review of Recent Clinical Research. *Pharmacogn. Res.* **2017**, *9*, 121–127. [CrossRef]
84. Boisard, S.; Le Ray, A.-M.; Gatto, J.; Aumond, M.-C.; Blanchard, P.; Derbré, S.; Flurin, C.; Richomme, P. Chemical composition, antioxidant and anti-AGEs activities of a French poplar type propolis. *J. Agric. Food Chem.* **2014**, *62*, 1344–1351. [CrossRef]
85. Fabris, S.; Bertelle, M.; Astafyeva, O.; Gregoris, E.; Zangrando, R.; Gambaro, A.; Lima, G.P.P.; Stevanato, R. *Antioxidant Properties and Chemical Composition Relationship of Europeans and Brazilians Propolis*; Scientific Research Publishing: Wuhan, China, 2013.
86. Socha, R.; Gałkowska, D.; Bugaj, M.; Juszczak, L. Phenolic composition and antioxidant activity of propolis from various regions of Poland. *Nat. Prod. Res.* **2015**, *29*, 416–422. [CrossRef] [PubMed]
87. Kumari, S.; Naik, P.; Vishma, B.; Salian, S.R.; Devkar, R.A.; Khan, S.; Mutalik, S.; Kalthur, G.; Adiga, S.K. Mitigating effect of Indian propolis against mitomycin C induced bone marrow toxicity. *Cytotechnology* **2016**, *68*, 1789–1800. [CrossRef] [PubMed]
88. Bonamigo, T.; Campos, J.F.; Alfredo, T.M.; Balestieri, J.B.P.; Cardoso, C.A.L.; Paredes-Gamero, E.J.; de Picoli Souza, K.; Dos Santos, E.L. Antioxidant, cytotoxic, and toxic activities of propolis from two native bees in Brazil: Scaptotrigona depilis and Melipona quadrifasciata anthidioides. *Oxidative Med. Cell. Longev.* **2017**, *2017*, 1038153. [CrossRef] [PubMed]
89. Özkök, D.; Silici, S. Antioxidant activities of honeybee products and their mixtures. *Food Sci. Biotechnol.* **2017**, *26*, 201–206. [CrossRef] [PubMed]
90. Kocot, J.; Kielczykowska, M.; Luchowska-Kocot, D.; Kurzepa, J.; Musik, I. Antioxidant Potential of Propolis, Bee Pollen, and Royal Jelly: Possible Medical Application. *Oxidative Med. Cell. Longev.* **2018**, *2018*, 7074209. [CrossRef]
91. Aghel, S.; Pouramir, M.; Moghadamnia, A.A.; Moslemi, D.; Molania, T.; Ghassemi, L.; Motallebnejad, M. Effect of Iranian propolis on salivary total antioxidant capacity in gamma-irradiated rats. *J. Dent. Res. Dent. Clin. Dent. Prospect.* **2014**, *8*, 235.
92. El-Sharkawy, H.M.; Anees, M.M.; Van Dyke, T.E. Propolis Improves Periodontal Status and Glycemic Control in Patients With Type 2 Diabetes Mellitus and Chronic Periodontitis: A Randomized Clinical Trial. *J. Periodontol.* **2016**, *87*, 1418–1426. [CrossRef]
93. Giammarinaro, E.; Marconcini, S.; Genovesi, A.; Poli, G.; Lorenzi, C.; Covani, U. Propolis as an adjuvant to non-surgical periodontal treatment: A clinical study with salivary anti-oxidant capacity assessment. *Minerva Stomatol.* **2018**, *67*, 183–188. [CrossRef]
94. Guo, H.; Kouzuma, Y.; Yonekura, M. Structures and properties of antioxidative peptides derived from royal jelly protein. *Food Chem.* **2009**, *113*, 238–245. [CrossRef]
95. Silici, S.; Ekmekcioglu, O.; Eraslan, G.; Demirtas, A. Antioxidative effect of royal jelly in cisplatin-induced testes damage. *Urology* **2009**, *74*, 545–551. [CrossRef]
96. Park, M.J.; Kim, B.Y.; Deng, Y.; Park, H.G.; Choi, Y.S.; Lee, K.S.; Jin, B.R. Antioxidant capacity of major royal jelly proteins of honeybee (*Apis mellifera*) royal jelly. *J. Asia-Pac. Entomol.* **2020**, *23*, 445–448. [CrossRef]
97. Liu, J.-R.; Yang, Y.-C.; Shi, L.-S.; Peng, C.-C. Antioxidant properties of royal jelly associated with larval age and time of harvest. *J. Agric. Food Chem.* **2008**, *56*, 11447–11452. [CrossRef] [PubMed]
98. Balkanska, R.; Marghitas, L.-A.; Pavel, C.I. Antioxidant activity and total polyphenol content of royal jelly from Bulgaria. *Int. J. Curr. Microbiol. Appl. Sci* **2017**, *6*, 578–585. [CrossRef]

99. Uçar, M.; Barlak, Y. The Antioxidant Activity of Water, DMSO and Methanol Extracts of Royal Jelly from Bursa Province in Turkey. *Pharm. Drug Regul. Aff. J.* **2019**, *2*, 1–5. [CrossRef]
100. Aparna, S.; Srirangarajan, S.; Malgi, V.; Setlur, K.P.; Shashidhar, R.; Setty, S.; Thakur, S. A Comparative Evaluation of the Antibacterial Efficacy of Honey In Vitro and Antiplaque Efficacy in a 4-Day Plaque Regrowth Model In Vivo: Preliminary Results. *J. Periodontol.* **2012**, *83*, 1116–1121. [CrossRef]
101. Moussa, A.; Noureddine, D.; Saad, A.; Abdelmelek, M.; Abdelkader, B. Antifungal activity of four honeys of different types from Algeria against pathogenic yeast: *Candida albicans* and *Rhodotorula* sp. *Asian Pac. J. Trop. Biomed.* **2012**, *2*, 554–557. [CrossRef]
102. Awad, O.G.A.-N.; Hamad, A.-M.H. Honey can help in herpes simplex gingivostomatitis in children: Prospective randomized double blind placebo controlled clinical trial. *Am. J. Otolaryngol.* **2018**, *39*, 759–763. [CrossRef]
103. Candiracci, M.; Piatti, E.; Dominguez-Barragán, M.; García-Antrás, D.; Morgado, B.; Ruano, D.; Gutiérrez, J.F.; Parrado, J.; Castaño, A.l. Anti-inflammatory activity of a honey flavonoid extract on lipopolysaccharide-activated N13 microglial cells. *J. Agric. Food Chem.* **2012**, *60*, 12304–12311. [CrossRef]
104. Ghashm, A.A.; Othman, N.H.; Khattak, M.N.; Ismail, N.M.; Saini, R. Antiproliferative effect of Tualang honey on oral squamous cell carcinoma and osteosarcoma cell lines. *BMC Complement. Altern. Med.* **2010**, *10*, 49. [CrossRef]
105. Skaba, D.; Morawiec, T.; Tanasiewicz, M.; Mertas, A.; Bobela, E.; Szliszka, E.; Skucha-Nowak, M.; Dawiec, M.; Yamamoto, R.; Ishiai, S. Influence of the toothpaste with brazilian ethanol extract propolis on the oral cavity health. *Evid.-Based Complement. Altern. Med.* **2013**, *2013*, 215391. [CrossRef]
106. Kujumgiev, A.; Tsvetkova, I.; Serkedjieva, Y.; Bankova, V.; Christov, R.; Popov, S. Antibacterial, antifungal and antiviral activity of propolis of different geographic origin. *J. Ethnopharmacol.* **1999**, *64*, 235–240. [CrossRef] [PubMed]
107. Schnitzler, P.; Neuner, A.; Nolkemper, S.; Zundel, C.; Nowack, H.; Sensch, K.H.; Reichling, J. Antiviral activity and mode of action of propolis extracts and selected compounds. *Phytother. Res.* **2010**, *24*, S20–S28. [CrossRef] [PubMed]
108. Bretz, W.A.; Paulino, N.; Nör, J.E.; Moreira, A. The effectiveness of propolis on gingivitis: A randomized controlled trial. *J. Altern. Complement. Med.* **2014**, *20*, 943–948. [CrossRef]
109. Kuo, Y.-Y.; Jim, W.-T.; Su, L.-C.; Chung, C.-J.; Lin, C.-Y.; Huo, C.; Tseng, J.-C.; Huang, S.-H.; Lai, C.-J.; Chen, B.-C. Caffeic acid phenethyl ester is a potential therapeutic agent for oral cancer. *Int. J. Mol. Sci.* **2015**, *16*, 10748–10766. [CrossRef]
110. Moghim, H.; Taghipour, S.; Kheiri, S.; Khabbazi, H.; Baradaran, A. Antifungal effects of iranian propolis extract and royal jelly against candida albicans in-vitro. *Int. J. Prev. Med.* **2021**, *12*, 163. [PubMed]
111. Kohno, K.; Okamoto, I.; Sano, O.; Arai, N.; Iwaki, K.; Ikeda, M.; Kurimoto, M. Royal jelly inhibits the production of proinflammatory cytokines by activated macrophages. *Biosci. Biotechnol. Biochem.* **2004**, *68*, 138–145. [CrossRef]
112. Izuta, H.; Chikaraishi, Y.; Shimazawa, M.; Mishima, S.; Hara, H. 10-Hydroxy-2-decenoic acid, a major fatty acid from royal jelly, inhibits VEGF-induced angiogenesis in human umbilical vein endothelial cells. *Evid.-Based Complement. Altern. Med.* **2009**, *6*, 489–494. [CrossRef]
113. Garcia, G.; Ramos, F.; Maldonado, J.; Fernandez, A.; Yáñez, J.; Hernandez, L.; Gaytán, P. Prevalence of two Entamoeba gingivalis ST1 and ST2-kamaktli subtypes in the human oral cavity under various conditions. *Parasitol. Res.* **2018**, *117*, 2941–2948. [CrossRef]
114. Teles, R.; Teles, F.; Frias-Lopez, J.; Paster, B.; Haffajee, A. Lessons learned and unlearned in periodontal microbiology. *Periodontology* **2000**, *62*, 95–162. [CrossRef]
115. Prabu, G.; Gnanamani, A.; Sadulla, S. Guaijaverin–a plant flavonoid as potential antiplaque agent against Streptococcus mutans. *J. Appl. Microbiol.* **2006**, *101*, 487–495. [CrossRef]
116. McGrath, L.J.; Becker-Dreps, S.; Pate, V.; Brookhart, M.A. Trends in antibiotic treatment of acute otitis media and treatment failure in children, 2000–2011. *PLoS ONE* **2013**, *8*, e81210. [CrossRef]
117. Hwang, A.Y.; Gums, J.G. The emergence and evolution of antimicrobial resistance: Impact on a global scale. *Bioorg. Med. Chem.* **2016**, *24*, 6440–6445. [CrossRef]
118. Brown, H.L.; Roberts, A.E.L.; Cooper, R.; Jenkins, R. A review of selected bee products as potential anti-bacterial, anti-fungal, and anti-viral agents. *Med. Res. Arch.* **2016**, *4*, 1–20.
119. Topcuoglu, N.; Ozan, F.; Ozyurt, M.; Kulekci, G. In Vitro antibacterial effects of glassionomer cement containing ethanolic extract of propolis on Streptococcus mutans. *Eur. J. Dent.* **2012**, *6*, 428–433. [CrossRef]
120. Dzidic, M.; Collado, M.C.; Abrahamsson, T.; Artacho, A.; Stensson, M.; Jenmalm, M.C.; Mira, A. Oral microbiome development during childhood: An ecological succession influenced by postnatal factors and associated with tooth decay. *ISME J.* **2018**, *12*, 2292–2306. [CrossRef]
121. Tanner, A.; Kressirer, C.A.; Faller, L.L. Understanding Caries From the Oral Microbiome Perspective. *J. Calif. Dent. Assoc.* **2016**, *44*, 437–446. [CrossRef]
122. Dustmann, J.H. Antibacterial effect of honey. *Apiacta* **1979**, *14*, 7–11.
123. Kiamco, M.M.; Atci, E.; Mohamed, A.; Call, D.R.; Beyenal, H. Hyperosmotic agents and antibiotics affect dissolved oxygen and PH concentration gradients in Staphylococcus aureus biofilms. *Appl. Environ. Microbiol.* **2017**, *83*, e02783-16. [CrossRef]
124. Bucekova, M.; Jardekova, L.; Juricova, V.; Bugarova, V.; Di Marco, G.; Gismondi, A.; Leonardi, D.; Farkasovska, J.; Godocikova, J.; Laho, M. Antibacterial activity of different blossom honeys: New findings. *Molecules* **2019**, *24*, 1573. [CrossRef]
125. Combarros-Fuertes, P.; Estevinho, L.M.; Dias, L.G.; Castro, J.M.; Tomás-Barberán, F.A.; Tornadijo, M.E.; Fresno-Baro, J.M. Bioactive components and antioxidant and antibacterial activities of different varieties of honey: A screening prior to clinical application. *J. Agric. Food Chem.* **2018**, *67*, 688–698. [CrossRef]

126. Molan, P.C. The antibacterial activity of honey: 1. The nature of the antibacterial activity. *Bee World* **1992**, *73*, 5–28. [CrossRef]
127. Brudzynski, K.; Abubaker, K.; Castle, A. Re-examining the role of hydrogen peroxide in bacteriostatic and bactericidal activities of honey. *Front. Microbiol.* **2011**, *2*, 213. [CrossRef]
128. Rabie, E.; Serem, J.C.; Oberholzer, H.M.; Gaspar, A.R.M.; Bester, M.J. How methylglyoxal kills bacteria: An ultrastructural study. *Ultrastruct. Pathol.* **2016**, *40*, 107–111. [CrossRef]
129. Terio, V.; Bozzo, G.; Ceci, E.; Savarino, A.E.; Barrasso, R.; Di Pinto, A.; Mottola, A.; Marchetti, P.; Tantillo, G.; Bonerba, E. Methylglyoxal (MGO) in Italian Honey. *Appl. Sci.* **2021**, *11*, 831. [CrossRef]
130. Mandal, M.D.; Mandal, S. Honey: Its medicinal property and antibacterial activity. *Asian Pac. J. Trop. Biomed.* **2011**, *1*, 154–160. [CrossRef]
131. Koochak, H.; Seyyednejad, S.M.; Motamedi, H. Preliminary study on the antibacterial activity of some medicinal plants of Khuzestan (Iran). *Asian Pac. J. Trop. Med.* **2010**, *3*, 180–184. [CrossRef]
132. Pieper, B. Honey-based dressings and wound care: An option for care in the United States. *J. Wound Ostomy Cont. Nurs.* **2009**, *36*, 60–66. [CrossRef]
133. Al-Waili, N.S. Investigating the antimicrobial activity of natural honey and its effects on the pathogenic bacterial infections of surgical wounds and conjunctiva. *J. Med. Food* **2004**, *7*, 210–222. [CrossRef]
134. Shiga, H.; Jo, A.; Terao, K.; Nakano, M.; Oshima, T.; Maeda, N. Decrease of halitosis by intake of Manuka honey. *Gen. Sess. IADR Barc.* **2010**, *14*, 234–238.
135. Patel, R.; Thaker, V.; Patel, V.; Shukla, P.; Bhatnagar, P.; Patel, A. In-Vitro study of changing antibiotic sensitivity and resistance by honey on gingival inflammation during orthodontic treatment a preliminary report. *Orthod. Cyber J.* **2010**, *20*, 3–8.
136. Eick, S.; Schäfer, G.; Kwieciński, J.; Atrott, J.; Henle, T.; Pfister, W. Honey—A potential agent against Porphyromonas gingivalis: An in vitro study. *BMC Oral Health* **2014**, *14*, 24. [CrossRef]
137. Boyanova, L.; Ilieva, J.; Gergova, G.; Vladimirov, B.; Nikolov, R.; Mitov, I. Honey and green/black tea consumption may reduce the risk of Helicobacter pylori infection. *Diagn. Microbiol. Infect. Dis.* **2015**, *82*, 85–86. [CrossRef] [PubMed]
138. Burdock, G. Review of the biological properties and toxicity of bee propolis (propolis). *Food Chem. Toxicol.* **1998**, *36*, 347–363. [CrossRef] [PubMed]
139. Bouchelaghem, S. Propolis characterization and antimicrobial activities against *Staphylococcus aureus* and *Candida albicans*: A review. *Saudi J. Biol. Sci.* **2022**, *29*, 1936–1946. [CrossRef]
140. Popova, M.P.; Bankova, V.S.; Bogdanov, S.; Tsvetkova, I.; Naydenski, C.; Marcazzan, G.L.; Sabatini, A.-G. Chemical characteristics of poplar type propolis of different geographic origin. *Apidologie* **2007**, *38*, 306–311. [CrossRef]
141. Mirzoeva, O.; Grishanin, R.; Calder, P. Antimicrobial action of propolis and some of its components: The effects on growth, membrane potential and motility of bacteria. *Microbiol. Res.* **1997**, *152*, 239–246. [CrossRef]
142. Koo, H.; Cury, J.A.; Rosalen, P.L.; Ambrosano, G.M.; Ikegaki, M.; Park, Y.K. Effect of a mouthrinse containing selected propolis on 3-day dental plaque accumulation and polysaccharide formation. *Caries Res.* **2002**, *36*, 445–448. [CrossRef]
143. Duailibe, S.A.d.C.; Gonçalves, A.G.; Ahid, F.J.M. Effect of a propolis extract on Streptococcus mutans counts in vivo. *J. Appl. Oral Sci.* **2007**, *15*, 420–423. [CrossRef]
144. do Amaral, R.C.; Gomes, R.T.; Rocha, W.M.S.; Lemos, S.; Abreu, R.; Santos, V.R. Periodontitis treatment with Brazilian green propolis gel. *Pharmacologyonline* **2006**, *3*, 336–341.
145. Chandna, P.; Adlakha, V.K.; Das, S.; Singh, S. Complementary and Alternative Medicine (CAM): A review of propolis in dentistry. *Technology* **2014**, *4*, 675–685.
146. Castro, S.L. Propolis: Biological and pharmacological activities. Therapeutic uses of this bee-product. *Annu. Rev. Biomed. Sci.* **2001**, *3*, 49–83. [CrossRef]
147. Coutinho, D.; Karibasappa, S.N.; Mehta, D.S. Royal jelly antimicrobial activity against periodontopathic bacteria. *J. Interdiscip. Dent.* **2018**, *8*, 18.
148. Sugiyama, T.; Takahashi, K.; Mori, H. Royal jelly acid, 10-hydroxy-trans-2-decenoic acid, as a modulator of the innate immune responses. *Endocr. Metab. Immune Disord.-Drug Targets (Former. Curr. Drug Targets-Immune Endocr. Metab. Disord.)* **2012**, *12*, 368–376. [CrossRef]
149. Terada, Y.; Narukawa, M.; Watanabe, T. Specific hydroxy fatty acids in royal jelly activate TRPA1. *J. Agric. Food Chem.* **2011**, *59*, 2627–2635. [CrossRef] [PubMed]
150. Khosla, A.; Gupta, S.J.; Jain, A.; Shetty, D.C.; Sharma, N. Evaluation and comparison of the antimicrobial activity of royal jelly—A holistic healer against periodontopathic bacteria: An in vitro study. *J. Indian Soc. Periodontol.* **2020**, *24*, 221–226. [CrossRef]
151. Baker, J.L.; Bor, B.; Agnello, M.; Shi, W.; He, X. Ecology of the oral microbiome: Beyond bacteria. *Trends Microbiol.* **2017**, *25*, 362–374. [CrossRef]
152. Sheikh, D.; Zaman, S.; Naqvi, S.; Sheikh, M.; Ali, G. Studies on the antimicrobial activity of honey. *Pak. J. Pharm. Sci.* **1995**, *8*, 51–62.
153. White, J.W., Jr.; Subers, M.H.; Schepartz, A.I. The identification of inhibine, the antibacterial factor in honey, as hydrogen peroxide and its origin in a honey glucose-oxidase system. *Biochim. Biophys. Acta (BBA)-Spec. Sect. Enzymol. Subj.* **1963**, *73*, 57–70. [CrossRef]
154. Bang, L.M.; Buntting, C.; Molan, P. The effect of dilution on the rate of hydrogen peroxide production in honey and its implications for wound healing. *J. Altern. Complement. Med.* **2003**, *9*, 267–273. [CrossRef]

155. Irish, J.; Carter, D.A.; Shokohi, T.; Blair, S.E. Honey has an antifungal effect against Candida species. *Med. Mycol.* **2006**, *44*, 289–291. [CrossRef]
156. English, H.; Pack, A.; Molan, P.C. The effects of manuka honey on plaque and gingivitis: A pilot study. *J. Int. Acad. Periodontol.* **2004**, *6*, 63–67.
157. Takaisi-Kikuni, N.B.; Schilcher, H. Electron microscopic and microcalorimetric investigations of the possible mechanism of the antibacterial action of a defined propolis provenance. *Planta Medica* **1994**, *60*, 222–227. [CrossRef] [PubMed]
158. Feres, M.; Figueiredo, L.C.; Barreto, I.; Coelho, M.; Araujo, M.; Cortelli, S.C. In vitro antimicrobial activity of plant extracts and propolis in saliva samples of healthy and periodontally-involved subjects. *J. Int. Acad. Periodontol.* **2005**, *7*, 90–96.
159. Santos, V.; Pimenta, F.; Aguiar, M.; Do Carmo, M.; Naves, M.; Mesquita, R. Oral candidiasis treatment with Brazilian ethanol propolis extract. *Phytother. Res. Int. J. Devoted Pharmacol. Toxicol. Eval. Nat. Prod. Deriv.* **2005**, *19*, 652–654. [CrossRef] [PubMed]
160. Bachanová, K.; Klaudiny, J.; Kopernický, J.; Šimúth, J. Identification of honeybee peptide active against *Paenibacillus larvae* larvae through bacterial growth-inhibition assay on polyacrylamide gel. *Apidologie* **2002**, *33*, 259–269. [CrossRef]
161. Bílikova, K.; Huang, S.-C.; Lin, I.-P.; Šimuth, J.; Peng, C.-C. Structure and antimicrobial activity relationship of royalisin, an antimicrobial peptide from royal jelly of Apis mellifera. *Peptides* **2015**, *68*, 190–196. [CrossRef] [PubMed]
162. Hetru, C. Antimicrobial peptides from insects. In *Molecular Mechanisms of Immune Responses in Insects*; Springer: Berlin/Heidelberg, Germany, 1998.
163. Fujiwara, S.; Imai, J.; Fujiwara, M.; Yaeshima, T.; Kawashima, T.; Kobayashi, K. A potent antibacterial protein in royal jelly. Purification and determination of the primary structure of royalisin. *J. Biol. Chem.* **1990**, *265*, 11333–11337. [CrossRef]
164. Melliou, E.; Chinou, I. Chemistry and bioactivity of royal jelly from Greece. *J. Agric. Food Chem.* **2005**, *53*, 8987–8992. [CrossRef]
165. Wang, J.; Gao, Y.; Zhao, F. Phage–bacteria interaction network in human oral microbiome. *Environ. Microbiol.* **2016**, *18*, 2143–2158. [CrossRef]
166. Sällberg, M. Oral viral infections of children. *Periodontol. 2000* **2009**, *49*, 87–95. [CrossRef]
167. Presti, R.M.; Handley, S.; Droit, L.; Ghannoum, M.; Jacobson, M.; Shiboski, C.H.; Webster-Cyriaque, J.; Brown, T.; Michael, T.Y.; Overton, E.T. Alterations in the oral microbiome in HIV-infected participants after ART administration are influenced by immune status. *AIDS* **2018**, *32*, 1279. [CrossRef]
168. Santosh, A.B.R.; Muddana, K. Viral infections of oral cavity. *J. Fam. Med. Prim. Care* **2020**, *9*, 36. [CrossRef]
169. Asai, D.; Nakashima, H. Pathogenic viruses commonly present in the oral cavity and relevant antiviral compounds derived from natural products. *Medicines* **2018**, *5*, 120. [CrossRef]
170. Nagi, R.; Patil, D.J.; Rakesh, N.; Jain, S.; Sahu, S. Natural agents in the management of oral mucositis in cancer patients-systematic review. *J. Oral Biol. Craniofac. Res.* **2018**, *8*, 245–254. [CrossRef]
171. Münstedt, K. Bee products and the treatment of blister-like lesions around the mouth, skin and genitalia caused by herpes viruses—A systematic review. *Complement. Ther. Med.* **2019**, *43*, 81–84. [CrossRef]
172. Pagani, L. Effects of propolis flavonoids on virus infectivity and replication. *Microbiologica* **1990**, *13*, 207–213.
173. Serkedjieva, J.; Manolova, N.; Bankova, V. Anti-influenza virus effect of some propolis constituents and their analogues (esters of substituted cinnamic acids). *J. Nat. Prod.* **1992**, *55*, 294–297. [CrossRef]
174. Woźniak, M.; Sip, A.; Mrówczyńska, L.; Broniarczyk, J.; Waśkiewicz, A.; Ratajczak, I. Biological Activity and Chemical Composition of Propolis from Various Regions of Poland. *Molecules* **2022**, *28*, 141. [CrossRef]
175. Hashemipour, M.A.; Tavakolineghad, Z.; Arabzadeh, S.; Iranmanesh, Z.; Nassab, S. Antiviral Activities of Honey, Royal Jelly, and Acyclovir Against HSV-1. *Wounds Compend. Clin. Res. Pract.* **2014**, *26*, 47–54.
176. Ingawale, D.K.; Mandlik, S.K.; Patel, S.S. An emphasis on molecular mechanisms of anti-inflammatory effects and glucocorticoid resistance. *J. Complement. Integr. Med.* **2015**, *12*, 1–13. [CrossRef]
177. Yang, D.; Elner, S.G.; Bian, Z.-M.; Till, G.O.; Petty, H.R.; Elner, V.M. Pro-inflammatory cytokines increase reactive oxygen species through mitochondria and NADPH oxidase in cultured RPE cells. *Exp. Eye Res.* **2007**, *85*, 462–472. [CrossRef]
178. Schieber, M.; Chandel, N.S. ROS function in redox signaling and oxidative stress. *Curr. Biol.* **2014**, *24*, R453–R462. [CrossRef]
179. Halim, D.S.; Mahanani, E.S.; Saini, R.; Omar, M.; bt Ibrahi, N.R.; Alam, M.K. A comparison study on the effectiveness of local honey and salicylate gel for treatment of minor recurrent aphtous stomatitis. *Int. Med. J.* **2013**, *20*, 770–772.
180. Song, J.J.; Twumasi-Ankrah, P.; Salcido, R. Systematic review and meta-analysis on the use of honey to protect from the effects of radiation-induced oral mucositis. *Adv. Ski. Wound Care* **2012**, *25*, 23–28. [CrossRef]
181. Li, H.; Wang, Q.; Ding, Y.; Bao, C.; Li, W. Mangiferin ameliorates Porphyromonas gingivalis-induced experimental periodontitis by inhibiting phosphorylation of nuclear factor-κB and Janus kinase 1–signal transducer and activator of transcription signaling pathways. *J. Periodontal Res.* **2017**, *52*, 1–7. [CrossRef]
182. Fuji, A.; Kobayashi, S.; Kuboyama, N.; Furukawa, Y.; Kaneko, Y.; Ishihama, S. Augmantation of wound healing by royal jelly in streptozocindiabetic rats. *J. Pharmacol.* **1990**, *53*, 331–337.
183. Okamoto, I.; Taniguchi, Y.; Kunikata, T.; Kohno, K.; Iwaki, K.; Ikeda, M.; Kurimoto, M. Major royal jelly protein 3 modulates immune responses in vitro and in vivo. *Life Sci.* **2003**, *73*, 2029–2045. [CrossRef]
184. Shillitoe, E.J. The microbiome of oral cancer. In *Critical Reviews™ in Oncogenesis*; Begell House: Danbury, CT, USA, 2018; Volume 23.
185. Tamura, T.; Fujii, A.; Kuboyama, N. Effects of royal jelly on experimental transplantable tumours. In Proceedings of the XXXth International Apicultural Congress, Nagoya, Japan, 10–16 October 1985; Constantinescu, C., Ed.; Apimondia Publ House: Bucharest, Romania, 1985; pp. 474–477.

186. Afrin, S.; Haneefa, S.M.; Fernandez-Cabezudo, M.J.; Giampieri, F.; Al-Ramadi, B.K.; Battino, M. Therapeutic and preventive properties of honey and its bioactive compounds in cancer: An evidence-based review. *Nutr. Res. Rev.* **2020**, *33*, 50–76. [CrossRef]
187. Saxena, S.; Gautam, S.; Maru, G.; Kawle, D.; Sharma, A. Suppression of error prone pathway is responsible for antimutagenic activity of honey. *Food Chem. Toxicol.* **2012**, *50*, 625–633. [CrossRef]
188. Meister, A. Glutathione deficiency produced by inhibition of its synthesis, and its reversal; applications in research and therapy. *Pharmacol. Ther.* **1991**, *51*, 155–194. [CrossRef]
189. Orsolic, N.; Tadic, Z.; Benkovic, V.; Horvat, A.; Lojkic, M.; Basic, I. Radioprotective effect of a water-soluble derivative of propolis in mice. *Mellifera* **2004**, *4*, 45–52.
190. Chung, L.C.; Chiang, K.C.; Feng, T.H.; Chang, K.S.; Chuang, S.T.; Chen, Y.J.; Tsui, K.H.; Lee, J.C.; Juang, H.H. Caffeic acid phenethyl ester upregulates N-myc downstream regulated gene 1 via ERK pathway to inhibit human oral cancer cell growth in vitro and in vivo. *Mol. Nutr. Food Res.* **2017**, *61*, 1600842. [CrossRef] [PubMed]
191. Taheri, J.B.; Azimi, S.; Rafieian, N.; Zanjani, H.A. Herbs in dentistry. *Int. Dent. J.* **2011**, *61*, 287–296. [CrossRef] [PubMed]
192. Tamura, T.; Fujii, A.; Kuboyama, N. Antitumor effects of royal jelly (RJ). *Nihon Yakurigaku Zasshi. Folia Pharmacol. Jpn.* **1987**, *89*, 73–80. [CrossRef] [PubMed]
193. Oršolić, N.; Terzić, S.; Šver, L.; Bašić, I. Honey-bee products in prevention and/or therapy of murine transplantable tumours. *J. Sci. Food Agric.* **2005**, *85*, 363–370. [CrossRef]
194. Aljaghwani, A.; Allemailem, K.S.; Aljaghwani, L.F.; Alrumaihi, F.; Joseph, R.J.; Khan, A.A.; Rahmani, A.H.; Almatroudi, A. Antimicrobial Effect of Different Types of Honey on Selected ATCC Bacterial Strains. *Pharmacogn. J.* **2021**, *13*, 217–225. [CrossRef]
195. Atwa, A.-D.A.; AbuShahba, R.Y.; Mostafa, M.; Hashem, M.I. Effect of honey in preventing gingivitis and dental caries in patients undergoing orthodontic treatment. *Saudi Dent. J.* **2014**, *26*, 108–114. [CrossRef]
196. Ramsay, E.I.; Rao, S.; Madathil, L.; Hegde, S.K.; Baliga-Rao, M.P.; George, T.; Baliga, M.S. Honey in oral health and care: A mini review. *J. Oral Biosci.* **2019**, *61*, 32–36. [CrossRef]
197. Beegum, N.; Nandan, N.; Vishwanathan, S. Honey the paradisiacal panacea-A Review. *J. Ayurveda Integr. Med. Sci.* **2019**, *4*, 273–280.
198. Sanghavi, T.; Shah, N.; Parekh, V.; Singbal, K. Evaluation and comparison of efficacy of three different storage media, coconut water, propolis, and oral rehydration solution, in maintaining the viability of periodontal ligament cells. *J. Conserv. Dent. JCD* **2013**, *16*, 71. [CrossRef]
199. Kamburoğlu, K.; Özen, T. Analgesic effect of Anatolian propolis in mice. *Agri* **2011**, *23*, 47–50.
200. Dhanesuan, N.; Srisuparbh, D.; Tiranathanagul, S.; Rungsiyanont, S. The In Vitro Effect of Royal Jelly, Apis mellifera, on Proliferation of Human Gingival, Periodontal Ligament Fibroblasts and Human Bone Cells. *Thai Pharm. Health Sci. J.* **2011**, *6*, 182–187.
201. Qamar, Z.; Alghonaim, M.; Almohana, S.; Almohana, A.; Zeeshan, T. Potential biochemical effects of honey in oral health care: A review. *Int. Food Res. J.* **2021**, *28*, 23–30. [CrossRef]
202. Singhal, R.; Siddibhavi, M.; Sankeshwari, R.; Patil, P.; Jalihal, S.; Ankola, A. Effectiveness of three mouthwashes—Manuka honey, Raw honey, and Chlorhexidine on plaque and gingival scores of 12-15-year-old school children: A randomized controlled field trial. *J. Indian Soc. Periodontol.* **2018**, *22*, 34–39. [CrossRef]
203. Beena, J.P.; Sahoo, P.; Konde, S.; Raj, N.S.; Kumar, N.C.; Agarwal, M. Manuka Honey: A Potent Cariostatic Agent—An in vitro Study. *Int. J. Clin. Pediatr. Dent.* **2018**, *11*, 105.
204. Xu, X.; Zhou, X.D.; Wu, C.D. Tea catechin epigallocatechin gallate inhibits Streptococcus mutans biofilm formation by suppressing gtf genes. *Arch. Oral Biol.* **2012**, *57*, 678–683. [CrossRef]
205. Koo, H.; Schobel, B.; Scott-Anne, K.; Watson, G.; Bowen, W.; Cury, J.; Rosalen, P.; Park, Y. Apigenin and tt-farnesol with fluoride effects on S. mutans biofilms and dental caries. *J. Dent. Res.* **2005**, *84*, 1016–1020. [CrossRef]
206. .Duarte, S.; Rosalen, P.L.; Hayacibara, M.F.; Cury, J.A.; Bowen, W.H.; Marquis, R.; Rehder, V.L.; Sartoratto, A.; Ikegaki, M.; Koo, H. The influence of a novel propolis on mutans streptococci biofilms and caries development in rats. *Arch. Oral Biol.* **2006**, *51*, 15–22. [CrossRef]
207. Hayacibara, M.F.; Koo, H.; Rosalen, P.L.; Duarte, S.; Franco, E.M.; Bowen, W.H.; Ikegaki, M.; Cury, J.A. In Vitro and In Vivo effects of isolated fractions of Brazilian propolis on caries development. *J. Ethnopharmacol.* **2005**, *101*, 110–115. [CrossRef]
208. Nam, S.; Choi, Y.; Jang, S.; Shim, Y.; Han, G. Antimicrobial activity of propolis on different oral bacteria. *Indian J. Sci. Technol.* **2016**, *9*, 1–4. [CrossRef]
209. Kuang, X.; Chen, V.; Xu, X. Novel approaches to the control of oral microbial biofilms. *BioMed Res. Int.* **2018**, *2018*, 6498932. [CrossRef]
210. Sricholpech, M.; Srisupabh, D. Royal jelly promotes the viability and proliferation of periodontal ligament fibroblasts in an in vitro tooth avulsion simulation. *Mahidol Dent. J.* **2015**, *35*, 47–56.
211. Scully, C.; Bagan, J. Oral squamous cell carcinoma: Overview of current understanding of aetiopathogenesis and clinical implications. *Oral Dis.* **2009**, *15*, 388–399. [CrossRef]
212. Borse, V.; Konwar, A.N.; Buragohain, P. Oral cancer diagnosis and perspectives in India. *Sens. Int.* **2020**, *1*, 100046. [CrossRef]
213. Al-Koshab, M.; Alabsi, A.M.; Bakri, M.M.; Naicker, M.S.; Seyedan, A. Chemopreventive activity of Tualang honey against oral squamous cell carcinoma—In Vivo. *Oral Surg. Oral Med. Oral Pathol. Oral Radiol.* **2020**, *129*, 484–492. [CrossRef]

214. Mahmood, R.; Asif, J.A.; Shahidan, W.N.S. Stingless-bee (Trigona itama) honey adversely impacts the growth of oral squamous cell carcinoma cell lines (HSC-2). *Eur. J. Integr. Med.* **2020**, *37*, 101162. [CrossRef]
215. Jaganathan, S.K.; Mandal, M. Antiproliferative effects of honey and of its polyphenols: A review. *J. Biomed. Biotechnol.* **2009**, *2009*, 830616. [CrossRef]
216. Dornelas, C.A.; Fechine-Jamacaru, F.V.; Albuquerque, I.L.; Magalhães, H.I.F.; Souza, A.J.S.D.; Alves, L.A.; Almeida, P.R.C.D.; Lemos, T.L.G.D.; Castro, J.D.V.D.; Moraes, M.E.A. Chemoprevention with green propolis green propolis extracted in L-lysine versus carcinogenesis promotion with L-lysine in N-Butyl-N-[4-hydroxybutyl] nitrosamine (BBN) induced rat bladder cancer. *Acta Cir. Bras.* **2012**, *27*, 185–192. [CrossRef]
217. Peng, C.-Y.; Yang, H.-W.; Chu, Y.-H.; Chang, Y.-C.; Hsieh, M.-J.; Chou, M.-Y.; Yeh, K.-T.; Lin, Y.-M.; Yang, S.-F.; Lin, C.-W. Caffeic Acid phenethyl ester inhibits oral cancer cell metastasis by regulating matrix metalloproteinase-2 and the mitogen-activated protein kinase pathway. *Evid.-Based Complement. Altern. Med.* **2012**, *2012*, 732578. [CrossRef]
218. Yanagita, M.; Kojima, Y.; Mori, K.; Yamada, S.; Murakami, S. Osteoinductive and anti-inflammatory effect of royal jelly on periodontal ligament cells. *Biomed. Res.* **2011**, *32*, 285–291. [CrossRef]
219. Drain, J.; Fleming, M.O. Palliative management of malodorous squamous cell carcinoma of the oral cavity with Manuka honey. *J. Wound Ostomy Cont. Nurs.* **2015**, *42*, 190–192. [CrossRef]
220. Lusby, P.; Coombes, A.; Wilkinson, J. Honey: A potent agent for wound healing? *J. WOCN* **2002**, *29*, 295–300. [CrossRef]
221. Kim, K.-E.; Kang, J.-O.; Park, Y.-D. Effect of mouthrinse containing propolis on oral malodor. *Int. J. Clin. Prev. Dent.* **2014**, *10*, 179–184. [CrossRef]
222. Biswal, B.M.; Zakaria, A.; Ahmad, N.M. Topical application of honey in the management of radiation mucositis. A preliminary study. *Support. Care Cancer* **2003**, *11*, 242–248. [CrossRef]
223. Raeessi, M.A.; Raeessi, N.; Panahi, Y.; Gharaie, H.; Davoudi, S.M.; Saadat, A.; Karimi Zarchi, A.A.; Raeessi, F.; Ahmadi, S.M.; Jalalian, H. "Coffee plus honey" versus "topical steroid" in the treatment of chemotherapy-induced oral mucositis: A randomised controlled trial. *BMC Complement. Altern. Med.* **2014**, *14*, 293. [CrossRef]
224. Davari, A.; Ataei, E.; Assarzadeh, H. Dentin hypersensitivity: Etiology, diagnosis and treatment; a literature review. *J. Dent.* **2013**, *14*, 136.
225. Erdem, O.; Gungormus, Z. The effect of royal jelly on oral mucositis in patients undergoing radiotherapy and chemotherapy. *Holist. Nurs. Pract.* **2014**, *28*, 242–246. [CrossRef]
226. Yamauchi, K.; Kogashiwa, Y.; Moro, Y.; Kohno, N. The effect of topical application of royal jelly on chemoradiotherapy-induced mucositis in head and neck cancer: A preliminary study. *Int. J. Otolaryngol.* **2014**, *2014*, 974967. [CrossRef]
227. Wojtaszek, C.; Kochis, L.; Cunningham, R. Nutrition Impact Systems in the Oncology Patient. *Oncol. Issues* **2002**, *17*, 15–17.
228. Charalambous, A.; Lambrinou, E.; Katodritis, N.; Vomvas, D.; Raftopoulos, V.; Georgiou, M.; Paikousis, L.; Charalambous, M. The effectiveness of thyme honey for the management of treatment-induced xerostomia in head and neck cancer patients: A feasibility randomized control trial. *Eur. J. Oncol. Nurs.* **2017**, *27*, 1–8. [CrossRef]
229. Sateriale, D.; Facchiano, S.; Colicchio, R.; Pagliuca, C.; Varricchio, E.; Paolucci, M.; Volpe, M.G.; Salvatore, P.; Pagliarulo, C. In Vitro synergy of polyphenolic extracts from honey, myrtle and pomegranate against oral pathogens, *S. mutans* and *R. dentocariosa*. *Front. Microbiol.* **2020**, *11*, 1465. [CrossRef]
230. Akber, A.; Ahmed, S.; Umer Hasan, S.M. A comparative study on the effect of propolis and dentine bonding agent in treating dentine hypersensitivity. *J. Pak. Med. Assoc.* **2022**, *72*, 2417–2421. [CrossRef]
231. Tavares, J.A.O.; da Silva, F.A.; Santos, T.M.L.; Caneppele, T.M.F.; Augusto, M.G. The effectiveness of propolis extract in reducing dentin hypersensitivity: A systematic review. *Arch. Oral Biol.* **2021**, *131*, 105248. [CrossRef]
232. Mehta, P.; Vimala, N.; Mandke, L. An insight into dentin desensitizing agents-In vivo study. *Indian J. Dent. Res.* **2013**, *24*, 571. [CrossRef]
233. Purra, A.R.; Mushtaq, M.; Acharya, S.R.; Saraswati, V. A comparative evaluation of propolis and 5.0% potassium nitrate as a dentine desensitizer: A clinical study. *J. Indian Soc. Periodontol.* **2014**, *18*, 466. [CrossRef]
234. Askari, M.; Yazdani, R. Comparison of two desensitizing agents for decreasing dentin hypersensitivity following periodontal surgeries: A randomized clinical trial. *Quintessence Int.* **2019**, *50*, 320–329.
235. Maity, S.; Priyadharshini, V.; Basavaraju, S. A comparative evaluation of propolis and light-cured ormocer-based desensitizer in reducing dentin hypersensitivity. *J. Indian Soc. Periodontol.* **2020**, *24*, 441. [CrossRef]
236. Nayak, P.A.; Nayak, U.A.; Mythili, R. Effect of Manuka honey, chlorhexidine gluconate and xylitol on the clinical levels of dental plaque. *Contemp. Clin. Dent.* **2010**, *1*, 214. [CrossRef]
237. Alibasyah, Z.M.; Saputri, D.; Alviana, V. The Comparison Between Dental Plaque Score Before and After Gargling with Tongra Original Honey 5% Solution (Study of Student in Dentistry of Syiah Kuala University). *Biomed. Pharmacol. J.* **2018**, *11*, 381–385. [CrossRef]
238. Nik Man, N.M.; Hassan, R.; Ang, C.Y.; Abdullah, A.D.; Mohd Radzi, M.A.; Sulaiman, S.A. Antileukemic Effect of Tualang Honey on Acute and Chronic Leukemia Cell Lines. *Biomed. Res. Int.* **2015**, *2015*, 307094. [CrossRef]
239. Yılmaz, N.; Nisbet, Ö.; Nisbet, C.; Ceylan, G.; Hoşgör, F.; Dede, Ö.D. Biochemical evaluation of the therapeutic effectiveness of honey in oral mucosal ulcers. *Bosn. J. Basic Med. Sci.* **2009**, *9*, 290. [CrossRef] [PubMed]
240. Nakao, R.; Senpuku, H.; Ohnishi, M.; Takai, H.; Ogata, Y. Effect of topical administration of propolis in chronic periodontitis. *Odontology* **2020**, *108*, 704–714. [CrossRef] [PubMed]

241. Bolouri, A.J.; Pakfetrat, A.; Tonkaboni, A.; Aledavood, S.A.; Najafi, M.F.; Delavarian, Z.; Shakeri, M.T.; Mohtashami, A. Preventing and therapeutic effect of propolis in radiotherapy induced mucositis of head and neck cancers: A triple-blind, randomized, placebo-controlled trial. *Iran. J. Cancer Prev.* **2015**, *8*, e4019.
242. Pina, G.d.; Lia, E.N.; Berretta, A.A.; Nascimento, A.P.; Torres, E.C.; Buszinski, A.F.; de Campos, T.A.; Coelho, E.B.; Martins, V.d.P. Efficacy of propolis on the denture stomatitis treatment in older adults: A multicentric randomized trial. *Evid.-Based Complement. Altern. Med.* **2017**, *2017*, 8971746. [CrossRef] [PubMed]
243. Netto, C.A.; Marcucci, M.C.; Paulino, N.; Anido-Anido, A.; Amore, R.; de Mendonça, S.; Neto, L.B.; Bretz, W.A. Effects of typified propolis on mutans streptococci and lactobacilli: A randomized clinical trial. *Braz. Dent. Sci.* **2013**, *16*, 31. [CrossRef] [PubMed]
244. Pereira, E.M.R.; da Silva, J.L.D.C.; Silva, F.F.; De Luca, M.P.; Lorentz, T.C.M.; Santos, V.R. Clinical evidence of the efficacy of a mouthwash containing propolis for the control of plaque and gingivitis: A phase II study. *Evid.-Based Complement. Altern. Med.* **2011**, *2011*, 750249. [CrossRef] [PubMed]
245. de Faveri, M.; Pupio, G.C.; Koo, H.; Bueno-Silva, B.; de Oliveira, K.M.; Figueiredo, L.C.; Rosalen, P.L.; Hayacibara, R.M.; Fujimaki, M. The effect of Brazilian propolis type-3 against oral microbiota and volatile sulfur compounds in subjects with morning breath malodor. *Clin. Oral Investig.* **2021**, *26*, 1531–1541. [CrossRef]
246. Barnett, M.L. The rationale for the daily use of an antimicrobial mouthrinse. *J. Am. Dent. Assoc.* **2006**, *137*, S16–S21. [CrossRef]
247. Colombo, A.; Tanner, A. The role of bacterial biofilms in dental caries and periodontal and peri-implant diseases: A historical perspective. *J. Dent. Res.* **2019**, *98*, 373–385. [CrossRef]
248. James, P.; Worthington, H.V.; Parnell, C.; Harding, M.; Lamont, T.; Cheung, A.; Whelton, H.; Riley, P. Chlorhexidine mouthrinse as an adjunctive treatment for gingival health. *Cochrane Database Syst. Rev.* **2017**, CD008676. [CrossRef]
249. Lim, K.-S.; Kam, P. Chlorhexidine-pharmacology and clinical applications. *Anaesth. Intensive Care* **2008**, *36*, 502–512. [CrossRef] [PubMed]
250. Osso, D.; Kanani, N. Antiseptic mouth rinses: An update on comparative effectiveness, risks and recommendations. *Am. Dent. Hyg. Assoc.* **2013**, *87*, 10–18.
251. Mude, G.; Pise, S.; Thombare, G. Formulation and evaluation of polyherbal toothpaste and comparative study with marketed formulations. *Int. J. Creat. Res. Thoughts* **2020**, *8*, 3796–3806.
252. Morawiec, T.; Dziedzic, A.; Niedzielska, I.; Mertas, A.; Tanasiewicz, M.; Skaba, D.; Kasperski, J.; Machorowska-Pieniążek, A.; Kucharzewski, M.; Szaniawska, K. The biological activity of propolis-containing toothpaste on oral health environment in patients who underwent implant-supported prosthodontic rehabilitation. *Evid.-Based Complement. Altern. Med.* **2013**, *2013*, 704947. [CrossRef] [PubMed]
253. Wiatrak, K.; Morawiec, T.; Rój, R.; Kownacki, P.; Nitecka-Buchta, A.; Niedzielski, D.; Wychowański, P.; Machorowska-Pieniążek, A.; Cholewka, A.; Baldi, D. Evaluation of Effectiveness of a Toothpaste Containing Tea Tree Oil and Ethanolic Extract of Propolis on the Improvement of Oral Health in Patients Using Removable Partial Dentures. *Molecules* **2021**, *26*, 4071. [CrossRef] [PubMed]
254. Victorino, F.R.; Bramante, C.M.; Watanabe, E.; Ito, I.Y.; Franco, S.L.; Hidalgo, M.M. Antibacterial activity of propolis-based toothpastes for endodontic treatment. *Braz. J. Pharm. Sci.* **2009**, *45*, 795–800. [CrossRef]
255. Li, J.; Luo, S.; Susha, Y.L. Propolis Toothpaste and Preparation Method Thereof. CN102283795A, 22 August 2011.
256. Zhang, Y.; Zhang, Y. Propolis Toothpaste and Preparation Method Thereof. CN110755355A, 6 August 2019.
257. Alviano, W.; Mendonça-Filho, R.; Alviano, D.; Bizzo, H.; Souto-Padrón, T.; Rodrigues, M.; Bolognese, A.; Alviano, C.; Souza, M. Antimicrobial activity of Croton cajucara Benth linalool-rich essential oil on artificial biofilms and planktonic microorganisms. *Oral Microbiol. Immunol.* **2005**, *20*, 101–105. [CrossRef]
258. Yoo, C.-S.; Chae, S.-K.; Lee, H.-G.; Kim, C.-H.; Kim, D.-K.; Kim, W.-G.; Kim, K.-E.; Ahn, H.-K.; Kim, J.-E.; Oh, Y.-J. The inhibition effects of cetylpyridinium chloride in dentifrices on plaque formation: Clinical test. *Int. J. Clin. Prev. Dent.* **2014**, *10*, 9–14. [CrossRef]
259. Halboub, E.; Al-Maweri, S.A.; Al-Wesabi, M.; Al-Kamel, A.; Shamala, A.; Al-Sharani, A.; Koppolu, P. Efficacy of propolis-based mouthwashes on dental plaque and gingival inflammation: A systematic review. *BMC Oral Health* **2020**, *20*, 198. [CrossRef]
260. Murray, M.; Worthington, H.; Blinkhorn, A. A study to investigate the effect of a propolis-containing mouthrinse on the inhibition of de novo plaque formation. *J. Clin. Periodontol.* **1997**, *24*, 796–798. [CrossRef]
261. Dehghani, M.; Abtahi, M.; Hasanzadeh, N.; Farahzad, Z.; Noori, M.; Noori, M. Effect of Propolis mouthwash on plaque and gingival indices over fixed orthodontic patients. *J. Clin. Exp. Dent.* **2019**, *11*, e244–e249. [CrossRef] [PubMed]
262. Santiago, K.B.; Piana, G.M.; Conti, B.J.; Cardoso, E.d.O.; Murbach Teles Andrade, B.F.; Zanutto, M.R.; Mores Rall, V.L.; Fernandes, A., Jr.; Sforcin, J.M. Microbiological control and antibacterial action of a propolis-containing mouthwash and control of dental plaque in humans. *Nat. Prod. Res.* **2018**, *32*, 1441–1445. [CrossRef] [PubMed]
263. Krishna, K.; Manjunath, S.; Shivanagendra, S.; Kumar, S.D.; Shekar, S.S. Health from the hive: 5% Propolis mouth wash as an adjunct in the treatment of chronic generalized gingivitis-a randomized controlled clinical trial. *Dentistry* **2019**, *9*, 1–6. [CrossRef]
264. Ke, W.; Duan, Y. Propolis Mouthwash and Preparation Method Thereof. CN104739738A, 27 December 2013.
265. Tulsani, S.G.; Chikkanarasaiah, N.; Siddaiah, S.B.; Krishnamurthy, N.H. The effect of Propolis and Xylitol chewing gums on salivary Streptococcus mutans count: A clinical trial. *Indian J. Dent. Res.* **2014**, *25*, 737. [CrossRef]
266. Liu, G. Compound Bee Product Chewable Tablet. CN107198189B, 26 June 2017.
267. Fan, S. Chewable Propolis Tablet and Production Technology Thereof. CN102326723A, 31 May 2011.
268. Ibrahim, P. Method for Producing Chewing Gum Using Propolis in Solid Resin Form. WO2020101601, 15 November 2018.

269. Liu, S.; Liu, Y. Brazil Green Propolis Toothpaste. CN107412138B, 6 September 2017.
270. Wang, J. Manuka Honey Toothpaste for Pregnant Women and Preparation Method of Toothpaste. CN105287328A, 23 September 2015.
271. Sung, K.H. A composition for Cleaning Oral Cavity Containing Water-Soluble Propolis Extract. KR100719306B1, 4 November 2004.
272. Feng, M. Hive Honey Chewing Gum and Preparation Method. CN107751533A, 10 October 2017.
273. Koca, I.; Koca, A.F. Poisoning by mad honey: A brief review. *Food Chem. Toxicol.* **2007**, *45*, 1315–1318. [CrossRef] [PubMed]
274. Silici, S.; Atayoglu, A.T. Mad honey intoxication: A systematic review on the 1199 cases. *Food Chem. Toxicol.* **2015**, *86*, 282–290. [CrossRef]
275. Edgar, J.A.; Roeder, E.; Molyneux, R.J. Honey from plants containing pyrrolizidine alkaloids: A potential threat to health. *J. Agric. Food Chem.* **2002**, *50*, 2719–2730. [CrossRef]
276. Fields, B.A.; Reeve, J.; Bartholomaeus, A.; Mueller, U. Human pharmacokinetic study of tutin in honey; a plant-derived neurotoxin. *Food Chem. Toxicol.* **2014**, *72*, 234–241. [CrossRef]
277. Larsen, L.; Joyce, N.I.; Sansom, C.E.; Cooney, J.M.; Jensen, D.J.; Perry, N.B. Sweet poisons: Honeys contaminated with glycosides of the neurotoxin tutin. *J. Nat. Prod.* **2015**, *78*, 1363–1369. [CrossRef]
278. Islam, M.N.; Khalil, M.I.; Islam, M.A.; Gan, S.H. Toxic compounds in honey. *J. Appl. Toxicol.* **2014**, *34*, 733–742. [CrossRef]
279. Fearnley, J. *Bee Propolis: Natural Healing from the Hive*; Souvenir: London, UK, 2001.
280. Walgrave, S.E.; Warshaw, E.M.; Glesne, L.A. Allergic contact dermatitis from propolis. *Dermatitis* **2005**, *16*, 209–215. [PubMed]
281. Rosmilah, M.; Shahnaz, M.; Patel, G.; Lock, J.; Rahman, D.; Masita, A.; Noormalin, A. Characterization of major allergens of royal jelly Apis mellifera. *Trop. Biomed.* **2008**, *25*, 243–251. [PubMed]

antioxidants

MDPI

Review

Honey: A Promising Therapeutic Supplement for the Prevention and Management of Osteoporosis and Breast Cancer

Monika Martiniakova [1,*,†], Veronika Kovacova [1,†], Vladimira Mondockova [2], Nina Zemanova [2], Martina Babikova [2], Roman Biro [1], Sona Ciernikova [3] and Radoslav Omelka [2,*]

1 Department of Zoology and Anthropology, Faculty of Natural Sciences and Informatics, Constantine the Philosopher University in Nitra, 949 01 Nitra, Slovakia
2 Department of Botany and Genetics, Faculty of Natural Sciences and Informatics, Constantine the Philosopher University in Nitra, 949 01 Nitra, Slovakia
3 Department of Genetics, Cancer Research Institute, Biomedical Research Center of Slovak Academy of Sciences, 845 05 Bratislava, Slovakia
* Correspondence: mmartiniakova@ukf.sk (M.M.); romelka@ukf.sk (R.O.); Tel.: +421-37-6408-737 (R.O.)
† These authors contributed equally to this work.

Abstract: Osteoporosis and breast cancer are serious diseases that have become a significant socioeconomic burden. There are biochemical associations between the two disorders in terms of the amended function of estrogen, receptor activator of nuclear factor kappa beta ligand, oxidative stress, inflammation, and lipid accumulation. Honey as a functional food with high antioxidant and anti-inflammatory properties can contribute to the prevention of various diseases. Its health benefits are mainly related to the content of polyphenols. This review aims to summarize the current knowledge from in vitro, animal, and human studies on the use of honey as a potential therapeutic agent for osteoporosis and breast cancer. Preclinical studies have revealed a beneficial impact of honey on both bone health (microstructure, strength, oxidative stress) and breast tissue health (breast cancer cell proliferation and apoptosis, tumor growth rate, and volume). The limited number of clinical trials, especially in osteoporosis, indicates the need for further research to evaluate the potential benefits of honey in the treatment. Clinical studies related to breast cancer have revealed that honey is effective in increasing blood cell counts, interleukin-3 levels, and quality of life. In summary, honey may serve as a prospective therapeutic supplement for bone and breast tissue health.

Keywords: honey; osteoporosis; breast cancer; associations; preclinical studies; clinical trials; prevention; management; treatment

Citation: Martiniakova, M.; Kovacova, V.; Mondockova, V.; Zemanova, N.; Babikova, M.; Biro, R.; Ciernikova, S.; Omelka, R. Honey: A Promising Therapeutic Supplement for the Prevention and Management of Osteoporosis and Breast Cancer. *Antioxidants* **2023**, *12*, 567. https://doi.org/10.3390/antiox12030567

Academic Editors: Josipa Vlainić and Ivana Tlak Gajger

Received: 7 February 2023
Revised: 19 February 2023
Accepted: 21 February 2023
Published: 24 February 2023

1. Introduction

Menopause is a biological process characterized by dysfunction of ovarian follicles and estrogen deficiency, oxidative stress, and inflammation, that together lead to different chronic disorders [1,2]. When the organism is exposed to high levels of oxidative stress following estrogen depletion, lipid accumulation also occurs [3]. Osteoporosis and breast cancer are considered serious diseases in which the aforementioned factors are involved and are currently becoming a significant socioeconomic burden worldwide.

Generally, postmenopausal osteoporosis is characterized by reduced bone mineral density (BMD) and increased risk of fragility fractures that are associated with significant pain, suffering, and disability [4]. Moreover, hip and vertebral fractures are consistent with significantly increased mortality [5]. It has been reported that a decrease in estrogen production represents a major cause of reduced bone mass [6–8]. During menopause, the osteoprotective effect of estrogen is weakened, leading to elevated expression of pro-inflammatory cytokines that promote osteoclastogenesis [9–12]. In general, estrogen regulates bone metabolism through two receptors: estrogen receptor alpha (ERα) and estrogen receptor beta (ERβ), with ERα being more dominant. Loss of estrogen also influences osteoblast progenitor cells via reduced ERα expression and lower responsiveness to mechanical stimulation [13]. Thus, estrogen deficiency not only directly affects the differentiation of precursor cells more toward active osteoclasts and less toward osteoblasts but can also influence their cellular energetics. Increased adiposity and inflammation after menopause can indirectly lead to bone loss as well [14].

Similar to bones, breast tissue is also dependent on estrogen [15]. In breast carcinogenesis, elevated exposure to estrogen is linked with early menarche, late menopause, obesity, and estrogen replacement therapy. High blood estrogen levels are able to increase the risk, incidence, and severity of breast malignancy in pre- and postmenopausal women [16]. In general, breast cancer is the second leading cause of cancer death in women, with a higher prevalence in postmenopausal women [1,17]. Consequently, postmenopausal women are at risk of morbidity and mortality, which are a combination of both diseases mentioned above.

Current pharmacological treatment for osteoporosis and breast cancer is often associated with adverse side effects; therefore, various natural therapeutic substances have been intensively studied to find an alternative and effective treatment method with less harmful impacts [17,18]. Honey and other bee products (e.g., royal jelly, propolis, bee bread, drone brood homogenate) are widely used as a functional food due to their high antioxidant and anti-inflammatory properties, which contribute to the prevention of various diseases, including diabetes, osteoporosis, cancer, reproductive disorders [19–23].

In general, honey is a sweet viscous liquid stored in combs after bees collect it from plants. It is produced by regurgitation, enzymatic activity, and evaporation of water in the hives. In addition to the source of nectar, bees also collect insect secretions (belonging to the genus *Rhynchota*) to produce honeydew honey [2,24,25]. Honey consists of at least 181 substances, mainly carbohydrates such as fructose (38%) and glucose (31%). It also contains enzymes, proteins, amino acids, polyphenols, vitamins, and minerals in lower quantities [26]. The content of polyphenols, which cover a wide spectrum of phytochemicals and are found in almost all types of natural honey, contributes to its health-promoting potential. Such polyphenols include flavonoids (e.g., quercetin, kaempferol, luteolin, hesperetin, chrysin, apigenin, galangin), phenolic acids (e.g., ellagic, caffeic, gallic, ferulic, benzoic, ascorbic), antioxidant enzymes (e.g., catalase, glucose oxidase, peroxidase) and carotenoids [27,28]. Most of these compounds interact with each other to create a range of synergistic antioxidant properties. Many studies revealed antioxidant, antibacterial, antiviral, immunomodulatory, anti-inflammatory, hypocholesterolemic, hypotensive, and antitumor impacts of honey [2,29]. The composition of a particular honey sample depends to a large extent on the nectar composition, the method of nectar collecting, environmental and seasonal factors, geographical origin, as well as storage conditions [30].

This review aims to summarize the current knowledge from preclinical and clinical studies regarding the use of honey as a potential therapeutic agent for osteoporosis and breast cancer due to their elevated incidence in postmenopausal women. Biochemical connections between the two disorders are also provided.

2. Biochemical Associations between Osteoporosis and Breast Cancer

Biochemical connections between osteoporosis and breast cancer include the amended function of receptor activator of nuclear factor kappa beta ligand (RANKL), estrogen, reactive oxygen species (ROS)-induced oxidative stress, chronic low-grade inflammation, and lipid accumulation [1,2]. A clearer understanding of the associations between these diseases can lead to the development of a therapeutic target for postmenopausal breast cancer patients. Figure 1 illustrates the influence of RANKL, estrogen, ROS, and inflammation on the development of osteoporosis and breast cancer.

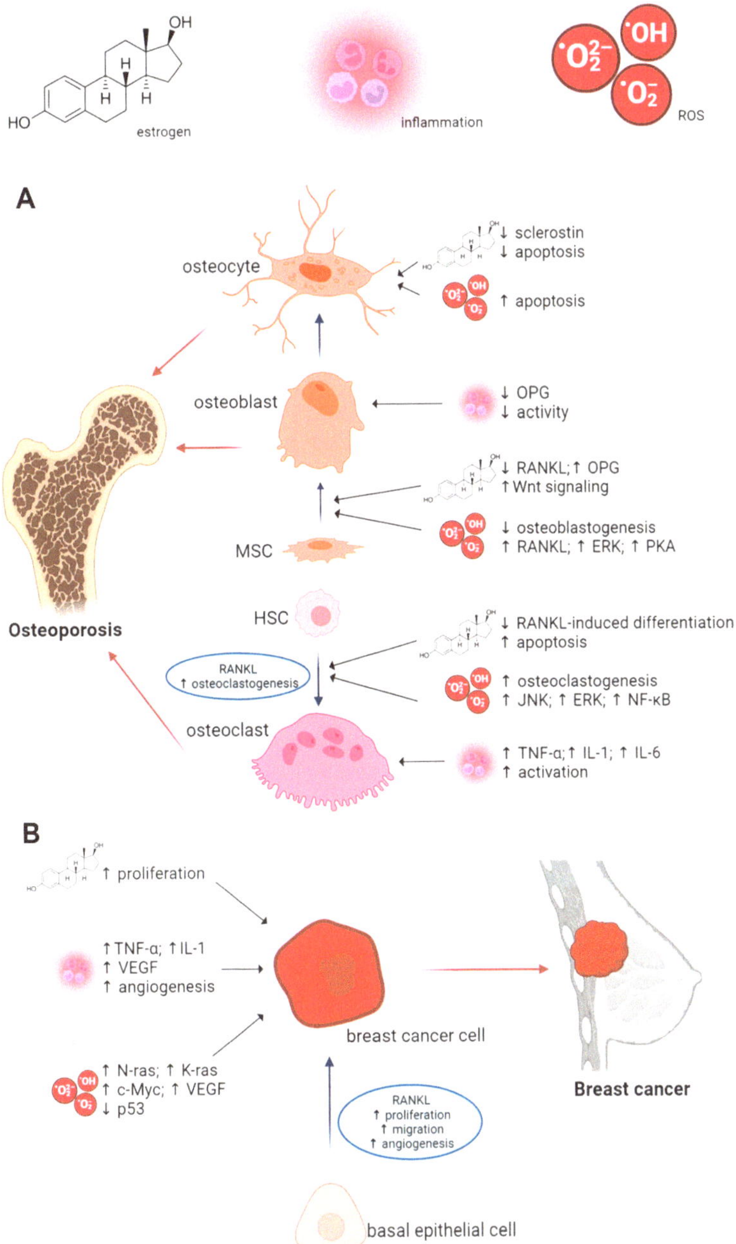

Figure 1. The impact of RANKL, estrogen, ROS, and inflammation on the development of osteoporosis (**A**) and breast cancer (**B**) (created with BioRender.com, https://www.biorender.com/, accessed on 27 January 2023). Abbreviations: c-Myc—c-myelocytomatosis oncogene product; ERK—extracellular signal-regulated kinase; HSC—hematopoietic stem cell; IL-1—interleukin 1; IL-6—interleukin 6; JNK—c-Jun N-terminal kinase; K-ras—Kirsten rat sarcoma viral oncogene homolog; MSC—mesenchymal stem cell; N-ras—neuroblastoma RAS viral oncogene homolog; NF-κB—nuclear factor kappa-B; OPG—osteoprotegerin; p53—tumor protein p53; PKA—protein kinase A; RANKL—receptor activator of nuclear factor kappa-B ligand; ROS—reactive oxygen species; TNF-α—tumor necrosis factor-alpha; VEGF—vascular endothelial growth factor; ↑—increased; ↓—decreased.

RANKL is an important cytokine that is a member of the tumor necrosis factor (TNF) family and is encoded by the tumor necrosis factor ligand super family 11 (*TNFSF11*) gene [31]. It plays an important role in human physiology by controlling the differentiation and activation of osteoclasts [32]. Generally, RANKL binds to the receptor activator of nuclear factor kappa beta (RANK) on osteoclast precursor cells. RANKL/RANK interaction subsequently activates nuclear factor kB (NF-kB) and supports the expression of other osteoclastogenic factors. Conversely, a soluble decoy receptor for RANKL-osteoprotegerin (OPG) prevents RANKL from binding to RANK. Therefore, RANKL/RANK/OPG system is considered a key mediator of osteoclastogenesis [18,33]. Moreover, RANKL/RANK pathway has been implicated in breast development as well as breast carcinogenesis. According to Fata et al. [34], lactating mammary gland did not develop cancer in RANK and RANKL receptor-deficient mice. In the study by Gonzalez-Suarez et al. [35], the development of mammary carcinogenesis was related to a higher expression of RANKL in 7,12-dimethylbenzeneanthracene (DMBA)-induced mice, with accelerated breast carcinogenesis identified in RANK-transgenic mice. RANKL also initiates the formation of pre-cancerous lesions and the metastatic process. Additionally, RANKL up-regulates the angiogenic process by stimulating the proliferation and survival of endothelial cells (Figure 1).

Both bone and breast tissues are dependent on estrogen. Moreover, high BMD can be associated with the risk of breast cancer [36]. The hormone estrogen is a key regulator of BMD [37], maintaining the balance between bone formation and bone resorption [38]. Specifically, estrogens stimulate osteoblast differentiation and activate Wnt signaling. They also have an indirect effect through suppression of RANKL and up-regulation of OPG, which ultimately inhibits osteoclastogenesis. Another mechanism for preventing bone resorption is the induction of apoptosis in osteoclasts. Furthermore, estrogens act at the osteocyte level since estrogen decreases sclerostin level and osteocyte apoptosis (Figure 1). Epidemiological and clinical evidence has shown that factors consistent with raised estrogen levels during a woman's lifetime (e.g., early menarche, late menopause, late first full-term pregnancy, obesity) are related to increased risk of breast cancer [39]. Estrogens are generally believed to induce breast cancer cell proliferation via the ER and serve as a transcription factor to regulate the expression of target genes encoding proteins with important biological functions [40]. The impact of estrogen on both aforementioned diseases documents the fact that women who develop ER-positive breast cancer at a relatively younger age and are treated with anti-estrogen drugs such as tamoxifen have an elevated risk of postmenopausal osteoporosis [41]. Due to the role of estrogen in breast cancer, aromatase inhibitors (inhibitors of the estrogen-metabolizing enzyme aromatase) are used in the treatment of postmenopausal individuals with ER-positive breast cancer, despite problems with bone fractures. Recently, researchers are examining the potential of denosumab, an anti-RANKL antibody, in preventing aromatase inhibitor-associated bone loss [42], which could provide major benefits for postmenopausal breast cancer patients.

Oxidative stress is a contributing factor in many chronic diseases, including osteoporosis and breast cancer [43,44]. ROS directly promote osteoclast formation in a process mediated by RANKL-RANK interaction and enhance bone resorption [45,46]. This signaling pathway includes redox-sensitive components such as tumor necrosis factor receptor-associated factor 6 (TRAF6), Rac1 (a member of the Rho-GTPase subfamily), and nicotinamide adenine dinucleotide phosphate oxidases (NOX) [47]. Moreover, ROS induce apoptosis of osteoblasts and osteocytes by activating numerous signaling pathways. Mitogen-activated protein kinases such as ERK and JNK are involved in this process (Figure 1). ROS also reduce osteoblast activity and differentiation, thus mineralization and osteogenesis [48]. Postmenopausal women are not only exposed to high levels of oxidative stress, but also to elevated levels of nitric oxide (NO) in erythrocytes [49]. NO can increase the ability of cytokines to stimulate osteoclast activity and enhances their inhibitory impacts on osteoblast growth [50,51]. Ultimately, bone formation prevails over bone resorption. In breast cancer, oxidative stress has been implicated in the initiation, promotion, and progression grades of breast carcinogenesis [52]. Mammary tissue is a complex combination of different cell types, including stromal and neoplastic cells [53]. In cancerous breast tissue, stromal fibroblasts acquire a phenotype characterized by raised levels of cytokines, growth factors, and metalloproteinases [54]. In the tumor microenvironment, an altered redox state in favor of pro-oxidants induces the formation of activated fibroblasts, leading to modifications of epithelial cells that support tumorigenesis [55]. Oxidative stress in the tumor microenvironment is also characterized by activated stromal cells that generate tumor-enhancing signals, thereby promoting tumor growth and vascularization [56]. Elevated ROS induce oncogenes and DNA damage, inhibit tumor suppressor genes, and can interfere with different signaling pathways (Figure 1).

Chronic age-related inflammation also plays an important role in the pathogenesis of osteoporosis by affecting bone remodeling [57]. In the presence of RANKL, pro-inflammatory cytokines such as tumor necrosis factor alpha (TNF-α), interleukins (IL)-1, and IL-6 cause the excessive formation of osteoclasts and simultaneously inhibit the activities of osteoblasts [58]. The aforementioned cytokines also stimulate osteoclast development and elevate the production of macrophage colony-stimulating factor (M-CSF) by bone marrow stromal cells (BMSC) [59,60]. They also suppress osteoblasts in releasing OPG [58]. According to several studies, raised levels of cytokine-mediated acute phase C-reactive protein [61], and pro-inflammatory cytokines, including IL-6 [62], IL-1β [63], and TNF-α [64] are found in breast cancer patients, documenting that breast cancer is associated with inflammation. Elevation of these cytokines has been linked with breast cancer invasiveness and has also been used as a prognostic factor in breast cancer patients [65]. Inflammatory cells such as macrophages play a role during tumor progression by stimulating angiogenesis via the production of pro-inflammatory cytokines and VEGF (Figure 1).

Obesity, one of the abnormalities of lipid metabolism, has been hypothesized to protect the skeleton by increasing BMD [1] through mechanical loading, which stimulates bone formation by reducing apoptosis and increasing the proliferation and differentiation of osteoblasts and osteocytes [66,67]. This mechanism is supposed to be controlled by the Wnt/β-catenin signaling pathway [68,69]. For this reason, bone mass increases as a compensatory mechanism to adapt to a greater load [70]. However, several researchers reported conflicting findings. According to Hsu et al. [71] and Pollock et al. [72], excess fat mass was associated with reduced total BMD and total bone mineral content. The link between obesity and 13 cancer types, including ER-positive postmenopausal breast cancer, was established by International Agency for Cancer Research [73]. Moreover, obesity was consistent with poor response outcomes in patients with ER-positive breast cancer [74]. Therefore, obesity presents a challenge in treating individuals with postmenopausal breast cancer who suffer from osteoporosis [75]. Targeting the metabolic pathways linked to estrogen production and immune surveillance modulation might represent an effective trend in breast cancer prevention and treatment [76]. Studies on estradiol depletion by aromatase inhibitors in subjects with postmenopausal breast cancer indicate that higher levels of aromatase activity associated with elevated adipose tissue mass, reduce the efficacy of aromatase inhibitor therapy [77]. Nowadays, bisphosphonates are used to prevent aromatase inhibitor-induced bone loss and improve survival in postmenopausal patients with ER-positive breast cancer [78,79].

3. Honey and Osteoporosis

Honey is able to protect the bone mainly due to antioxidant and anti-inflammatory properties, primarily through its content of polyphenols, which act on several signaling pathways, resulting in anabolic and antiresorptive effects [2]. From the group of polyphenols, the anti-osteoporotic impact of quercetin, kaempferol, and luteolin was recorded [4]. In addition, vitamin D3 and its hydroxyderivatives with antioxidant properties were also detected in honey [80,81]. Vitamin D3 supplementation was found to have protective effects on the inhibition of bone loss and BMD in both experimental animals and postmenopausal women [82–84].

According to Zaid et al. [85], the thickness of trabecular bone was elevated in ovariectomized (OVX) rats receiving Tualang honey (a type of Malaysian honey that is especially produced by the rock bee) at the dose of 0.2 g/kg/day for 2 weeks compared to OVX rats fed calcium [86]. Additionally, identical Tualang honey administration (0.2 g/kg/day for 2 weeks) significantly increased BMD in OVX rats [87]. The study of Husniati Y et al. [88] showed that daily consumption of Tualang honey (20 mg/day for 4 months) in postmenopausal women resulted in similar bone densitometry findings as in individuals with hormone replacement therapy. Moreover, Shafin et al. [89] revealed that postmenopausal women consuming Tualang honey (20 g for 16 weeks) had comparable blood oxidative stress (e.g., glutathione peroxidase, catalase, superoxide dismutase) levels to those receiving estrogen-progestin therapy. The aforementioned beneficial effects of Tualang honey can be attributed to the highest content of kaempferol, quercetin, ellagic acid, gallic acid, hesperetin, and catechin among different types of honey, indicating its highest antioxidant potential [90–92].

According to Kamaruzzaman et al. [93], the administration of Kelulut honey (a type of Malaysian honey that is mainly produced by stingless bumblebees) at doses of 200 mg/kg/day and 400 mg/kg/day for 2 months alleviated glucocorticoid-induced osteoporosis through its antioxidant activity in rats. Significantly elevated bone volume/tissue volume, trabecular number, osteoblast surface, superoxide dismutase level and decreased osteoclast surface and malondialdehyde levels were determined in osteoporotic rats fed this type of honey. The impact of Kelulut honey supplementation (1 g/kg for 8 weeks) on the bone health of rats with metabolic syndrome was investigated by Ekeuku et al. [94]. Oxidative stress and chronic low-grade inflammation present in metabolic syndrome are known to play a major role in osteoporosis induction or bone loss [19]. Rats receiving Kelulut honey showed a significant reduction in osteoclast surface compared to the control group, other bone parameters did not differ between the two groups [94]. However, Ramli et al. [24] report that honey has a strong potential to be used in the management of metabolic syndrome and related osteoporosis by exerting anti-obesity, hypolipidemic, antidiabetic, and hypotensive activities.

Yudaniayanti et al. [30] examined the impact of honey supplements on bone strength in OVX rats. These authors determined significantly increased bone strength in OVX rats receiving honey (2 g/kg and 4 g/kg for 12 weeks) in comparison with the untreated group. According to Hasib et al. [95], honey administration (1 g/kg, 2 g/kg, and 4 g/kg for 2 weeks) had a favorable effect on osteoporotic fracture healing in rat femur by promoting osteoblastogenesis. The pro-osteoblastic influence of honey was documented by an enhanced level of alkaline phosphatase (ALP) in the serum.

Abu-Serie et al. [96] revealed the ameliorative impact of a combined extract of Greek thyme (*Thymus vulgaris*) and honey on hydrocortisone-induced osteoporosis in rat bone cells through modulation of bone turnover, oxidative stress, and inflammation. Moreover, a stronger anti-osteoporotic effect of the combined extract was recorded compared to a commonly used bisphosphonate drug (alendronate).

Interestingly, Manuka honey (a type of New Zealand honey with antimicrobial and antioxidant capacities) was used as an antibacterial agent incorporated into a biopolymer coating based on corn protein zein to evaluate the combined effects of bioactive glass and Manuka honey in a new type of scaffold. According to the results of Arango-Ospina et al. [97], Manuka honey and zein coatings imparted antibacterial properties and excellent mechanical properties to bioactive glass bone tissue scaffolds.

From the information mentioned above it is clear that honey may serve as a promising therapeutic supplement for the prevention and management of osteoporosis. Anyway, more scientific or epidemiological evidence is needed for the use of any type of honey in the treatment of postmenopausal osteoporosis in women due to the limited number of clinical trials. Summary data from the aforementioned research is presented in Table 1.

Table 1. Preclinical and clinical studies on the anti-osteoporotic potential of honey.

Research Models	Applied Treatment	Obtained Results	References
OVX Rats	Tualang honey; 0.2 g/kg/day/2 weeks	↑BV/TV ↑Tb.Th ↑Tb.N ↓Tb.Sp	[85]
OVX Rats	Tualang honey; 0.2, 1.0, and 2.0 g/kg/2 weeks	Tibia: ↑BMD	[87]
OVX Rats	Apis dorsata honey; 2 and 4 g/kg/12 weeks	↑Bone strength	[30]
Rats	Kelulut honey; 200 and 400 mg/kg/day/2 months	↑BV/TV ↑Tb.N ↓Tb.Sp ↑SOD activity ↓MDA activity	[93]
Rats	Kelulut honey 1 g/kg/8 weeks	↓Oc.S/BS ↓OS/BS	[94]
Rats	Apis melifera honey; 1, 2, and 4 g/kg/2 weeks	↑ALP	[95]
Bone cells Rat/HC-induced bone damage	Greek thyme + honey	↓ROS ↓Lipid peroxidation Synergistic improving effect on parameters of bone turnover	[96]
Postmenopausal women (n = 39)	Tualang honey; 20 mg/day/4 months	No difference in BMD and cardiovascular risk between honey and HRT groups	[88]
Postmenopausal women (n = 78)	Tualang honey; 20 g/day/16 weeks	↓Blood oxidative stress	[89]

ALP—alkaline phosphatase; HRT—hormone replacement therapy; BMD—bone mineral density; BV/TV—bone volume per tissue volume; MDA—malondialdehyde; Oc.S/BS—osteoclast surface/bone surface; OS/BS—osteoid surface/bone surface; OVX—ovariectomized; ROS—reactive oxygen species; SOD—superoxide dismutase; Tb.Th—trabecular thickness, Tb.N—trabecular number; Tb.Sp—trabecular separation; ↑—increased; ↓—decreased.

4. Honey and Breast Cancer

Honey as a potential preventive and therapeutic supplement is currently gaining attention in cancer research. Various studies have been reported to investigate the anticancer benefits of different types of honey from different origins. The anticancer activity of honey has been demonstrated against various cancer cell lines and tissues, such as breast, prostate, colorectal, endometrial, and renal [98–106]. In general, the chemo-preventive properties of honey are consistent with its bioactive compounds, mostly quercetin, luteolin, chrysin, and esters of caffeic [107]. Although the exact mechanism is still unclear, some studies revealed the interference of bioactive compounds with anti-proliferative [108], antioxidant [109], and pro-apoptotic cell-signaling pathways [110]. Choi et al. [111] documented the anti-proliferative effects of quercetin in the human breast cancer cell line MCF-7 by inhibiting cell cycle progression via transient accumulation in the M phase followed by G2 arrest. Moreover, quercetin treatment activated apoptosis in MCF-7 cells via the p38MAPK signaling pathway [112]. Kim et al. [113] detected melatonin and its metabolites in honey, which possess strong free radical scavenging properties [114]. However, high concentrations of melatonin can induce the production of ROS, leading to apoptosis in a variety of cancers [115–117]. In general, favorable impacts of honey against breast cancer have been proven in both preclinical and clinical studies.

Tualang honey has been found to induce apoptosis of MDA-MB-231 and MCF-7 breast cancer cells through activation of the mitochondrial apoptotic pathway by elevating caspase-3/7 and caspase-9 and reducing mitochondrial membrane potential [102]. Moreover, Tualang honey combined with tamoxifen enhanced the anticancer activity of tamoxifen, activated multiple caspase enzymes, and increased mitochondrial membrane depolarization, leading to a breast cancer cell (MCF-7 and MDA-MB-231) apoptosis [118]. Tualang honey with tamoxifen can therefore be used as an alternative for the treatment of breast cancer, thereby reducing the required dose of tamoxifen and subsequently eliminating the side effects of tamoxifen. According to Kadir et al. [119], the growth of DMBA-induced mammary tumors was inhibited by Tualang honey administration (0.2–2 g/kg for 150 days) in rats. Additionally, vascular endothelial growth factor (VEGF), a pro-angiogenic factor, was reduced in honey-supplemented rats. In the study of Zakaria et al. [120], elevated levels of alanine aminotransferase were determined in postmenopausal women with breast cancer compared to those consuming Tualang honey (20 g/day for 12 weeks). Moreover, an increase in creatinine levels, leukocyte, and platelet counts was observed in the honey-treated group. In a clinical trial by Hizan et al. [121], the combination of Tualang honey with the aromatase inhibitor anastrozole lowered background parenchyma enhancement (a correlate of cancer relapse) more efficiently than anastrozole treatment alone (42% vs. 10% reduction) in patients with ER-positive breast cancer.

The anti-proliferative impact of Manuka honey was determined in MDA-MB-231 and MCF-7 breast cancer cells and was time- and dose-dependent. Moreover, the IL-6/STAT3 signaling pathway was highlighted as one of the first potential targets for Manuka honey-induced breast cancer cell suppression [122]. In another study, Aryappalli et al. [123] found that inhibition of tyrosine-phosphorylated STAT3 in breast cancer cells by Manuka honey is mediated by selective antagonism of the IL-6 receptor. Ahmed et al. [124] revealed that supplementation with both Manuka and Tualang kinds of honey (1.0 g/kg for 120 days) was able to reduce tumor volume, numbers, weight, and growth rate in the 1-methyl-1-nitrosourea (MNU)-induced breast cancer in rats. In addition, a higher expression of pro-apoptotic proteins and lowered expression of anti-apoptotic proteins were recorded. These types of honey administered orally exhibit anticancer effects by modulating the immune system and activating the intrinsic apoptotic pathway.

Greek honey extract (pine, thyme, and fir) reduced the viability of MCF-7 breast cancer cells [100,125], while thyme honey inhibited the progression of MCF-7 cells by suppressing estrogenic impacts [100]. Anatolian honey with different botanical origins (pine, chestnut, and cedar) produced stronger inhibitory effects on MDA-MB-231, MCF-7, and SKBR3 breast cancer cells in a time- and dose-dependent manner [126]. In MCF-7 and MDA-MB-231 cancer cells, the aforementioned types of honey suppressed breast cancer through the IL-6/STAT3 signaling pathway.

Kurniawan et al. [127] examined the impact of apis Dorsata honey (two tablespoons orally, 3 times/day for 15 days) on IL-3 (multi-potential hematopoietic growth factor) levels in breast cancer patients undergoing chemotherapy. These authors determined increased levels of IL-3 in the honey-treated group compared to the control group. The effect of Dorsata honey on IL-6 (breast cancer metastases factor) levels and T lymphocytes in post-chemotherapy breast cancer individuals was investigated by Syam et al. [107]. It has been found that there is a significant increase in the levels of T lymphocytes, which can indirectly enhance the immune system and inhibit tumor cell growth in honey-treated patients with breast cancer. The results also showed that Dorsata honey consumption did not affect IL-6 levels in contrast to the Manuka honey, where differences were noted.

The ability of honey to mitigate the chemo- and radiotherapy-induced oral mucositis (OM) was documented in numerous studies that mainly involved patients with head and neck cancers [128]. The studies on honey-treatment toxicity associations are limited in breast cancer but a pilot randomized trial comprising breast cancer patients receiving doxorubicin and cyclophosphamide reported the clinical efficacy of propolis plus bicarbonate in OM prevention [129]. According to Aghamohammadi et al. [130], a mixture of honey (30 g) and cinnamon (4 g) powder administered to breast cancer patients three times a day for 1 week led to a significant improvement in overall health and quality of life after the treatment.

Although honey supplementation has been associated with breast cancer modification in most of the experimental studies mentioned above, further experiments (especially animal studies and prospective randomized clinical trials) are still needed to evaluate the potential usefulness of honey as a therapeutic supplement in prevention and management of breast cancer. Table 2 provides summary data from the aforementioned studies.

Table 2. Preclinical and clinical studies on honey's potential against breast cancer.

Research Models	Applied Treatment	Obtained Results	References
Cells MCF-7; MDA-MB-231 HeLa	Tualang honey; 1–10%/72 h	↑Cytotoxicity ↑Cell death ↑Apoptosis ↓Δψm ↑Caspase-3/7 and -9	[102]
Cells MCF-7; MDA-MB-231	Tualang honey; 10%/6, 24, 48, and 72 h	↑Apoptosis ↑Caspase-3/7 and -9 ↓TAM-induced adverse effects	[118]
Cells MDA-MB-231; MDA-MB-435; MCF-7	Manuka honey; 0.3–1.25 %/24–72 h	↓Viability of cancer cells ↑Caspase-dependent apoptosis ↑Bax protein expression ↑Apoptosis ↓IL-6/STAT3 signaling pathway	[122]
Cells MCF-7	Greek honey extract	↓Viability of MCF-7 cells	[125]
Cells MCF-7	Fir honey extract 0.2–125 μg/ml	↑Viability of MCF-7 cells	[100]
Cells MCF7, SKBR3, and MDAMB-231	Chestnut, pine, cedar, multifloral honey; 1, 2.5, 5, 7.5, and 10 μg/mL/ 24, 48, and 72 h	↑Cytotoxic effect	[126]
Rats	Tualang honey; 0.2, 1.0, and 2.0 g/kg/day/150 days	↓Tumor development ↓Tumor mean size ↓VEGF protein	[119]
Rats	Tualang honey, Manuka honey 1.0 g/kg/day/120 days	↓Cancer masses ↓Tumor size, weight, and multiplicity ↓Growth rate ↑Expression of pro-apoptotic proteins (Apaf-1, Caspase-9, IFN-γ, IFNGR1, and p53) ↓Expression of anti-apoptotic proteins (TNF-α, COX-2, and Bcl-xL 1)	[124]
Postmenopausal breast cancer women (*n* = 72)	Tualang honey; 20 g/day/12 weeks	↓Alanine aminotransferase levels ↑Creatinine levels ↑Leukocyte counts ↑Platelet counts	[120]
Postmenopausal breast cancer women (*n* = 40)	Tualang honey; 20 g/day/6 months	↓BPE	[121]
Adult women with breast cancer (*n* = 30)	Dorsata honey; 15 mL/3 times daily/15 days	↑IL-3	[127]
Adult women with breast cancer (*n* = 30)	Dorsata honey; 15 mL/3 times daily/15 days	↑T lymphocytes levels No differences in IL-6 level	[107]
Adult women with breast cancer (*n* = 117)	Honey + cinnamon powder; 30 g + 4 g/3 times daily/1 week	↑Overall quality of life	[130]

Δψm—mitochondrial membrane potential; Apaf-1—apoptotic protease activating factor-1; Bcl-xL 1—B-cell lymphoma-extra large; BPE—background parenchymal enhancement; COX-2—cyclooxygenase-2; HeLa—cervical carcinoma; IFN-γ—interferon gamma; IFNGR1—interferon gamma receptor 1; IL-3—interleukin 3; IL-6/STAT3—interleukin-6/tyrosine-phosphorylated; MCF-7, MDA-MB-231, MDA-MB-435—human breast adenocarcinoma cell line; p53—tumor protein; SKBR-3—human breast cancer cell line; STAT3—signal transducer and activator of transcription 3; TAM—tamoxifen; TNF-α—tumor necrosis factor alpha; VEGF—vascular endothelial growth factor; ↑—increased; ↓—decreased.

5. Conclusions

Nowadays, the administration of dietary supplements and functional food intake in standard care of osteoporotic and oncological patients is gaining more attention. Honey is one of the oldest organic natural substances used for medical purposes. Many studies have pointed to the antioxidant, antibacterial, antiviral, immunomodulatory, anti-inflammatory, hypocholesterolemic, hypotensive, and antitumor impacts of honey, making it beneficial for human health.

In this review, the current knowledge from in vitro, animal, and human studies concerning the use of honey as a potential therapeutic supplement for osteoporosis and breast cancer is presented, due to their increasing incidence in postmenopausal women. Preclinical studies related to osteoporosis have reported favorable effects of honey on cortical and trabecular bone microstructure, bone strength, and oxidative stress. The limited number of clinical trials suggests the need for further research to evaluate the potential benefits of honey in the treatment of postmenopausal osteoporosis. In relation to breast cancer, in vitro experiments revealed the anti-proliferative and pro-apoptotic impact of honey on breast cancer cells, as well as their increased apoptosis. Animal studies have shown that honey reduces the number, growth rate, volume, and tumor weight. Findings from clinical trials reported its immunomodulatory properties showing that honey is effective in increasing leukocyte and platelet counts, IL-3 levels, and quality of life. In this context, the potential role of honey and its oligosaccharides as prebiotics for specific beneficial bacteria might be examined in future clinical studies.

In conclusion, we can state that honey represents a prospective therapeutic supplement for bone and breast tissue health. However, several issues need to be addressed before administration, including the presence of allergens or pesticides, antibiotics, and contaminants. Since the existing differences among honey types, precise identification and quantification of bioactive compound content should be provided in detail. In addition, clinical studies published so far are limited by a small sample size without the involvement of all different ethnicities, a single dose of honey and often a short duration of experiments and different parameters analyzed. Therefore, further clinical trials should also be aimed at eliminating these shortcomings. Importantly, large-scale placebo-controlled clinical studies concerning nutrigenomics are highly warranted to evaluate the effects of honey with its bioactive components on global gene and protein expression.

Author Contributions: Conceptualization, M.M. and R.O.; methodology, M.M., V.K. and R.O.; formal analysis, V.M., N.Z., M.B. and R.B; writing-original draft preparation, M.M., V.K., S.C. and R.O.; writing-review and editing, M.M., S.C. and R.O.; visualization, V.M., N.Z., M.B. and R.B.; supervision, M.M. and R.O.; funding acquisition, M.M. and R.O. All authors have read and agreed to the published version of the manuscript.

Funding: This work was supported by the Ministry of Education, Science, Research and Sport of the Slovak Republic, grant number VEGA 1/0444/20 and KEGA 012UKF-4/2023.

Institutional Review Board Statement: Not applicable.

Informed Consent Statement: Not applicable.

Data Availability Statement: Not applicable.

Conflicts of Interest: The authors declare no conflict of interest.

References

1. Muhammad, A.; Mada, S.B.; Malami, I.; Forcados, G.E.; Erukainure, O.L.; Sani, H.; Abubakar, I.B. Postmenopausal osteoporosis and breast cancer: The biochemical links and beneficial effects of functional foods. *Biomed. Pharmacother.* **2018**, *107*, 571–582. [CrossRef] [PubMed]
2. Kamaruzzaman, M.A.; Chin, K.-Y.; Mohd Ramli, E.S. A Review of Potential Beneficial Effects of Honey on Bone Health. *Evid. Based Complement. Altern. Med.* **2019**, *2019*, e8543618. [CrossRef] [PubMed]
3. Badeau, M.; Adlercreutz, H.; Kaihovaara, P.; Tikkanen, M.J. Estrogen A-ring structure and antioxidative effect on lipoproteins. *J. Steroid Biochem. Mol. Biol.* **2005**, *96*, 271–278. [CrossRef] [PubMed]

4. Martiniakova, M.; Babikova, M.; Mondockova, V.; Blahova, J.; Kovacova, V.; Omelka, R. The Role of Macronutrients, Micronutrients and Flavonoid Polyphenols in the Prevention and Treatment of Osteoporosis. *Nutrients* **2022**, *14*, 523. [CrossRef] [PubMed]
5. Teng, G.G.; Curtis, J.R.; Saag, K.G. Mortality and osteoporotic fractures: Is the link causal, and is it modifiable? *Clin. Exp. Rheumatol.* **2008**, *26*, S125–S137. [PubMed]
6. Eteraf-Oskouei, T.; Najafi, M. Traditional and Modern Uses of Natural Honey in Human Diseases: A Review. *Iran. J. Basic Med. Sci.* **2013**, *16*, 731–742. [PubMed]
7. Ahmed, S.; Sulaiman, S.A.; Baig, A.A.; Ibrahim, M.; Liaqat, S.; Fatima, S.; Jabeen, S.; Shamim, N.; Othman, N.H. Honey as a Potential Natural Antioxidant Medicine: An Insight into Its Molecular Mechanisms of Action. *Oxidative Med. Cell. Longev.* **2018**, *2018*, e8367846. [CrossRef]
8. Kassim, M.; Achoui, M.; Mustafa, M.R.; Mohd, M.A.; Yusoff, K.M. Ellagic acid, phenolic acids, and flavonoids in Malaysian honey extracts demonstrate in vitro anti-inflammatory activity. *Nutr. Res.* **2010**, *30*, 650–659. [CrossRef]
9. Hussein, S.Z.; Mohd Yusoff, K.; Makpol, S.; Mohd Yusof, Y.A. Gelam Honey Inhibits the Production of Proinflammatory, Mediators NO, PGE(2), TNF-α, and IL-6 in Carrageenan-Induced Acute Paw Edema in Rats. *Evid. Based Complement. Altern. Med.* **2012**, *2012*, 109636. [CrossRef]
10. Owoyele, B.V.; Adenekan, O.T.; Soladoye, A.O. Effects of honey on inflammation and nitric oxide production in Wistar rats. *Zhong Xi Yi Jie He Xue Bao J. Chin. Integr. Med.* **2011**, *9*, 447–452. [CrossRef]
11. Afroz, R.; Tanvir, E.M.; Paul, S.; Bhoumik, N.C.; Gan, S.H.; Khalil, M.I. DNA Damage Inhibition Properties of Sundarban Honey and its Phenolic Composition. *J. Food Biochem.* **2016**, *40*, 436–445. [CrossRef]
12. Hussein, S.Z.; Mohd Yusoff, K.; Makpol, S.; Mohd Yusof, Y.A. Gelam honey attenuates carrageenan-induced rat paw inflammation via NF-κB pathway. *PLoS ONE* **2013**, *8*, e72365. [CrossRef]
13. Desroches, S.; Lamarche, B. The evolving definitions and increasing prevalence of the metabolic syndrome. *Appl. Physiol. Nutr. Metab.* **2007**, *32*, 23–32. [CrossRef] [PubMed]
14. Thapa, S.; Nandy, A.; Rendina-Ruedy, E. Endocrinal metabolic regulation on the skeletal system in post-menopausal women. *Front. Physiol.* **2022**, *13*, 1052429. [CrossRef]
15. Nestel, P.; Lyu, R.; Low, L.P.; Sheu, W.H.-H.; Nitiyanant, W.; Saito, I.; Tan, C.E. Metabolic syndrome: Recent prevalence in East and Southeast Asian populations. *Asia Pac. J. Clin. Nutr.* **2007**, *16*, 362–367.
16. Lear, S.A.; Gasevic, D. Ethnicity and Metabolic Syndrome: Implications for Assessment, Management and Prevention. *Nutrients* **2019**, *12*, 15. [CrossRef] [PubMed]
17. Qadirifard, M.S.; Fathabadi, A.; Hajishah, H.; Gholami, K.; Abbasi, M.; Sami, N.; Zo, M.M.; Kadkhodaei, F.; Sina, M.; Ansari, A.; et al. Anti-breast cancer potential of honey: A narrative review. *OncoReview* **2022**, *12*, 5–15. [CrossRef]
18. Martiniakova, M.; Babikova, M.; Omelka, R. Pharmacological agents and natural compounds: Available treatments for osteoporosis. *J. Physiol. Pharmacol.* **2020**, *71*, 307–320. [CrossRef]
19. Chin, K.-Y.; Wong, S.K.; Ekeuku, S.O.; Pang, K.-L. Relationship Between Metabolic Syndrome and Bone Health—An Evaluation of Epidemiological Studies and Mechanisms Involved. *Diabetes Metab. Syndr. Obes.* **2020**, *13*, 3667–3690. [CrossRef] [PubMed]
20. Luchese, R.H.; Prudêncio, E.R.; Guerra, A.F.; Luchese, R.H.; Prudêncio, E.R.; Guerra, A.F. *Honey as a Functional Food*; IntechOpen: Rijeka, Croatia, 2017; ISBN 978-953-51-2880-9.
21. Blahova, J.; Martiniakova, M.; Babikova, M.; Kovacova, V.; Mondockova, V.; Omelka, R. Pharmaceutical Drugs and Natural Therapeutic Products for the Treatment of Type 2 Diabetes Mellitus. *Pharmaceuticals* **2021**, *14*, 806. [CrossRef]
22. Sidor, E.; Dżugan, M. Drone Brood Homogenate as Natural Remedy for Treating Health Care Problem: A Scientific and Practical Approach. *Molecules* **2020**, *25*, 5699. [CrossRef] [PubMed]
23. Martiniakova, M.; Blahova, J.; Kovacova, V.; Babikova, M.; Mondockova, V.; Kalafova, A.; Capcarova, M.; Omelka, R. Bee Bread Can Alleviate Lipid Abnormalities and Impaired Bone Morphology in Obese Zucker Diabetic Rats. *Molecules* **2021**, *26*, 2616. [CrossRef] [PubMed]
24. Mohd Ramli, E.S.; Sukalingam, K.; Kamaruzzaman, M.A.; Soelaiman, I.N.; Pang, K.-L.; Chin, K.-Y. Direct and Indirect Effect of Honey as a Functional Food Against Metabolic Syndrome and Its Skeletal Complications. *Diabetes Metab. Syndr. Obes.* **2021**, *14*, 241–256. [CrossRef] [PubMed]
25. Tomczyk, M.; Bocian, A.; Sidor, E.; Miłek, M.; Zaguła, G.; Dżugan, M. The Use of HPTLC and SDS-PAGE Methods for Coniferous Honeydew Honey Fingerprinting Compiled with Mineral Content and Antioxidant Activity. *Molecules* **2022**, *27*, 720. [CrossRef]
26. Alvarez-Suarez, J.M.; Tulipani, S.; Romandini, S.; Bertoli, E.; Battino, M. Contribution of honey in nutrition and human health: A review. *Mediterr. J. Nutr. Metab.* **2010**, *3*, 15–23. [CrossRef]
27. da Silva, P.M.; Gauche, C.; Gonzaga, L.V.; Costa, A.C.O.; Fett, R. Honey: Chemical composition, stability and authenticity. *Food Chem.* **2016**, *196*, 309–323. [CrossRef]
28. Cianciosi, D.; Forbes-Hernández, T.Y.; Afrin, S.; Gasparrini, M.; Reboredo-Rodriguez, P.; Manna, P.P.; Zhang, J.; Bravo Lamas, L.; Martínez Flórez, S.; Agudo Toyos, P.; et al. Phenolic Compounds in Honey and Their Associated Health Benefits: A Review. *Molecules* **2018**, *23*, 2322. [CrossRef]
29. Miguel, M.; Antunes, M.; Faleiro, M. Honey as a Complementary Medicine. *Integr. Med. Insights* **2017**, *12*, 1178633717702869. [CrossRef]

30. Yudaniayanti, I.S.; Primarizky, H.; Nangoi, L. The effects of honey (Apis dorsata) supplements on increased bone strength in ovariectomized rat as animal model of osteoporosis. *AIP Conf. Proc.* **2018**, *1945*, 020004. [CrossRef]
31. Boman, A.; Kokkonen, H.; Ärlestig, L.; Berglin, E.; Rantapää-Dahlqvist, S. Receptor activator of nuclear factor kappa-B ligand (RANKL) but not sclerostin or gene polymorphisms is related to joint destruction in early rheumatoid arthritis. *Clin. Rheumatol.* **2017**, *36*, 1005–1012. [CrossRef]
32. Hofbauer, L.C.; Rachner, T.D.; Hamann, C. From bone to breast and back—The bone cytokine RANKL and breast cancer. *Breast Cancer Res.* **2011**, *13*, 107. [CrossRef]
33. Park, J.H.; Lee, N.K.; Lee, S.Y. Current Understanding of RANK Signaling in Osteoclast Differentiation and Maturation. *Mol. Cells* **2017**, *40*, 706–713. [CrossRef]
34. Fata, J.E.; Kong, Y.Y.; Li, J.; Sasaki, T.; Irie-Sasaki, J.; Moorehead, R.A.; Elliott, R.; Scully, S.; Voura, E.B.; Lacey, D.L.; et al. The osteoclast differentiation factor osteoprotegerin-ligand is essential for mammary gland development. *Cell* **2000**, *103*, 41–50. [CrossRef]
35. Gonzalez-Suarez, E.; Jacob, A.P.; Jones, J.; Miller, R.; Roudier-Meyer, M.P.; Erwert, R.; Pinkas, J.; Branstetter, D.; Dougall, W.C. RANK ligand mediates progestin-induced mammary epithelial proliferation and carcinogenesis. *Nature* **2010**, *468*, 103–107. [CrossRef]
36. Trémollieres, F.A. Screening for osteoporosis after breast cancer: For whom, why and when. *Maturitas* **2014**, *79*, 343–348. [CrossRef] [PubMed]
37. Cauley, J.A. Estrogen and bone health in men and women. *Steroids* **2015**, *99*, 11–15. [CrossRef] [PubMed]
38. Manolagas, S.C. From estrogen-centric to aging and oxidative stress: A revised perspective of the pathogenesis of osteoporosis. *Endocr. Rev.* **2010**, *31*, 266–300. [CrossRef] [PubMed]
39. Yue, W.; Wang, J.-P.; Li, Y.; Fan, P.; Liu, G.; Zhang, N.; Conaway, M.; Wang, H.; Korach, K.S.; Bocchinfuso, W.; et al. Effects of estrogen on breast cancer development: Role of estrogen receptor independent mechanisms. *Int. J. Cancer* **2010**, *127*, 1748–1757. [CrossRef] [PubMed]
40. Wen, C.; Wu, L.; Fu, L.; Wang, B.; Zhou, H. Unifying mechanism in the initiation of breast cancer by metabolism of estrogen (Review). *Mol. Med. Rep.* **2017**, *16*, 1001–1006. [CrossRef] [PubMed]
41. Taxel, P.; Choksi, P.; Poznak, C.V. The Management of Osteoporosis in Breast Cancer Survivors. *Maturitas* **2012**, *73*, 275–279. [CrossRef] [PubMed]
42. Trémollieres, F.A.; Ceausu, I.; Depypere, H.; Lambrinoudaki, I.; Mueck, A.; Pérez-López, F.R.; van der Schouw, Y.T.; Senturk, L.M.; Simoncini, T.; Stevenson, J.C.; et al. Osteoporosis management in patients with breast cancer: EMAS position statement. *Maturitas* **2017**, *95*, 65–71. [CrossRef]
43. Mirończuk-Chodakowska, I.; Witkowska, A.M.; Zujko, M.E. Endogenous non-enzymatic antioxidants in the human body. *Adv. Med. Sci.* **2018**, *63*, 68–78. [CrossRef]
44. Rao, L.G.; Rao, A.V.; Rao, L.G.; Rao, A.V. *Oxidative Stress and Antioxidants in the Risk of Osteoporosis—Role of the Antioxidants Lycopene and Polyphenols*; IntechOpen: Rijeka, Croatia, 2013; ISBN 978-953-51-1066-8.
45. Callaway, D.A.; Jiang, J.X. Reactive oxygen species and oxidative stress in osteoclastogenesis, skeletal aging and bone diseases. *J. Bone Miner. Metab.* **2015**, *33*, 359–370. [CrossRef] [PubMed]
46. Li, D.-Z.; Zhang, Q.-X.; Dong, X.-X.; Li, H.-D.; Ma, X. Treatment with hydrogen molecules prevents RANKL-induced osteoclast differentiation associated with inhibition of ROS formation and inactivation of MAPK, AKT and NF-kappa B pathways in murine RAW264.7 cells. *J. Bone Miner. Metab.* **2014**, *32*, 494–504. [CrossRef] [PubMed]
47. Kanzaki, H.; Shinohara, F.; Kanako, I.; Yamaguchi, Y.; Fukaya, S.; Miyamoto, Y.; Wada, S.; Nakamura, Y. Molecular regulatory mechanisms of osteoclastogenesis through cytoprotective enzymes. *Redox Biol.* **2016**, *8*, 186–191. [CrossRef]
48. Zhang, B.; Xie, Q.; Quan, Y.; Pan, X.; Liao, D. Reactive oxygen species induce cell death via Akt signaling in rat osteoblast-like cell line ROS 17/2.8. *Toxicol. Ind. Health* **2015**, *31*, 1236–1242. [CrossRef] [PubMed]
49. Ozgocmen, S.; Kaya, H.; Fadillioglu, E.; Aydogan, R.; Yilmaz, Z. Role of antioxidant systems, lipid peroxidation, and nitric oxide in postmenopausal osteoporosis. *Mol. Cell. Biochem.* **2007**, *295*, 45–52. [CrossRef]
50. van't Hof, R.J.; Ralston, S.H. Cytokine-induced nitric oxide inhibits bone resorption by inducing apoptosis of osteoclast progenitors and suppressing osteoclast activity. *J. Bone Miner. Res.* **1997**, *12*, 1797–1804. [CrossRef]
51. MacPherson, H.; Noble, B.S.; Ralston, S.H. Expression and functional role of nitric oxide synthase isoforms in human osteoblast-like cells. *Bone* **1999**, *24*, 179–185. [CrossRef] [PubMed]
52. Hecht, F.; Pessoa, C.F.; Gentile, L.B.; Rosenthal, D.; Carvalho, D.P.; Fortunato, R.S. The role of oxidative stress on breast cancer development and therapy. *Tumour Biol.* **2016**, *37*, 4281–4291. [CrossRef] [PubMed]
53. Yu, T.; Di, G. Role of tumor microenvironment in triple-negative breast cancer and its prognostic significance. *Chin. J. Cancer Res.* **2017**, *29*, 237–252. [CrossRef]
54. Chen, Y.; Zeng, C.; Zhan, Y.; Wang, H.; Jiang, X.; Li, W. Aberrant low expression of p85α in stromal fibroblasts promotes breast cancer cell metastasis through exosome-mediated paracrine Wnt10b. *Oncogene* **2017**, *36*, 4692–4705. [CrossRef] [PubMed]
55. Bussard, K.M.; Mutkus, L.; Stumpf, K.; Gomez-Manzano, C.; Marini, F.C. Tumor-associated stromal cells as key contributors to the tumor microenvironment. *Breast Cancer Res.* **2016**, *18*, 84. [CrossRef] [PubMed]
56. Jezierska-Drutel, A.; Rosenzweig, S.A.; Neumann, C.A. Role of oxidative stress and the microenvironment in breast cancer development and progression. *Adv. Cancer Res.* **2013**, *119*, 107–125. [CrossRef] [PubMed]

57. Flegar, D.; Šućur, A.; Markotić, A.; Kovačić, N.; Grčević, D. Osteoporosis in the view of osteoimmunology: Common feature underlined by different pathogenic mechanisms. *Period. Biol.* **2015**, *117*, 35–43.
58. Pietschmann, P.; Mechtcheriakova, D.; Meshcheryakova, A.; Föger-Samwald, U.; Ellinger, I. Immunology of Osteoporosis: A Mini-Review. *Gerontology* **2016**, *62*, 128–137. [CrossRef]
59. Tencerova, M.; Kassem, M. The Bone Marrow-Derived Stromal Cells: Commitment and Regulation of Adipogenesis. *Front. Endocrinol* **2016**, *7*, 127. [CrossRef] [PubMed]
60. Souza, P.P.C.; Lerner, U.H. The role of cytokines in inflammatory bone loss. *Immunol. Investig.* **2013**, *42*, 555–622. [CrossRef]
61. Kaur, R.P.; Rubal; Banipal, R.P.S.; Vashistha, R.; Dhiman, M.; Munshi, A. Association of elevated levels of C-reactive protein with breast cancer, breast cancer subtypes, and poor outcome. *Curr. Probl. Cancer* **2019**, *43*, 123–129. [CrossRef] [PubMed]
62. Kozłowski, L.; Zakrzewska, I.; Tokajuk, P.; Wojtukiewicz, M.Z. Concentration of interleukin-6 (IL-6), interleukin-8 (IL-8) and interleukin-10 (IL-10) in blood serum of breast cancer patients. *Rocz. Akad. Med. Bialymst.* **2003**, *48*, 82–84.
63. Nicolini, A.; Carpi, A.; Rossi, G. Cytokines in breast cancer. *Cytokine Growth Factor Rev.* **2006**, *17*, 325–337. [CrossRef] [PubMed]
64. García-Tuñón, I.; Ricote, M.; Ruiz, A.; Fraile, B.; Paniagua, R.; Royuela, M. Role of tumor necrosis factor-alpha and its receptors in human benign breast lesions and tumors (in situ and infiltrative). *Cancer Sci.* **2006**, *97*, 1044–1049. [CrossRef] [PubMed]
65. Zakaria, Z.; Hamid, W.Z.W.A.; Mohamed, M. Tualang honey supplementation improves inflammatory and bone markers among postmenopausal breast cancer patients : A randomised controlled trial. *Sains Malays.* **2021**, *50*, 1971–1985. [CrossRef]
66. Kopelman, P.G. Obesity as a medical problem. *Nature* **2000**, *404*, 635–643. [CrossRef]
67. Villareal, D.T.; Apovian, C.M.; Kushner, R.F.; Klein, S.; American Society for Nutrition; NAASO, The Obesity Society. Obesity in older adults: Technical review and position statement of the American Society for Nutrition and NAASO, The Obesity Society. *Am. J. Clin. Nutr.* **2005**, *82*, 923–934. [CrossRef]
68. Bonewald, L.F.; Johnson, M.L. Osteocytes, mechanosensing and Wnt signaling. *Bone* **2008**, *42*, 606–615. [CrossRef]
69. Sawakami, K.; Robling, A.G.; Ai, M.; Pitner, N.D.; Liu, D.; Warden, S.J.; Li, J.; Maye, P.; Rowe, D.W.; Duncan, R.L.; et al. The Wnt co-receptor LRP5 is essential for skeletal mechanotransduction but not for the anabolic bone response to parathyroid hormone treatment. *J. Biol. Chem.* **2006**, *281*, 23698–23711. [CrossRef]
70. Ehrlich, P.J.; Lanyon, L.E. Mechanical strain and bone cell function: A review. *Osteoporos. Int.* **2002**, *13*, 688–700. [CrossRef]
71. Hsu, Y.-H.; Venners, S.A.; Terwedow, H.A.; Feng, Y.; Niu, T.; Li, Z.; Laird, N.; Brain, J.D.; Cummings, S.R.; Bouxsein, M.L.; et al. Relation of body composition, fat mass, and serum lipids to osteoporotic fractures and bone mineral density in Chinese men and women. *Am. J. Clin. Nutr.* **2006**, *83*, 146–154. [CrossRef]
72. Pollock, N.K.; Laing, E.M.; Baile, C.A.; Hamrick, M.W.; Hall, D.B.; Lewis, R.D. Is adiposity advantageous for bone strength? A peripheral quantitative computed tomography study in late adolescent females. *Am. J. Clin. Nutr.* **2007**, *86*, 1530–1538. [CrossRef]
73. Lauby-Secretan, B.; Scoccianti, C.; Loomis, D.; Grosse, Y.; Bianchini, F.; Straif, K.; International Agency for Research on Cancer Handbook Working Group. Body Fatness and Cancer–Viewpoint of the IARC Working Group. *N. Engl. J. Med.* **2016**, *375*, 794–798. [CrossRef] [PubMed]
74. Ewertz, M.; Jensen, M.-B.; Gunnarsdóttir, K.Á.; Højris, I.; Jakobsen, E.H.; Nielsen, D.; Stenbygaard, L.E.; Tange, U.B.; Cold, S. Effect of obesity on prognosis after early-stage breast cancer. *J. Clin. Oncol.* **2011**, *29*, 25–31. [CrossRef] [PubMed]
75. Cheung, A.M.; Heisey, R.; Srighanthan, J. Breast cancer and osteoporosis. *Curr. Opin. Endocrinol. Diabetes Obes.* **2013**, *20*, 532–538. [CrossRef] [PubMed]
76. Brown, K.A. Metabolic pathways in obesity-related breast cancer. *Nat. Rev. Endocrinol.* **2021**, *17*, 350–363. [CrossRef]
77. Pfeiler, G.; Königsberg, R.; Hadji, P.; Fitzal, F.; Maroske, M.; Dressel-Ban, G.; Zellinger, J.; Exner, R.; Seifert, M.; Singer, C.; et al. Impact of body mass index on estradiol depletion by aromatase inhibitors in postmenopausal women with early breast cancer. *Br. J. Cancer* **2013**, *109*, 1522–1527. [CrossRef]
78. Fournier, A.; Mesrine, S.; Gelot, A.; Fagherazzi, G.; Baglietto, L.; Clavel-Chapelon, F.; Boutron-Ruault, M.-C.; Chabbert-Buffet, N. Use of Bisphosphonates and Risk of Breast Cancer in a French Cohort of Postmenopausal Women. *J. Clin. Oncol.* **2017**, *35*, 3230–3239. [CrossRef]
79. Rennert, G.; Pinchev, M.; Gronich, N.; Saliba, W.; Flugelman, A.; Lavi, I.; Goldberg, H.; Fried, G.; Steiner, M.; Bitterman, A.; et al. Oral Bisphosphonates and Improved Survival of Breast Cancer. *Clin. Cancer Res.* **2017**, *23*, 1684–1689. [CrossRef]
80. Kim, T.-K.; Atigadda, V.; Brzeminski, P.; Fabisiak, A.; Tang, E.K.Y.; Tuckey, R.C.; Slominski, A.T. Detection of 7-Dehydrocholesterol and Vitamin D3 Derivatives in Honey. *Molecules* **2020**, *25*, 2583. [CrossRef]
81. Slominski, A.T.; Chaiprasongsuk, A.; Janjetovic, Z.; Kim, T.-K.; Stefan, J.; Slominski, R.M.; Hanumanthu, V.S.; Raman, C.; Qayyum, S.; Song, Y.; et al. Photoprotective Properties of Vitamin D and Lumisterol Hydroxyderivatives. *Cell Biochem. Biophys.* **2020**, *78*, 165–180. [CrossRef]
82. Omelka, R.; Martiniakova, M.; Svik, K.; Slovak, L.; Payer, J.; Oppenbergerova, I.; Kovacova, V.; Babikova, M.; Soltesova-Prnova, M. The effects of eggshell calcium (Biomin H®) and its combinations with alfacalcidol (1α-hydroxyvitamin D3) and menaquinone-7 (vitamin K2) on ovariectomy-induced bone loss in a rat model of osteoporosis. *J. Anim. Physiol. Anim. Nutr.* **2021**, *105*, 336–344. [CrossRef]
83. Kazemian, E.; Pourali, A.; Sedaghat, F.; Karimi, M.; Basirat, V.; Sajadi Hezaveh, Z.; Davoodi, S.H.; Holick, M.F. Effect of supplemental vitamin D3 on bone mineral density: A systematic review and meta-analysis. *Nutr. Rev.* **2022**, nuac068. [CrossRef]
84. Iwamoto, J.; Takeda, T.; Ichimura, S. Effect of combined administration of vitamin D3 and vitamin K2 on bone mineral density of the lumbar spine in postmenopausal women with osteoporosis. *J. Orthop. Sci.* **2000**, *5*, 546–551. [CrossRef]

85. Zaid, S.S.M.; Sulaiman, S.A.; Othman, N.H.; Soelaiman, I.-N.; Shuid, A.N.; Mohamad, N.; Muhamad, N. Protective effects of Tualang honey on bone structure in experimental postmenopausal rats. *Clinics* **2012**, *67*, 779–784. [CrossRef]
86. Ahmed, S.; Othman, N.H. Review of the Medicinal Effects of Tualang Honey and a Comparison with Manuka Honey. *Malays. J. Med. Sci.* **2013**, *20*, 6–13. [PubMed]
87. Zaid, S.S.; Sulaiman, S.A.; Sirajudeen, K.N.; Othman, N.H. The effects of tualang honey on female reproductive organs, tibia bone and hormonal profile in ovariectomised rats—Animal model for menopause. *BMC Complement. Altern. Med.* **2010**, *10*, 82. [CrossRef] [PubMed]
88. Husniati, Y.L.; Hazlina, N.H.N.; Amrah, S.S.; Idiana, H.I.; Azidah, A.K.; Norhayati, M.N.; Bahari, I.S.; Rahimah, Z.; Shaniza, S.N.; Juhara, H.; et al. The effects of Tualang Honey on postmenopausal women. In Proceedings of the Journal of ApiProduct and ApiMedical Science, Kota Bharu, Malaysia, 13–15 January 2010; Volume 2, p. 33.
89. Shafin, N.; Othman, Z.; Zakaria, R.; Nik Hussain, N.H. Tualang Honey Supplementation Reduces Blood Oxidative Stress Levels/Activities in Postmenopausal Women. *Int. Sch. Res. Not.* **2014**, *2014*, e364836. [CrossRef]
90. Hussein, S.Z.; Yusoff, K.M.; Makpol, S.; Yusof, Y.A.M. Antioxidant Capacities and Total Phenolic Contents Increase with Gamma Irradiation in Two Types of Malaysian Honey. *Molecules* **2011**, *16*, 6378–6395. [CrossRef] [PubMed]
91. Mohd Effendy, N.; Mohamed, N.; Muhammad, N.; Mohamad, I.N.; Shuid, A.N. The Effects of Tualang Honey on Bone Metabolism of Postmenopausal Women. *Evid. Based Complement. Altern. Med.* **2012**, *2012*, e938574. [CrossRef]
92. Moniruzzaman, M.; Khalil, M.I.; Sulaiman, S.A.; Gan, S.H. Physicochemical and antioxidant properties of Malaysian honeys produced by Apis cerana, Apis dorsata and Apis mellifera. *BMC Complement. Altern. Med.* **2013**, *13*, 43. [CrossRef] [PubMed]
93. Kamaruzzaman, M.A.; Thanu, A.; Yusof, M.R.; Soelaiman, I.N.; Ramli, E.S. Kelulut honey ameliorates glucocorticoid induced osteoporosis via its antioxidant activity in rats. *Asian Pac. J. Trop. Biomed.* **2019**, *9*, 493. [CrossRef]
94. Ekeuku, S.O.; Chin, K.-Y.; Ramli, N.Z.; Zarkasi, K.A.; Ahmad, F. Effect of Kelulut honey supplementation on bone health in male rats on high-carbohydrate high-fat diet. *Trop. J. Pharm. Res.* **2021**, *20*, 1185–1192. [CrossRef]
95. Hasib, A.; Wahjuningrum, D.A.; Ibrahim, M.H.R.; Kurniawan, H.J.; Ernawati, R.; Hadinoto, M.E.K.; Mooduto, L. ALP (Alkaline Phosphatase) Expression in Simple Fracture Incident in Rat (*Rattus norvegicus*) Femur Bone Supplemented by Apis Mellifera Honey. *J. Int. Dent. Med. Res.* **2020**, *13*, 887–891.
96. Abu-Serie, M.M.; Habashy, N.H. The ameliorating effect of the combined extract from Greek Thymus vulgaris and bee's honey on the hydrocortisone-induced osteoporosis in rat bone cells via modulating the bone turnover, oxidative stress, and inflammation. *RSC Adv.* **2018**, *8*, 28341–28355. [CrossRef] [PubMed]
97. Arango-Ospina, M.; Lasch, K.; Weidinger, J.; Boccaccini, A.R. Manuka Honey and Zein Coatings Impart Bioactive Glass Bone Tissue Scaffolds Antibacterial Properties and Superior Mechanical Properties. *Front. Mater.* **2021**, *7*, 610889. [CrossRef]
98. Tomasin, R.; Gomes-Marcondes, M.C.C. Oral administration of Aloe vera and honey reduces Walker tumour growth by decreasing cell proliferation and increasing apoptosis in tumour tissue. *Phytother. Res.* **2011**, *25*, 619–623. [CrossRef]
99. Fukuda, M.; Kobayashi, K.; Hirono, Y.; Miyagawa, M.; Ishida, T.; Ejiogu, E.C.; Sawai, M.; Pinkerton, K.E.; Takeuchi, M. Jungle honey enhances immune function and antitumor activity. *Evid. Based Complement. Altern. Med.* **2011**, *2011*, 908743. [CrossRef] [PubMed]
100. Tsiapara, A.V.; Jaakkola, M.; Chinou, I.; Graikou, K.; Tolonen, T.; Virtanen, V.; Moutsatsou, P. Bioactivity of Greek honey extracts on breast cancer (MCF-7), prostate cancer (PC-3) and endometrial cancer (Ishikawa) cells: Profile analysis of extracts. *Food Chem.* **2009**, *116*, 702–708. [CrossRef]
101. Gribel, N.V.; Pashinskiĭ, V.G. The antitumor properties of honey. *Vopr. Onkol.* **1990**, *36*, 704–709.
102. Fauzi, A.N.; Norazmi, M.N.; Yaacob, N.S. Tualang honey induces apoptosis and disrupts the mitochondrial membrane potential of human breast and cervical cancer cell lines. *Food Chem. Toxicol.* **2011**, *49*, 871–878. [CrossRef]
103. Jaganathan, S.K. Honey Constituents and their apoptotic effect in colon cancer cells. *J. ApiProduct ApiMedical Sci.* **2009**, *1*, 29–36. [CrossRef]
104. Samarghandian, S.; Afshari, J.T.; Davoodi, S. Honey induces apoptosis in renal cell carcinoma. *Pharmacogn. Mag.* **2011**, *7*, 46–52. [CrossRef]
105. Ahmed, S.; Othman, N.H. Honey as a Potential Natural Anticancer Agent: A Review of Its Mechanisms. *Evid. Based Complement. Altern. Med.* **2013**, *2013*, e829070. [CrossRef]
106. Afrin, S.; Haneefa, S.M.; Fernandez-Cabezudo, M.J.; Giampieri, F.; Al-Ramadi, B.K.; Battino, M. Therapeutic and preventive properties of honey and its bioactive compounds in cancer: An evidence-based review. *Nutr. Res. Rev.* **2020**, *33*, 50–76. [CrossRef]
107. Syam, Y.; Prihantono, P.; Majid, S.; Sjattar, E.L.; Kana, M.; Usman, A.N. The effect of apis Dorsata honey as a complementary therapy to interleukin-6 (IL-6) levels and T lymphocytes of post-chemotherapy breast cancer patients. *Breast Dis.* **2021**, *40*, S97–S101. [CrossRef]
108. Jaganathan, S.K.; Mandal, M. Antiproliferative effects of honey and of its polyphenols: A review. *J. Biomed. Biotechnol.* **2009**, *2009*, 830616. [CrossRef]
109. Valko, M.; Leibfritz, D.; Moncol, J.; Cronin, M.T.D.; Mazur, M.; Telser, J. Free radicals and antioxidants in normal physiological functions and human disease. *Int. J. Biochem. Cell Biol.* **2007**, *39*, 44–84. [CrossRef] [PubMed]
110. Jaganathan, S.K.; Mandal, M. Involvement of non-protein thiols, mitochondrial dysfunction, reactive oxygen species and p53 in honey-induced apoptosis. *Investig. New Drugs* **2010**, *28*, 624–633. [CrossRef] [PubMed]

111. Choi, J.A.; Kim, J.Y.; Lee, J.Y.; Kang, C.M.; Kwon, H.J.; Yoo, Y.D.; Kim, T.W.; Lee, Y.S.; Lee, S.J. Induction of cell cycle arrest and apoptosis in human breast cancer cells by quercetin. *Int. J. Oncol.* **2001**, *19*, 837–844. [CrossRef] [PubMed]
112. Ranganathan, S.; Halagowder, D.; Sivasithambaram, N.D. Quercetin Suppresses Twist to Induce Apoptosis in MCF-7 Breast Cancer Cells. *PLoS ONE* **2015**, *10*, e0141370. [CrossRef] [PubMed]
113. Kim, T.-K.; Fabisiak, A.; Brzeminski, P.; Reiter, R.J.; Slominski, A.T. Serotonin, melatonin and their precursors and metabolites and vitamin D3 derivatives in honey. *Melatonin Res.* **2022**, *5*, 374–380. [CrossRef]
114. Florido, J.; Rodriguez-Santana, C.; Martinez-Ruiz, L.; López-Rodríguez, A.; Acuña-Castroviejo, D.; Rusanova, I.; Escames, G. Understanding the Mechanism of Action of Melatonin, Which Induces ROS Production in Cancer Cells. *Antioxidants* **2022**, *11*, 1621. [CrossRef] [PubMed]
115. Moreira, A.J.; Ordoñez, R.; Cerski, C.T.; Picada, J.N.; García-Palomo, A.; Marroni, N.P.; Mauriz, J.L.; González-Gallego, J. Melatonin Activates Endoplasmic Reticulum Stress and Apoptosis in Rats with Diethylnitrosamine-Induced Hepatocarcinogenesis. *PLoS ONE* **2015**, *10*, e0144517. [CrossRef] [PubMed]
116. Yun, S.-M.; Woo, S.H.; Oh, S.T.; Hong, S.-E.; Choe, T.-B.; Ye, S.-K.; Kim, E.-K.; Seong, M.K.; Kim, H.-A.; Noh, W.C.; et al. Melatonin enhances arsenic trioxide-induced cell death via sustained upregulation of Redd1 expression in breast cancer cells. *Mol. Cell Endocrinol.* **2016**, *422*, 64–73. [CrossRef]
117. Guerra-Librero, A.; Fernandez-Gil, B.I.; Florido, J.; Martinez-Ruiz, L.; Rodríguez-Santana, C.; Shen, Y.-Q.; García-Verdugo, J.M.; López-Rodríguez, A.; Rusanova, I.; Quiñones-Hinojosa, A.; et al. Melatonin Targets Metabolism in Head and Neck Cancer Cells by Regulating Mitochondrial Structure and Function. *Antioxidants* **2021**, *10*, 603. [CrossRef] [PubMed]
118. Yaacob, N.S.; Nengsih, A.; Norazmi, M.N. Tualang honey promotes apoptotic cell death induced by tamoxifen in breast cancer cell lines. *Evid. Based Complement. Altern. Med.* **2013**, *2013*, 989841. [CrossRef]
119. Kadir, E.A.; Sulaiman, S.A.; Yahya, N.K.; Othman, N.H. Inhibitory effects of Tualang Honey on experimental breast cancer in rats: A preliminary study. *Asian Pac. J. Cancer Prev.* **2013**, *14*, 2249–2254. [CrossRef]
120. Zakaria, Z.; Zainal Abidin, Z.F.; Gan, S.H.; Wan Abdul Hamid, W.Z.; Mohamed, M. Effects of honey supplementation on safety profiles among postmenopausal breast cancer patients. *J. Taibah Univ. Med. Sci.* **2018**, *13*, 535–540. [CrossRef]
121. Hizan, N.S.; Hassan, N.H.M.; Haron, J.; Abubakar, M.B.; Mahdi, N.M.N.; Gan, S.H. Tualang honey adjunct with anastrozole improve parenchyma enhancement of breast tissue in breast cancer patients: A randomized controlled trial. *Integr. Med. Res.* **2018**, *7*, 322–327. [CrossRef]
122. Aryappalli, P.; Al-Qubaisi, S.S.; Attoub, S.; George, J.A.; Arafat, K.; Ramadi, K.B.; Mohamed, Y.A.; Al-Dhaheri, M.M.; Al-Sbiei, A.; Fernandez-Cabezudo, M.J.; et al. The IL-6/STAT3 Signaling Pathway Is an Early Target of Manuka Honey-Induced Suppression of Human Breast Cancer Cells. *Front. Oncol.* **2017**, *7*, 167. [CrossRef]
123. Aryappalli, P.; Shabbiri, K.; Masad, R.J.; Al-Marri, R.H.; Haneefa, S.M.; Mohamed, Y.A.; Arafat, K.; Attoub, S.; Cabral-Marques, O.; Ramadi, K.B.; et al. Inhibition of Tyrosine-Phosphorylated STAT3 in Human Breast and Lung Cancer Cells by Manuka Honey is Mediated by Selective Antagonism of the IL-6 Receptor. *Int. J. Mol. Sci.* **2019**, *20*, 4340. [CrossRef]
124. Ahmed, S.; Sulaiman, S.A.; Othman, N.H. Oral Administration of Tualang and Manuka Honeys Modulates Breast Cancer Progression in Sprague-Dawley Rats Model. *Evid. Based Complement. Altern. Med.* **2017**, *2017*, 5904361. [CrossRef]
125. Spilioti, E.; Jaakkola, M.; Tolonen, T.; Lipponen, M.; Virtanen, V.; Chinou, I.; Kassi, E.; Karabournioti, S.; Moutsatsou, P. Phenolic Acid Composition, Antiatherogenic and Anticancer Potential of Honeys Derived from Various Regions in Greece. *PLoS ONE* **2014**, *9*, e94860. [CrossRef] [PubMed]
126. Seyhan, M.F.; Yılmaz, E.; Timirci-Kahraman, Ö.; Saygılı, N.; Kısakesen, H.İ.; Eronat, A.P.; Ceviz, A.B.; Bilgiç Gazioğlu, S.; Yılmaz-Aydoğan, H.; Öztürk, O. Anatolian honey is not only sweet but can also protect from breast cancer: Elixir for women from artemis to present. *IUBMB Life* **2017**, *69*, 677–688. [CrossRef]
127. Kurniawanl, A.; Sampepajungl, D.; Syamsul, S.A.; Prihantonol, P. Effectiveness of Dorsata Honey Suplement on Interleukin-3 Levels in Breast Cancer Patients Who Underwent Chemotherapy. *Indian J. Public Health Res. Dev.* **2020**, *11*, 947–952. [CrossRef]
128. Yang, C.; Gong, G.; Jin, E.; Han, X.; Zhuo, Y.; Yang, S.; Song, B.; Zhang, Y.; Piao, C. Topical application of honey in the management of chemo/radiotherapy-induced oral mucositis: A systematic review and network meta-analysis. *Int. J. Nurs. Stud.* **2019**, *89*, 80–87. [CrossRef] [PubMed]
129. Piredda, M.; Facchinetti, G.; Biagioli, V.; Giannarelli, D.; Armento, G.; Tonini, G.; De Marinis, M.G. Propolis in the prevention of oral mucositis in breast cancer patients receiving adjuvant chemotherapy: A pilot randomised controlled trial. *Eur. J. Cancer Care* **2017**, *26*, e12757. [CrossRef]
130. Aghamohammadi, D.; Fakhari, S.; Bilehjani, E.; Hassanzadeh, S. The effects of honey and cinnamon mixture on improving the quality of life in breast cancer. *Crescent J. Med. Biol. Sci.* **2017**, *4*, 74–79.

antioxidants

MDPI

Article

Antioxidant Activity and Phenolic Compound Identification and Quantification in Western Australian Honeys

Ivan Lozada Lawag [1,2], Md Khairul Islam [1,2], Tomislav Sostaric [2], Lee Yong Lim [2], Katherine Hammer [1,3] and Cornelia Locher [1,2,*]

1 Cooperative Research Centre for Honey Bee Products Limited (CRC HBP), The University of Western Australia, Agriculture North M085, Crawley, WA 6009, Australia
2 Division of Pharmacy, School of Allied Health, The University of Western Australia, Curnow Building M315, Crawley, WA 6009, Australia
3 School of Biomedical Sciences, The University of Western Australia, L Block QEII Medical Centre, Monash Ave., Crawley, WA 6009, Australia
* Correspondence: connie.locher@uwa.edu.au

Abstract: This study reports on the total phenolic content and antioxidant activity as well as the phenolic compounds that are present in *Calothamnus* spp. (Red Bell), *Agonis flexuosa* (Coastal Peppermint), *Corymbia calophylla* (Marri) and *Eucalyptus marginata* (Jarrah) honeys from Western Australia. The honey's total phenolic content (TPC) was determined using a modified Folin–Ciocalteu assay, while their total antioxidant activity was determined using FRAP and DPPH assays. Phenolic constituents were identified using a High Performance Thin-Layer Chromatography (HTPLC)-derived phenolic database, and the identified phenolic compounds were quantified using HPTLC. Finally, constituents that contribute to the honeys' antioxidant activity were identified using a DPPH-HPTLC bioautography assay. Based on the results, *Calothamnus* spp. honey (n = 8) was found to contain the highest (59.4 ± 7.91 mg GAE/100 g) TPC, followed by *Eucalyptus marginata* honey (50.58 ± 3.76 mg GAE/100 g), *Agonis flexuosa* honey (36.08 ± 4.2 mg GAE/100 g) and *Corymbia calophylla* honey (29.15 ± 5.46 mg GAE/100 g). In the FRAP assay, *Calothamnus* spp. honey also had the highest activity (9.24 ± 1.68 mmol Fe^{2+}/kg), followed by *Eucalyptus marginata* honey (mmol Fe^{2+}/kg), whereas *Agonis flexuosa* (5.45 ± 1.64 mmol Fe^{2+}/kg) and *Corymbia calophylla* honeys (4.48 ± 0.82 mmol Fe^{2+}/kg) had comparable FRAP activity. In the DPPH assay, when the mean values were compared, it was found that *Calothamnus* spp. honey again had the highest activity (3.88 ± 0.96 mmol TE/kg) while the mean DPPH antioxidant activity of *Eucalyptus marginata*, *Agonis flexuosa*, and *Corymbia calophylla* honeys were comparable. Kojic acid and epigallocatechin gallate were found in all honeys, whilst other constituents (e.g., m-coumaric acid, lumichrome, gallic acid, taxifolin, luteolin, epicatechin, hesperitin, eudesmic acid, syringic acid, protocatechuic acid, t-cinnamic acid, o-anisic acid) were only identified in some of the honeys. DPPH-HPTLC bioautography demonstrated that most of the identified compounds possess antioxidant activity, except for t-cinnamic acid, eudesmic acid, o-anisic acid, and lumichrome.

Keywords: HPTLC; HPTLC-DPPH; HPTLC-derived database; DPPH; FRAP; TPC; Folin–Ciocalteu assay; phenolics; *Eucalyptus marginata* (Jarrah); *Corymbia calophylla* (Marri); *Calothamnus* spp. (Red Bell); *Agonis flexuosa* (Coastal Peppermint); honey; colour hue

Citation: Lawag, I.L.; Islam, M.K.; Sostaric, T.; Lim, L.Y.; Hammer, K.; Locher, C. Antioxidant Activity and Phenolic Compound Identification and Quantification in Western Australian Honeys. *Antioxidants* **2023**, *12*, 189. https://doi.org/10.3390/antiox12010189

Academic Editors: Josipa Vlainić and Ivana Tlak Gajger

Received: 25 November 2022
Revised: 9 January 2023
Accepted: 9 January 2023
Published: 12 January 2023

1. Introduction

Next to antibacterial activity, the antioxidant activity of honey has attracted considerable interest in recent years because of its association with anti-inflammatory, anti-cancer and also anti-aging effects [1]. Commonly, the determination of the antioxidant activity of honey involves the use of several popular colorimetric assays, such as the measurement of total phenolic content (TPC) [2–6], total flavonoid content (TFC) [5–7], free radical scavenging activity using the 2,2-diphenyl-1-picrylhydrazyl (DPPH) assay [4–9], or measuring the ferric reducing antioxidant power (FRAP) [2,8,10] which is also known as Trolox equivalent antioxidant capacity (TEAC). High antioxidant potential in these assays is usually observed for samples with high phenolic and flavonoid content [11]. Thus, the variance in antioxidant properties among honeys from different floral and geographical origins is mainly due to the difference in the composition of their polyphenolic fraction.

Although they are present in honey in only small amounts, phenolic compounds are well studied due to their biological activities [1,12] and their influence on honeys' organoleptic characteristics [1,13–17]. They have also been identified as potential chemical markers for quality assurance and in authenticating the geographical and botanical origin of honeys [18–21]. For example, kaempferol is seen as a key marker for rosemary honey [22], naringenin, caffeic acid and hesperetin for citrus blossom honey [23,24], safflomin for safflower honey [16], ellagic acid for heather honey, caffeic, p-coumaric and ferulic acids for chestnut honey [25] and quercetin for sunflower honey [21]. Similarly, pinocembrin, pinobanksin and chrysin are not only characteristic flavonoids present in propolis but have also been found in many European honeys [21]. The identification of such marker compounds in honey is, however, challenging because they are present at only low concentrations and their isolation and successful chemical identification is strongly dependent on the respective extraction and analysis methods employed [13,26–28].

Based on a comprehensive review of 130 research papers, it was found that 161 phenolic compounds have been identified in honey to date [29], most of which belong to the class of hydroxycinnamic acid derivatives, hydroxybenzoic acid derivatives, flavonols, flavones and flavanones [29]. High-Performance Liquid Chromatography (HPLC) coupled with diode array detection (DAD) appears to be the most commonly employed technique for the qualitative and quantitative analysis of phenolic compounds, followed by Liquid Chromatography-Mass Spectrometry (LC-MS).

Recently, a novel High Performance Thin-Layer Chromatography (HPTLC)-derived database has also been developed and successfully used for the identification of phenolic compounds in Manuka honey [30]. An important advantage of HPTLC over other chromatographic techniques is that it can be paired with post-chromatographic derivatization, even with biochemical reagents. An example is HPTLC-DPPH analysis, which allows visualization and quantification of the antioxidant activity of individual compounds in the chromatographically separated mixture [31]. This method has already been employed in honey analysis and has been demonstrated to be useful in visualizing honey constituents with antioxidant properties [32–34].

Western Australia (WA) is home to 8 of Australia's 15 biodiversity hotspots, which are characterised by a high percentage of endemic flora, (https://www.dcceew.gov.au/science-research/australias-biological-resources/access-resources/wa, accessed on 22 December 2022) therefore, most of the plants foraged by bees (*Apis mellifera*) are unique only to the State. Based on unpublished data by the Cooperative Research Centre for Honey Bee Products (a program established and supported under the Australian Government's Cooperative Research Centres Program), honey samples collected (437 samples) across Western Australia were mostly monofloral in nature and belonged to 48 different botanical species from 10 different families, a majority of which belonged to Myrtaceae (34) and Proteaceae (7) families. Myrtaceae include the genus *Eucalyptus* which represents the most abundant type of trees in the State, along with trees and shrubs of the genus *Melaleuca*, and shrubs of the genus *Calothamnus*. Myrtaceae is the most important plant family foraged by bees. Proteaceae is another important honey producing family and includes trees and shrubs from the genus *Banksia* and *Grevilla* [35]. Table S1 summarises the identity, botanical origin, and families of the honeys collected in WA. Among the honeys produced in the state, honeys from *Eucalyptus marginata* (Jarrah, Myrtaceae), *Corymbia calophylla* (Marri, Myrtaceae), and *Agonis flexuosa* (Coastal Peppermint, Myrtaceae) and a shrub honey harvested from *Calothamnus* spp. (Red Bell, Myrtaceae) are considered iconic and are popular amongst consumers.

The level of information available on the chemical composition and bioactivity of various honeys of different floral origin varies considerably. While some, such as New

Zealand Manuka honey derived from *Leptospermum scoparium*, have attracted considerable academic and commercial interests [36], research data on other honeys is scant. This is undoubtedly the case for honeys derived from Western Australian (WA) floral sources. This presents a significant gap in current knowledge that is addressed in this study. For Western Australian monofloral honeys such as *Eucalyptus marginata* (Jarrah), *Corymbia calophylla* (Marri), *Calothamnus* spp. (Red Bell) and *Agonis flexuosa* (Coastal Peppermint), very few studies have yet focused on their antioxidant properties [37,38] and the constituents that contribute to this activity. For example, to date, only some phenolic constituents have been reported for Jarrah honey [39]. To address these knowledge gaps, the aims of this study were to determine the total phenolic content and the total antioxidant activity of the said Western Australian honeys using a modified Folin–Ciocalteu as well as FRAP and DPPH assays, to identify and quantify phenolic constituents in these honeys using a HPTLC-derived phenolic database and to determine the contribution of various constituents to the overall antioxidant activity of the honeys using an HPTLC-DPPH assay. The findings of this research will assist WA's honey industry in selecting appropriate floral sources that might lead to high-value products.

2. Materials and Methods

2.1. Chemicals and Reagents

The chemicals and reagents used in this study were sourced as follows: Folin and Ciocalteu's phenol reagent 2N, (F9252-1L), 2,4,6-tris(2-pyridyl)-1,3,5-triazine (TPTZ, 3682-35-7), iron (III) chloride hexahydrate (10025-77-1), iron (II) sulphate heptahydrate (7782-63-0), trolox (53188-07-1), fructose (57-48-7), and maltose (6363-53-7) from Sigma Aldrich Truganina, Australia; vanillin (121-33-5) from Sigma-Aldrich, St. Louis, MO, USA; anhydrous magnesium sulfate (7487-88-9), anhydrous sodium carbonate (497-19-8), aminoethyl diphenylborinate (524-95-8), glucose (50-99-7) sucrose (57-50-1), ethanol (64-17-5), from Chem Supply, Port Adelaide, South Australia, Australia; toluene from APS Chemicals, Sydney, New South Wales, Australia; naringenin (98%, 67604-48-2) from Alfa Aesar, Heysham, Lancashire, UK; anhydrous sodium acetate (127-09-3), glacial acetic acid (64-19-7), ethyl acetate (141-78-6), and formic acid (64-18-6) from Ajax Finechem, Wollongong, New South Wales, Australia; hydrochloric acid (7647-01-0) from Asia Pacific Specialty Chemicals Limited, Seven Hills, New South Wales, Australia; 2,2-diphenyl-1-picrylhydrazyl (DPPH, 1898-66-4) from Fluka AG, Buchs, St. Gallen, Switzerland; 3,4,5-trihydroxybenzoic acid (149-91-7) from Ajax Chemicals Ltd. Sydney, New South Wales, Australia; dichloromethane (75-09-2), acetonitrile (75-05-8), concentrated sulfuric acid (7664-93-9) and HPTLC Silica gel 60 F254 Plates 10 $\times$ 20 cm from Merck KGaA, Darmstadt, Hesse, Germany; PEG (25322-68-3) from PharmAust Manufacturing, Welshpool, Western Australia, Australia; Methanol (CH_3OH, B.n. 19758725, 67-56-1) from Scharlau, Barcelona, Catalonia, Spain.

Phenolic compounds and other standards that were included in the HPTLC database were chosen based on an extensive review of phenolic compounds reported in honey [29] and were purchased from Ajax Finechem Pvt. Ltd., (Sydney, New South Wales, Australia), AK Scientific, Inc. (Union City, CA, USA), Alfa Aesar (Heysham, Lancashire, UK), Angene International Ltd. (Nanjing, China), Chem Supply Australia Pty Ltd. (Port Adelaide, Australia), Combi-Blocks Inc., (San Diego, CA, USA), Wuhan ChemFaces Biochemical Co., Ltd. (Wuhan, China), Sigma Aldrich (Castle Hill, Australia), and Sigma-Aldrich (St. Louis, MO, USA) [30].

2.2. Honey Samples

Calothamnus spp. honey (Red Bell, Myrtaceae, n = 8), *Agonis flexuosa* honey (Coastal Peppermint, Myrtaceae, n = 5), *Corymbia calophylla* honey (Marri, Myrtaceae, n = 13), and *Eucalyptus marginata* honey (Jarrah, Myrtaceae, n = 6) were purchased from different suppliers in Western Australia (WA) (see Table S1). Figure 1 shows the geographical locations from where the honeys were collected.

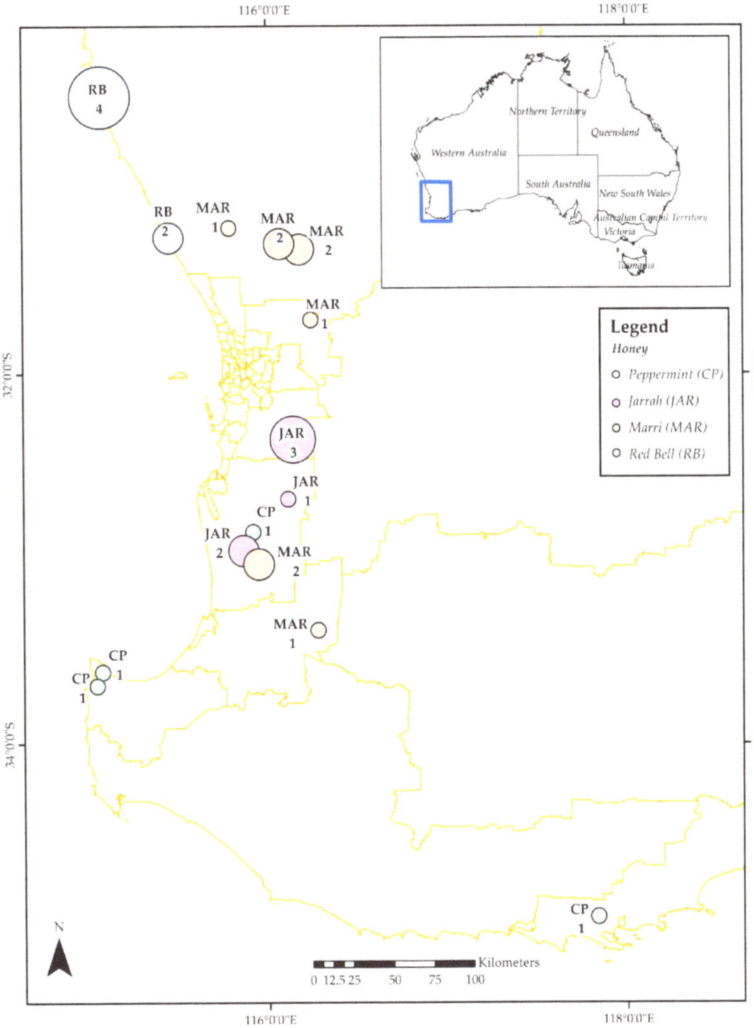

Figure 1. Collection sites of honeys used in this study (samples shown for which this information was available). Map generated using ARCGIS Version 10.8. Redlands, CA, USA (Note: Numbers correspond to the number of samples of honey collected from each specific location; Source: https://www.abs.gov.au/statistics/standards/australian-statistical-geography-standard-asgs-edition-3/jul2021-jun2026/access-and-downloads/digital-boundary-files, accessed on 10 November 2022).

Individual honey samples were authenticated based on their HPTLC fingerprints following established protocols (see Figures S1–S4) [37,38,40] and based on this authentication, a pooled sample for each honey was prepared by mixing equal amounts of each individual sample from the same floral source. It was deemed that such a pooled sample would better reflect the typical chemical composition of a honey rather than analysing an individual honey with a chemical profile that specifically mirrors its unique location, time of collection and processing [40]. Therefore, the pooled honey samples were used in this study for constituent identification and quantification.

An artificial honey solution was prepared by mixing 21.625 g of fructose, 18.125 g of glucose, 1.000 g of maltose, 0.750 g of sucrose and 8.500 g of water [41].

2.3. Preparation of Honey Samples

For the total phenolic content analysis and antioxidant analyses, individual honeys were prepared in triplicates as 20% *w/v* aqueous solutions while for phenolic identification and quantification experiments, pooled honeys were extracted using an organic solvent. The extraction process involved adding 1 g of each pooled honey sample to 2 mL deionised water in stoppered glass test tubes followed by vortex mixing. The resulting solution was then extracted three times with 5 mL dichloromethane and acetonitrile (1:1, *v/v*). The combined organic extracts were dried using anhydrous $MgSO_4$, filtered, and evaporated to dryness using a heating block (Stuart SBHCONC/1 Sample Concentrator) set at 35 °C. The organic honey extracts were stored at 4 °C until analysis for which they were reconstituted with 100 μL methanol.

2.4. Determination of Total Phenolic Content (TPC)

The TPC assay was performed based on the methodology described by Liberato et al. with minor modifications [42]. This protocol has previously been employed in the analysis of the TPC of some Western Australian bee products [37,38,41,43].

In brief, 200 μL of aqueous honey solution (20%, *w/v*) or 100 μL of gallic acid standards (0.06 mg/mL to 0.18 mg/mL) spiked with 100 μL of artificial honey solution (40%, *w/v*) [39] were reacted with 1 mL of diluted Folin–Ciocalteu reagent (1 mL of Folin–Ciocalteu reagent in 30 mL deionised water). After 5 min, 800 μL of 0.75% Na_2CO_3 was added and allowed to react for 2 h, excluded from light. Sample absorbance at 760 nm was then measured (Carry 60 Bio UV–Vis spectrophotometer) using 100 μL of deionised water spiked with 100 μL of artificial honey solution (40%, *w/v*), 1 mL of Folin–Ciocalteu reagent and 800 μL of 0.75% Na_2CO_3 as a blank. The analysis was carried out in triplicate and the mean result for each sample was expressed as mg gallic acid equivalent (GAE) per 100 g of honey.

$$TPC\ Value\ of\ Sample\ (mg\ Gallic\ Acid\ ) = \frac{(\Delta Abs - intercept)}{slope} \quad (1)$$

2.5. Determination of Antioxidant Activity Using the Ferric Reducing Antioxidant Power (FRAP) Assay

The FRAP assay, which is based on the reduction of ferric 2,4,6-tris(2-pyridyl)-1,3,5-triazine [Fe(III)-TPTZ] to ferrous complex at low pH followed by a spectrophotometric analysis, was performed according to the protocol described by Almeida et al. [44] with minor modifications. This protocol has previously been used in our laboratory to determine the FRAP activity of various bee products [37,38,41,43].

In brief, a 1:1:10 (*v/v/v*) ratio of the FRAP reagent was prepared by mixing 10 mM TPTZ (dissolved in 40 mM HCl), 20 mM aqueous $FeCl_3 \cdot 6H_2O$ and 300 mM aqueous acetate buffer (pH 3.6). The reagent mixture was freshly prepared prior to each experiment and incubated at 37 °C prior to use. Ferrous sulphate ($FeSO_4 \cdot 7H_2O$) standards ranging from 200 μM to 1200 μM, along with the standard concentration of 600 μM which was used as a positive control, were freshly prepared prior to each experiment and stored on ice.

A total of 20 μL of honey solution or standards were mixed with 180 μL of FRAP reagent in a 96-well microplate (Greiner Bio-One 96-well Microplate Flat Bottom), and the absorbance of the reaction mixture after 30 min of incubation at 37 °C was determined at 620 nm (BMG Labtech POLARstar Optima Microplate Reader). The FRAP antioxidant activity was determined based on the interpolation of the standard curve and expressed as mmol Fe^{2+} equivalent (FE)/kg of honey (mean of triplicate results).

$$FRAP\ Value\ of\ Sample\ (\mu M\ Fe\ (II)) = \frac{(\Delta Abs - intercept)}{slope} \quad (2)$$

2.6. Determination of Antioxidant Activity Using the 2,2-Diphenyl-1-Picrylhydrazyl (DPPH) Radical Scavenging Assay

The DPPH assay in this study was based on the protocol described by Karabagias et al. [45] with minor modifications [37,38,41,43]. The radical 2,2-diphenyl-1-picrylhydrazyl (DPPH) is purple in colour and decays to yellow in the presence of antioxidants. The resulting change can be captured at 520 nm. The DPPH reagent mixture was prepared using a ratio of 19:10 (v/v) of 0.130 mM methanolic DPPH solution and 100 mM pH 5.5 aqueous $NaC_2H_3O_2$ buffer. Aqueous Trolox solutions with concentrations ranging from 100–600 μM (pH adjusted to pH 7.0) were used to derive the calibration curve, with the 400 μM standard also serving as a positive control.

A total of 10 μL of aqueous honey solution or standards were placed in a 96-well microplate, followed by 290 μL of DPPH reagent, and then mixed. The reaction mixture was kept in the dark and the absorbance was measured at 520 nm after 120 min using a microplate reader (Greiner Bio-One 96-well Microplate Flat Bottom). The mean radical scavenging activity of triplicate samples of honey solutions or standards was expressed as Trolox Equivalent (TE), calculated based on the interpolation of the standard curve, and for the honey samples then also expressed as μmol Trolox equivalent per kg of honey.

$$DPPH\ Value\ of\ Sample\ (\mu M\ Trolox) = \frac{(\Delta Abs - intercept)}{slope} \quad (3)$$

2.7. Phenolic Constituent Identification in Honey

The identification of phenolic honey constituents was performed using a validated HPTLC based database of phenolic compounds. In brief, honeys were first fingerprinted using HPTLC under various conditions and the resulting data (i.e., Rf values, colour hues, UV-Vis and fluorescence λmax and λmin prior to derivatisation, UV-Vis and fluorescence λmax after derivatisation) were matched with standards included in the database [30]. Potential matches were confirmed by spectral overlay analysis [30].

In this study, as additional confirmation of correct identification, a mixture of the identified compounds (7.4 μL, for concentrations see Table 1) in each honey was used to over-spot the respective neat honey extract (7 μL). A corresponding increase in the absorbance of the respective honey extract bands was seen as confirmation of the correct identification.

Table 1. Key parameters for the phenolic standards applied in the quantification experiments.

Compound and Code	Purity (%)	Supplier	Mobile Phase	Concentration (μg/mL)	Rf	UV Detection (nm)	R^2	% Accuracy (n = 3)
Luteolin–Lut (**1**)	97	C	MPB	12.5	0.375	352	0.998	97.4
Hesperetin–Hesp (**2**)	95	C	MPB	12.5	0.527	290	0.999	98.9
Taxifolin–Tax (**3**)	95	E	MPB	5	0.335	292	0.998	98.5
Epicatechin–Epi (**4**)	98	D	MPB	50	0.176	281	0.990	97.8
Epigallocatechin gallate–EGCG (**5**)	98	B	MPB	50	0.065	282	0.996	99.3
2,3,4-Trihydroxy benzoic acid–2,3,4-THBA (**6**)	98	B	MPA	25	0.589	267	1.000	98.4
Eudesmic acid–EudA (**7**)	98	C	MPB	50	0.478	264	0.991	95.8
Gallic acid–GA (**8**)	NI	F	MPB	25	0.27	272	0.997	98.9
o-Anisic acid–o-AA (**9**)	NI	A	MPB	25	0.44	299	1.000	98.9
Protocatechuic acid–ProA (**10**)	98	C	MPB	25	0.377	295	0.997	97.7
Syringic acid–SyrA (**11**)	98	C	MPB	25	0.395	277	0.998	96.3

Table 1. *Cont.*

Compound and Code	Purity (%)	Supplier	Mobile Phase	Concentration (μg/mL)	Rf	UV Detection (nm)	R^2	% Accuracy (n = 3)
m-Coumaric acid–m-CoA (**12**)	98	C	MPB	25	0.467	280	0.996	95.2
t-Cinnamic acid–TCA (**13**)	99	C	MPB	12.5	0.557	279	0.994	97.2
Kojic acid–KA (**14**)	NI	A	MPB	25	0.13	277	0.995	97.0
Lumichrome–Lum (**15**)	NI	A	MPB	10	0.266	357	0.993	97.5

Legend: NI-No information, Suppliers: A = Sigma Aldrich (Castle Hill, NSW, Australia), B = Angene International Ltd. (Nanjing, China), C = Combi-Blocks Inc., (San Diego, CA, USA), D = Wuhan ChemFaces Biochemical Co., Ltd. (Wuhan, Hubei, China), E = AK Scientific, Inc. (Union City, CA, USA), F = Ajax Finechem Pvt. Ltd., (Sydney, NSW, Australia), MPA—toluene: ethyl acetate: formic acid (2:8:1, *v*/*v*/*v*), MPB—toluene: ethyl acetate: formic acid (6:5:1, *v*/*v*/*v*).

The CAMAG HPTLC system (Muttenz, Switzerland) used in this study consisted of a CAMAG TLC visualizer 2, Linomat V semi-automatic sample applicator, and ADC2 automated development chamber, a TLC scanner IV, a derivatiser, and a TLC plate heater III. The system was operated by VisionCATS Version 3.1 software, which controls all chromatographic operations and analyses.

In order to perform the phenolic compound identification, honey extracts were subjected to the same HPTLC conditions used to establish the database (Table 2) using two solvent systems: (a) MPA, consisting of toluene: ethyl acetate: formic acid (2:8:1, *v*/*v*/*v*), and (b) MPB, consisting of toluene: ethyl acetate: formic acid (6:5:1, *v*/*v*/*v*) [44,45], as well as two different derivatising reagents, natural product-polyethylene glycol reagent (NP-PEG) and vanillin-sulfuric acid reagent (VSA).

Table 2. Conditions used in performing the HPTLC analysis.

Name	Solvent System	Derivatising Agent
DB-1A	MPA	NP-PEG
DB-1B	MPA	VSA
DB-2A	MPB	NP-PEG
DB-2B	MPB	VSA

Naringenin (0.5 mg/mL in methanol), with an application volume of 4 μL, was used as HPTLC reference standard, and for all honey extracts a volume of 7 μL was used. All samples were applied as 8 mm bands, 8 mm from the bottom of the HPTLC plate at a rate of 150 nLs^{-1} (aided by liquid nitrogen at a pressure of 10,000 mmHg). The chromatographic separation was performed on 20 × 10 cm HPTLC plates (glass-backed silica gel 60 F_{254} plates) in an automated twin trough development chamber activated with $MgCl_2{\cdot}6H_2O$ at 33–38% relative humidity. Saturation pads were used to saturate the system for 15 min and plates were preconditioned with the mobile phase for 5 min, and then developed automatically to a distance of 70 mm at room temperature before being automatically dried for 5 min. Photo-documentations under 254 nm, 366 nm, and white light in transmittance mode (T) were performed on the developed plates in order to detect the separated honey constituents. From this information corresponding peak profiles were generated, and major peaks automatically determined by the software.

The scanning of individual major bands in the honey extracts was carried out using the TLC Scanner 4 in both UV-Vis mode (190–900 nm) and fluorescence mode (190–380 nm) with the following settings: Dimension set at 5 × 0.2 mm (micro), optimisation set for maximum resolution, scanning speed 20 nm/s and use of K400 optical filter. Deuterium (190–380 nm) and tungsten (380–900 nm) were used as lamps and the scans in fluorescence excitation mode were set at 380 < /400 nm and the emissions were observed at 190–270 nm.

Three spectral scans were performed for each sample, prior to and after derivatisation with each of the derivatisation reagents used.

To perform the derivatisation of the plates with NP-PEG reagent, plates were first sprayed with 3 mL of 1% NP reagent using a green nozzle at level 3 and then allowed to dry for 5 min at 40 °C. The plates were then sprayed again, this time with 5% PEG reagent using a blue nozzle at level 2, dried for 5 min at 40 °C and the resulting image was captured at 366 nm [29]. To derivatise using VSA reagent, plates were sprayed with 3 mL of 1% vanillin sulphuric acid reagent using a yellow nozzle at level 3, and then heated for 3 min at 115 °C for 3 min, and after cooling for 2 min, the plates were visualised at 366 nm and T white light.

A system suitability test (SST) was performed for each plate analysis as a quality control step. This was performed by utilising the Rf and the minimum height of the reference sample (naringenin) prior to derivatisation at 254 nm and only those plates that passed the set threshold of ±0.05 for the Rf and the minimum height for MPA (Rf 0.690, minimum height 0.108) and MPB (Rf 0.550, minimum height 0.120) were used in the qualitative and quantitative analysis.

2.8. Quantification of Phenolic Compounds in Honey

The same chromatographic instrumentation and parameters as described in Section 2.7 were employed in the quantification of the identified phenolic compounds in the various honey samples. Standard concentrations, application volumes, derivatisation and scanning conditions were optimised. The optimised application volumes for the various standards ranged from 5.0 to 9.8 µL (1.2 µL interval) and each compound was quantified at its specific λmax using the evaluation feature of the VisionCATS software. Table 1 summarises the key parameters for the standards used in the quantification experiments.

2.9. HPTLC-DPPH Antioxidant Activity

The same chromatographic instrumentation and parameters as described in Section 2.7 were also employed to perform the HPTLC-DPPH analysis for antioxidant activity in the honey extracts and their respective matched constituents. Seven microliters of each honey extract were used for the analysis alongside the standards in varying volumes. After development, the plates were derivatised with 3 mL of 0.4% DPPH solution (1:1 ratio of methanol and water) using the yellow nozzle and sprayed at level 1 [32,33]. Plate images were obtained at transmittance in white light after 1 h, 2 h and 3 h. Peak profiles at 517 nm were also generated and from these the Rf values of the respective peaks were generated. Each band's colour in the form of RGB values was determined and then converted into corresponding hue values [30]. Compounds that possess antioxidant activity will quench the DPPH radical either by electron transfer or hydrogen atom transfer through radical attack, which is observed as a discoloration at 517 nm due to the formation of 2,2-diphenyl-1-hydrazine or a substituted analogue hydrazine [33]. Gallic acid was used as positive control, its quenching activity resulting in a maximum hue value of 40° (yellow colour). All obtained hue values were calculated using previously reported formula [30]. The DPPH radical scavenging activity (% DPPH RSA) of a band of interest was calculated as follows:

$$\%\ \text{DPPH RSA} = \left(\frac{\Delta H^{\circ}B}{H^{\circ}P \rightarrow 40^{\circ}}\right) * 100 \quad (4)$$

where: $H^{\circ}P \rightarrow 40^{\circ}$–Hue values (°) of unreacted DPPH on the plate (n = 10), $\Delta H^{\circ}B$-hue values (°) of the bands up to 40° (Note: Hue = 40° or yellow was the maximum hue value of the gallic acid).

The respective band's % DPPH RSA was then categorised as indicated in Table 3.

Table 3. Categories of antioxidant activity for individual bands based on DPPH % RSA.

% DPPH RSA	Category	Inference
0.0%	0	No activity
1.0–33.3%	+	Low activity
33.4–66.6%	++	Medium activity
66.7–100.0%	+++	High activity

The DPPH antioxidant activity of luteolin, epicatechin, epigallocatechin gallate, gallic acid, protocatechuic acid, m-coumaric acid and kojic acid was analysed at varying concentrations to validate the bioautographic analysis. Furthermore, the DPPH antioxidant activity of the matched compounds was determined at low and high concentrations to determine their inherent antioxidant activity.

2.10. Statistical Analysis

Analysis of variance (ANOVA) was performed using Graphpad Prism 9 (GraphPad Software, San Diego, CA, USA) in order to determine whether there was a significant difference in the total phenolic content, FRAP activity, and DPPH antioxidant activity of different honeys. Tukey's post hoc comparisons were used to identify differences between the groups ($p < 0.05$).

3. Results

3.1. Total Phenolic Content

Table 4 shows the average total phenolic content for *Calothamnus* spp. (Red Bell), *Agonis flexuosa* (Coastal Peppermint), *Corymbia calophylla* (Marri), and *Eucalyptus marginata* (Jarrah) honeys. Individual sample values, expressed as mg GAE/100 g of honey and based on the 32 samples tested, ranged from 18.91 (Marri honey) to 75.56 (Red Bell honey), with an overall average of 59.4. Individual TPC values for each investigated honey are shown in Table S2 (Supplementary Materials). The average TPC value for Red Bell honey ($n = 8$) was found to be the highest (59.4 ± 7.91 mg GAE/100 g), followed by Jarrah honey (50.58 ± 3.76 mg GAE/100 g), Coastal Peppermint honey (36.08 ± 4.2 mg GAE/100 g) and Marri honey (29.15 ± 5.46 mg GAE/100 g). The average TPC of the four honeys differed significantly when analysed using One way ANOVA ($p < 0.0001$). Tukey's post hoc analysis demonstrated that Red Bell honey had higher TPC than the other three honeys (Coastal Peppermint and Marri honey ($p = <0.0001$), Jarrah honey ($p = 0.0407$) while Jarrah honey also showed higher TPC compared to Coastal Peppermint ($p = 0.0016$) and Marri ($p = <0.0001$) honeys). No difference, however, was observed when the mean TPC values of Coastal Peppermint and Marri honeys were compared (see Figure 2).

Table 4. Total phenolic content and antioxidant activity of different Western Australian honeys.

Assay	Honey	Mean	Range	Minimum	Maximum
TPC (mg GAE/100 g)	*Calothamnus* spp. (Red Bell, *n* = 8)	59.4 ± 7.91	27.48	48.09	75.56
	Agonis flexuosa (Coastal Peppermint, *n* = 5)	36.08 ± 4.20	12.66	30.05	42.71
	Corymbia calophylla (Marri, *n* = 13)	29.15 ± 5.46	17.7	18.91	36.61
	Eucalyptus marginata (Jarrah, *n* = 6)	50.58 ± 3.76	9.85	46.24	56.09
FRAP (mmol Fe^{2+}/kg)	*Calothamnus* spp. (Red Bell, *n* = 8)	9.24 ± 1.68	4.90	6.76	11.66
	Agonis flexuosa (Coastal Peppermint, *n* = 5)	5.45 ± 1.64	4.72	3.69	8.41
	Corymbia calophylla (Marri, *n* = 13)	4.48 ± 0.82	3.05	3.47	6.52
	Eucalyptus marginata (Jarrah, *n* = 6)	6.83 ± 1.38	3.61	5.20	8.81
DPPH mmol TE/kg	*Calothamnus* spp. (Red Bell, *n* = 8)	3.88 ± 0.96	2.95	2.45	5.41
	Agonis flexuosa (Coastal Peppermint, *n* = 5)	2.04 ± 0.57	1.61	1.45	3.06
	Corymbia calophylla (Marri, *n* = 13)	1.76 ± 0.58	2.35	1.01	3.36
	Eucalyptus marginata (Jarrah, *n* = 6)	2.3 ± 0.76	2.08	1.67	3.75

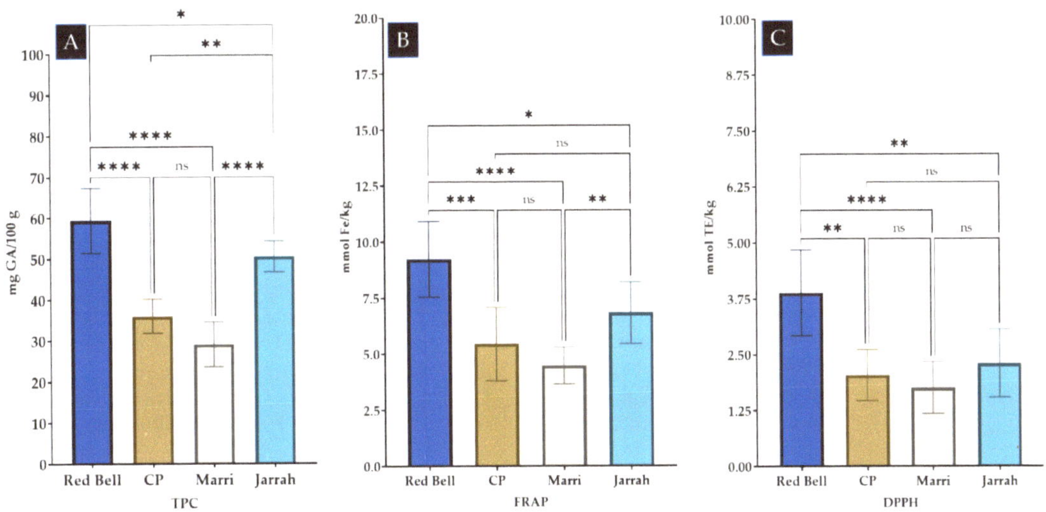

Figure 2. Comparison of the TPC (**A**), FRAP (**B**), and DPPH (**C**) of *Calothamnus* spp. (Red Bell), *Agonis flexuosa* (Coastal Peppermint, CP), *Corymbia calophylla* (Marri), and *Eucalyptus marginata* (Jarrah) honey. (Tukey post-hoc comparison: ns (not significant) = $p > 0.05$, * = $p < 0.05$, ** = $p < 0.005$, *** = $p < 0.0005$, **** = $p < 0.0001$).

3.2. Ferric Reducing Antioxidant Power (FRAP) Assay

Table 4 shows the average FRAP antioxidant activity of the investigated Western Australian honeys, expressed as mmol Fe^{2+} equivalent/kg. Based on the analysis of the 32 individual samples tested, mean FRAP activity was 6.26 and ranged from 3.47 (*Corymbia calophylla* (Marri) honey) to 11.66 (*Calothamnus* spp. (Red Bell) honey). The FRAP antioxidant activity of individual honeys is shown in Table S2 (Supplementary Materials). When the means of each honey type were analysed, it was found that *Calothamnus* spp. (Red Bell) honey had the highest activity, followed by *Eucalyptus marginata* (Jarrah), whereas *Agonis flexuosa* (Coastal Peppermint) and *Corymbia calophylla* (Marri) honey had comparable FRAP activity. One way ANOVA analysis demonstrated a significant difference (p = <0.0001) between the means of the honeys and post hoc analysis showed that the average Red Bell honey's FRAP activity was higher than that of Jarrah honey (p = 0.01030), Coastal

Peppermint honey (p = 0.0001), and also Marri honey (p = <0.0001). The FRAP activity of Jarrah honey was also found to be higher than that of Marri honey (p = 0.0058), whereas Coastal Peppermint and Marri honeys had comparable average FRAP antioxidant activities (p = 0.5061) (see Figure 2). In line with findings reported by others [46–49], a high correlation (0.912) was observed between FRAP antioxidant activity and TPC, indicating that the antioxidant activity of these honeys is strongly related to their phenolic constituents.

3.3. *2,2-Diphenyl-1-Picrylhydrazyl (DPPH) Radical Scavenging Assay*

Table 4 shows the average DPPH radical scavenging activity of the investigated Western Australian honeys, expressed as mmol TE/kg honey. Based on the results of the analysed 32 samples, a mean radical scavenging activity of 2.44 was found, ranging from 1.01 (*Corymbia calophylla* (Marri) honey) to 5.41 (*Calothamnus* spp. (Red Bell) honey). The DPPH radical scavenging activity of individual honey samples is shown in Table S2 (Supplementary Materials). When the mean values were compared, it was found that Red Bell honey had the highest activity. One way ANOVA analysis found a significant difference (p = 0.0001) amongst the means of the different honeys and post hoc analysis demonstrated that the mean DPPH antioxidant activity of Red Bell honey was higher when compared to *Eucalyptus marginata* (Jarrah) (p = 0.0026), *Agonis flexuosa* (Coastal Peppermint) (p = 0.0053), and Marri honey (p = <0.0001). Jarrah, Coastal Peppermint and Marri honey, have, however, comparable DPPH radical scavenging activities (p = >0.05) (see Figure 2). A high correlation (0.832) between DPPH antioxidant activity and TPC values of the individual honeys was observed, confirming that phenolic constituents contribute to honey's antioxidant activity. Furthermore, a high correlation (0.948) between DPPH and FRAP antioxidant activity was also observed.

3.4. *Phenolic Compound Identification*

The phenolic compound identification was carried out based on a previously reported database filtering approach [30]. The summary of the data (as described in Section 2.7) used to determine the identity of various phenolic constituents in the four investigated pooled honey samples is shown in Tables S3–S18 (Supplementary Materials). In addition, the identified candidate compounds for each significant band in the four different Western Australian honeys are shown in Table 5 along with correlations and percent match data based on the spectral overlays of four different UV-Vis spectra of the unknown and the candidate match compounds (254 nm and 366 nm prior to derivatisation, and 366 nm after derivatisation with VSA and NP-PEG reagents).

Based on the results obtained using database 1A and 1B (Figure 3A), the compound at Rf 0.570 in *Calothamnus* spp. (Red Bell) honey was identified as protocatechuic acid (**10**) as shown by the similarity of the spectral overlays of the unknown band and the standard when analysing their UV-Vis spectra prior to derivatisation (Figure 4A,B), after derivatisation with NP-PEG (Figure 4C,D), and also after derivatisation with VSA (Figure 4E,F). The unknown band at Rf 0.423 in Red Bell honey was identified as epigallocatechin gallate (**5**), and the unknown band at Rf 0.226 as kojic acid (**14**). By employing database 2A and 2B (Figure 3B), which utilised a less polar solvent, the unknown band at Rf 0.550 in Red Bell honey was identified as t-cinnamic acid (**13**), the band at Rf 0.380 as protocatechuic acid (**10**), the band at Rf 0.270 as gallic acid (**8**), and the band at Rf 0.115 as kojic acid (**14**) (see Figure 5 for structures).

Table 5. Match compounds, correlations, and % similarity of match compounds identified in Western Australian honey (Note: Compound codes are based on a previously published phenolic database paper [30]).

Honey	Data-Base	Rf	Name and Code	Rf	UV DEV	%	UV NP	%	UV VS	%	Match
Calothamnus spp. (Red Bell)	1A and 1B	0.630									none
		0.570	Daidzein (**31**)	0.600	0.926	51.2	0.290	24.2	0.841	64.1	Protocatechuic acid
			3,5-DHBA (**39**)	0.594	0.744	23.1	0.476	38.0	0.819	47.0	
			Protocatechuic acid (**55**)	0.577	0.998	100.0	0.978	75.8	0.707	61.2	
			Vanillic acid (**59**)	0.623	0.993	45.5	−0.127	15.4	−0.117	42.7	
		0.455									none
		0.423	EGCG (**29**)	0.407	0.430	57.1	0.711	19.5	0.812	52.7	EGCG
		0.382									none
		0.327									none
		0.299									none
		0.226	Kojic Acid (**105**)	0.287	0.952	39.7	0.583	33.8	−0.072	13.1	Kojic Acid
		0.178									none
		0.110									none
		0.078									none
		0.050									none
		0.020									none
	2A and 2B	0.550	m-Toluic Acid, (**51**)	0.577	0.817	18.7	−0.574	1.2	−0.684	1.99	t-Cinnamic acid
			o-Toluic Acid, (**53**)	0.591	0.791	18.7	0.606	31.5	−0.805	1.99	
			t-Cinnamic acid, (**75**)	0.557	0.963	28.6	−0.376	7.6	0.584	5.98	
		0.515									none
		0.465									none
		0.410									none
		0.335									none
		0.185									none
		0.380	Protocatechuic acid (**55**)	0.377	0.996	65.9	0.987	45.4	0.814	29.1	Protocatechuic acid
		0.270	Gallic acid (**44**)	0.270	0.965	45.0	0.750	40.6	0.580	14.0	Gallic acid
		0.115	Kojic Acid (**105**)	0.130	0.999	50.8	0.779	47.4	0.650	11.1	Kojic Acid
		0.075									none
		0.028									none
Agonis flexuosa (Coastal Peppermint)	1A and 1B	0.692									none
		0.615	Benzoic acid (**40**)	0.663	0.892	26.4	0.647	96.7	0.364	11.7	Syringic acid
			Methyl syringate (**48**)	0.610	0.936	44.0	0.829	25.4	0.738	70.5	
			Syringic acid (**58**)	0.577	0.941	45.1	0.859	26.2	0.797	80.8	
			m-Coumaric acid (**67**)	0.633	0.848	39.6	0.921	46.7	−0.806	3.2	
		0.588	Luteolin (**15**)	0.584	0.758	68.1	0.597	34.7	0.331	16.8	Luteolin
		0.500	Epicatechin (**27**)	0.499	0.646	17.6	0.748	31.4	0.284	16.8	Epicatechin
		0.460	Lumichrome (**107**)	0.464	0.837	67.5	0.832	32.7	0.253	20.2	Lumichrome
		0.380	EGCG (**29**)	0.407	0.804	61.5	0.713	19.5	0.442	15.1	EGCG
		0.265	Kojic Acid (**105**)	0.287	0.979	43.7	0.729	29.8	0.715	31.1	Kojic Acid
		0.195									none
		0.140									none
		0.050									none
	2A and 2B	0.475	2,3,4-TMBA (**37**)	0.453	0.883	23.1	0.631	17.1	−0.620	8.3	m-Coumaric acid
			Eudesmic acid (**43**)	0.478	0.970	28.6	0.793	29.5	0.363	3.4	
			Methyl syringate (**48**)	0.471	0.947	40.7	0.805	25.1	0.363	3.4	
			p-Hydroxybenzoic acid (**54**)	0.462	0.784	20.9	0.596	10.0	0.589	14.2	
			m-Coumaric acid (**67**)	0.467	0.866	80.2	0.906	24.3	0.918	7.7	
		0.450									none
		0.415	Syringic acid (**58**)	0.395	0.946	34.1	0.967	24.2	0.739	18.5	Syringic acid
			p-HPAA, HPAAD (**82**)	0.427	0.742	13.2	0.702	11.0	−0.827	19.9	
			DL-p-HPLA, HPLAD (**84**)	0.444	0.777	15.4	0.750	14.3	0.490	29.9	
		0.375	Luteolin (**8**)	0.372	0.911	71.3	0.597	66.5	0.781	28.5	Luteolin
		0.325									none
		0.266	Lumichrome (**107**)	0.266	0.613	63.4	0.638	12.7	0.796	33.3	Lumichrome
		0.235									none
		0.180	Epicatechin (**27**)	0.176	0.727	20.9	0.817	34.3	−0.278	16.8	Epicatechin
		0.105	Kojic Acid (**105**)	0.13	0.962	46.0	0.612	35.1	0.664	46.4	Kojic Acid
		0.090	EGCG (**29**)	0.065	0.852	58.2	0.577	42.2	−0.028	21.7	EGCG
		0.050									none

Table 5. *Cont.*

Honey	Data-Base	Rf	Name and Code	Rf	UV DEV	%	UV NP	%	UV VS	%	Match
Corymbia calophylla (Marri)	1A and 1B	0.697									none
		0.620	2,3,4-TMBA (**54**)	0.637	0.915	21.3	−0.634	12.9	−0.092	78.3	Eudesmic acid
			Eudesmic acid (**94**)	0.602	0.983	46.7	−0.339	26.7	0.161	32.8	
			p-Hydroxybenzoic acid (**33**)	0.663	0.846	20.5	−0.711	12.9	−0.064	70.1	
		0.600									None
		0.550	Luteolin (**8**)	0.584	0.431	17.1	0.639	90.6	0.500	36.5	Luteolin
		0.475	Epicatechin (**26**)	0.520	0.800	42.9	0.862	12.4	0.765	39.3	Epicatechin
		0.375	EGCG (**29**)	0.407	0.791	79.1	0.679	14.3	0.653	45.6	EGCG
		0.250	Kojic Acid (**105**)	0.241	0.926	98.9	0.711	23.5	−0.061	25.9	Kojic Acid
		0.259									none
		0.216									none
		0.186									none
		0.156									none
		0.115									none
		0.105									none
		0.052									none
		0.025									none
		0.012									none
	2A and 2B	0.470	2,3,4-TMBA (**37**)	0.453	0.977	30.8	0.858	16.8	0.564	11.7	Eudesmic acid
			Eudesmic acid (**43**)	0.478	0.941	65.9	0.981	44.6	0.334	6.0	
			p-HBA (**54**)	0.462	0.955	27.5	0.796	19.8	0.564	14.2	
		0.426	Methyl syringate (**48**)	0.471	0.335	18.7	0.820	22.6	0.211	17.9	m-Coumaric acid
			o-Anisic acid (**52**)	0.440	−0.075	23.1	0.604	16.5	0.644	26.5	
			Syringic acid (**58**)	0.395	0.350	18.7	0.841	23.3	0.751	20.5	
			m-Coumaric acid (**67**)	0.467	0.180	68.1	0.918	42.1	0.758	8.0	
			DL-p-HPLA (**84**)	0.444	0.319	9.9	0.641	11.3	0.601	36.5	
		0.390	Luteolin (**8**)	0.372	0.574	23.9	0.033	31.5	0.887	34.2	Luteolin
		0.346									none
		0.300	Taxifolin (**25**)	0.335	0.758	29.5	0.759	70.1	0.661	8.5	Taxifolin
		0.270	Gallic acid (**44**)	0.270	0.906	71.6	0.655	37.5	0.632	16.5	Gallic acid
		0.186									none
		0.150	Epicatechin (**27**)	0.176	0.774	19.8	0.751	35.7	−0.345	35.9	Epicatechin
		0.110	Kojic Acid (**105**)	0.130	0.960	38.9	0.339	32.5	0.633	27.6	Kojic Acid
Eucalyptus marginata (Jarrah)	1A and 1B	0.633	Genistein (**33**)	0.633	0.959	59.6	0.769	31.6	0.432	40.5	2,3,4-THBA
			2,3,4-THBA (**36**)	0.589	0.979	65.1	0.901	76.9	0.900	50.1	
			p-HBA (**54**)	0.637	0.885	24.5	−0.653	8.8	0.707	55.8	
		0.562									none
		0.538									none
		0.471	Lumichrome (**107**)	0.471	0.278	59.7	0.090	18.7	0.592	50.5	Lumichrome
		0.371	EGCG (**29**)	0.371	0.940	41.3	0.165	25.4	0.357	31.0	EGCG
		0.320									none
		0.250	Kojic Acid (**105**)	0.241	0.970	34.1	0.760	44.3	0.128	45.9	Kojic Acid
		0.196									none
		0.117									none
		0.072									none
	2A and 2B	0.525	Hesperetin (**18**)	0.520	0.598	29.8	0.821	40.6	−0.369	32.2	Hesperetin
		0.470	Methyl syringate (**48**)	0.471	0.912	35.5	0.346	36.8	−0.060	16.0	m-Coumaric acid
			m-Coumaric acid (**67**)	0.467	0.944	76.0	0.334	21.8	0.909	7.7	
			p-HPAA (**82**)	0.427	0.709	14.0	0.137	18.0	0.867	8.8	
			DL-p-HPLA (**84**)	0.444	0.690	14.0	0.627	60.2	−0.722	23.1	
			Phloretic acid (**87**)	0.46	0.704	14.0	0.641	37.6	−0.538	3.1	
		0.420	o-Anisic acid (**52**)	0.44	0.729	25.2	0.689	21.8	0.524	19.9	o-Anisic acid
		0.375									none
		0.320	Taxifolin (**25**)	0.335	0.600	63.9	0.800	71.8	0.661	8.5	Taxifolin
		0.270	Lumichrome (**107**)	0.266	0.136	35.1	0.594	12.7	0.745	64.1	Lumichrome
		0.130	Kojic Acid (**105**)	0.13	0.942	53.4	0.969	82.0	0.695	87.9	Kojic Acid
		0.090	EGCG (**29**)	0.065	0.978	47.7	0.806	49.6	−0.478	14.6	EGCG

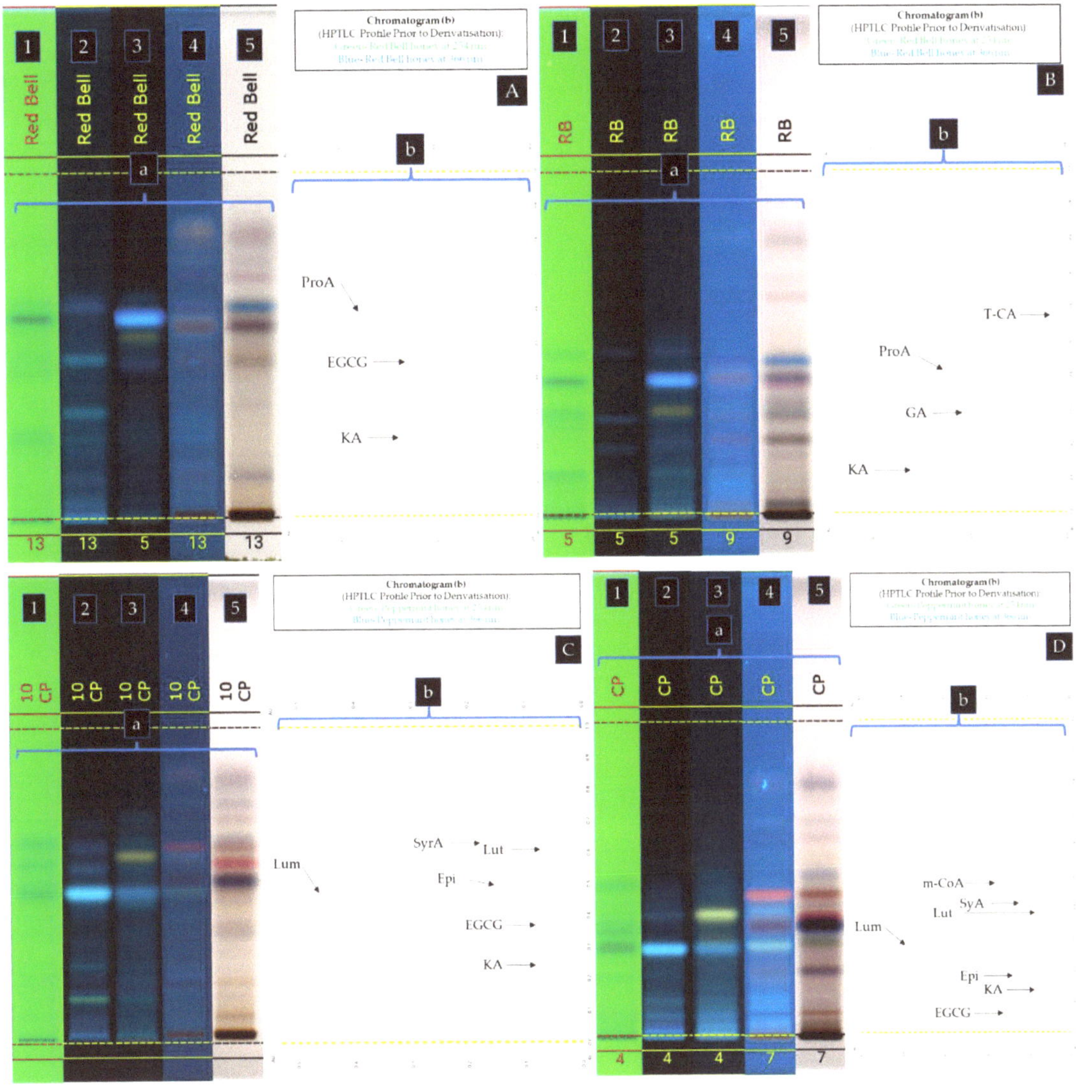

Figure 3. HPTLC Profile of *Calothamnus* spp. (Red Bell) honey (**A**,**B**) and *Agonis flexuosa* (Coastal Peppermint) honey (**C**,**D**) using MPA (**A**,**C**), and MPB (**B**,**D**). (**a**) Plate images obtained under the following light conditions: 254 nm prior to derivatisation (**1**), 366 nm prior to derivatisation (**2**), 366 nm after derivatisation with NP-PEG (**3**), 366 nm after derivatisation with VSA (**4**), transmittance in white light after derivatisation with VSA (**5**); (**b**) Chromatograms prior to derivatisation obtained at 254 nm and 366 nm.

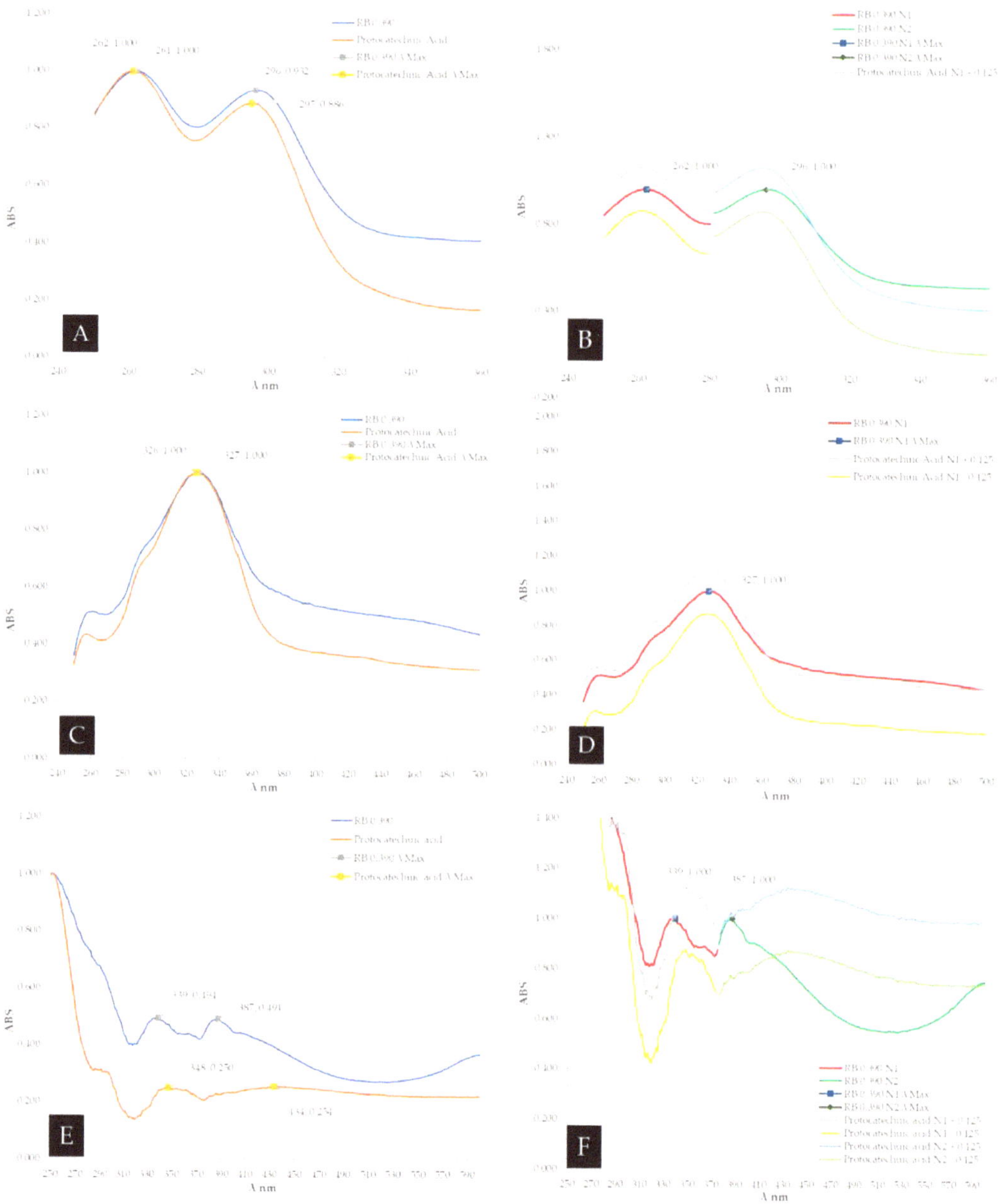

Figure 4. Spectra overlay of unknown band at Rf 0.390 in *Calothamnus* spp. (Red Bell) honey vs. protocatechuic acid (**10**) using MPB. (**A**)—UV-Vis spectra and (**B**)—overlay of the ±0.125 AU comparison prior to derivatisation, (**C**)—UV-Vis spectra, (**D**)—overlay of the ±0.125 AU comparison after derivatisation with NP-PEG, (**E**)—UV-Vis spectra, (**F**)—overlay of the ±0.125 AU comparison after derivatisation with VSA.

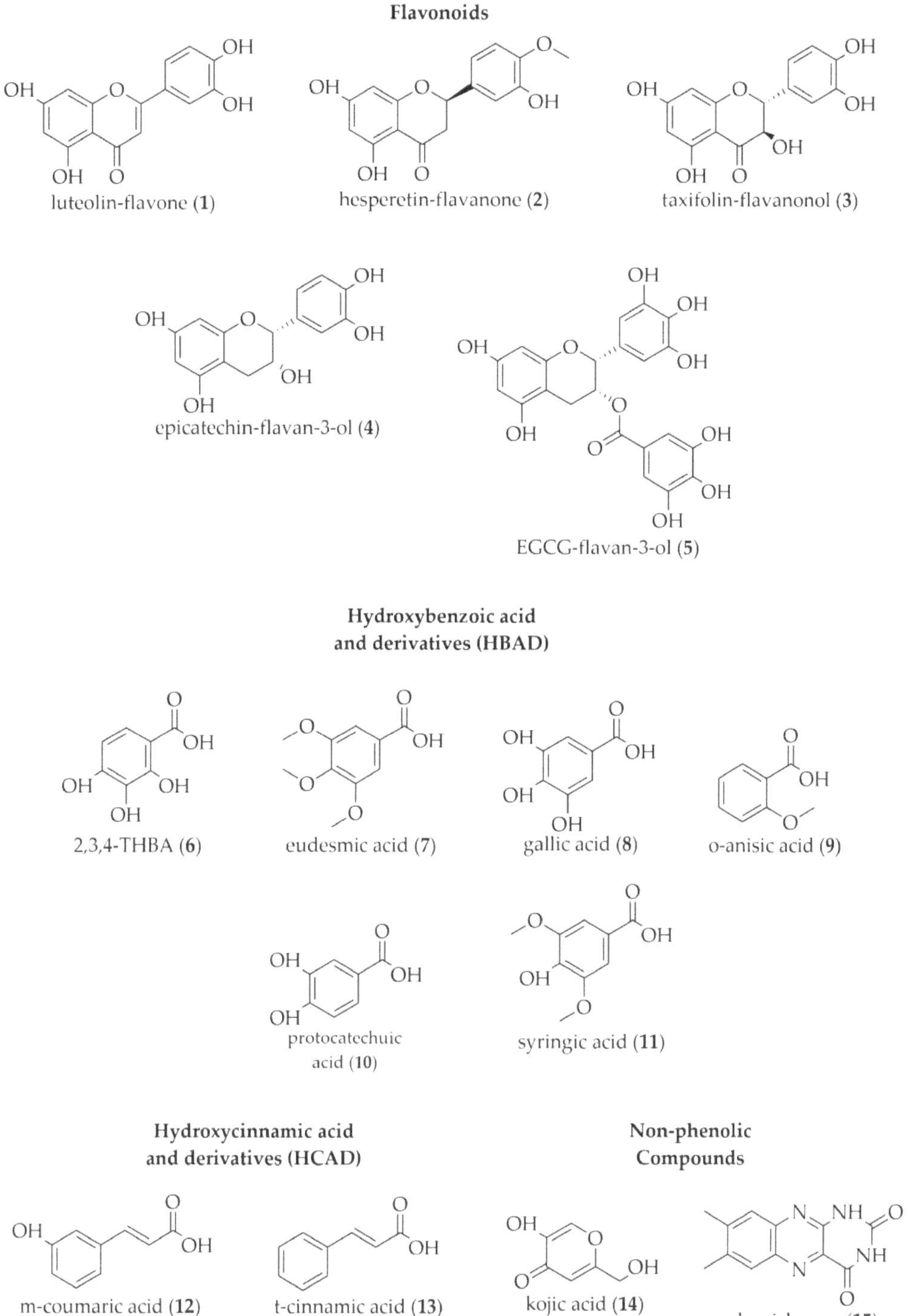

Figure 5. Structures of the compounds identified in Western Australian honeys (generated using ChemDraw version 20.1.1, PerkinElmer Informatics, Inc., Waltham, MA, USA).

All other compounds reported here were identified in the three other honey samples in the same manner. Table 2 summarises the identified honey constituents. Figure 3C,D shows the identified compounds in *Agonis flexuosa* (Coastal Peppermint) honey. Figure 6A,B summarises the identified compounds in *Corymbia calophylla* (Marri) honey, while Figure 6C,D summarises the identified compounds in *Eucalyptus marginata* (Jarrah) honey.

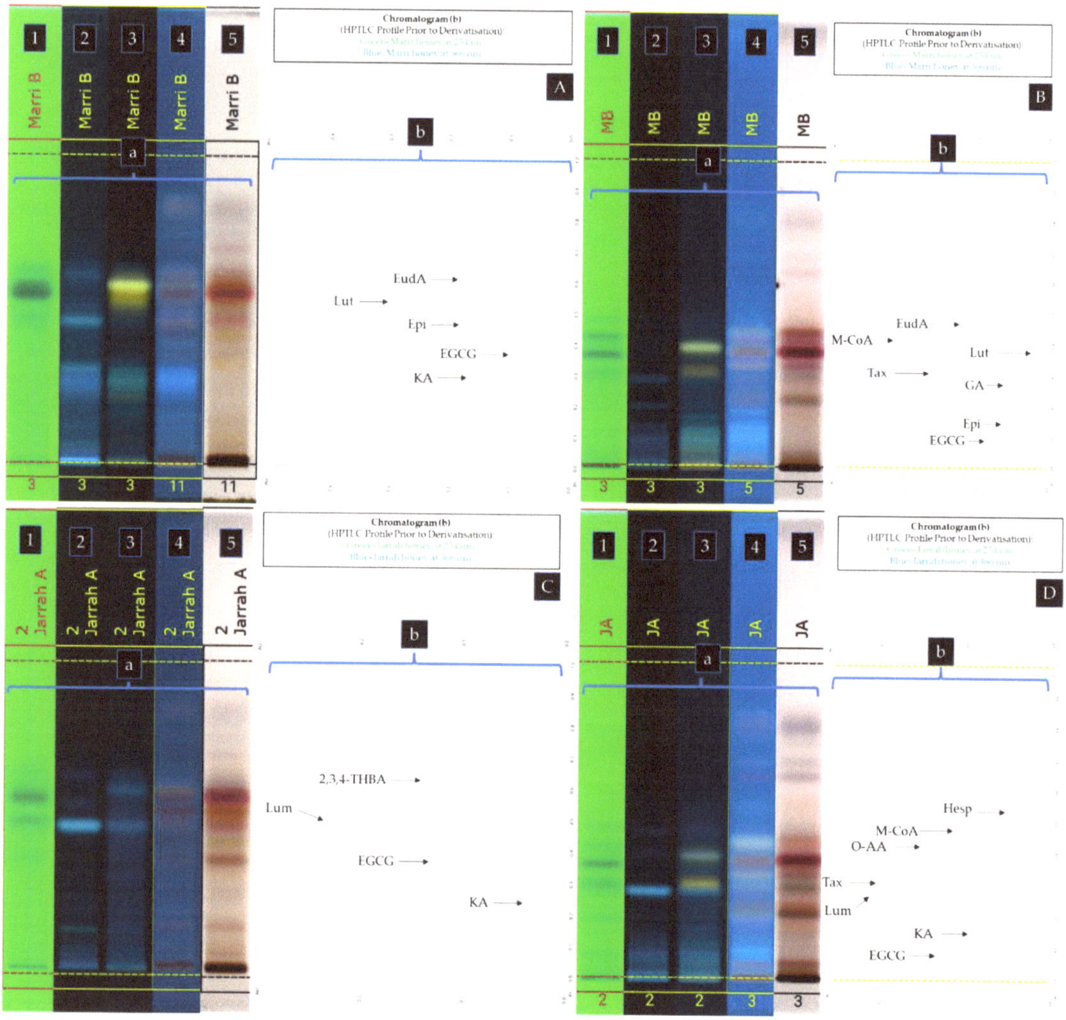

Figure 6. HPTLC Profile of *Corymbia calophylla* (Marri) honey (**A**,**B**) and *Eucalyptus marginata* (Jarrah) honey (**C**,**D**) using mobile phase A (**A**,**C**), and mobile phase B (**B**,**D**). Plate images (**a**) obtained under the following light conditions: 254 nm prior to derivatisation (**1**), 366 nm prior to derivatisation (**2**), 366 nm after derivatised with NP-PEG (**3**), 366 nm after derivatisation with VSA (**4**), transmittance in white light after derivatisation with VSA (**5**) and chromatograms (**b**) prior to derivatisation obtained at 254 nm and 366 nm.

A comparison between the peak profile of each honey and the respective honey overspotted with a mixture of its identified constituents was also used to further confirm the phenolic compound determination. For confirmation, scans were performed, for example, at each specific λmax of each identified compound in *Calothamnus* spp. (Red Bell) honey

(Figure 7A–D) using databases 2A and 2B and based on this analysis, the Rf of the matched compounds were found to be similar to that of the identified bands in the honey. Moreover, an increase in the absorbance confirmed the presence of the compounds in the honey. Profile comparisons for the other investigated honeys are included in the Supplementary Materials (Figures S5–S11).

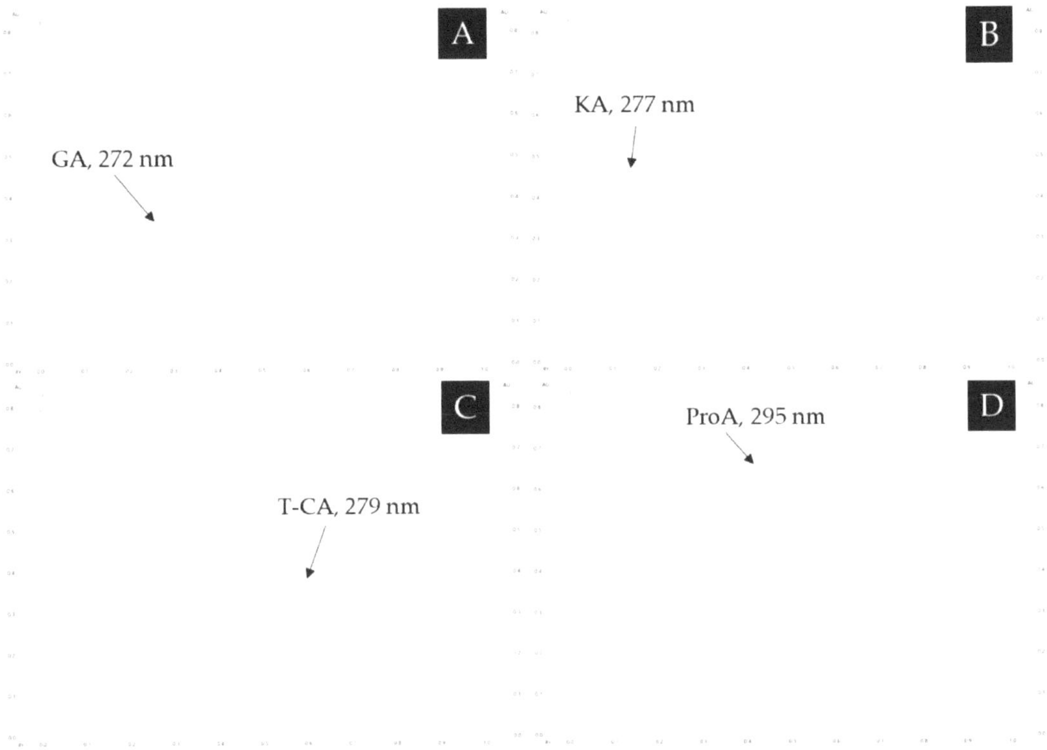

Figure 7. (**A–D**) Peak profile comparison of *Calothamnus* spp. (Red Bell) honey (green) and *Calothamnus* spp. (Red Bell) honey spiked with the identified compounds based on Database 2A and 2B (blue) scanned at the λmax of each specific compound prior to derivatisation.

3.5. Phenolic Compound Quantification

Optimised parameters, such as standard concentrations, application volumes, mode in obtaining the profile/chromatogram, and derivatisation for quantification of the identified phenolic compounds in the four pooled honey samples, are shown in Table S19 (Supplementary Materials). Based on the findings of the optimisation, it was concluded that standard concentrations ranging from 5 µg/mL to 50 µg/mL, application volumes ranging from 5.0 to 9.8 µL (1.2 µL interval), peak profiles obtained by scanning the plate at the respective specific λmax, and the absence of any derivatisation constituted the best approach for quick and accurate quantification of phenolic compounds in the honey matrices. Linearity was observed to be greater than 0.99 for each standard and the percent recovery ranged from 95.2 to 102.6%. Table 2 details the standard concentrations, linearity, and % recovery of each identified constituent that was used in the quantification experiment. Furthermore, a sample of an HPTLC plate and its corresponding peak profile used in the quantification of phenolic compounds in *Calothamnus* spp. (Red Bell) honey is shown in Figure 8A–C.

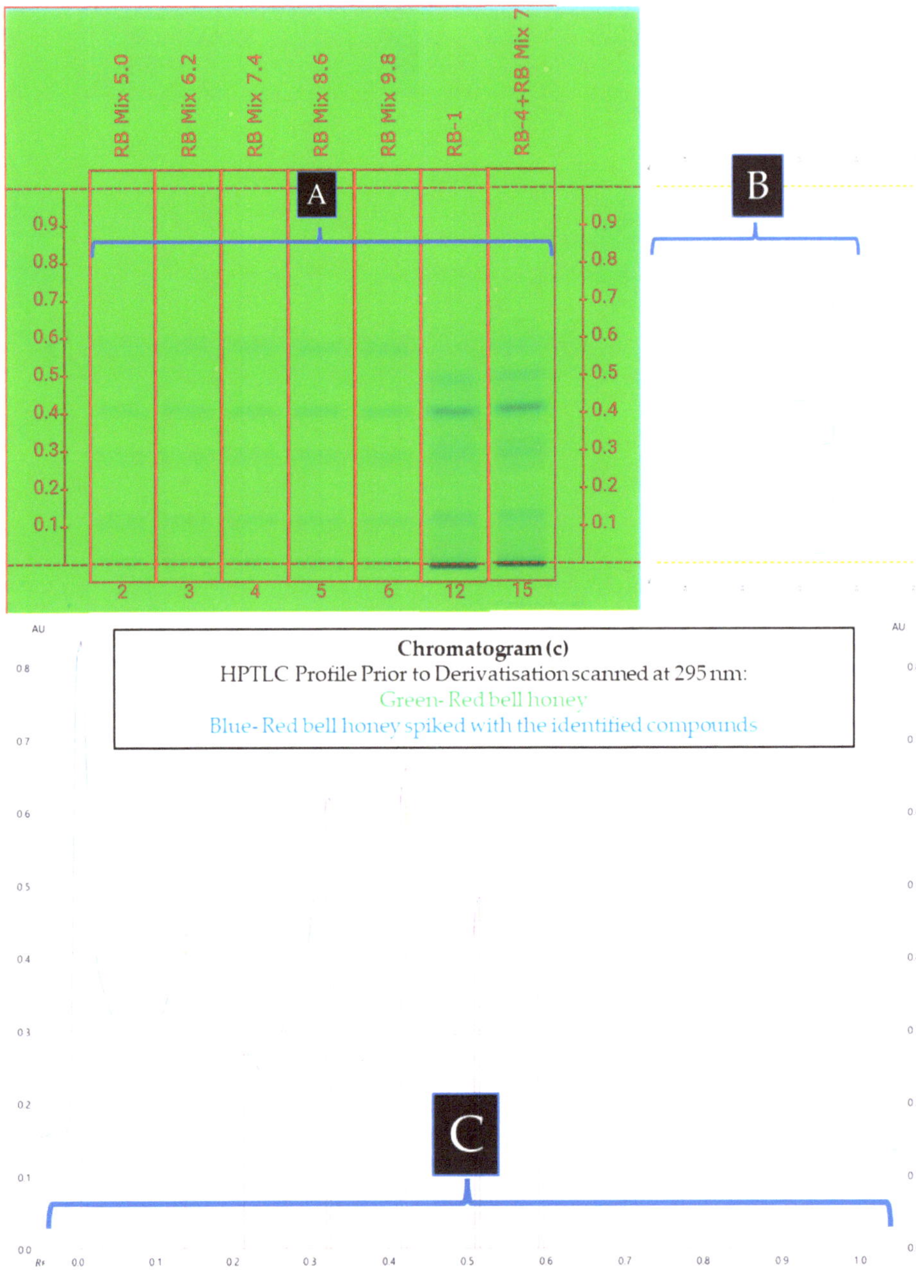

Figure 8. (**A**) HPTLC Images of the compound mixture of identified compounds in *Calothamnus* spp. (Red Bell) honey using various application volumes (Tracks 2–6) as compared to Red Bell honey (Track 12), and Red Bell honey spiked with the mixture of identified compounds (Track 15); (**B**) peak profile of compounds identified in *Calothamnus* spp. (Red Bell) honey; (**C**) peak profile of Red Bell honey (green), and Red Bell honey spiked with the identified compound mixture (blue) scanned at 295 nm using MPB.

By utilising the optimised conditions, the concentration of the compounds identified in honey ranged from 0.003 μg/g (t-cinnamic acid (**13**)) to 13.49 μg/g (2,3,4-trihydroxy benzoic acid (**6**)) (see Table 6 for the specific quantities).

Table 6. Quantity of Specific Phenolic Constituents (in μg/g, $n = 3$) identified in different Western Australian Honeys.

Compound (see Figure 5 for Structures)	*Calothamnus* spp. (Red Bell)	*Agonis flexuosa* (Coastal Peppermint)	*Corymbia calophylla* (Marri)	*Eucalyptus marginata* (Jarrah)
Luteolin (**1**)	-	1.14 ± 0.00	1.50 ± 0.01	-
Hesperitin (**2**)	-	-	-	0.62 ± 0.01
Taxifolin (**3**)	-	-	1.40 ± 0.01	1.34 ± 0.00
Epicatechin (**4**)	-	6.90 ± 0.03	2.40 ± 0.03	
EGCG (**5**)	3.81 ± 0.01	2.45 ± 0.04	5.11 ± 0.03	5.61 ± 0.04
2,3,4-THBA (**6**)	-	-	-	13.49 ± 0.16
Eudesmic acid (**7**)	-	-	3.25 ± 0.02	-
Gallic acid (**8**)	1.64 ± 0.00	-	5.84 ± 0.00	-
o-Anisic acid (**9**)		-	-	3.52 ± 0.04
Protocatechuic acid (**10**)	5.09 ± 0.02	-	-	-
Syringic acid (**11**)	-	1.47 ± 0.02	-	-
m-Coumaric acid (**12**)	-	0.58 ± 0.02	0.54 ± 0.02	0.72 ± 0.01
t-Cinnamic acid (**13**)	0.003 ± 0.00	-	-	-
Kojic acid (**14**)	3.64 ± 0.00	2.88 ± 0.02	1.32 ± 0.01	0.64 ± 0.01
Lumichrome (**15**)	-	1.94 ± 0.01	-	1.03 ± 0.00

Protocatechuic acid (**10**) was found to be the most abundant constituent in *Calothamnus* spp. (Red Bell) honey, followed by epigallocatechin gallate (**5**), kojic acid (**14**), gallic acid (**8**), and t-cinnamic acid (**13**). In the case of *Agonis flexuosa* (Coastal Peppermint) honey, epicatechin (**4**) was found to be the most abundant, followed by kojic acid, epigallocatechin gallate, lumichrome (**15**), syringic acid (**11**), luteolin (**1**), and m-coumaric acid (**12**). For *Corymbia calophylla* (Marri) honey, gallic acid was found to be the most abundant compound, followed by epigallocatechin gallate, eudesmic acid (**7**), epicatechin, luteolin, taxifolin (**3**), kojic acid, and m-coumaric acid (**12**). 2,3,4-Trihydroxy benzoic acid (**6**) was found to be the most abundant compound in *Eucalyptus marginata* (Jarrah) honey, followed by epicatechin, o-anisic acid (**9**), taxofolin, lumichrome, m-coumaric acid, kojic acid, and hesperitin (**2**) (see Figure 5 for structures).

3.6. HPTLC DPPH Assay

The DPPH-HPTLC assay was carried out to determine which constituents contributed to the respective honey's overall antioxidant activity. Previously, the HPTLC-DPPH assay was developed to determine the DPPH antioxidant activity after 1 h of exposure to the reagent, the corresponding peak profiles were obtained using white light, and the antioxidant activity was expressed as mg GAE/100 g of honey [32]. In this experiment, incubation time as well as the mode for peak profile generation were optimised (Figure S12, Supplementary Materials). It was found that the colour of the unreacted DPPH on the plate degraded by 14.7% after 2 h and by 19.3% after 3 h. It was also observed that a significant decrease in the absorbances of the test compounds was observed after 2 and 3 h of incubation time as compared to 1 h (Figure S8, Supplementary Materials). Furthermore, naringenin (used as reference standard) showed a DPPH radical scavenging activity (% DPPH RSA) of 66.5% after 1 h, 81.4% after 2 h, and 81.9% after 3 h of incubation. DPPH scavenging activity is generally evaluated at the point when the absorbance remains constant [50]. Because of this, in this study the photo-documentations and the recording of the corresponding peak profiles were carried out after 2 h in order to allow sufficient time for compounds to react with the DPPH reagent but not too long that the reagent would autodegrade and produce false positive results. Furthermore, it was also observed that peak profiles generated using a scan at 517 nm were more sensitive in determining the reaction of the individual bands as compared to profiles generated using white light, and therefore scanning at 517 nm was adopted for all analyses in this study (Figure S12, Supplementary Materials).

The use of a change in hue to determine the DPPH antioxidant activity was also validated. Based on the findings (Table 7), all standards showed an increase in % DPPH RSA which correlated with increases in sample concentration, indicating that hue values can be a very useful tool in describing the antioxidant activity of a particular compound.

Table 7. Colour and % DPPH RSA of various compounds tested for validation of the HPTLC-DPPH analysis.

Sample	Conc. (µg/mL)	H°					%RSA					R^2
		Volume Application (µL)					Volume Application (µL)					
		5	6.2	7.4	8.6	9.8	5	6.2	7.4	8.6	9.8	
Plate (unreacted DPPH)	0.0	335.0	335.0	335.0	335.0	335.0	0	0	0	0	0	NA
Luteolin (**1**)	12.5	347.9	358.5	3.8	11.6	15.9	19.8	36.2	44.3	56.3	62.9	0.979
Epicatechin (**4**)	50.0	22.7	29.2	32.3	35.6	36.2	73.4	83.4	88.2	93.2	94.2	0.916
Epigallocatechin gallate (**5**)	50.0	27.0	29.3	33.2	35.0	36.5	80.0	83.5	89.5	92.3	94.6	0.972
Gallic acid (**8**)	25.0	23.0	31.5	34.0	35.9	38.3	73.8	86.9	90.8	93.7	97.4	0.883
Protocatechuic acid (**10**)	25.0	28.8	32.7	34.6	34.8	35.0	82.8	88.8	91.7	92.0	92.3	0.768
m-Coumaric acid (**12**)	25.0	349.9	352.6	353.9	354.1	359.2	22.9	27.1	29.1	29.4	37.2	0.984
Kojic acid (**14**)	25.0	1.5	6.3	10.6	12.7	14.3	40.8	48.2	54.8	58.0	60.5	0.950

The DPPH antioxidant activity of the four pooled and extracted Western Australian honeys along with the phenolic compounds that were previously identified in each honey were analysed in the HPTLC-DPPH assay using two solvent systems (MPA and MPB). Figure 8 shows the DPPH-HPTLC plate of Red Bell honey. The images of the other honeys are shown in Figures S13–S15.

Calothamnus spp. (Red Bell) honey was found to have nine bands with DPPH antioxidant activity of which the band at Rf 0.390 had very high activity, a medium activity band was found at Rf 0.115, and low activity bands at Rf 0.505, 0.450, 0.281, 0.246, 0.207, and 0.174. A total of nine bands were also observed to be antioxidant in the case of *Agonis flexuosa* (Coastal Peppermint) honey, where the bands at Rf 0.473, 0.395, 0.352, 0.279, 0.187, 0.114, 0.100, and 0.090 were all found to be low in activity. *Corymbia calophylla* (Marri) honey presented 11 antioxidant bands, of which a medium active antioxidant band was found at Rf 0.391, while bands at Rf 0.484, 0.444, 0.391, 0.313, 0.275, 0.26, 0.212, 0.179, 0.146, 0.105 were all low in activity. *Eucalyptus marginata* (Jarrah) honey, on the other hand, showed 12 antioxidant bands, of which the band at Rf 0.391 had medium activity and low activity bands were observed at Rf 0.530, 0.455, 0.322, 0.282, 0.254, 0.22, 0.189, 0.147, 0.117, 0.100, 0.083, and 0.024. The average antioxidant band activity of each honey (% AVE) was also calculated based on the total % DPPH RSA over the total number of antioxidant bands. It was found that *Calothamnus* spp. (Red Bell) honey had an average of 33.6% DPPH RSA, *Eucalyptus marginata* (Jarrah) honey had an average of 21.1%, *Agonis flexuosa* (Coastal Peppermint) honey an average of 18.4%, and *Corymbia calophylla* (Marri) honey an average of 18.2%, a trend which is similar to that reported for the total DPPH antioxidant activity of each individual honey (Table 8).

Table 8. Percentage DPPH RSA antioxidant activity of individual bands in *Calothamnus* spp. (Red Bell), *Agonis flexuosa* (Coastal Peppermint), *Corymbia calophylla* (Marri) and *Eucalyptus marginata* (Jarrah) honey along with their corresponding matched compounds.

Sample	Rf	Match Compounds	H°	% RSA	Category	% AVE
Baseline	NA	NA	335.0	0.0	0	0
Gallic acid (8)	NA	NA	38.3	97.4	+++	97.4
Calothamnus spp. (Red Bell)	0.562	t-CA	335.0	0.0	0	33.6
	0.505	-	342.6	11.7	+	
	0.450	-	344.6	14.8	+	
	0.390	ProA	37.9	96.8	+++	
	0.281	GA	339.1	6.3	+	
	0.246	-	337.2	3.4	+	
	0.207	-	340.6	8.6	+	
	0.174	-	338.7	5.7	+	
	0.115	KA	14.5	60.8	++	
	0.000	-	36.3	94.3	+++	
Agonis flexuosa (Coastal Peppermint)	0.473	m-CoA	342.3	11.2	+	18.4
	0.420	SyrA	335.0	0.0	0	
	0.395	Lut	344.4	14.5	+	
	0.352	-	336.9	2.9	+	
	0.279	Lum	337.1	3.2	+	
	0.187	Epi	337.6	4.0	+	
	0.114	-	337.7	4.2	+	
	0.100	KA	344.8	15.1	+	
	0.090	EGCG	344.8	15.1	+	
	0.000	-	37.3	95.8	+++	
Corymbia calophylla (Marri)	0.484	EudA	336.1	1.7	+	18.2
	0.444	m-CoA	340.9	9.1	+	
	0.391	Lut	7.6	50.2	++	
	0.313	Tax	342.5	11.5	+	
	0.275	GA	337.4	3.7	+	
	0.260	-	337.2	3.4	+	
	0.212	-	337.6	4.0	+	
	0.179	Epi	337.6	4.0	+	
	0.146	-	339.1	6.3	+	
	0.105	KA	343.9	13.7	+	
	0.00	-	35.2	92.6	+++	
Eucalyptus marginata (Jarrah)	0.530	Hesp	335.0	0.0	0	21.1
	0.455	o-AA	344.3	14.3	+	
	0.420	m-CoA	335.0	0.0	0	
	0.391	-	8.3	51.2	++	
	0.322	Tax	336.1	1.7	+	
	0.282	Lum	336.3	2.0	+	
	0.254	-	339.1	6.3	+	
	0.220	-	336.6	2.5	+	
	0.189	-	336.6	2.5	+	
	0.147	-	337.6	4.0	+	
	0.117	-	339.4	6.8	+	
	0.100	KA	351.6	25.5	+	
	0.083	EGCG	350.3	23.5	+	
	0.024	-	350.8	24.3	+	
	0.00	-	32.3	88.2	+++	

The DPPH antioxidant activity of the compounds identified in each pooled honey sample was also determined at a low concentration to mimic the concentration of the compounds in each honey and also at a high concentration in order to determine whether the activity is based on its concentration in the honey or an inherent antioxidant activity of the constituent (see Table 9). Based on the data generated, it was found that most compounds were antioxidant with the exception of eudesmic acid, o-anisic acid, t-cinnamic acid, and lumichrome, which remained inactive even when analysed at a higher concentration (see Figure 5 for structures).

Table 9. Colour and % DPPH RSA of matched compounds at a higher and lower application volume (Note: See Table 3 for Antioxidant Category and Inference).

Compound and Code	Sample Applied (ng)	H°	% RSA	Category	Sample Applied (ng)	H°	% RSA	Category	No. of OH	Bors Criteria	Remarks
background (plate)	0	336.6	0.0	0	0	336.6	0.0	0	NA	NA	NA
Luteolin (**1**)	87.5	336.7	2.6	+	350	20.3	69.7	++	NA	1	Active
Hesperetin (**2**)	87.5	334.0	−1.5	0	700	356.9	31.2	+	3	None	Active
Taxifolin (**3**)	35	340.6	8.6	+	140	2.4	42.2	++	NA	1,3	Active
Epicatechin (**4**)	350	33.1	89.4	+++	700	37.1	95.5	+++	NA	1	Active
Epigallocatechin gallate (**5**)	350	28.7	82.6	+++	700	36.8	95.1	+++	NA	1	Active
2,3,4-Trihydroxy benzoic acid (**6**)	175	345	15.4	+	700	28.9	85.8	+	3	NA	Active
Eudesmic acid (**7**)	350	333.7	−2.0	0	1400	336.6	0.0	0	0	NA	Inactive
Gallic acid (**8**)	175	354.1	29.4	+	700	38.2	97.2	+++	3	NA	Active
o-Anisic acid (**9**)	175	333.7	−2.0	0	700	334.5	−3.2	0	0	NA	Inactive
Protocatechuic acid (**10**)	175	10.4	54.5	++	700	34.6	91.7	+++	2	NA	Active
Syringic acid (**11**)	175	340.6	8.6	+	700	31.8	87.4	+++	1	NA	Active
m-Coumaric acid (**12**)	175	336.6	2.5	+	700	352.9	25.1	+	1	NA	Active
t-Cinnamic acid (**13**)	87.5	334.0	−1.5	0	700	337.2	0.9	+	0	NA	Inactive
Kojic acid (**14**)	175	344.6	14.8	+	700	11.6	56.3	++	1	NA	Active
Lumichrome (**15**)	35	333.7	−2.0	0	140	336.6	0.0	0	0	NA	Inactive

For *Calothamnus* spp. (Red Bell) honey, one of the identified constituents, t-cinnamic acid at Rf 0.562, was observed to be inactive, which was consistent with the finding that the t-cinnamic acid (**13**) standard did not possess any DPPH antioxidant activity. The other compounds in Red Bell honey, identified as protocatechuic acid (**10**) at Rf 0.390, gallic acid (**8**) at Rf 0.281, and kojic acid (**14**) at Rf 0.115, were found to be the dominant antioxidants in the honey (Figure 9B).

For *Agonis flexuosa* (Coastal Peppermint) honey, the respective quantities present for the bands at Rf 0.473 (m-coumaric acid (**12**)), at Rf 0.420 (syringic acid (**11**)), at Rf 0.395 (luteolin (**1**)), at Rf 0.187 (epicatechin (**4**)), at Rf 0.100 (kojic acid (**14**)), and at Rf 0.090 (epigallocatechin gallate (**5**)), showed the expected antioxidant activity based on the calibrated antioxidant activity of the standards, except for the constituent at Rf 0.279 (identified as lumichrome (**15**)) which showed an unexpected result since lumichrome standard itself was found to be inactive, indicating that there might be a constituent that was co-eluting with lumichrome at this Rf value which might cause the honey band at this Rf to show antioxidant activity (Figure S13, Supplementary Materials).

In the case of *Corymbia calophylla* (Marri) honey, the bands at Rf 0.484 (eudesmic acid (**7**)), Rf 0.444 (m-coumaric acid), Rf 0.391 (luteolin), Rf 0.313 (taxifolin (**3**)), Rf 0.275 (gallic acid (**8**)), Rf 0.179 (epicatechin), and Rf 0.105 (kojic acid) all showed the expected antioxidant activity consistent with that of the calibrated antioxidant activity of the standards (Figure S14, Supplementary Materials).

For *Eucalyptus marginata* (Jarrah) honey, the band at Rf 0.530, identified as hesperitin (**2**), was found to be consistent in its behaviour with the analysed activity of the corresponding standard, which was found to be inactive at low concentration. The band at Rf 0.455 (o-anisisc acid (**9**)) was found to be active, whereas the corresponding standard showed no activity, even at higher concentration, implying that this honey band has a co-eluting constituent which causes a low level of antioxidant activity. The band at Rf 0.282 was identified as lumichrome (**15**), which showed a very low level of activity similar to the standard, which was found to be inactive in the investigated concentrations. The activity of the compounds at Rf 0.420 (m-coumaric acid (**12**)), Rf 0.322 (taxifolin (**3**)), Rf 0.100 (kojic acid (**14**)), and Rf 0.083 (epigallocatechin gallate (**5**)) were found to be consistent with the activity of the respective standards (Figure S15, Supplementary Materials).

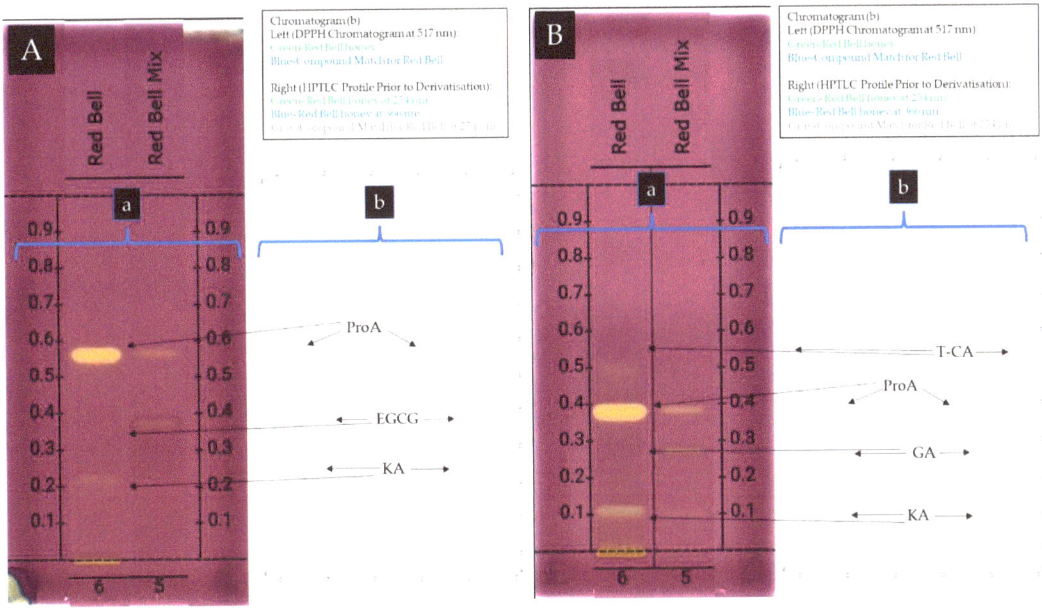

Figure 9. HPTLC-DPPH plate image (a) of *Calothamnus* spp. (Red Bell) honey after development in MPA (**A**) and after development in MPB (**B**) recorded with transmission white light, and comparison of the peak profiles of *Calothamnus* spp. (Red Bell) honey (green) and Red Bell honey spiked with the identified compounds (blue) after derivatisation with DPPH reagent and scanning at 517 nm (b-left) and comparison of the profiles of *Calothamnus* spp. (Red Bell) honey obtained at 254 nm (green) and 366 nm (blue) prior to derivatisation and the profile of *Calothamnus* spp. (Red Bell) honey spiked with the identified compounds (gray) obtained at 254 nm prior to derivatisation (b-right).

4. Discussion

The data obtained in the TPC assay were consistent with previous studies where Red Bell honey had shown higher phenolic content than nine other monofloral honeys from Western Australia [37]. By using the same conditions for the assay, Manuka honey from Australia and New Zealand [43] was found to have a TPC of 35.08 mg GAE/100 g (minimum 22.6, maximum 66.3) indicating that *Calothamnus* spp. (Red Bell) honey and also *Eucalyptus marginata* (Jarrah) honey have higher TPC than Manuka honey, which is generally seen as a honey with high antioxidant activity [51]. By comparing the findings of this study with TPC data for other monofloral honeys across the globe, TPC values of 18.9 $\pm$ 3.82 to 23.7 $\pm$ 4.37 GAE/100 g [52] were reported for some Romanian monofloral honeys, while Mexican monofloral honeys had TPC range of 18.02 $\pm$ 0.49 to 102.77 $\pm$ 1.29 GAE/100 g [53], Czech and Slovak honeys had TPC between 54.0 $\pm$ 1.7 and 254.2 $\pm$ 1.4 GAE/100 g [54], and Brazilian honeys were reported to have TPC between 13.3 and 100 GAE/100 g [55]. The TPC values obtained in this study are lower in comparison. The assay used in this study was, however, a modified Folin–Ciocalteu assay in which the concentration of the sodium carbonate solution was optimised in such a way that sugar interference was muted, as sugars were also observed to react with the reagent leading to an overestimation of TPC without this modification [41].

Similar to the generated TPC values, the FRAP activity of *Calothamnus* spp. (Red Bell) honey was also observed to be higher compared to the other investigated Western Australian honeys [37] and also when compared to that of Manuka honey (2.88 to 10.72 mmol Fe^{2+}/kg) [43]. By comparing the FRAP activity with that of other monofloral honeys from across the globe, Bangladeshi monofloral honeys were reported to have FRAP activity of 1.00–8.00 mmol Fe^{2+}/kg [56], the FRAP activity of Oak honeydew honey from Croatia was reported to be 4.8 mmol Fe^{2+}/kg [57], Polish monofloral honeys were reported to have between 1.00 and 7.00 mmol Fe^{2+}/kg FRAP activity [58], and Thai monofloral honeys 0.61 to 4.34 mmol Fe^{2+}/kg [59]. Compared to these findings, the honeys from Western Australia investigated in this study showed a higher FRAP activity.

The DPPH radical scavenging activity of *Calothamnus* spp. (Red Bell) honey was also observed to be higher compared to the other investigated Western Australian honeys [37] and when compared to that of Manuka honey (mean = 1.98, range of 0.56 to 4.35 mmol TE/kg) [43]. Polish monofloral honeys were reported to have 0.20 to 1.20 mmol TE/kg DPPH activity [58], Oak honeydew honey from Croatia was reported to have a DPPH activity of 4.5 mmol TE/kg [57], and Thai monofloral honeys of 0.107 to 1.224 mmol TE/kg [59]. These values were lower compared to the DPPH radical scavenging activity of the investigated WA honeys.

High correlations between TPC values, FRAP and DPPH antioxidant activity were observed in this study, consistent with other reports [46–49]. DPPH and FRAP assays were chosen to express the total antioxidant activities of honey because the application of multiple assays can be helpful in reflecting the antioxidant properties of honeys more accurately than a single assay can do [60]. DPPH and FRAP assays have been widely used to determine the antioxidant activity of various plant extracts and food products since they use stable free radicals and the determination of antioxidant capacity is simple, quick and easy to perform, results are readily validated, accurate, and highly reproducible and the reagents are inexpensive and easy to prepare [61,62]. The TPC assay was employed to confirm that the antioxidant assay can be attributed to the phenolic compounds present in honey as it has been found that high antioxidant potential in FRAP and DPPH assays is usually observed for samples with high phenolic and flavonoid content [11].

By employing the HPTLC database to identify the phenolic constituents in the honeys, kojic acid (**14**) and epigallocatechin gallate (**5**) were found in all investigated honeys. m-Coumaric acid (**12**) was present in most honeys except *Calothamnus* spp. (Red Bell) honey. Lumichrome (**15**) was identified in *Agonis flexuosa* (Coastal Peppermint) and *Eucalyptus marginata* (Jarrah) honey, gallic acid (**8**) was found in both Red Bell and *Corymbia calophylla* (Marri) honey, taxifolin (**3**) was only found in Marri and Jarrah honey, while luteolin (**1**) and epicatechin (**4**) were only found in Coastal Peppermint and Marri honey. Hesperitin (**2**) was only identified in Jarrah honey, eudesmic acid (**7**) only in Marri honey, syringic acid (**11**) only in Coastal Peppermint honey, and protocatechuic acid (**10**) and t-cinnamic acid (**13**) only in Red Bell honey. Compounds that were only identified in a specific honey might in the future potentially be used as biomarkers for that honey.

The HPTLC-based database for phenolic compound identification was previously employed in the analysis of Manuka honey where kojic acid, gallic acid, epigallocatechin gallate, lumichrome, 2,3,4-trihydroxy benzoic acid, and o-anisic acid were also identified. However, leptosperine, mandelic acid, lepteridine, methyl syringate, salicylic acid, and benzoic acid were only found in Manuka honey [30]. This implies that there are some compounds that are ubiquitous in honeys while others are unique and can only be found in a certain honey. Since all the honeys investigated in this study along with Manuka honey belong to the plant family Myrtaceae, it can be speculated that this might explain some of the overlaps in the compounds identified in the four honey types.

To date, reports on the presence and concentration of phenolic compounds in honeys originating from Western Australia has been very scant. Prior to this study, only for Jarrah honey had some compounds been reported. Using HPLC-ESI-MS/MS analysis, Anand et al. in 2019 were able to quantify quercetin, hesperitin, cinnamic acid, methyl syringate, rutin, sinapic acid, ferulic acid, p-coumaric acid, phenyllactic acid, syringic acid, caffeic acid, vanillic acid, chlorogenic acid, p-hydroxybenzoic acid, protocatechuic acid, and gallic acid [39]. By employing the HPTLC-based database in this study, only hesperitin was identified from the compounds reported by Anand et al., which can be attributed to a number of reasons: Firstly, Anand et al. (2019) utilised a Strata-X cartridge solid phase extraction which was eluted with acidified water (pH-2), and then with methanol prior to their analysis [39]. This study, however, employed a liquid–liquid solvent extraction using dichloromethane and methanol (1:1) as a solvent system.

The solvent system used in the development of the HPTLC plates was toluene: ethyl acetate: formic acid 6:5:1 (MPB) which has frequently been used to fingerprint honeys [32,33,40,63,64]. A more polar solvent system, toluene: ethyl acetate: formic acid 2:8:1 (MPA), was also utilised in order to identify compounds of higher polarity. A better separation in the bands in honey was observed with MPB, however, it was found to be unable to fully develop all honey constituents as seen by dark bands on the baseline of the plate prior to derivatisation, and after derivatisation with NP-PEG and VSA reagents (Figures 2 and 5), as well as after derivatisation with DPPH reagent (Figure 9). It is a recommendation that another solvent system with higher polarity is also used in the future in order to identify those more polar compounds that were not fully captured by the solvent systems used in the current study.

This research utilised pooled honey samples, as the composition of such a pooled sample will be more representative of the typical chemical composition of the respective honey compared to the analysis of a randomly chosen single sample. Specifically, eight samples were pooled to represent the Red Bell honey used in this study, six samples each were pooled for Coastal Peppermint and Jarrah honey, and 13 samples were used to represent Marri honey.

An HPTLC-DPPH assay was previously employed in the qualitative and quantitative analysis of the antioxidant fingerprints of honeys [32–34]. Islam et al. in 2020 and 2021 utilised the method for the quantification of antioxidant band activities for various Australian honeys. However, the analysis was performed using dichloromethane as an extraction solvent, the incubation time was set to only 1 h, and peak profiles were obtained with white light [32,33]. In this study, however, a more polar extraction solvent was used (dichloromethane: methanol 1:1 v/v) which led to the observation of more antioxidant bands. The incubation time was also optimised as it was found that 1 h was not enough for some phenolic compounds to fully react with the DPPH reagent (Figure S8, Supplementary Materials). Longer incubation times of 2 h and 3 h were also tested and it was found that 2 h is the optimum time for the multiple types of polyphenols present in honey to react with the reagent but not long enough for the DPPH reagent to autodegrade.

The findings of the HPTLC-DPPH assay for honeys are often expressed in a qualitative manner by presenting active bands that showed a discoloration of the DPPH reagent [34]. In some instances, the antioxidant band activity was also quantified, expressed as mg GAE/100 g of honey [32,33]. Quantification of individual bands is, however, challenging given that some antioxidant bands are very low in absorbance and might thus be below the limit of detection of this quantification method.

In this study, the colours of the unreacted DPPH reagent and the colours of the active bands in the analysed samples, converted into hue values, were compared and from this, their DPPH radical scavenging activity was calculated. The HPTLC software usually provides colour information in the form of RGB values which can be converted into hue values (based on the hue, saturation, and brightness (HSB) colour space) [30]. The use of hues in expressing the colour of a particular band was found to be very helpful in the early stage of identification of an unknown sample using the HPTLC-derived database where, upon the use of a suitable derivatisation agent, a discrimination of one compound group from another based on colour was possible [30]. It was found in this study that colour captured in the form of hue values can also be used to express the results of the HPTLC-DPPH assay (Tables 6–8). Various antioxidant compounds were tested, and the findings demonstrated that hues varied according to the sample concentration that was applied. However, linear regression did not reach 0.99 indicating that the current parameters used in this study are only able to describe the antioxidant results in a semi-quantitative manner, expressed here in inferences ranging from + to +++. More optimisation is required in order to use the method for full quantification of the antioxidant activity of individual bands in an unknown sample.

The DPPH antioxidant activities of the individual constituents that were identified in each honey were also determined as a mixture at low concentration to mimic the concentrations that were quantified in honey and also at higher concentration (Table 7) to determine whether its activity is based on its concentration in the honey or inherent antioxidant activity of the constituent. All identified compounds except eudesmic acid (**7**), o-anisic acid (**9**), t-cinnamic acid (**13**), and lumichrome (**15**) showed activity towards the DPPH reagent. The inactive compounds (Table 7) lack a hydroxyl group in the phenolic ring that can react with the DPPH reagent [65,66] indicating that not all phenolic compounds are antioxidant.

The reaction of compounds with DPPH is governed by the reagent's steric accessibility indicating that smaller molecules have greater access to the radical site as compared to larger molecules [50]. This explains why flavonoids tend to react slower compared to smaller molecules like simple phenolic acids. The reactivity of flavonoids with DPPH on the other hand is dictated by the so-called Bors criteria (Figure 10).

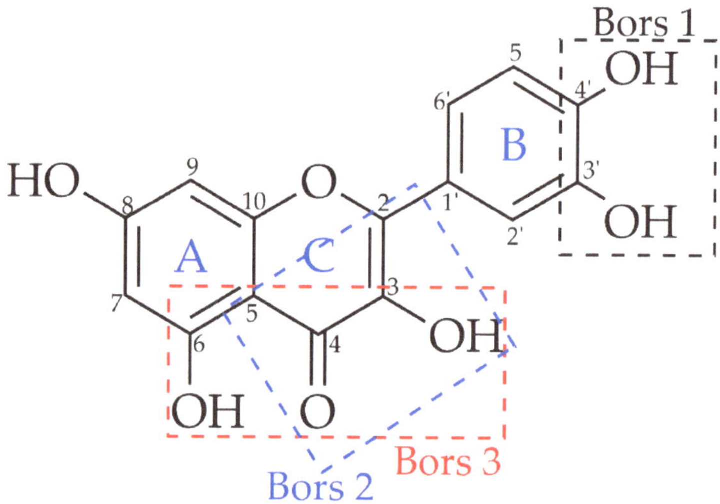

Figure 10. Bors Criteria to describe flavonoid activity (adapted from Platzer et al. [65]).

The first criterion is the presence of a catechol group on Ring B (Bors 1), which increases the stability of the resulting antioxidant radical. The second is the presence of a 2,3 double bond combined with a 4-oxo group on Ring C (Bors 2), which facilitates electron delocalization. The third is the presence of OH groups at positions 3 and 5 in combination with a 4-oxo group, which enables electron delocalization via hydrogen bonds (Bors 3) [65,66]. Among the flavonoids identified in this study, taxifolin (**3**) possesses Bors 1 and 2 criteria, confirmed by very high radical scavenging activity even when analysed at a lower concentration. Luteolin (**1**), epicatechin (**4**), and epigallocatechin gallate (**5**) all possess the Bors 1 criterion, while hesperetin (**2**) does not possess any, which explains why it has shown only a very weak radical scavenging activity. The trends seen in the antioxidant activity of the different phenolic compounds investigated in this study were consistent with the trends that were previously reported [65–67]. The HPTLC-DPPH assay has thus been demonstrated to be a very powerful tool in the identification of antioxidant constituents. However, DPPH or a similarly structured radical does not exist in a biological or food system [50] and it is therefore suggested that a more biochemically relevant antioxidant model should be used in future studies.

This is the first report on the use of band colours as a basis of expressing antioxidant activity in samples, which demonstrates that colour values derived from HPTLC analysis can also be used to (semi-quantitatively) express antioxidant activity in addition to more traditional quantification (using generated standard curves) that HPTLC can also perform. While this presents a novel analytical angle to HPTLC-DPPH analysis, some limitations need to be acknowledged. Given the very general nature of the DPPH assay and its common use in natural product research as a screening tool for antioxidant activity, a qualitative (i.e., active or inactive) or semi-quantitative (i.e., activity ranges from + to +++) approach might suffice in many instances. This can be achieved, as illustrated in this study, by expressing antioxidant activity of individual honey bands as % RSA, which is a widely accepted way of expressing antioxidant activity. However, should a fully quantitative result be the aim, more optimisation is needed, specifically to determine the concentration range of each match compound that yields linear regression equal to or greater than 0.99.

In recent years High Performance Thin Layer Chromatography (HPTLC) has emerged as a very versatile tool for various aspects of honey analysis. It can, for example, be used to identify and quantify various sugars in honey [68,69] and with this can also be used to identify post-harvest sugar adulterations [70]. It is also applied to identify and quantify the presence of hydroxymethyl furfural (HMF) in honey, which is a marker for excessive heat treatment-associated degradation and thus reduced honey quality [71–73]. HPTLC in combination with DPPH derivatisation has also been successfully used to visualise and quantify (as gallic acid equivalents) antioxidant honey constituents [32,33]. Moreover, the HPTLC analysis of organic honey extracts has been demonstrated to yield unique signatures that are reflective of a honey's floral origin and can thus be used for honey authentication [40,63,64]. This study contributes to the growing body of literature that demonstrates the versatility of HPTLC in the analysis of honey. The identity of some phenolic constituents in the four investigated Western Australian honeys was revealed using a HPTLC-based database along with their quantification, also using HPTLC. Moreover, compounds that contribute to these honeys' antioxidant activity could be identified and semi-quantified using a modification of the previously published HPTLC-DPPH analysis protocol.

5. Conclusions

This study investigated the antioxidant activity of four Western Australian honeys, *Calothamnus* spp. (Red Bell), *Agonis flexuosa* (Coastal Peppermint), *Corymbia calophylla* (Marri) and *Eucalyptus marginata* (Jarrah) honey. It was found that Red Bell honey has the highest total phenolic content, followed by Jarrah, Coastal Peppermint, and Marri honey. The same trends were observed for their respective FRAP and DPPH antioxidant activities.

t-Cinnamic acid, protocatechuic acid, gallic acid, epigallocatechin gallate, and kojic acid were identified and quantified in Red Bell honey. For Coastal Peppermint honey, the presence of syringic acid, m-coumaric acid, luteolin, epicatechin, lumichrome, and kojic acid was determined and quantified. Eudesmic acid, epicatechin, epigallocatechin gallate, luteolin, gallic acid, kojic acid, m-coumaric acid, and taxifolin were identified and quantified in Marri honey, and hesperitin, o-anisic acid, taxifolin, kojic acid, m-coumaric acid, lumichrome, epigallocatechin gallate, kojic acid, and 2,3,4-trihydroxy benzoic acid in Jarrah honey.

HPTLC-DPPH bioautography was also carried out to determine which honey constituents contribute to the respective honey's antioxidant activity using a novel method of analysis based on the changes of hues on reaction with the DPPH reagent. This change in hue was used to determine the % RSA of each active band. The method was able to identify the individual bands that contribute to the honeys' overall antioxidant activity. Based on the findings of this analysis, most identified compounds showed antioxidant activity except for t-cinnamic acid, lumichrome, o-anisic acid, and eudesmic acid due to the absence of hydroxyl groups in their benzene ring.

As most analyses were carried out using HPTLC, the study was also able to demonstrate the versatility of this instrumentation in the analysis of various aspects of honey chemistry and bioactivity.

Supplementary Materials: The following supporting information can be downloaded at: https://www.mdpi.com/article/10.3390/antiox12010189/s1, Table S1: Table S1 summarises the identity, botanical origin, and families of honeys collected as part of a study on Western Australia honeys; Table S2: Honey Sample Collection and Floral Information, TPC, FRAP, and DPPH Antioxidant Activity, Table S3: Summary of the data used to determine the identity of the unknown bands in *Calothamnus* spp. (Red Bell) honey (Database 1A), Table S4: Summary of the data used to determine the identity of the unknown bands in *Calothamnus* spp. (Red Bell) honey (Database 1B), Table S5: Summary of the data used to determine the identity of the unknown bands in *Calothamnus* spp. (Red Bell) honey (Database 2A), Table S6: Summary of the data used to determine the identity of the unknown bands in *Calothamnus* spp. (Red Bell) honey (Database 2B), Table S7: Summary of the data used to determine the identity of the unknown bands in *Agonis flexuosa* (Coastal Peppermint) honey (Database 1A), Table S8: Summary of the data used to determine the identity of the unknown bands in *Agonis flexuosa* (Coastal Peppermint) honey (Database 1B), Table S9: Summary of the data used to determine the identity of the unknown bands in *Agonis flexuosa* (Coastal Peppermint) honey (Database 2A), Table S10: Summary of the data used to determine the identity of the unknown bands in *Agonis flexuosa* (Coastal Peppermint) honey (Database 2B), Table S11: Summary of the data used to determine the identity of the unknown bands in *Corymbia calophylla* (Marri) honey (Database 1A), Table S12: Summary of the data used to determine the identity of the unknown bands in *Corymbia calophylla* (Marri) honey (Database 1B), Table S13: Summary of the data used to determine the identity of the unknown bands in *Corymbia calophylla* (Marri) honey (Database 2A), Table S14: Summary of the data used to determine the identity of the unknown bands in *Corymbia calophylla* (Marri) honey (Database 2B), Table S15: Summary of the data used to determine the identity of the unknown bands in *Eucalyptus marginata* (Jarrah) honey (Database 1A), Table S16: Summary of the data used to determine the identity of the unknown bands in *Eucalyptus marginata* (Jarrah) honey (Database 1B), Table S17: Summary of the data used to determine the identity of the unknown bands in *Eucalyptus marginata* (Jarrah) honey (Database 2A), Table S18: Summary of the data used to determine the identity of the unknown bands in *Eucalyptus marginata* (Jarrah) honey (Database 2B); Table S19: Parameters used in optimising the quantification of phenolic compounds in honey, Figure S1. HPTLC fingerprint patterns for various samples of *Calothamnus* spp. (Red bell, n = 8), Figure S2. HPTLC fingerprint patterns for various samples of *Agonis flexuosa* (Coastal Peppermint, n = 5), Figure S3. HPTLC fingerprint patterns for various samples of *Corymbia calophylla* (Marri, n = 13), Figure S4. HPTLC fingerprint patterns for various samples of *Eucalyptus marginata* (Jarrah, n = 6), Figure S5A–C. Profile comparison of *Calothamnus* spp. (Red bell) honey (green) and *Calothamnus* spp. (Red bell) honey spiked with the identified compounds based on database 1A and 1B (blue) scanned at the λmax of each specific compounds prior to derivatization, Figure S6A–E. Profile comparison of *Agonis flexuosa* (Coastal Peppermint) Honey (green) and *Agonis flexuosa* (Coastal Peppermint) honey spiked with the identified compounds based on database 1A and 1B (blue) scanned at the λmax of each specific compounds prior to derivatization, Figure S7A–F. Profile comparison of *Agonis flexuosa* (Coastal Peppermint) honey (green) and *Agonis flexuosa* (Coastal Peppermint) honey spiked with the identified compounds based on database 2A and2B (blue) scanned at the λmax of each specific compounds prior to derivatization, Figure S8A–E. Profile comparison of *Corymbia calophylla* (Marri) honey (green) and *Corymbia calophylla* (Marri) spiked with the identified compounds based on database 1A and 1B (blue) scanned at the λmax of each specific compounds prior to derivatization, Figure S9A–G. Profile comparison of *Corymbia calophylla* (Marri) honey (green) and *Corymbia calophylla* (Marri) spiked with the identified compounds based on database 2A and 2B (blue) scanned at the λmax of each specific compounds prior to derivatization, Figure S10A–D. Profile comparison of *Eucalyptus marginata* (Jarrah) honey (green) and *Eucalyptus marginata* (Jarrah) spiked with the identified compounds based on database 1A and 1B (blue) scanned at the λmax of each specific compounds prior to derivatization, Figure S11A–G. Profile comparison of *Eucalyptus marginata* (Jarrah) honey (green) and *Eucalyptus marginata* (Jarrah) spiked with the identified compounds based on database 2A and 2B (blue) scanned at the λmax of each specific compounds prior to derivatization, Figure S12A,B. Comparison of the profiles of compounds identified in *Corymbia calophylla* (Marri) honey after derivatised with DPPH reagent and obtained after 1 h with transmittance in white light (green) vs. scanned at 517 nm (A) and comparison of the profiles of compounds identified in *Corymbia calophylla* (Marri) honey after being derivatised with DPPH reagent scanned at 517 nm and taken at 1 h (green), 2 h (blue), 3 h (grey) (B) developed using mobile phase 1B, Figure S13A–D. HPTLC plate image (a) of *Agonis flexuosa* (Coastal Peppermint) honey after derivatised with DPPH reagent and developed using mobile phase A (A,B) and developed using mobile phase B (C,D) obtained with transmission in white light, and comparison of the profiles of *Agonis flexuosa* (Coastal Peppermint) honey (green) and *Agonis flexuosa* (Coastal Peppermint) honey spiked with the identified compounds (blue) after being derivatised with DPPH reagent obtained at 517 nm (b-left) and comparison of the profiles of *Agonis flexuosa* (Coastal Peppermint) honey obtained at 254 nm (green) and 366 nm (blue) prior to derivatisation and the profile of *Agonis flexuosa* (Coastal Peppermint) honey spiked with the identified compounds (grey) obtained at 277 nm prior to derivatisation (b right), Figure S14A–C. HPTLC plate image (a) of *Corymbia calophylla* (Marri) honey after being derivatised with DPPH reagent and developed using mobile phase A (A) and developed using mobile phase B (B,C) obtained with transmission in white light, and comparison of the profiles of *Corymbia calophylla* (Marri) honey (green) and *Corymbia calophylla* (Marri) honey spiked with the identified compounds (blue) after being derivatised with DPPH reagent obtained at 517 nm (b-left) and comparison of the profiles of *Corymbia calophylla* (Marri) honey obtained at 254 nm (green) and 366 nm (blue) prior to derivatisation and the profile of *Corymbia calophylla* (Marri) honey spiked with the identified compounds (grey) obtained at 277 nm prior to derivatisation (b right), Figure S15A–D. HPTLC plate image (a) of *Eucalyptus marginata* (Jarrah) honey after being derivatised with DPPH reagent and developed using mobile phase A (A and B) and developed using mobile phase B (C,D) obtained with transmission in white light, and comparison of the profiles of *Eucalyptus marginata* (Jarrah) honey (green) and *Eucalyptus marginata* (Jarrah) honey spiked with the identified compounds (blue) after being derivatised with DPPH reagent obtained at 517 nm (b-left) and comparison of the profiles of *Eucalyptus marginata* (Jarrah) honey obtained at 254 nm (green) and 366 nm (blue) prior to derivatisation and the profile of *Eucalyptus marginata* (Jarrah) honey spiked with the identified compounds (grey) obtained at 277 nm prior to derivatisation (b right).

Author Contributions: Conceptualisation, I.L.L. and C.L.; methodology, I.L.L., M.K.I., T.S. and C.L.; formal analysis, I.L.L.; software, I.L.L., M.K.I. and T.S.; writing—original draft preparation, I.L.L.; writing—review and editing, C.L. and L.Y.L.; supervision, C.L., L.Y.L. and K.H.; project administration, C.L.; funding acquisition, C.L. All authors have read and agreed to the published version of the manuscript.

Funding: This research was funded by the Cooperative Research Centre for Honey Bee Products (CRC HBP) Project 33.

Institutional Review Board Statement: Not applicable.

Informed Consent Statement: Not applicable.

Data Availability Statement: No new data were created or analysed in this study. Data sharing is not applicable to this article.

Acknowledgments: The authors would like to thank Ranee Joshi of the Centre for Exploration Targeting, School of Earth Sciences, University of Western Australia, for her technical support and supervision in generating the map used for this paper.

Conflicts of Interest: The authors declare no conflict of interest.

Sample Availability: Samples of the compounds and honeys are available from the authors.

References

1. Nguyen, H.T.L.; Panyoyai, N.; Kasapis, S.; Pang, E.; Mantri, N. Honey and Its Role in Relieving Multiple Facets of Atherosclerosis. *Nutrients* **2019**, *11*, 167. [CrossRef] [PubMed]
2. Borycka, K.; Grabek-Lejko, D.; Kasprzyk, I. Antioxidant and antibacterial properties of commercial bee pollen products. *J. Apic. Res.* **2016**, *54*, 491–502. [CrossRef]
3. Ozkok, A. Determination of antioxidant capacities, chemical composition and melissopalynological characterization of *Verbascum* spp. and *Euphorbia* spp. honeys. *Fresenius Environ. Bull.* **2019**, *28*, 2644–2649.
4. Ceksteryte, V.; Kurtinaitiene, B.; Venskutonis, P.R.; Pukalskas, A.; Kazernaviciute, R.; Balzekas, J. Evaluation of antioxidant activity and flavonoid composition in differently preserved bee products. *Czech J. Food Sci.* **2016**, *34*, 133–142. [CrossRef]

5. Nayik, G.A.; Nanda, V. A chemometric approach to evaluate the phenolic compounds, antioxidant activity and mineral content of different unifloral honey types from Kashmir, India. *LWT—Food Sci. Technol.* **2016**, *74*, 504–513. [CrossRef]
6. Cinkmanis, I.; Dimins, F.; Mikelsone, V. Influence of Lyophilization and convective type drying on antioxidant properties, total phenols and flavonoids in pollens. In Proceedings of the 11th Baltic Conference on Food Science and Technology "Food Science and Technology in a Changing World", Jelgava, Latvia, 27–28 April 2017; pp. 201–203.
7. Ozkan, K.; Sagcan, N.; Ozulku, G.; Sagdic, O.; Toker, O.S.; Muz, M.N. Bioactive and bioaccessibility characteristics of honeybee pollens collected from different regions of Turkey. *J. Food Meas. Charact.* **2017**, *12*, 581–587. [CrossRef]
8. Bakour, M.; Fernandes, Â.; Barros, L.; Sokovic, M.; Ferreira, I.C.F.R.; Lyoussi, B. Bee bread as a functional product: Chemical composition and bioactive properties. *LWT* **2019**, *109*, 276–282. [CrossRef]
9. Yang, Y.; Zhang, J.-l.; Zhou, Q.; Wang, L.; Huang, W.; Wang, R.-d. Effect of ultrasonic and ball-milling treatment on cell wall, nutrients, and antioxidant capacity of rose (*Rosa rugosa*) bee pollen, and identification of bioactive components. *J. Sci. Food Agric.* **2019**, *99*, 5350–5357. [CrossRef]
10. Gorjanovic, S.Z.; Alvarez-Suarez, J.M.; Novakovic, M.M.; Pastor, F.T.; Pezo, L.; Battino, M.; Suznjevic, D.Z. Comparative analysis of antioxidant activity of honey of different floral sources using recently developed polarographic and various spectrophotometric assays. *J. Food Compos. Anal.* **2013**, *30*, 13–18. [CrossRef]
11. Khalil, M.I.; Moniruzzaman, M.; Boukraâ, L.; Benhanifia, M.; Islam, M.A.; Islam, M.N.; Sulaiman, S.A.; Gan, S.H. Physicochemical and antioxidant properties of algerian honey. *Molecules* **2012**, *17*, 11199–11215. [CrossRef]
12. Sant'ana, L.D.O.; Buarque Ferreira, A.B.; Lorenzon, M.C.A.; Berbara, R.L.L.; Castro, R.N. Correlation of Total Phenolic and Flavonoid Contents of Brazilian Honeys with Colour and Antioxidant Capacity. *Int. J. Food Prop.* **2014**, *17*, 65–76. [CrossRef]
13. Stanek, N.; Teper, D.; Kafarski, P.; Jasicka-Misiak, I. Authentication of phacelia honeys (*Phacelia tanacetifolia*) based on a combination of HPLC and HPTLC analyses as well as spectrophotometric measurements. *LWT—Food Sci. Technol.* **2019**, *107*, 199–207. [CrossRef]
14. Michalkiewicz, A.; Biesaga, M.; Pyrzynska, K. Solid-phase extraction procedure for determination of phenolic acids and some flavonols in honey. *J. Chromatogr. A* **2008**, *1187*, 18–24. [CrossRef] [PubMed]
15. Gheldof, N.; Wang, X.-H.; Engeseth, N.J. Identification and Quantification of Antioxidant Components of Honeys from Various Floral Sources. *J. Agric. Food Chem.* **2002**, *50*, 5870–5877. [CrossRef] [PubMed]
16. Zhao, L.; Ren, C.; Xue, X.; Lu, H.; Wang, K.; Wu, L. Safflomin A: A novel chemical marker for *Carthamus tinctorius* L. (Safflower) monofloral honey. *Food Chem.* **2022**, *366*, 130584. [CrossRef] [PubMed]
17. Demir Kanbur, E.; Yuksek, T.; Atamov, V.; Ozcelik, A.E. A comparison of the physicochemical properties of chestnut and highland honey: The case of Senoz Valley in the Rize province of Turkey. *Food Chem.* **2021**, *345*, 128864. [CrossRef] [PubMed]
18. Das, A.; Datta, S.; Mukherjee, S.; Bose, S.; Ghosh, S.; Dhar, P. Evaluation of antioxidative, antibacterial and probiotic growth stimulatory activities of Sesamum indicum honey containing phenolic compounds and lignans. *LWT—Food Sci. Technol.* **2015**, *61*, 244–250. [CrossRef]
19. Badjah Hadj Ahmed, A.Y.; Wabaidur, S.M.; Siddiqui, M.R.; Alothman, Z.A.; Obeid, M.S.; Khan, M.R.; Al-Tamrah, S.A. Simultaneous determination of twenty-five polyphenols in multifloral and cactus honeys using solid-phase extraction and high-performance liquid chromatography with photodiode array detection. *Eur. Food Res. Technol.* **2016**, *242*, 943–952. [CrossRef]
20. Martos, I.; Ferreres, F.; Tomás-Barberán, F.A. Identification of flavonoid markers for the botanical origin of Eucalyptus honey. *J. Agric. Food Chem.* **2000**, *48*, 1498–1502. [CrossRef]
21. Tomás-Barberán, F.A.; Martos, I.; Ferreres, F.; Radovic, B.S.; Anklam, E. HPLC flavonoid profiles as markers for the botanical origin of European unifloral honeys. *J. Sci. Food Agric.* **2001**, *81*, 485–496. [CrossRef]
22. Yao, L.; Datta, N.; Tomás-Barberán, F.A.; Ferreres, F.; Martos, I.; Singanusong, R. Flavonoids, phenolic acids and abscisic acid in Australian and New Zealand *Leptospermum* honeys. *Food Chem.* **2003**, *81*, 159–168. [CrossRef]
23. Ferreres, F.; García-Viguera, C.; Tomás-Lorente, F.; Tomás-Barberán, F.A. Hesperetin: A marker of the floral origin of citrus honey. *J. Sci. Food Agric.* **1993**, *61*, 121–123. [CrossRef]
24. Escriche, I.; Kadar, M.; Juan-Borrás, M.; Domenech, E. Using flavonoids, phenolic compounds and headspace volatile profile for botanical authentication of lemon and orange honeys. *Food Res. Int.* **2011**, *44*, 1504–1513. [CrossRef]
25. Andrade, P.; Ferreres, F.; Amaral, M.T. Analysis of Honey Phenolic Acids by HPLC, Its Application to Honey Botanical Characterization. *J. Liq. Chromatogr. Relat. Technol.* **1997**, *20*, 2281–2288. [CrossRef]
26. Pyrzynska, K.; Biesaga, M. Analysis of phenolic acids and flavonoids in honey. *TrAC Trends Anal. Chem.* **2009**, *28*, 893–902. [CrossRef]
27. Ciulu, M.; Spano, N.; Pilo, M.I.; Sanna, G. Recent Advances in the Analysis of Phenolic Compounds in Unifloral Honeys. *Molecules* **2016**, *21*, 451. [CrossRef]
28. Kečkeš, S.; Gašić, U.; Veličković, T.Ć.; Milojković-Opsenica, D.; Natić, M.; Tešić, Ž. The determination of phenolic profiles of Serbian unifloral honeys using ultra-high-performance liquid chromatography/high resolution accurate mass spectrometry. *Food Chem.* **2013**, *138*, 32–40. [CrossRef]
29. Lawag, I.L.; Lim, L.-Y.; Joshi, R.; Hammer, K.A.; Locher, C. A Comprehensive Survey of Phenolic Constituents Reported in Monofloral Honeys around the Globe. *Foods* **2022**, *11*, 1152. [CrossRef]
30. Lawag, I.L.; Sostaric, T.; Lim, L.Y.; Hammer, K.; Locher, C. The Development and Application of a HPTLC-Derived Database for the Identification of Phenolics in Honey. *Molecules* **2022**, *27*, 6651. [CrossRef]

31. Roussel, J.M.; Bardot, V.; Berthomier, L.; Cotte, C.; Dubourdeaux, M.; Holowacz, S.; Bernard-Savary, P. Application of the Life Cycle Management of Analytical methods concept to a HPTLC-DPPH assay method for acteoside content in industrial extracts of *Plantago lanceolata* L. *J. Chromatogr. B* **2021**, *1181*, 122923. [CrossRef]
32. Islam, M.K.; Sostaric, T.; Lim, L.Y.; Hammer, K.; Locher, C. Development and validation of an HPTLC–DPPH assay and its application to the analysis of honey. *JPC—J. Planar Chromatogr.—Mod. TLC* **2020**, *33*, 301–311. [CrossRef]
33. Islam, M.K.; Sostaric, T.; Lim, L.Y.; Hammer, K.; Locher, C. Antioxidant HPTLC-DPPH Fingerprinting of Honeys and Tracking of Antioxidant Constituents upon Thermal Exposure. *Foods* **2021**, *10*, 357. [CrossRef] [PubMed]
34. Morlock, G.E.; Belay, A.; Heil, J.; Mehl, A.; Borck, H. Effect-Directed Profiling of Monofloral Honeys from Ethiopia by High-Performance Thin-Layer Chromatography and High-Resolution Mass Spectrometry. *Molecules* **2022**, *27*, 3541. [CrossRef] [PubMed]
35. Smith, F.G. *Honey Plants in Western Australia*; Bulletin 3618; Department of Primary Industries and Regional Development: Perth, Australia, 1969.
36. Lin, B.; Loomes, K.M.; Prijic, G.; Schlothauer, R.; Stephens, J.M. Lepteridine as a unique fluorescent marker for the authentication of manuka honey. *Food Chem.* **2017**, *225*, 175–180. [CrossRef] [PubMed]
37. Green, K.J.; Islam, M.K.; Lawag, I.; Locher, C.; Hammer, K.A. Honeys derived from plants of the coastal sandplains of Western Australia: Antibacterial and antioxidant activity, and other characteristics. *J. Apic. Res.* **2022**, 1–14. [CrossRef]
38. Islam, M.K.; Lawag, I.L.; Green, K.J.; Sostaric, T.; Hammer, K.A.; Lim, L.Y.; Locher, C. An investigation of the suitability of melissopalynology to authenticate Jarrah honey. *Curr. Res. Food Sci.* **2022**, *5*, 506–514. [CrossRef]
39. Anand, S.; Deighton, M.; Livanos, G.; Morrison, P.D.; Pang, E.C.K.; Mantri, N. Antimicrobial Activity of Agastache Honey and Characterization of Its Bioactive Compounds in Comparison with Important Commercial Honeys. *Front. Microbiol.* **2019**, *10*, 263. [CrossRef]
40. Islam, M.K.; Sostaric, T.; Lim, L.Y.; Hammer, K.; Locher, C. Development of an HPTLC-based dynamic reference standard for the analysis of complex natural products using Jarrah honey as test sample. *PLoS ONE* **2021**, *16*, e0254857. [CrossRef]
41. Lawag, I.L.; Yoo, O.; Lim, L.Y.; Hammer, K.; Locher, C. Optimisation of Bee Pollen Extraction to Maximise Extractable Antioxidant Constituents. *Antioxidants* **2021**, *10*, 1113. [CrossRef]
42. Liberato Mda, C.; de Morais, S.M.; Siqueira, S.M.; de Menezes, J.E.; Ramos, D.N.; Machado, L.K.; Magalhaes, I.L. Phenolic content and antioxidant and antiacetylcholinesterase properties of honeys from different floral origins. *J. Med. Food* **2011**, *14*, 658–663. [CrossRef]
43. Green, K.J.; Lawag, I.L.; Locher, C.; Hammer, K.A. Correlation of the antibacterial activity of commercial manuka and Leptospermum honeys from Australia and New Zealand with methylglyoxal content and other physicochemical characteristics. *PLoS ONE* **2022**, *17*, e0272376. [CrossRef]
44. Almeida, A.M.M.d.; Oliveira, M.B.S.; Costa, J.G.d.; Valentim, I.B.; Goulart, M.O.F. Antioxidant Capacity, Physicochemical and Floral Characterization of Honeys from the Northeast of Brazil. *Rev. Virtual Quím.* **2016**, *8*, 57–77. [CrossRef]
45. Karabagias, I.K.; Louppis, A.P.; Karabournioti, S.; Kontakos, S.; Papastephanou, C.; Kontominas, M.G. Characterization and geographical discrimination of commercial Citrus spp. honeys produced in different Mediterranean countries based on minerals, volatile compounds and physicochemical parameters, using chemometrics. *Food Chem.* **2017**, *217*, 445–455. [CrossRef] [PubMed]
46. Gül, A.; Pehlivan, T. Antioxidant activities of some monofloral honey types produced across Turkey. *Saudi J. Biol. Sci.* **2018**, *25*, 1056–1065. [CrossRef]
47. Lewoyehu, M.; Amare, M. Comparative evaluation of analytical methods for determining the antioxidant activities of honey: A review. *Cogent Food Agric.* **2019**, *5*, 1685059. [CrossRef]
48. Akgün, N.; Çelik, Ö.F.; Kelebekli, L. Physicochemical properties, total phenolic content, and antioxidant activity of chestnut, rhododendron, acacia and multifloral honey. *J. Food Meas. Charact.* **2021**, *15*, 3501–3508. [CrossRef]
49. Cianciosi, D.; Forbes-Hernández, T.Y.; Giampieri, F.; Zhang, J.; Ansary, J.; Pacetti, M.; Quiles, J.L.; Simal-Gandara, J.; Battino, M. Effect of In vitro Gastrointestinal Digestion on the Bioaccessibility of Phenolic Compounds and Antioxidant Activity of Manuka Honey. *eFood* **2020**, *1*, 85–93. [CrossRef]
50. Gülçin, I. Antioxidants and antioxidant methods: An updated overview. *Arch. Toxicol.* **2020**, *94*, 651–715. [CrossRef]
51. Hunter, M.; Ghildyal, R.; D'Cunha, N.M.; Gouws, C.; Georgousopoulou, E.N.; Naumovski, N. The bioactive, antioxidant, antibacterial, and physicochemical properties of a range of commercially available Australian honeys. *Curr. Res. Food Sci.* **2021**, *4*, 532–542. [CrossRef]
52. Pauliuc, D.; Dranca, F.; Oroian, M. Antioxidant Activity, Total Phenolic Content, Individual Phenolics and Physicochemical Parameters Suitability for Romanian Honey Authentication. *Foods* **2020**, *9*, 306. [CrossRef]
53. Hernández-Fuentes, A.; Chávez-Borges, D.; Cenobio-Galindo, A.J.; Zepeda-Velázquez, A.; Figueira, A.; Jiménez-Alvarado, R.; Campos-Montiel, R. Characterization of total phenol and flavonoid contents, colour, functional properties from honey samples with different floral origins. *Int. J. Food Stud.* **2021**, *10*, 346–358. [CrossRef]
54. Škrovánková, S.; Snopek, L.; Mlček, J.; Volaříková, E. Bioactive compounds evaluation in different types of Czech and Slovak honeys. *Potravin. Slovak J. Food Sci.* **2019**, *13*, 94–99. [CrossRef]
55. Nascimento, K.S.D.; Gasparotto Sattler, J.A.; Lauer Macedo, L.F.; Serna González, C.V.; Pereira de Melo, I.L.; da Silva Araújo, E.; Granato, D.; Sattler, A.; de Almeida-Muradian, L.B. Phenolic compounds, antioxidant capacity and physicochemical properties of Brazilian *Apis mellifera* honeys. *LWT—Food Sci. Technol.* **2018**, *91*, 85–94. [CrossRef]

56. Moniruzzaman, M.; Yung An, C.; Azlan, S.A.B.M.; Sulaiman, S.A.; Rao, P.V.; Hawlader, M.N.I.; Gan, S.H. Identification of phenolic acids and flavonoids in monofloral honey from Bangladesh by high performance liquid chromatography: Determination of antioxidant capacity. *BioMed Res. Int.* **2014**, *2014*, 737490. [CrossRef]
57. Jerkovic, I.; Marijanović, Z. Oak (*Quercus frainetto* Ten.) Honeydew Honey—Approach to Screening of Volatile Organic Composition and Antioxidant Capacity (DPPH and FRAP Assay). *Molecules* **2010**, *15*, 3744–3756. [CrossRef]
58. Kus, P.M.; Szweda, P.; Jerkovic, I.; Tuberoso, C.I. Activity of Polish unifloral honeys against pathogenic bacteria and its correlation with colour, phenolic content, antioxidant capacity and other parameters. *Lett. Appl. Microbiol.* **2016**, *62*, 269–276. [CrossRef]
59. Bundit, T.; Anothai, T.; Pattaramart, P.; Roongpet, T.; Chuleeporn, S. Comparison of antioxidant contents of Thai Honeys to Manuka Honey. *Malays. J. Nutr.* **2016**, *22*, 413–420.
60. Moniruzzaman, M.; Khalil, M.I.; Sulaiman, S.A.; Gan, S.H. Advances in the analytical methods for determining the antioxidant properties of honey: A review. *Afr. J. Tradit. Complement. Altern. Med.* **2012**, *9*, 36–42. [CrossRef]
61. Kedare, S.B.; Singh, R.P. Genesis and development of DPPH method of antioxidant assay. *J. Food Sci. Technol.* **2011**, *48*, 412–422. [CrossRef]
62. Hayes, W.A.; Mills, D.S.; Neville, R.F.; Kiddie, J.; Collins, L.M. Determination of the molar extinction coefficient for the ferric reducing/antioxidant power assay. *Anal. Biochem.* **2011**, *416*, 202–205. [CrossRef]
63. Locher, C.; Tang, E.; Neumann, J.; Sostaric, T. High-performance thin-layer chromatography profiling of Jarrah and Manuka honeys. *JPC—J. Planar Chromatogr.—Mod. TLC* **2018**, *31*, 181–189. [CrossRef]
64. Locher, C.; Neumann, J.; Sostaric, T. Authentication of honeys of different floral origins via high-performance thin-layer chromatographic fingerprinting. *J. Planar Chromatogr.—Mod. TLC* **2017**, *30*, 57–62. [CrossRef]
65. Platzer, M.; Kiese, S.; Herfellner, T.; Schweiggert-Weisz, U.; Eisner, P. How Does the Phenol Structure Influence the Results of the Folin-Ciocalteu Assay? *Antioxidants* **2021**, *10*, 811. [CrossRef] [PubMed]
66. Platzer, M.; Kiese, S.; Tybussek, T.; Herfellner, T.; Schneider, F.; Schweiggert-Weisz, U.; Eisner, P. Radical Scavenging Mechanisms of Phenolic Compounds: A Quantitative Structure-Property Relationship (QSPR) Study. *Front. Nutr.* **2022**, *9*, 882458. [CrossRef]
67. Mathew, S.; Abraham, T.E.; Zakaria, Z.A. Reactivity of phenolic compounds towards free radicals under in vitro conditions. *J. Food Sci. Technol.* **2015**, *52*, 5790–5798. [CrossRef] [PubMed]
68. Islam, M.K.; Sostaric, T.; Lim, L.Y.; Hammer, K.; Locher, C. Sugar Profiling of Honeys for Authentication and Detection of Adulterants Using High-Performance Thin Layer Chromatography. *Molecules* **2020**, *25*, 5289. [CrossRef]
69. Islam, M.K.; Sostaric, T.; Lim, L.Y.; Hammer, K.; Locher, C. A validated method for the quantitative determination of sugars in honey using high-performance thin-layer chromatography. *JPC—J. Planar Chromatogr.—Mod. TLC* **2020**, *33*, 489–499. [CrossRef]
70. Islam, M.K.; Vinsen, K.; Sostaric, T.; Lim, L.Y.; Locher, C. Detection of syrup adulterants in manuka and jarrah honey using HPTLC-multivariate data analysis. *PeerJ* **2021**, *9*, e12186. [CrossRef]
71. Shapla, U.M.; Solayman, M.; Alam, N.; Khalil, M.I.; Gan, S.H. 5-Hydroxymethylfurfural (HMF) levels in honey and other food products: Effects on bees and human health. *Chem. Cent. J.* **2018**, *12*, 35. [CrossRef]
72. Gregorc, A.; Jurišić, S.; Sampson, B. Hydroxymethylfurfural Affects Caged Honey Bees (*Apis mellifera carnica*). *Diversity* **2020**, *12*, 18. [CrossRef]
73. Biluca, F.C.; Della Betta, F.; De Oliveira, G.P.; Pereira, L.M.; Gonzaga, L.V.; Costa, A.C.O.; Fett, R. 5-HMF and carbohydrates content in stingless bee honey by CE before and after thermal treatment. *Food Chem.* **2014**, *159*, 244–249. [CrossRef] [PubMed]

antioxidants

Article

The Antioxidant and Prebiotic Activities of Mixtures Honey/Biomimetic NaDES and Polyphenols Show Differences between Honeysuckle and Raspberry Extracts

Luminița Dimitriu [1,2], Diana Constantinescu-Aruxandei [1,*], Daniel Preda [1,3], Ionuț Moraru [4], Narcisa Elena Băbeanu [2] and Florin Oancea [1,2,*]

1 Bioproducts Team, Bioresources Department, National Institute for Research & Development in Chemistry and Petrochemistry—ICECHIM, Splaiul Independent, ei No. 202, Sector 6, 060021 Bucharest, Romania; luminita.dimitriu@icechim.ro (L.D.); daniel.preda@stud.chimie.upb.ro (D.P.)
2 Faculty of Biotechnologies, University of Agronomic Sciences and Veterinary Medicine of Bucharest, Marasti Blv., No. 59, Sector 1, 011464 Bucharest, Romania; narcisa.babeanu@biotehnologii.usamv.ro
3 Department of Analytical Chemistry and Environmental Engineering, Faculty of Chemical Engineering and Biotechnologies, University Politehnica Bucharest, Str. Gheorghe Polizu nr. 1-7, Sector 1, 011061 Bucharest, Romania
4 Medica Laboratories, Str. Frasinului nr. 11, 075100 Otopeni, Romania; ionut.moraru@pro-natura.ro
* Correspondence: diana.constantinescu@icechim.ro (D.C.-A.); florin.oancea@icechim.ro (F.O.); Tel.: +40-21-316-3071 (D.C.-A. & F.O.)

Abstract: In our previous research, we demonstrated that honey and its biomimetic natural deep eutectic solvent (NaDES) modulate the antioxidant activity (AOA) of the raspberry extract (RE). In this study, we evaluated the AOA behaviour of the mixture honey/NaDES–honeysuckle (*Lonicera caprifolium*, LFL) extract and compared it with the mixture honey/NaDES–RE. These two extracts have similar major flavonoids and hydroxycinnamic acid compounds but differ in their total content and the presence of anthocyanins in RE. Therefore, it was of interest to see if the modulation of the LFL polyphenols by honey/NaDES was similar to that of RE. We also evaluated the prebiotic activity of these mixtures and individual components on *Limosilactobacillus reuteri* DSM 20016. Although honey/NaDES modulated the AOA of both extracts, from synergism to antagonism, the modulation was different between the two extracts for some AOA activities. Honey/NaDES mixtures enriched with LFL and RE did not show significant differences in bacterial growth stimulation. However, at a concentration of 45 mg/mL, the honey -LFL mixture exhibited a higher effect compared to the honey–RE mixture. The antioxidant and prebiotic properties of mixtures between honey and polyphenol-rich extracts are determined by multiple interactions in complex chemical systems.

Keywords: honey; biomimetic natural deep eutectic solvents; *Lonicera caprifolium*; lactic acid production; *Limosilactobacillus reuteri*

check for updates

Citation: Dimitriu, L.; Constantinescu-Aruxandei, D.; Preda, D.; Moraru, I.; Băbeanu, N.E.; Oancea, F. The Antioxidant and Prebiotic Activities of Mixtures Honey/Biomimetic NaDES and Polyphenols Show Differences between Honeysuckle and Raspberry Extracts. *Antioxidants* **2023**, *12*, 1678. https://doi.org/10.3390/antiox12091678

Academic Editor: Volker Böhm

Received: 15 July 2023
Revised: 17 August 2023
Accepted: 25 August 2023
Published: 28 August 2023

1. Introduction

Honey, being a natural product, displays a diverse range of biological activities such as antioxidant, antimicrobial, anti-inflammatory, cytoprotective, prebiotic, and postbiotic [1]. These diverse ranges of biological activities arise from the complex composition of honey, which includes carbohydrates, amino acids, phenolic compounds, minerals, enzymes, and electrolytes [2]. Furthermore, honey exhibits characteristics similar to those of a natural deep eutectic solvent (NaDES) due to the intermolecular interactions between its monosaccharides and disaccharides, as well as the hydrogen bonds formed between them [3–5]. Initially, NaDESs were introduced in green chemistry as a viable and eco-friendly alternative to conventional organic solvents. Their distinct properties, including bioavailability, biodegradability, and cost-effectiveness, captured the attention and motivated researchers to assess their potential applicability in the food sector for creating innovative functional food products. Moreover, their relatively more challenging removal after extraction has contributed to their investigation as functional ingredients in the food industry [6–8]. Additionally, other characteristics such as water activity, pH, antimicrobial activity, and enzyme interactions play an essential role in the storage and stabilization of food compounds when utilizing NaDES in food applications [9,10]. Honey possesses notable prebiotic properties, making it beneficial for the growth and activity of beneficial gut bacteria [11–13]. Prebiotics are non-digestible substances that selectively promote the growth and activity of beneficial microorganisms in the gastrointestinal tract. The prebiotic effects of honey are primarily attributed to its carbohydrate composition, especially oligosaccharides [14].

The antioxidant activity of honey is attributed to its rich content of phenolic acids, flavonoids, and other phenolic compounds. The antioxidant activity of honey helps protect cells from oxidative damage by neutralizing harmful free radicals [15–17]. On the other hand, the prebiotic activity of honey refers to its ability to selectively promote the growth and activity of beneficial gut bacteria. There appears to be an interplay between the antioxidant and prebiotic activities of honey. The presence of healthy gut microbiota is essential for the effective absorption and utilization of dietary antioxidants. Beneficial gut bacteria can metabolize certain components of honey, releasing bioactive compounds that contribute to its antioxidant potential. In turn, the antioxidants present in honey can help protect the gut microbiota from oxidative stress, maintaining a balanced microbial community [14,18]. In this case, a good solution would be to improve the biological and especially antioxidant properties of honey.

Plants are a rich source of bioactive compounds, including polyphenols which have various properties (antioxidant, antimicrobial, prebiotic, and others) with different applications in human health and industry [19]. Honeysuckle (*Lonicera caprifolium*) is a perennial flowering plant native to Europe and belongs to the family *Caprifoliaceae*. The honeysuckle flowers have a history of use in traditional herbal medicine due to their antibacterial, antioxidant, and antiviral activities [20]. While several species of the Lonicera genus, such as *L. japonica* (Japanese honeysuckle) and others, have been extensively studied and utilized in traditional medicine and cosmetics, the chemical composition of European honeysuckle (*L. caprifolium*) has received less research attention.

In our recent research [5], we formulated a NaDES that mimics the composition of honey by incorporating essential sugars found in honey (glucose, fructose, and sucrose). The NaDES derived from this formulation was analyzed, comparing its structural and physicochemical properties with honey. Our findings revealed that the honey-biomimetic NaDES closely resembled honey in terms of its characteristic features. Within the same study, we improved the antioxidant potential of honey and its biomimetic DES by incorporating dried raspberry extract and standard polyphenols found in the raspberry extract (caffeic acid and epicatechin) and evaluated the interaction in terms of antioxidant activity between them (between honey/NaDES and polyphenols). The main scope of analysing honey in comparison to NaDES was to understand better the behaviour of honey-polyphenols mixtures and test if the polysaccharides composition and interactions were sufficient to explain this behavior. A better understanding of honey properties could also help to design an edible or biocompatible biomimetic product based on its NaDES characteristics.

In this study, we aimed to explore the antioxidant activity of formulations between honey/biomimetic NaDES and polyphenols extracted from honeysuckle flowers. Building on previous research, we investigated how incorporating dried extract of honeysuckle flowers could enhance the antioxidant activity (AOA) of honey and its biomimetic NaDES. Our focus was on evaluating the AOA of this new mixture and understanding the interactions between its components.

Furthermore, we compared the AOA behaviour of these honey/NaDES mixtures with those enriched with raspberry extract to determine their relative efficacy. Alongside this, we conducted an assessment of the prebiotic activity of these mixtures and their individual components using the strain *Limosilactobacillus reuteri* DSM 20016. To evaluate prebiotic activity, we measured growth activity and determined L-lactic acid production as metabolites during fermentation.

2. Materials and Methods

2.1. Materials

Fresh honeysuckle flowers (*Lonicera caprifolium, family Caprifoliaceae*) were harvested in Bucharest area, identified based on their morphological characteristics, and a herbarium voucher with the number [USAMV B 4093] was deposited in the herbarium of USAMV Bucharest. These flowers and multifloral honey (RomHoney Group, Iasi, Romania) were used in this work. The multifloral honey was prepared by mixing 1/3 rapeseed honey with 1/3 sunflower honey and 1/3 meadow honey. The honeysuckle flowers were dried at room temperature and were ground to a fine powder using an electrical grinder. The following chemicals were used: ethanol 96% (Reactivul București Srl, Bucharest, Romania), D(+)-Glucose anhydrous extra pure, D(−)-Fructose, extra pure, D(+) Saccharose, reagent grade (Scharlau, Barcelona, Spain) Trolox 97% (Acros Organics, Thermo Fisher Scientific, Pittburghs, PA, USA), 2,2-Diphenyl-1-picrylhydrazyl (Sigma-Aldrich, Merck Group, Darmstad, Germany), 2,20-Azino-bis(3-ethylbenzothiazo-line-6-sulfonic acid) diammonium salt, 98%, 2,4,6-tri (2-pyridyl-1,3,5-triazine) 98% (Alfa Aesar, Kandel, Germany), Folin Ciocalteu's phenol reagent, Iron chloride (III) (Merck, Darmstadt, Germany), hydrochloric acid, acetic acid (Chimopar Srl, Bucharest, Romania), sodium acetate, MRS broth and aga (Scharlau, Barcelona, Spain), HPLC standards: ferulic acid, p-coumaric acid, caffeic acid, quercetin dihydrate (Sigma-Aldrich, Merck Group, Darmstad, Germany), syringic acid, luteolin, (+)-rutin trihydrate, (Alfa Aesar, Haverhill, MA, USA), chlorogenic acid, myricetin (Cayman Chemical, Ann Arbor, MI, USA), apigenin, (−) epicatechin (Roth, Karlsruhe, Germany), and kaempferol (Cayman Chemical, Ann Arbor, MI, USA). K-DLATE kit for D-/L-Lactic Acid (D-/L-Lactate) (Rapid) Assay Kit (Megazyme, Wicklow, Ireland).

2.2. Hydroalcoholic Extraction of Polyphenols from Honeysuckle Flower

The polyphenols extraction from honeysuckle flowers (*L. caprifolium*) was performed according to the method described by [21]. The polyphenol compounds were extracted using an ultrasound-assisted method in 61% (*v*/*v*) ethanol, in the ratio 1:20 (plant material/solvent) and 30 min of reaction in an ultrasound bath. The supernatant was removed after centrifugation at 7350 rcf for 30 min, and extraction was repeated in the same condition described below.

2.3. Analysis of Polyphenolic Content of Honeysuckle Extract

2.3.1. Total Polyphenol Content

The total polyphenol content (TPC) of the honeysuckle flower extract was measured spectrophotometrically by the Folin–Ciocalteau assay according to [22]. The method involved mixing 0.01 mL of honeysuckle extract or standard solutions of gallic acid with 0.09 mL double-distilled water (ddH_2O) and 0.010 mL of Folin–Ciocalteu reagent. After 5 min of reaction, 0.1 mL of 7% Na_2CO_3 and 0.04 mL ddH_2O were added to the mixture and incubated at room temperature for 60 min. The absorbance of solutions was measured at 765 nm using a plate reader (CLARIOstar, BMG LABTECH, Ortenberg, Germany). The calibration curve was performed using different concentrations of gallic acid in 70% (*v*/*v*) of ethanol. The range of gallic acid concentrations used was between 5 and 30 μg/mL. The results of TPC were calculated and reported as mg gallic acid equivalent/100 g dry weight of the sample (mg GAE/100 g DW).

2.3.2. Total Hydroxycinnamic Acid Content

The total hydroxycinnamic acid content (HAT) of honeysuckle flower extract was quantified spectrophotometrically according to the method adapted from the European Pharmacopoeia [23]. The method involved mixing 0.025 mL of honeysuckle extract or standard solutions of chlorogenic acid with 0.05 mL of 0.5 M HCl and with 0.05 mL of a solution composed of 1% (*w*/*v*) $NaNO_2$ and 1% (*w*/*v*) Na_2MoO_4, followed by 0.05 mL of 8.5% NaOH and 0.07 mL ddH_2O. The absorbance of solutions was measured at 524 nm. The calibration curve was performed using different concentrations (0–50 μg/mL) of chlorogenic acid in 70% (*v*/*v*) of ethanol. The HAT of the honeysuckle flower was expressed as mg chlorogenic acid equivalent/100 g DW of the sample (ChaE mg/100 g DW).

2.3.3. Total Flavonoid Content

Determination of the total flavonoid content (TFC) of the honeysuckle flower extract extracts was performed using the aluminium chloride/sodium acetate method according to [24] with some modifications. The method involved mixing 0.1 mL of honeysuckle extract or standard solution of quercetin with 0.1 mL of 10% CH3COONa and 0.12 mL of 2.5% $AlCl_3$ as well as 0.68 mL of ddH_2O were added to the mixture. The absorbance of the mixture was measured at 430 nm after 45 min. The results of TFC were calculated as quercetin equivalent mg/100 g DW of the sample (QE mg/100 g DW).

2.3.4. Total Anthocyanin Content

The total anthocyanin content (TAC) was determined using the pH differential method [25]. In brief, the absorbance of 2.5 × diluted sample in 25 mM potassium chloride buffer at pH 1 and 0.4 M sodium acetate buffer at pH 4.5 was measured after 30 min of incubation at room temperature, at 520 nm and 700 nm using a UV-VIS-NIR spectrophotometer (Ocean Optics, Orlando, FL, USA). The TAC was calculated by the following equation: TAC = $(\Delta As \times MW \times DF \times V \times 1000)/(\varepsilon \times L \times m)$, where ΔAs—difference of the absorbance of the sample at pH 1 and pH 4.5, DF is the dilution factor, L—optical pathlength (1 cm), V—the volume of the extracts (L), ε—molar absorptivity coefficient and MW—the molecular weight of cyanidin 3-glucoside ($\varepsilon = 26{,}900\ M^{-1}\ cm^{-1}$ and MW = 449.2 g/mol), $\Delta As = (A_{520} - A_{700})pH1.0 - (A_{520} - A_{700})pH4.5$. The result was expressed as milligrams of cyanidin 3-glucoside equivalent per 100 g of dry weight (DW) of the sample (mg cya, 3-Gluequivalent/100 g DW).

2.3.5. HPLC Analysis

The high-pressure liquid chromatography (HPLC) analysis of polyphenolic compounds from honeysuckle flower extract was carried out on Dionex Ultimate 3000 system (Thermo Fisher Scientific, Waltham, MA, USA) equipped with VWD-3100 detector. Data processing and analysis were performed by Chromelleon 7.0 software (Thermo Fisher Scientific, Waltham).

HPLC Analysis of Phenolic Acids. The phenolic acids content and composition of honeysuckle flower extract were determined by HPLC analysis according to a method described by [25] with some modifications. Chromatographic separation was performed on a Luna Omega 5 μm Polar C18 100 Å column (250 mm × 4.6 mm) (Phenomenex, Torrance, CA, USA). The mobile phase consisted of an aqueous solution with 0.1% formic acid (solvent A) and methanol (solvent B). The total runtime of the method was 40 min with the following gradient elution program: 0–25 min. 5% B/95% A, 25–33 min. 30% B/70% A, 34–40 min. 5% B/95% A. The analysis was conducted at a constant flow rate of 1.25 mL × min^{-1}, and the injection volume was set to 10 μL. The phenolic acids were detected at λ = 280 nm.

The identification of the phenolic acids involved comparing them with standards for each identified compound based on the retention time of standards. Quantification was accomplished by creating calibration curves for each determined compound using the standards. These calibration curves exhibited excellent linearity ($R^2 = 0.9996$) when plotting peak area against concentration and were in the range of 18.125–1000 μg/mL.

HPLC Analysis of Flavonoids. The composition and quantification of flavonoids from the extract of honeysuckle flower were determined by HPLC analysis according to the method described by [26]. The separation of flavonoids was performed on an Omega 5 μm Polar C18 100 Å column (250 mm × 4.6 mm) (Phenomenex, Torrance). The method involved using a gradient elution of two solvents: Methanol (solvent A) and 0.5% H_3PO_4

(solvent B). The gradient elution program was set as follows: 0–10 min 15% A/85% B, 15–25 min 85% A/15% B, 25–30 min. 60% A/40% B. The flow rate of the mobile phase was 1.5 mL/min, and the column temperature was 25 °C to detect flavonoids at 280 nm.

Flavonoids were detected and measured by correlating their retention time and spectral properties with established standards through the utilization of a calibration curve.

2.4. *Preparation of a Mixture of Honey/GFSw with Honeysuckle Extract*

The biomimetic NaDES, abbreviated from this point onwards GFSw (glucose/fructose/sucrose/water—the components of NaDES), was prepared by the method described in our previous work [5]. Honey/GFSw mixtures enriched with dried *L. caprifolium* extract were prepared in the same way as the mixtures with a raspberry extract from our previous study [5]. The honeysuckle flower extract was divided into three equal fractions, and each fraction was then concentrated to dryness (E_CD) at 40 °C using a semi-automated evaporation system called MultiVap54 (Lab tech, Sorisole, Italy). Two of the fractions E_CD were resuspended in honey (H) and, respectively, in its biomimetic NaDES named GFSw from this point onwards, at a ratio of 1:20 (*w*/*w*), resulting in the honey-honeysuckle mixture sample (H_LFL) and GFSw_LFL. The last fraction of E_CD was resuspended in 70% ethanol solution at the same ratio as in honey/GFSw (1:20 *w*/*v*), resulting in the LFL sample. The E_CD was dissolved in honey/GFSw, subjecting it to an ultrasonic bath, ensuring thorough mixing, and allowing the polyphenols to diffuse into the honey/GFSw overnight.

2.5. *Antioxidant Activity*

For the assessment of antioxidant activity (AOA), the H/GFSw and H_LFL/GFSw_LFL samples were dissolved in 70% (*v*/*v*) ethanol at a concentration of 0.2 g/mL (*w*/*v*). Four spectrophotometric methods, namely radical scavenging activity (ABTS and DPPH) and reducing antioxidant power (CUPRAC and FRAP), were employed to measure the AOA of the samples.

The AOA analysis was conducted at various concentrations of the samples, and calibration curves of the samples were generated for each method. The concentration values of LFL were individually tested and matched with the concentrations of LFL present in mixtures containing honey/GFSw. The final concentrations used in the final testing encompassed a range of 2 to 200 mg/mL for honey or GFSw and their mixtures and 0.1 to 10 mg/mL for LFL.

2.5.1. Radical Scavenging Activity by the DPPH Assay

The radical scavenging activity of the samples was tested by the DPPH method as described by [27] with slight modifications. To 100 μL of the sample, 100 μL of 0.3 mM DPPH solution dissolved in 99.6% (*v*/*v*) ethanol. The samples were incubated in the dark at room temperature for 30 min. The absorbance of the reaction mixture was measured at 517 nm using a UV-Vis plate reader (CLARIOstar, BMG LABTECH, Ortenberg, Germany).

2.5.2. Radical Scavenging Activity by the ABTS Assay

The ABTS cation scavenging activity of the samples was evaluated using the method adopted by [28]. ABTS radical cation solution was produced by mixing 7 mM ABTS in H_2Od.d. and 2.45 mM potassium persulfate solution. This solution was left for 12–16 h before being used in the dark at room temperature. Before use, the $ABTS^+$ solution was diluted with 96% ethanol to obtain an absorbance of 0.700 ± 0.04 at 734 nm. An aliquot of 0.02 mL of the sample was added to 0.180 mL of diluted $ABTS^+$ solution, and the absorbance was read at 734 nm after 30 min of incubation in the dark at room temperature.

2.5.3. Ferric-Ion Reducing Antioxidant Power (FRAP) Assay

The FRAP assay was performed according to the procedure described by [29] with slight modifications. The FRAP reagent was composed of 0.3 M acetate buffer at pH 3.6, 0.01 M TPTZ (solubilized in 0.04 mM HCl), and 0.02 M $FeCl_3$ solution in the ratio 10:1:1, which was warmed at 37 °C before to use. The method involved mixing 15 µL of sample/standard solution of Trolox with 285 µL freshly prepared FRAP reagent. The reaction mixtures were incubated at 37 °C in the dark for 30 min, and the absorbance was measured at 593 nm. The calibration curve was developed using different concentrations (50–450 µg/mL) of Trolox in 70% (*v*/*v*) of ethanol.

2.5.4. Cupric-Ion Reducing Antioxidant Capacity (CUPRAC) Assay

The CUPRAC method was performed according to the adopted procedure described by [30]. An aliquot of 10 µL of the samples/standard solutions of Trolox was mixed with 30 µL $CuSO_4$ (5 mM), 30 µL neocuproine (3.75 mM), and 280 µL distilled H_2Od.d. The absorbance was measured at 450 nm after 30 min of incubation at room temperature in the dark. The calibration curve was made from a stock solution of 10 mM Trolox in 70% ethanol, with a concentration interval of 0.25–2 mM Trolox.

2.6. Evaluation of Interaction between Honey/GFSw and Honeysuckle Extract in Terms of AOA

In order to evaluate the interaction between honey/GFSw and honeysuckle extract (H_GFSw_LFL) in terms of AOA and also to compare it to the AOA behaviour of H/GFSw_RE, we first used the procedure described in our previous research [5]. This procedure involved the determination of the combination index, isobolograms, the dose–response curve of each compound, and the evaluation of theoretical and experimental AOA of the samples. The combination index (CI) of mixture H_LFL/GFSw_LFL was calculated based on the ratio of the concentration of each compound when combined in the mixture($Cc_{1,c}$ and $Cc_{2,c}$) to the concentration when used separately ($Cc_{1,s}$ and $Cc_{2,s}$) to achieve the same effect as observed in the mixture [31]:

$$CI = \frac{C_{c1,\,c}}{C_{c1,s}} + \frac{C_{c2,c}}{C_{c2,\,s}}$$

where Cc_1 means the concentration of H/GFSW, and Cc_2—is the concentration of extract of honeysuckle (LFL). Isobolomic analysis was the graphical representation of the same data.

The CI and isobolomic analysis of the samples in the case of FRAP and CUPRAC methods was expressed as the effective concentration of the samples at 1 mM Trolox (EC 1 mM Trolox, mg/mL).

In the case of ABTS and DPPH methods of AOA, for evaluation of the CI and isobolograms of the samples, the values of IC_{50} and IC_{20} (50% and 20% inhibitory concentration of the substrate) were used. The IC_{50} and IC_{20} values were calculated based on the median-effect equation, transforming the non-linear equation for the dose–response curve into a linear one:

$$log\frac{f_i}{f_u} = m \times log(conc.) + n$$

f_i and f_u are inhibited and uninhibited fractions of the reaction, m—the slope and n—respectively intercept of the curve.

f_u—is inhibited fraction of the substrate by antioxidant sample and was calculated as follows:

$$f_i = \frac{(A_0 - A_{blank\_0}) - (A_s - A_{blank\_s})}{(A_0 - A_{blank_0})} \times 100\%$$

where A_0 is the absorbance of the substrate (DPPH or ABTS reagent), A_{blank_0}—absorbance of the blank of the substrate (solvent), A_s—absorbance of the sample, A_{blank_s}—absorbance of the blank of the sample (sample without substrate).

f_u—is the uninhibited fraction of the reaction and was calculated as $f_u = 100 - f_i$.

Another way to evaluate the interactions between polyphenols and honey/GFSw was by plotting concentration–dependent curves of the theoretical and experimental antioxidant activities. In the case of FRAP and CUPRAC methods, the theoretical AOA was calculated by addition (absorbance of H/GFSw + absorbance of LFL). In the case of ABTS and DPPH methods, which were non-linear dose–response curves, the theoretical AOA of the samples was calculated by the Webb equation: $100 - ((100 - f_{i,C1}) \times (100 - f_{i,C2}))$, where $f_{i,C1}$ and

$f_{i,C2}$ are the inhibited fraction of compound 1 (C1) and compound 2 (C2), respectively, when analysed separately.

For comparison of LFL and RE within the experimental concentration range, we generated dose–response curves, Dose–Response Matrix, and 2D representation of Synergy Score for DPPH and ABTS using the SynergyFinder R package [32]. We conducted an analysis of the interactions between the components within H_ LFL, GFSw_LFL mixtures from this study, and H_RE and GFSw_RE from our previous study [5]. We generated Dose–response curves for LFL and RE in the absence and presence of honey/GFSw, as well as for honey and GFSw. The common concentration interval and concentration values between LFL and RE were chosen. The objective of this analysis was to compare the degree of inhibition of DPPH and ABTS radicals, both the experimental and those simulated by SynergyFinder. However, the FRAP and CUPRAC methods could not be analysed through the SynergyFinder R package as these methods do not result in an inhibition percent.

2.7. Prebiotic Activity

The prebiotic activity of the mixtures of honey/GFSw enriched with dried plant extract (raspberry and honeysuckle flowers) was assessed by the evaluation of the growth-promoting activity of the samples on the strain of *Limosilactobacillus reuteri* DSM 20016 and determination of L-lactic acid content as a metabolite produced during of sample fermentation. The results were compared to simple honey/GFSw, plant extract at the same concentration as in the mixture, and the control (C+) of strain. The statistics were conducted between extract (RE, LFL) and C+, and between honey/GFSw mixtures and simple honey/GFSw.

2.7.1. Probiotic Growth-Promoting

The evaluation of the growth-promoting of the samples was performed according to the method described by [33,34] with some modifications. The probiotic strain of *L. reuteri* DSM 20016 was obtained from Leibniz Institute DSMZ-German Collection of Microorganisms and Cell Cultures GmbH (DSMZ, Braunschweig, Germany). The strain was stored in a cryotube with 25% (*v*/*v*) glycerol solution at −90 °C. Before experiments, the probiotic was activated by being inoculated in MRS broth for 48 h at 30 °C in Oxoid™ AnaeroJar™ 2.5 L (Thermo Scientific™) and after was subcultured on MRS agar plate under the same conditions for preparing the probiotic inoculum (0.5 McFarland) in sterile saline solution (0.8% NaCl).

The stock solution of the samples (50 mg/mL for H/GFSw and their mixtures and 2.25 mg/mL for LFL and RE) was prepared by solubilisation in MRS broth and sterile filtration through sterile 0.22 µm PES filters. The test itself was carried out in Eppendorf tubes by making dilutions in MRS to obtain 5 test concentrations between 1–45 mg/mL (for H/GFSw and their mixtures) and 0.05–2.25 mg/mL for plant extract (LFL and RE). The concentration of the samples was calculated for the final volume in the test tube after adding 10% of probiotic inoculum. The control sample (C+, which means the control of the strain *L. reuteri* without any supplements) was prepared in the same way as the samples by adding 10% of probiotic inoculum in the medium MRS broth. The samples were incubated for 48 h at 30 °C in Oxoid™ AnaeroJar™, and the absorbance of the samples was measured at 600 nm in 96-well plates using a plate reader after carefully and thoroughly mixing the Eppendorf tube. The samples were stored in the freezer at −20 °C for further analysis of lactic acid content.

The growth-promoting effect of the samples was calculated as follows: $(A_s - A_{blank\_s})/A_c - A_{blank\_c}) \times 100$, where A_s—absorbance of the sample after incubation time, A_c—absorbance of control samples of the strain, A_{blank_s}—absorbance of the blank of the sample (before incubated time), A_{blank_c}—absorbance of the blank of the control of the strain before incubated time. The results are expressed as percent bacterial growth.

2.7.2. L-Lactic Acid Content

The L-lactic acid content produced during cultivation of *L. reuteri* in the presence of the samples tested was determined enzymatically using the commercial kit–K-DLATE kit for D-/L-Lactic Acid (D-/L-Lactate) (Rapid) Assay Kit (Megazyme, International Ireland Ltd., Wicklow, Ireland). Before analysis, the samples were centrifugated at 1470 rcf for 10 min, and the supernatant of the samples was analysed according to the manufacturer's kit protocol.

2.8. Statistical Analysis

Statistical analysis for prebiotic activity was performed using IBM ®SPSS® Statistics, version 26 (IBM SPSS Corp., Armonk, NY, USA). All assays were carried out in triplicate, and the results are expressed as mean values ± standard deviation (SD). One-way analysis of variance (ANOVA) was used to determine if significant differences exist between the tested samples of honey and GFSw with or without plant extract (honeysuckle and raspberry extract) and vice-versa. The homogeneity of variance was tested by Levene's test. To explore the significant difference between group means, Tukey's honestly significant difference (HSD) test was performed for homoscedastic groups and Games–Howell for heteroscedastic groups.

For the isobolographic analysis of the AOA, we computed 95% confidence intervals. These intervals were determined by subtracting and adding the value of 1.96 times the standard deviation (SD) divided by the square root of the number of measurement replicates (n) from the mean of results (mean ± 1.96 × SD/sqrt (n)). In this study, the number of replicates was three (n = 3) for all cases.

3. Results

3.1. Screening of Polyphenolic Compounds from the Honeysuckle Flower

The polyphenolic profile of honeysuckle flowers was evaluated by several methods: total polyphenols content (TPC), total flavonoids content (TFC), total hydroxycinnamic acid content (HAT), total anthocyanin content (TAC), and HPLC assays and the results are summarized in Tables 1 and 2.

Table 1. The results of TPC, TFC, HAT, and TAC of *L. caprifolium*.

Sample	TPC GAE mg/100 g DW	TFC QE mg/100 g DW	HAT ChaE mg/100 g DW	TAC, mg cya, 3-Glu Equivalent/100 g DW
L. caprifolium flower	651.79 ± 5.11	64.56 ± 2.12	587.38 ± 1.19	4.926 ± 0.011

GAE—gallic acid equivalent, QE—quercetin equivalent, ChaE—chlorogenic acid equivalent, 3-Gluequivalent—cyaniding 3-glucoside, DW—dried weight.

Table 2. The polyphenol compound from honeysuckle flowers by HPLC analysis.

Polyphenols	*Lonicera caprifolium* Flowers mg/g DW
Phenolic acids	
Caffeic acid	36.54 ± 0.04
Ferulic acid	1.72 ± 0.02
p-coumaric acid	0.46 ± 0.001
Chlorogenic acid	2.45 ± 0. 11
Flavonoids	
Epicatechin	2.83 ± 0.02
Apigenin	1.47 ± 0.007

The phenolic acids and flavonoids found and identified in honeysuckle flowers by HPLC analysis after ultrasound-assisted extraction were caffeic acid (RT–23.62 min), chlorogenic acid (RT–22.50 min), ferulic acid (RT–30.68 min), p-coumaric acid (RT–33.91 min), epicatechin (RT–14.61 min) and apigenin (RT–17.20 min) The chromatograms illustrating the polyphenolic compounds found in honeysuckle flowers can be found in the Supplementary Materials (Figures S1 and S2). Caffeic acid, 36.54 ± 0.04 mg/g DW, and epicatechin, 2.83 ± 0.02 mg/g DW had the highest content among polyphenols.

3.2. Evaluation of the Interaction between Honey/GFSw and Extract of Honeysuckle Flowers in Terms of AOA

The honey mixture with honeysuckle extract (H_LFL) showed much higher antioxidant activity (AOA) compared to commercial honey at all concentrations tested (2–200 mg/mL) as determined by all the methods of AOA (FRAP, CUPRAC, DPPH, ABTS) evaluated. Figures S3–S6 in the Supplementary Material provides a detailed comparison of the AOA values of the analysed samples, illustrated by the dose–response curves for the AOA of the sample. We can see that the slope and intercept values of H_LFL were much higher than the values for simple honey. Similar behaviour can be observed for the mixture between GFSw and LFL (GFSw_LFL).

In an intention to evaluate the interactions between LFL and honey/GFSw and to compare them with our previous data reported for the raspberry extract [5], the combination index (CI) was calculated (Table 3) based on the calibration curves of the samples, and the isobolograms were plotted—Figure 1 DPPH and ABTS) and Figure 2 (FRAP and CUPRAC).

Table 3. The combination index of honey (H_LFL) and its biomimetic DES (GFSw_LFL) mixtures with honeysuckle extract.

AOA Method	H_LFL	GFSw_LFL
FRAP	0.86 ±0.038	0.85 ± 0.029
CUPRAC	1.287 ± 0.020	1.195 ± 0.075
DPPH IC_{50}	1.157 ± 0.036	1.346 ± 0.19
DPPH IC_{20}	1.158 ± 0.038	1.536 ± 0.071
ABTS IC_{50}	1.026 ± 0.015	1.093 ± 0.039
ABTS IC_{20}	1.028 ± 0.014	1.093 ± 0.02

H_LFL—mixture of honey with honeysuckle flowers extract, GFSw_LFL—mixture of the NaDES GFSw—honeysuckle flowers extract. The terms IC_{50} and IC_{20} represent the analysis of concentration at 50% and 20% substrate inhibition, respectively.

In accordance with our previous study [5], we have categorized the CI values for ease of comparison as follows: 0.5–0.7 indicates strong synergism, 0.7–0.9 denotes moderate synergism, 0.9–1.1 implies nearly additive behaviour, 1.1–1.5 signifies moderate antagonism, 1.5–2 indicates moderate to strong antagonism, and CI > 2 represents strong antagonism.

The interaction between LFL and honey ranged from moderate synergism (CI = 0.86 ± 0.04 for FRAP) to nearly additive behaviour (CI = 1.03 ± 0.02 for ABTS IC50 and 1.03 ± 0.01 for ABTS IC20) and moderate antagonism (CI = 1.16 ± 0.04 for DPPH IC50 and 1.16 ± 0.04 for DPPH IC20, CI = 1.29 ± 0.020 for CUPRAC). GFSw with LFL exhibited similar behaviour in the case of FRAP (CI = 0.85 ± 0.03), ABTS IC50 and ABTS IC20 (CI = 1.09 ± 0.04 and 1.09 ± 0.02, respectively), and CUPRAC (CI = 1.35 ± 0.19). However, some differences in behaviour were observed in terms of DPPH, the CI being higher than in the case of H_LFL. The DPPH CI values of GFSw_LFL indicated a moderate antagonism feature (CI = 1.35 ± 0.19 for DPPH IC50) and moderate to strong antagonism (CI = 1.54 ± 0.07 for DPPH IC20).

The behaviour of H_LFL is very similar to the behaviour of GFSw_LFL in terms of CI and ranges from synergism, additive, and antagonism (0.85 ± 0.029 FRAP for GFSw_LFL and 1.536 ± 0.071 DPPH IC20 for GFSw_LFL). The only synergic behaviour was obtained in the case of FRAP.

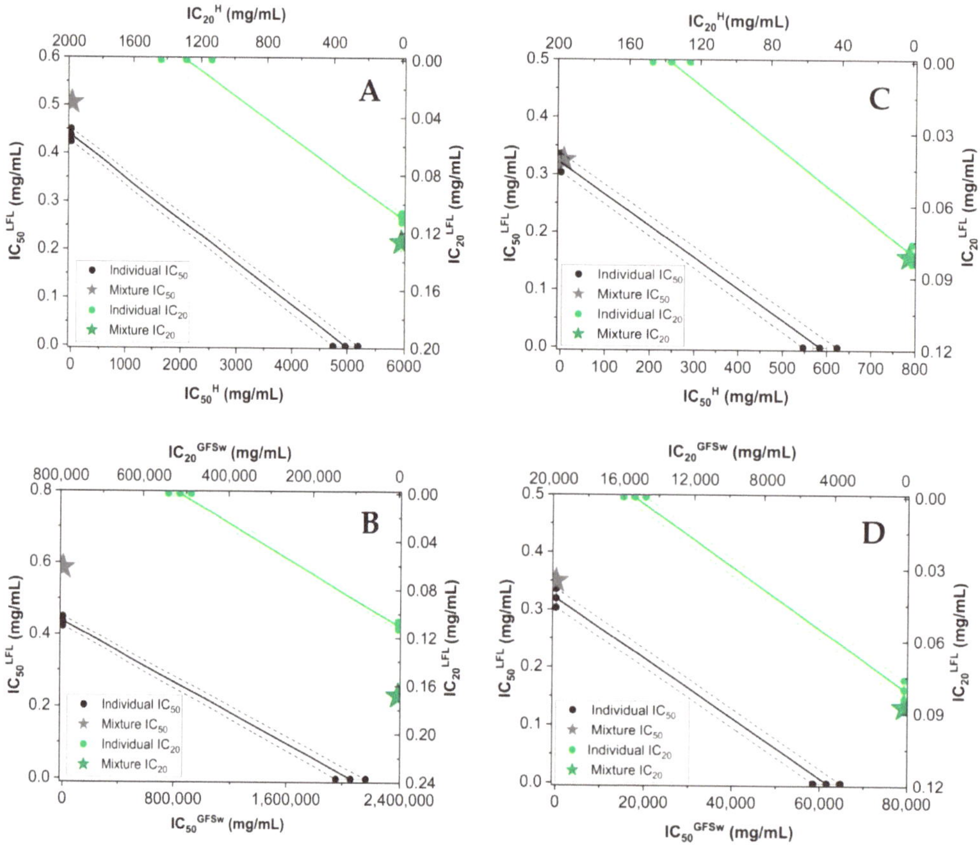

Figure 1. Isobolograms of honey (H) and its biomimetic NaDES (GFSw), and honeysuckle flowers extract (LFL) in terms of AOA by DPPH (**A**,**B**), and ABTS (**C**,**D**); IC_{50} (half-maximal inhibitory concentration) and IC_{20} (inhibitory concentration at 20% substrate inhibition); dashed lines indicate the 95% confidence intervals.

The isobologram is a graphical representation of the interactions between two compounds of a mixture and represents the effect of separate compounds when they are in a mixture.

The x- and y-axes on the graph represent the concentrations of the compounds in the mixture, specifically honey (H)/GFSw and honeysuckle extract (LFL). In Figure 1, the black and green circles on the graph represent the effects of IC_{50} (half-maximal inhibitory concentration) and, respectively, IC_{20} (20% inhibitory concentration) of H/GFSw and LFL when each of the two components is used individually. Before analyzing the interactions between the compounds graphically, an additive line is drawn between the two compounds (between the black or green circles). The concentrations of the components in the mixture that gave the same result (IC_{50}, IC_{20}) were plotted as stars. If the position of the mixture lies above the additive line, it indicates antagonism. If it lies below the additive line, it indicates synergism. If it lies on the additive line, it represents an additive effect.

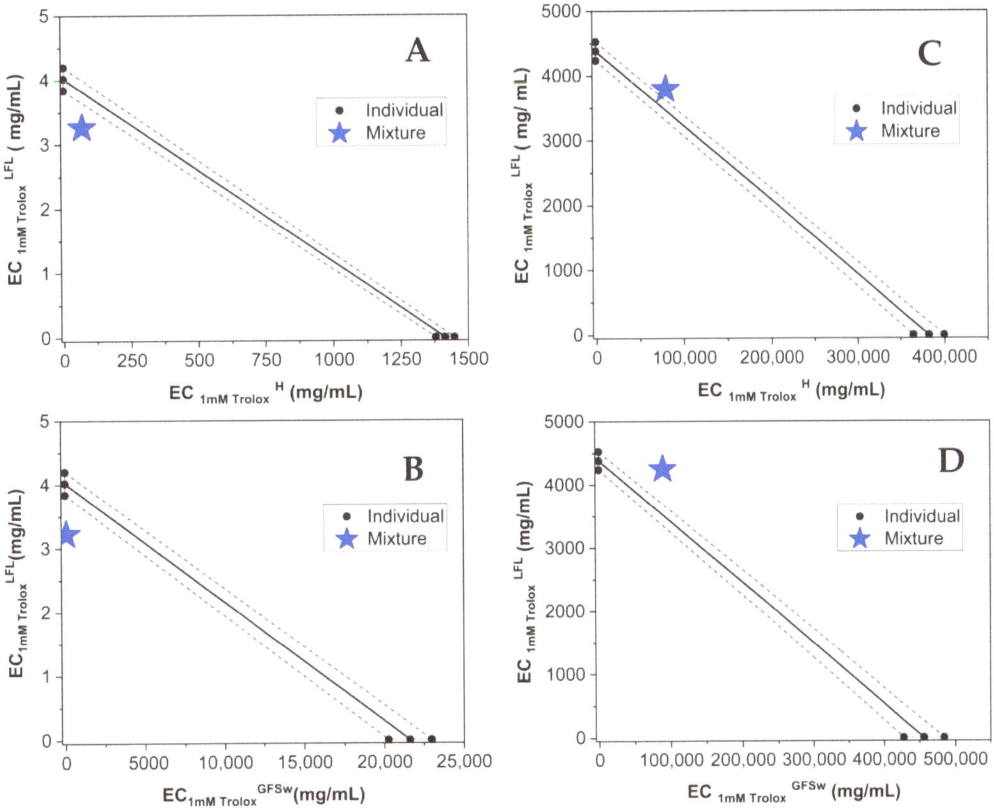

Figure 2. Isobolograms of honey (H) and its biomimetic NaDES (GFSw), and honeysuckle flowers extract (LFL) in the therm of AOA by FRAP (**A**,**B**) and CUPRAC (**C**,**D**) methods, CE- effective concentration at 1 mM Trolox equivalent of the sample. Each value is accompanied by error bars representing three measurements. Dashed lines indicate the 95% confidence intervals.

As can be observed in Figure 1, the mixtures of honey with LFL and GFSw with LFL exhibited similar behaviour. Furthermore, no significant difference between the behaviour of the mixture at IC_{50} and IC_{20} was observed, and the data corroborated with the CI results from Table 3.

Figure 2 illustrates the isobolograms of the correlation between honey (H)/GFSw and LFL regarding AOA using the FRAP and CUPRAC methods. It is evident from the figure that the mixture of honey and GFSw with LFL displayed comparable behaviour, which correlated with the CI data from Table 3.

To further assess the interactions between polyphenols and honey/GFSw, concentration–dependent curves of both theoretical and experimental AOA (Figures 3 and 4) were plotted. This provided an additional means of evaluating the extent and nature of these interactions.

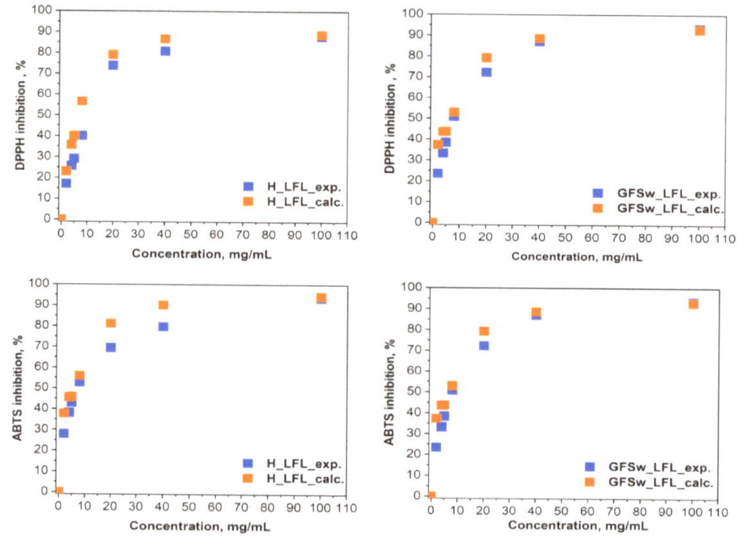

Figure 3. Webb analysis of experimental (H_LFL/GFSw_LFL_exp.) and theoretical (H_LFL/GFSw_LFL_calc.) AOA in the mixture of honey (H)/GFSw with the honeysuckle extract (LFL).

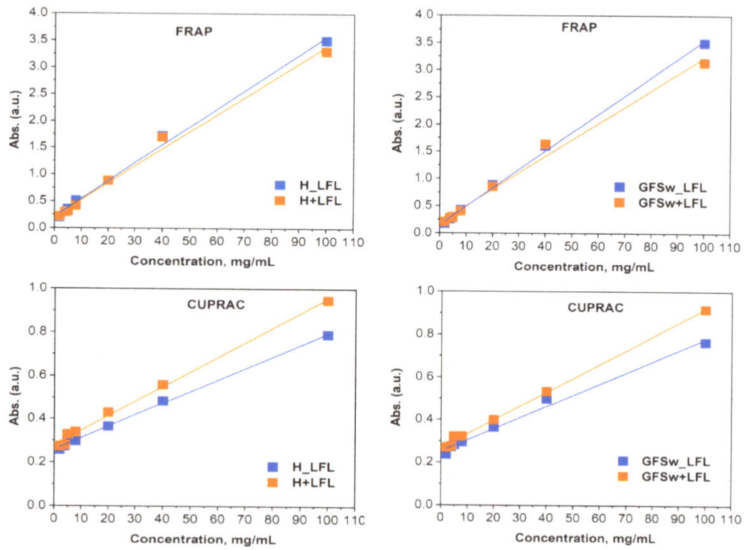

Figure 4. Evaluation of the concentration dependence of experimental (H_LFL/GFSw_LFL) and theoretical (H+LFL/GFSw + LFL) AOA in the mixture of honey(H)/GFSw with the honeysuckle extract (LFL).

As previously stated, the theoretical AOA for the FRAP and CUPRAC methods was obtained by summing the absorbance values of H/GFSw and LFL. However, for the ABTS and DPPH methods, which exhibited nonlinear concentration–dependent curves, the theoretical AOA was calculated using the Webb equation, taking into account the inhibited and uninhibited fractions.

By examining the relative positioning of the concentration-dependence curves for the theoretical versus the experimental AOA of the analyzed samples, we can assess the modulation of polyphenols in honey and GFSw. If these curves overlap, it indicates an additive effect between the components. If thc theoretical (calculated) curve lies below the experimental curve, it suggests synergism, whereas if the theoretical curve is higher than the experimental curve, it indicates antagonism.

The observed trend indicates that the AOA exhibited a linear relationship with concentration for the FRAP and CUPRAC methods, and a sigmoidal relationship was observed for the DPPH and ABTS methods.

For a more in-depth comparison between LFL and RE, we analysed the DPPH and ABTS data using SynergyFinder R package. The dose–response Curves from Figures S7 and S8 confirm that LFL has a higher AOA than RE. The curves of LFL start to saturate at the maximum concentration tested, with a final inhibition of approx. 80%, but the curves of RE are still on the ascendent trend, reaching approx. 50–60% inhibition at the same concentration. We showed in our previous work [5] that the saturation takes place at RE mixture concentrations higher than 100 mg/mL (5 mg/mL RE extract).

Based on the dose–response curves of the sample analysed (in the case of samples with RE, the dose–response curves based on the data from our previous article [5] were used), the values of IC_{50} (for DPPH and ABTS methods) and Trolox Equivalent Antioxidant Capacity (TEAC coefficient) were calculated and are shown in Table 4 (from the AOA as a function of extract concentration in the mixture) and Table 5 (from the AOA as a function of total mixture concentration).

Table 4. Quantitative data of the antioxidant activity as a function of RE and LFL concentrations.

Methods	RE	RE in H_RE Mixture	RE in GFSw_RE Mixture	LFL	LFL in H_LFL Mixture	LFL in GFSw_LFL Mixture
DPPH IC_{50} (mg/mL)	3.89 ± 0.12^{c}	1.94 ± 0.13^{b}	3.79 ± 0.21^{c}	0.43 ± 0.011^{a}	0.50 ± 0.02^{a}	0.67 ± 0.15^{a}
ABTS IC_{50} (mg/mL)	2.25 ± 0.11^{b}	4.68 ± 0.24^{c}	5.98 ± 0.95^{d}	0.32 ± 0.01^{a}	0.32 ± 0.01^{a}	0.35 ± 0.02^{a}
$TEAC_{FRAP}$	94.72 ± 1.45^{a}	121.81 ± 6.71^{b}	93.12 ± 3.11^{a}	263.81 ± 15.41^{c}	293.41 ± 0.80^{d}	296.58 ± 0.31^{d}
$TEAC_{CUPRAC}$	0.1 ± 0.004^{a}	0.1 ± 0.013^{a}	0.1 ± 0.009^{a}	0.24 ± 0.012^{b}	0.23 ± 0.039^{b}	0.22 ± 0.0006^{b}

Different letters show statistically different differences (±error bars, $\sigma < 0.05$, n = 3).

Table 5. Quantitative data of the antioxidant activity as a function of total mixture concentrations.

Methods	H_RE	GFSw_RE	H_LFL	GFSw_LFL
DPPH IC_{50} (mg/mL)	40.76 ± 2.89^{b}	79.58 ± 4.45^{c}	10.59 ± 0.48^{a}	14.09 ± 1.18^{a}
ABTS IC_{50} (mg/mL)	98.28 ± 2.96^{b}	125.49 ± 11.56^{c}	6.81 ± 0.16^{a}	7.34 ± 0.22^{a}
$TEAC_{FRAP}$	6.07 ± 0.34^{b}	4.63 ± 0.16^{a}	14.59 ± 0.026^{c}	14.77 ± 0.025^{c}
$TEAC_{CUPRAC}$	0.0061 ± 0.0006^{a}	0.0049 ± 0.0004^{a}	0.0114 ± 0.0019^{b}	0.0112 ± 0.00012^{b}

Different letters show statistically different differences (±error bars, $\sigma < 0.05$, n = 3).

The IC_{50} value is a measure of the concentration of a substance needed to inhibit a specific biological or biochemical activity by 50%. In the context of antioxidant activity, a lower IC_{50} value indicates a stronger antioxidant capacity, as it implies that a lower concentration of the sample is required to achieve the same inhibitory effect.

The results in Table 4 show that LFL had an almost 10× lower IC50 and higher TEAC than RE, both applied alone or in a mixture. Honeysuckle flower extract demonstrated consistently superior AOA performance compared to raspberry extract in all methodologies examined. Honey improved the AOA only in the case of DPPH and FRAP of RE, which correlates with the synergism reported previously [5], and FRAP of LFL, which correlates with the small synergism observed at higher concentrations for FRAP of H_LFL in Figure 4. GFSw improved the AOA only in the case of FRAP of LFL, which correlated with the small synergism in Figure 4. All the other mixtures showed similar or lower AOA compared with the extract itself. In the case of RE, the AOA was higher in mixtures with honey than with GFSw, except for CUPRAC, where the AOA is the same. In the case of LFL, honey and GFSw behaved similarly.

A similar trend as in Table 4 is observed when the total mixture concentration is used (Table 5).

The Dose–Response Matrix (DRM) and Loewe Synergy score (LSS) for DPPH generated by SynergyFinder are shown in Figures 5 and 6. The points that form the diagonal represent the experimental points, and the rest of the combinations represent predicted behaviour generated by the software. The following observations can be drawn from DRM: all DRM have very similar patterns, except H_RE and H_CA, which present more significant differences and look similar one to the other; the H_LFL and GFSw_LFL are almost identical; the earlier saturation and higher AOA of H_LFL compared to H_RE is also predicted for other combinations of concentrations.

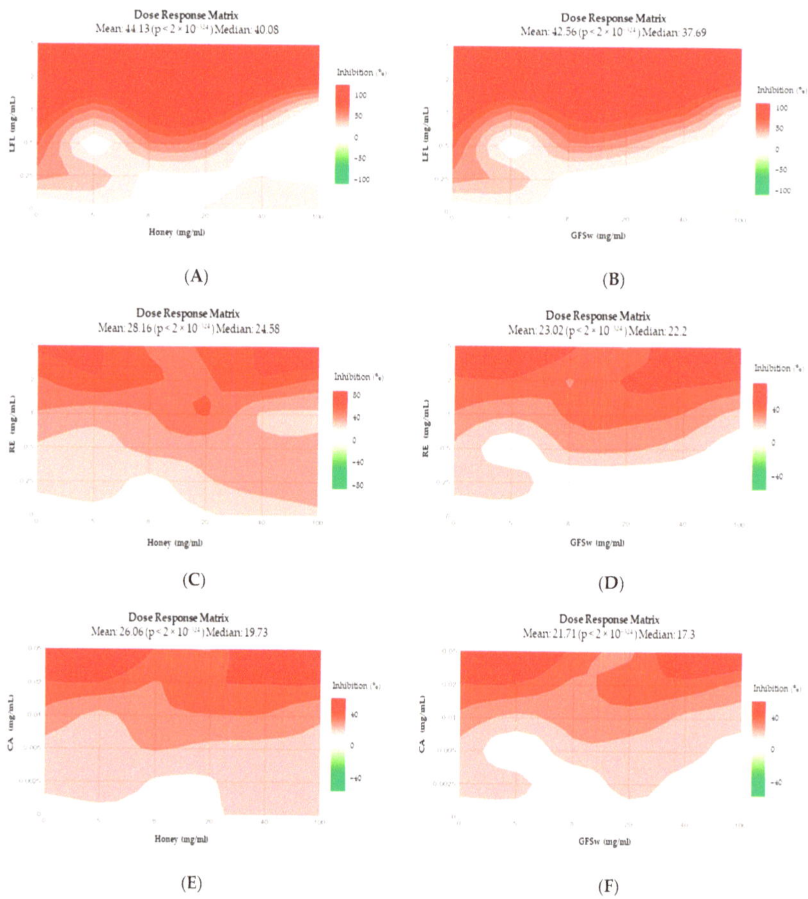

Figure 5. *Cont.*

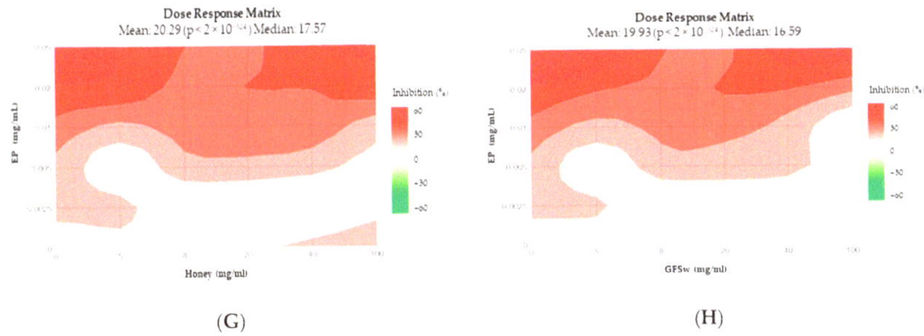

Figure 5. Dose–Response Matrix of DPPH, generated by SynergyFinder for (**A**) H_LFL; (**B**) GFSw_LFL; (**C**) H_RE; (**D**) GFSw_RE; (**E**) H_CA; (**F**) GFSw_CA; (**G**) H_EP; (**H**) GFSw_EP.

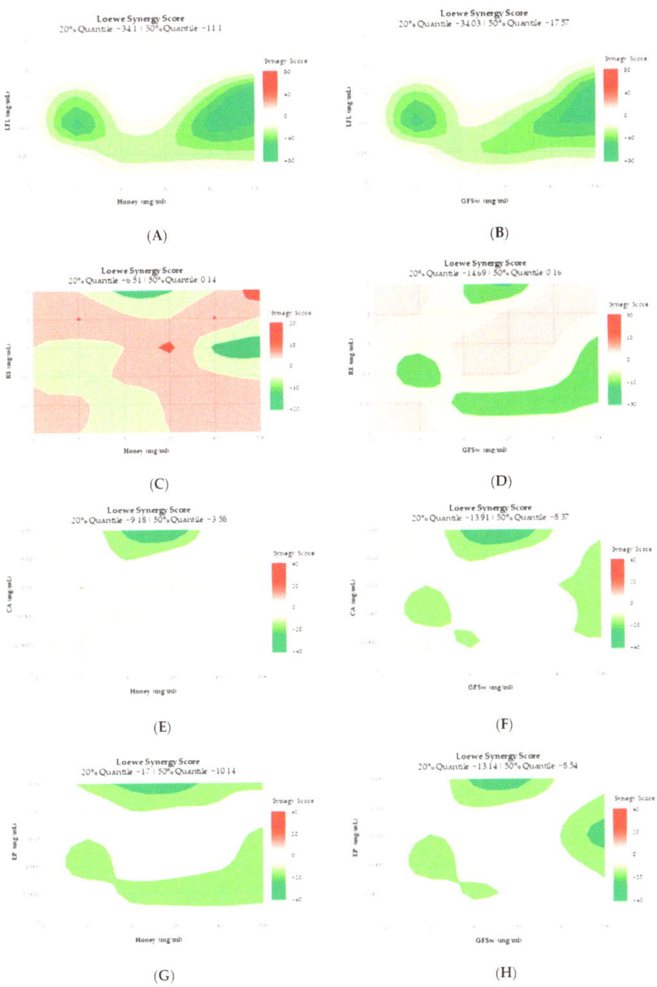

Figure 6. Loewe Synergy Score of DPPH generated by SynergyFinder for: (**A**) H_LFL; (**B**) GFSw_LFL; (**C**) H_RE; (**D**) GFSw_RE; (**E**) H_CA; (**F**) GFSw_CA; (**G**) H_EP; (**H**) GFSw_EP. Red represents synergism, and green represents antagonism.

The LSS confirms the prevalence of synergism for RE and antagonism for LFL for diagonal combinations and predicts the same difference in other combinations. The synergism pattern of H_LFL and GFSw_LFL are again almost identical, while there are some differences in the case of RE.

In the case of ABTS, a similar difference between LFL and RE is observed in DRM (Figure 7). In this case, H_RE resembles not only H_CA but also H_EP, and GFSw induces changes that give patterns similar to that of H_LFL and GFSw_LFL, which resemble very much, just as in the case of DPPH.

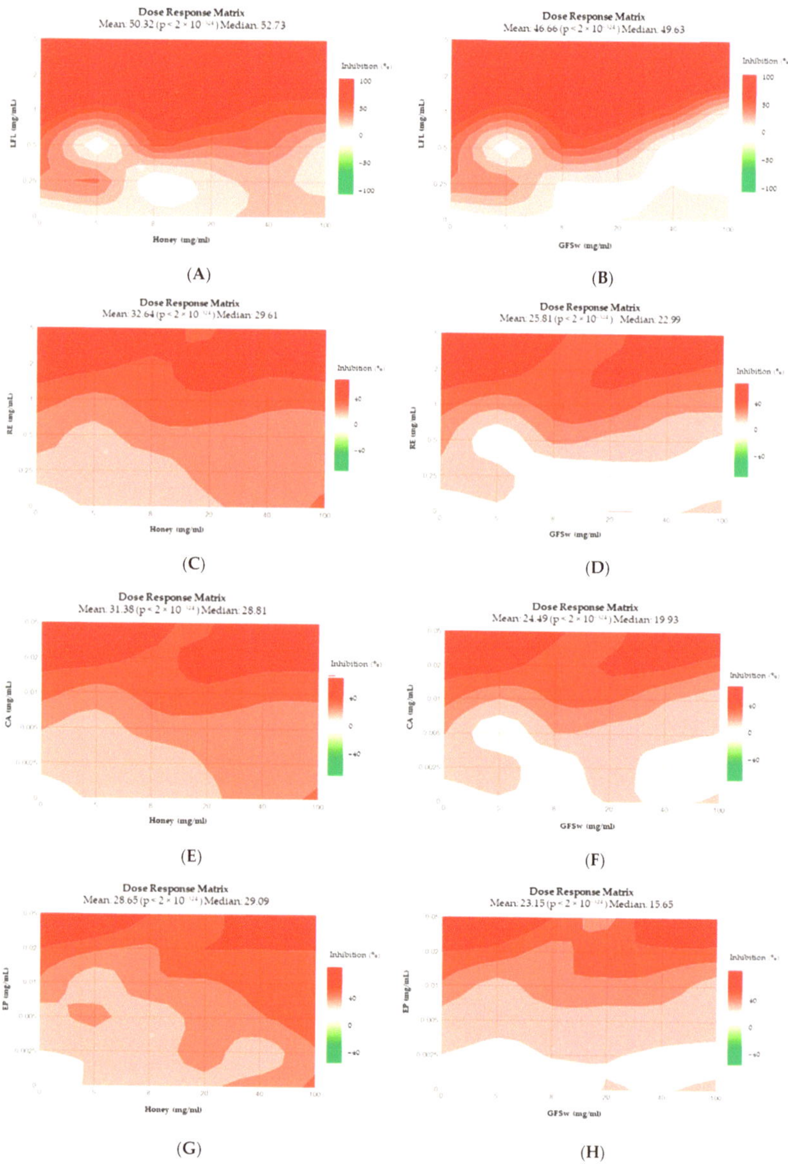

Figure 7. Dose–Response Matrix of ABTS, generated by SynergyFinder for (**A**) H_LFL; (**B**) GFSw_LFL; (**C**) H_RE; (**D**) GFSw_RE; (**E**) H_CA; (**F**) GFSw_CA; (**G**) H_EP; (**H**) GFSw_EP.

The LSS of ABTS confirms the predominant antagonist behavior for the experimental points (diagonal) and predicts similar behavior at other concentrations (Figure 8).

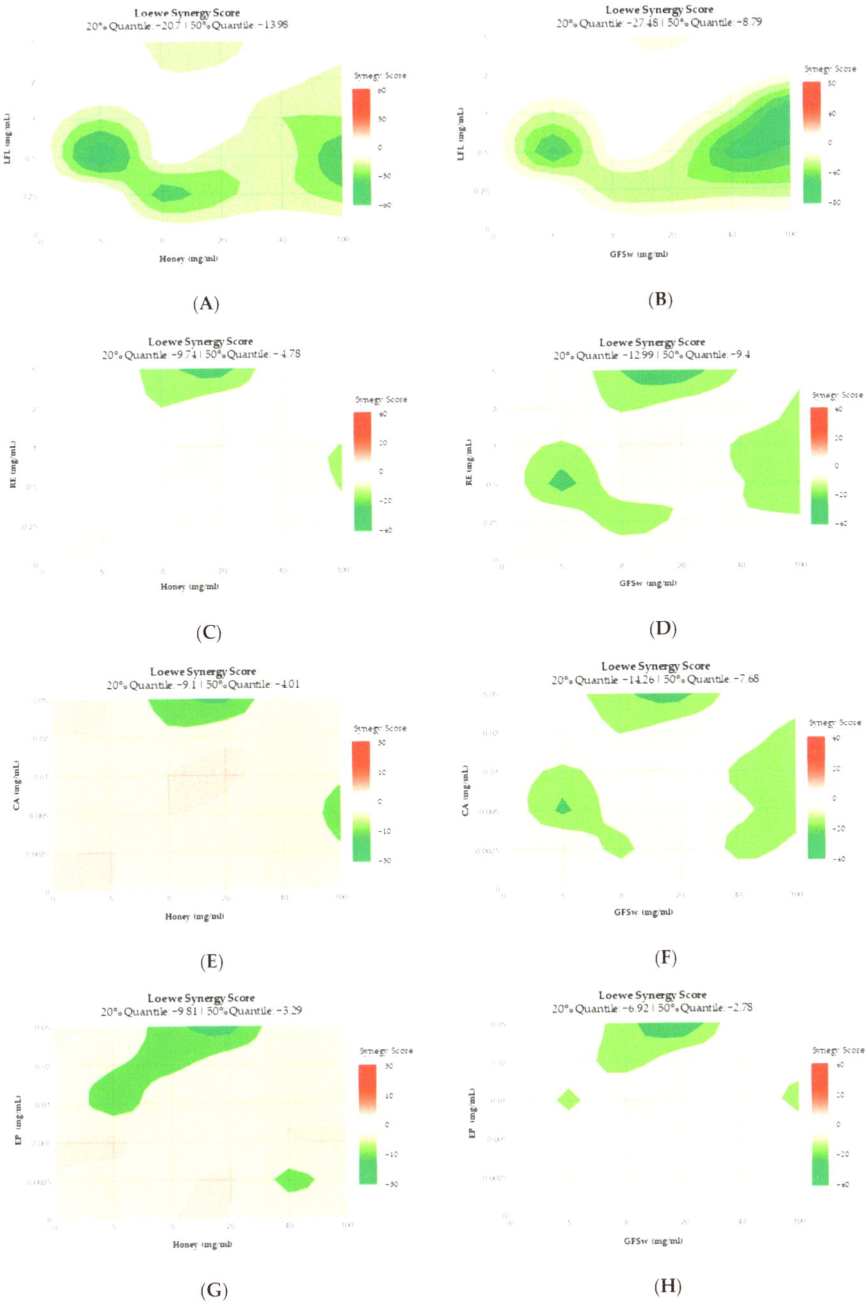

Figure 8. Loewe Synergy Score of ABTS generated by SynergyFinder for: (**A**) H_LFL; (**B**) GFSw_LFL; (**C**) H_RE; (**D**) GFSw_RE; (**E**) H_CA; (**F**) GFSw_CA; (**G**) H_EP; (**H**) GFSw_EP. Red represents synergism and green antagonism.

3.3. Prebiotic Activity

The prebiotic activity of the mixtures of honey (H)/GFSw enriched with dried plant extracts (raspberry fruits and honeysuckle flowers) was evaluated by assessing their growth-promoting effects on the *Limosilactobacillus reuteri* DSM 20,016 strain and measuring the production of L-lactic acid as a metabolite during sample fermentation. In order to assess whether the improved antioxidant activity of honey also improves the prebiotic activity, five concentrations ranging from 45–1 mg/mL of H/GFSw and their mixtures with the two extracts (LFL and RE) were tested. The individual extracts were tested at concentrations of 2.25–0.05 mg/mL corresponding to their respective concentrations in the mixtures—Figure 9.

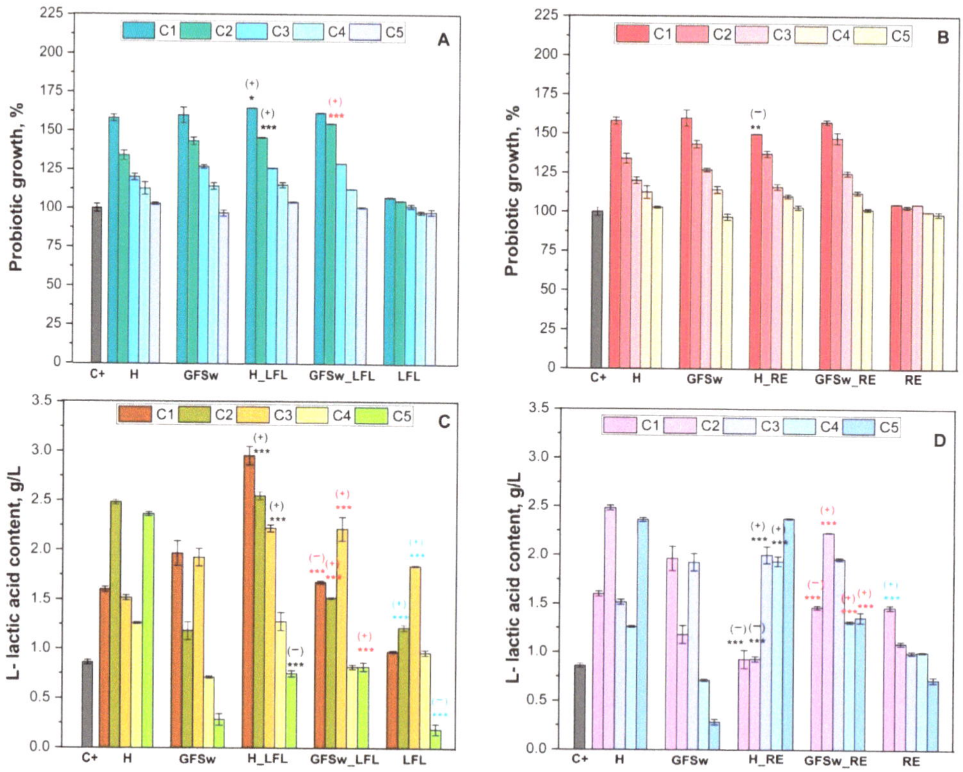

Figure 9. Probiotic growth-promoting (**A**,**B**) and L- lactic acid content (**C**,**D**) of honey (H), its biomimetic DES (GFSw), honey enhanced with honeysuckle flowers extract (H_LFL), and with raspberry extract (H_RE), GFSw enhanced with honeysuckle flowers extract (GFSw_LFL), and with raspberry extract (GFSw_RE), honeysuckle flowers extract (LFL) and raspeberry extract (RE), C1—45 mg/mL; C2—25 mg/mL; C3—10 mg/mL, C4—5 mg/mL, C5—1 mg/mL for H, GFSw, H_LFL/H_RE, GFSw_LFL/GFSw_RE and C1 − C5—2.25–0.05 mg/mL for LFL and RE, ± error bars, $\alpha < 0.05$, n = 3, *—σ between 0.05 and 0.01, **—σ between 0.01 and 0.001, ***—$\sigma < 0.001$; Black stars indicate statistically significant values oh H_LFL/H_RE compared to H, Red stars indicate statistically significant values of GFS compared to GFSw_LFL/GFSw_RE, blue stars indicate statistically significant values of LFL/RE compared to C+ (the strain of *L. reuteri* without any supplements); (+)—prebiotic activity; (−) inhibition.

As observed in Figure 9, the majority of samples exhibited a positive impact on bacterial growth compared to the control (C+, which is the control sample of the strain

L. reuteri without any supplements), except GFSw at lower concentrations (1 mg/mL–96.76 ± 1.93%) and the LFL extract at low concentrations (1 mg/mL–97.29 ± 0.94%, and 5 mg/mL–97.63 ± 1.77%). Moreover, the highest bacterial growth was observed for the mixture of honey/GFSw with LFL compared to simple honey/GFSw, particularly at a concentration of 25 mg/mL (statistically significant differences with a σ-value of 0.000). The growth percentages were 145.38 ± 0.29% for H_LFL compared to 134.08 ± 3.10% for simple honey, and 154.69 ± 0.13% for GFSw_LFL compared to 143.38 ± 2.36% for GFSw. Furthermore, GFSw samples exhibited a slightly stronger influence on the growth of the *L. reuteri* strain compared to honey at all tested concentrations, except for the lowest concentration tested of 1 mg/mL, but all honey concentrations showed prebiotic activity. GFSw at the lowest concentration tested of 1 mg/mL had a slight inhibitory effect. The growth percentages from the highest to the lowest concentration tested were 160.02% to 96.76% for GFSw and 158.08% to 102.96% for honey.

At the highest concentration (45 mg/mL), the mixture of honey with raspberry extract (RE) demonstrated a lower prebiotic activity in terms of bacterial growth (149.93 ± 0.12%) when compared to simple honey (158.08 ± 2.19%). Conversely, slightly higher probiotic growth is observed for the mixture of honey/GFSw with RE compared to simple honey/GFSw, at 25 mg/mL (137.29 ± 1.98 for H_RE in comparison with 134.08 ± 3.09% for simple honey and 147.32 ± 3.53 compared to 143.38 ± 2.36% for GFSw), but the differences are not statistically significant. Overall, RE does not induce a prebiotic effect neither in the absence nor in the presence of honey/GFSw at the concentrations tested.

According to our data, honey mixtures with honeysuckle extracts exert a slightly more positive effect on bacterial growth than those with raspberry extract, the results being in the range of 164.63 ± 0.12–104.13 ± 0.26% for H_LFL and 149.93 ± 0.12–102.94 ± 1.43% for H_RE. Similar data were observed in the case of the GFSw mixture, the value of bacterial growth being in the range of 161.514 ± 0.17–100.55 ± 0.32% for GFSw_LFL and 157.47 ± 1.42–101.67 ± 0.82% for GFSw_RE.

Upon observation, it is evident that both the mixtures and individual samples of honey and GFSw exhibit concentration–dependent effects on the probiotic growth, where the observed effect diminishes as the concentration decreases.

Regarding L-lactic acid production, the effects are diverse due to the complex metabolic interactions between polyphenols and the carbohydrate metabolism in heterofermentative lactic bacteria (like the used *L. reuteri* DSM 20016) under anaerobic conditions (Figure 9C,D). The individual extracts, LFL and RE, behaved relatively differently. The maximum positive effect of LFL compared to C+ was at the median LFL concentration of 10 mg/mL (1.85 ± 0.00 g/L versus 0.86 ± 0.02 g/L L-lactic acid, respectively), followed by 25 mg/mL LFL (1.22 ± 0.03 g/L L-lactic acid). Both values were statistically significant. At the lowest LFL concentration tested, 1 mg/mL, there was a significant inhibition of L-lactic acid production (0.19 ± 0.05 g/L) compared to C+. The other two LFL concentrations, 45 and 5 mg/mL, did not have a significant effect compared to C+ (Figure 9C).

Most of the RE concentrations tested had a positive effect on the L-lactic acid production except the lowest RE concentration, 1 mg/mL, which had a slight but not statistically significant inhibitory effect (0.72 ± 0.03 g/L L-lactic acid) compared to control C+. The only statistically significant positive effect compared to C+ was at the highest RE concentration of 45 mg/mL (1.47 ± 0.02 g/L L-lactic acid).

A statistically significant difference was observed in the L-lactic acid content between H_LFL and simple honey at tested concentrations of 45 mg/mL, 10 mg/mL, and 1 mg/mL (Figure 9C). The first two concentrations exhibited a positive trend, with higher L-lactic acid content (2.96 ± 0.09 g/L at 45 mg/mL of H_LFL and 2.22 ± 0.04 g/L at 10 mg/mL of H_LFL) compared to simple honey fermentation (1.60 ± 0.03 g/L L-lactic acid at 45 mg/mL of H, and 1.52 ± 0.03 g/L L-lactic acid at 10 mg/mL of H). At the concentration of 1 mg/mL H_LFL, the L-lactic acid content was lower (0.75 ± 0.03 g/L) than that observed during H fermentation (2.36 ± 0.02 g/L).

In the case of the mixture of honey with raspberry extract, higher L-lactic acid content than in the case of simple honey was observed at the tested concentrations 10 and 5 mg/mL (2.01 ± 0.08 g/L for H_RE and 1.52 ± 0.03 g/L for H at 10 mg/mL, 1.93 ± 0.05 g/L for H_RE and 1.26 ± 0.01 g/L for H at 5 mg/mL).

According to our data, the highest positive effect of GFSw_LFL on the L-lactic acid content produced by *L. reuteri* compared to simple GFSw was at 10 mg/mL (2.22 ± 0.12 g/L L-lactic acid for GFSw_LFL compared to 1.93 ± 0.09 g/L L-lactic acid for GSFw). Other statistically significant positive effects on the L-lactic acid content were obtained at 25 and 1 mg/mL GFSw_LFL compared to GFSw. At the highest GFSw_LFL concentration tested, 45 mg/mL, there was an inhibition of L-lactic acid production (Figure 9C). The highest effect of GFSw_RE in comparison to GFSw was at 25 mg/mL tested concentration (2.24 ± 0.003 g/L for GFSw_RE and 1.18 ± 0.092 g/L for GFSw). The only GFSw_RE concentration that inhibited the production of L-lactic acid was the highest concentration of the mixture, 45 mg/mL (Figure 9D).

4. Discussion

Based on the Information found in previous studies [35–38], the concentration of phenolic compounds is influenced by various factors such as plant species, cultivars, environmental conditions, storage, extraction methods, and analysis techniques. Consequently, the reported concentrations of phenolic compounds can differ significantly across different scientific articles, and it is difficult to make direct comparisons. In our study, the value of TPC from honeysuckle was in the range of some literature data [21,36,38]. The TPC value was lower compared to the results reported by [38]–87.48 ± 6.32 mg GAE/g and by other researchers [36], who extracted the polyphenols in water (40.18 mg GAE/g) and in ethanol (5.25 mg GAE/g). The TPC values were higher compared to those obtained by our research group (392.093–1741.05 µg GAE/g DW) in another study [21]. This is due to repeated extraction from the same substrate. The value of TFC in our study was much lower than those revealed by [38] –52.51 mg CAE/g, but they expressed the TFC results as catechine equivalent (CAE) in comparison to our result (QE—quercetin equivalent). We tested catechine at the same concentrations as quercetin and found it to have much lower activity than quercetin; therefore, more catechin is needed to have the same AOA as 1 quercetin equivalent, which could explain the difference. Moreover, [38] employed a different methodology to determine the flavonoid content, involving the use of $AlCl_3$, $NaNO_2$, and NaOH. Additionally, the extraction of phenolic compounds in their study was conducted using a solution containing acetone, water, and acetic acid (70:29.5:0.5, *v*/*v*/*v*).

The results of TPC, TFC, and HAT (651.79 ± 5.11 GAE mg/100 g DW, 64.56 ± 2.12 QE mg/100 g DW, 587.38 ± 1.19 ChaE mg/100 g DW) were substantially higher in comparison to the raspberry extract (282 ± 10.72 GAE mg/100 g DW, 29.88 ± 1.05 QE mg/g DW, and 57.92 ± 2.92 Chae mg/100 g DW) from our previous results [5].

Chlorogenic acid and caffeic acid have also been identified in honeysuckle flowers by other authors [35–37]. The content of chlorogenic acid was lower than the values obtained by [35]–33.12 ± 0.25–48.84 ± 0. 04 µg/mg (depending on the growth stage of flowers), and our value was higher compared to the results reported by [21]—1331 µg/g. The content of caffeic acid in our case was lower than the results released by [35] 0.01–0.07 µg/mg for *L. japonica* and [39], which obtained a value of 0.195 ± 0.002 g/100 g. As can be seen, the values differ very much between studies.

As mentioned previously, the objective of this study was to build upon prior research by exploring the impact of incorporating a different polyphenol extract into honey or its biomimetic natural deep eutectic solvent (NaDES) and comparing it with the AOA behaviour of mixtures enriched with the raspberry extract from our previous study [5].

There is a correlation between the antioxidant activity of the samples (honey and plant extract) and the concentrations and profiles of the analyzed polyphenols, as determined by colourimetric tests (TPC, TFC, HAT) and HPLC analysis. In the case of the samples enriched with honeysuckle flower extract, the antioxidant activity (AOA) had to be analyzed at lower concentrations, ranging from 2 to 100 mg/mL, and the AOA of samples enriched with raspberry extract was analyzed at concentrations ranging from 5 to 200 mg/mL.

Additionally, in some cases, the samples with *L. caprifolium* extract exceed the detection limit of the method and instrument (absorbance at 200 mg/mL exceeds 3.5)–particularly in the case of the FRAP method, for which additional optimization of the method will be necessary in the future, such as the decrease of substrate concentration. Furthermore, at elevated concentrations (200 mg/mL for H_LFL/GFSw_LFL or 10 mg/mL for LFL), the dose–response curves for FRAP and CUPRAC methods deviated from linearity. Also, the AOA of the sample with LFL shows inhibition at 200 mg/mL when measured using the ABTS and DPPH methods (Figures S3 and S4 from Supplementary Material).

These issues can be attributed to the higher concentrations of active compounds extracted from *L. caprifolium* flowers compared to raspberry, as also observed in the TPC, HAT, TFC, and HPLC analyses.

As observed in Table 4, there are differences in terms of antioxidant activity of samples enriched with raspberry extract and samples enriched with honeysuckle flower extract. The samples with honeysuckle extract showed higher AOA than those with raspberry extract. We can conclude that there is a relationship between the polyphenol content and antioxidant activity, as the polyphenol content was higher in LFL than in RE by all the methods tested.

As mentioned previously, the incorporation of dried honeysuckle flower extract (LFL) into honey or its biomimetic NaDES, GFSw, increased the antioxidant activity of both honey and GFSw. To evaluate the modulation behaviour of LFL by honey and GFSw, the combination index (CI) was calculated.

The two extracts share some similar main compounds (caffeic acid–CA and epicatechin–EP), but they are different in the total polyphenolic content. In our previous study, we investigated the individual CA and EP as well [5]. It seems that the behaviour of the honeysuckle extract is similar to the behaviour of these tested polyphenols in certain cases. For example, in the case of the FRAP method, the AOA feature of LFL in honey was similar to CA in honey (CI = 0.866 $\pm$ 0.021), both of them exhibiting moderate synergistic effects. This similarity between LFL and CA is also observed for the CUPRAC and DPPH IC_{50} methods, the AOA behaviour of CA being moderate antagonism with CI value 1.43 $\pm$ 0.02 for CUPRAC and 1.17 $\pm$ 0.08 for DPPH IC_{50}. In the case of epicatechin and its behaviour in honey or GFSw, a similarity to LFL is observed in the case of DPPH IC_{20} (CI = 1.4 $\pm$ 0.11). The similarities in terms of antioxidant activity between LFL and CA are likely attributed to the higher content of CA in LFL (36.54 $\pm$ 0.04 mg/g DW)).

In comparison to raspberry extract, which exhibited varying behaviour depending on the tested concentrations, the AOA behaviour of honeysuckle extract demonstrated minimal variation. This can be observed in DPPH and ABTS assays at both 50% and 20% inhibition substrate, where the combination index (CI) values were nearly identical.

The interactions between LFL and GFSw are similar to RE and GFSw interactions in the case of the CUPRAC method, RE exerting moderate antagonistic behaviour (CI = 1.409 $\pm$ 0.023) and ABTS IC_{20} with nearly additive feature (CI = 1.011 $\pm$ 0.079).

In accordance with our previous study [5], we codified the CI intervals as follows: 0.5–0.7, which indicates strong synergism as (+2), 0.7–0.9, which denotes moderate synergism as (+1), 0.9–1.1 which implies nearly additive behaviour as (0), 1.1–1.5, which signifies moderate antagonism as (-1), 1.5–2, which indicates moderate to strong antagonism as (-2), and CI > 2, which represents strong antagonism as (-3).

Figure 10 reveals that the antioxidant behaviour (AOA) of polyphenols, including the two plant extracts and the polyphenol standards, in honey and GFSw exhibited a similar tendency, as indicated by a similar colour code. The strong antagonism values (-3) are assigned to dark blue and strong synergism(+2) to light blue. Out of the total 24 cases analyzed, 14 cases, accounting for approximately 58%, demonstrated that honey and GFSw (biomimetic NaDES) behaved similarly. From Figure 10, it is easy to see that LFL induced a more homogeneous behaviour than RE. It is also suggested that, in fact, the

AOA of the extracts is a result of the combined effects of different polyphenolic species, to which inter-polyphenolic interactions probably contribute. The quantitative values of the qualitative representation from Figure 10 can be found in Table S1.

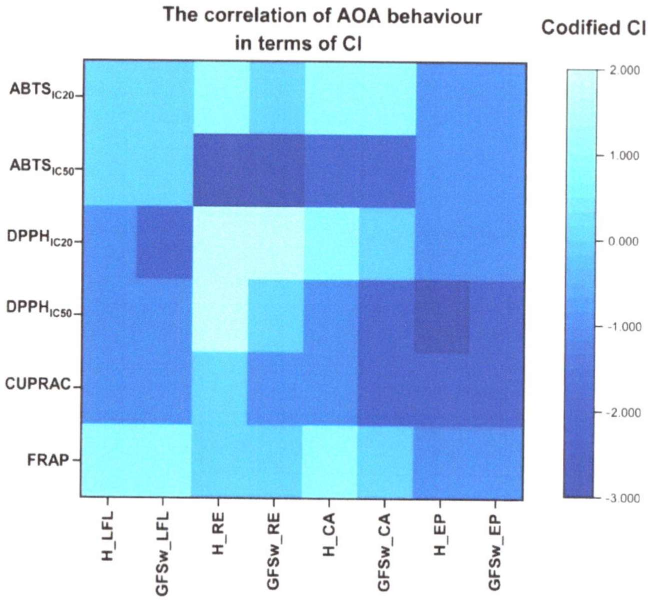

Figure 10. The Heatmap of the correlation of AOA in terms of CI.

Honey and polyphenols are both well-known for their prebiotic properties [14,40,41], which means they can support the growth and activity of beneficial bacteria in the gut. The prebiotic properties of honey are attributed to the presence of oligosaccharides, short-chain carbohydrates that are not fully digested in the upper gastrointestinal tract. Instead, they reach the colon intact, where they can exert prebiotic effects. Despite honey being primarily composed of simple sugars that are quickly absorbed in the small intestine, there are also di-, tri-, and oligosaccharides present in smaller quantities. These low-weight polysaccharides are likely to resist degradation by host enzymes, allowing them to reach the lower gut and contribute to the prebiotic effects of honey [13,14,42]. There are numerous studies on the prebiotic activity of honey [13,34,43–45]. In most cases, honey exerted positive effects on probiotic growth, and the prebiotic activity of the analysed substrate is often influenced by the concentration of substances they are exposed to. For example, in the study reported by [34], they tested two concentrations (1% and 2%) of several types of honey on five probiotic strains (*Lactobacillus acidophilus*, *Lactiplantibacillus plantarum*, *Lactobacillus gasseri*, *Lacticaseibacillus rhamnosus*, and *Lacticaseibacillus casei*,), the highest prebiotic activity expressed in terms of bacterial growth was observed at 2% honey in the case of *L. plantarum* (6× greater than control) and *L. acidophilus* (4× higher than control). In another study [45], different levels of active Manuka Factor (AMF:0.5, 10, 15, 20) were tested on the growth of the *strain Limosilactobacillus reuteri* DPC16, and it was observed that the highest biomass of probiotic substrate was obtained at AMF20 (4.77 mg/mL) after 36 h of incubation under anaerobic conditions compared to control (2.23 mg/mL).

In the context of our study, where honey and mixtures of honey/GFSw enriched with dried plant extracts were tested for prebiotic activity, it was observed that almost all concentrations of these mixtures had a positive impact on the growth of the probiotic strain *Limosilactobacillus reuteri* DSM 20016. Only the lowest concentration tested, 1 mg/mL did not show prebiotic activity. The higher the concentration of honey or the mixtures, the greater the positive effect on the probiotic growth. The data show that the prebiotic effects manifest after a certain concentration of compounds.

The prebiotic effect was mainly induced by honey, so probably by the saccharides present in it. The extracts showed a moderate prebiotic effect, statistically significant only in the case of LFL at the highest concentrations tested. As the higher the LFL extract, the higher the prebiotic effect, higher extract concentrations should be tested until reaching a plateau or an inhibition.

Based on our findings, the honey/GFS_W mixtures enriched with honeysuckle or raspberry did not show significant differences at most of the tested concentrations. The most significant difference was at the tested concentration of 45 mg/mL; the honey mixture with LFL exhibited a more pronounced effect (164.63 ± 0.01%) compared to the honey and RE mixture (149.93 ± 0.12%) on the bacterial growth (p-value = 0.019, <0.05). H_LFL had a prebiotic effect, and H_RE had an inhibitory effect compared to H. The higher content of hydroxycinnamic acids and lower anthocyanin content that we determined in the LFL extract compared to RE [5] could be involved in this effect, but more studies are needed to understand the mechanism behind this difference.

Most concentrations of honey and GFSw had positive effects on the production of L-lactic acid, as the lactic acid bacteria (LAB) metabolize sugars into lactic acid. The dependence of H/GFSw concentration presented an apparent hormetic behaviour nevertheless. In the case of the extracts, the maximum effect of LFL on the L-lactic acid content at lower concentrations than RE is probably related to the higher polyphenols content of LFL compared to RE. The mixtures presented a less-evident hormetic effect than H/GFSw, and there are two differences worth mentioning: (1) the trend of H_LFL is opposite (increasing effect with the concentration) to the trend of H_RE (decreasing effect with the concentration); (2) both GFSw_LFL and GFSw_RE differ from the corresponding honey mixtures, the first having the maximum effect at intermediate concentration tested (10 and 25 mg/mL, respectively). This suggests that honey interacts differently with LFL and RE, probably due to other compounds present in honey than saccharides.

Hydroxycinnamic acids are used as external electron acceptors by heterofermentative lactic acid bacteria [46], therefore decreasing the production of lactic acid, that is, the product of NAD(P)H reoxidation by using pyruvic acid as external electron acceptors [47]. However, other polyphenols exert different effects on lactic acid bacteria being metabolysed by several different enzyme classes besides the reductases, e.g., esterase and decarboxylases used for hydroxycinnamic acids [48]. The complex effects of polyphenols on lactic acid production require more investigation.

5. Conclusions

Although honey/GFSw modulated the AOA of both extracts, from synergism to antagonism, the modulation was different between the two extracts for some AOA activities, which could be explained by the differences between compositions in polyphenols of the two tested plant extracts. The effects are specific to complex chemical systems, wherein the biological and biochemical activities are determined by multiple interactions. The honeysuckle flower extract (LFL) has higher prebiotic activity than the raspberry extract. The effect on lactic acid production follows a hormetic behavior.

Supplementary Materials: The following supporting information can be downloaded at https://www.mdpi.com/article/10.3390/antiox12091678/s1, Figure S1. HPLC chromatogram of phenolic acids from honeysuckle flower extract; Figure S2. HPLC chromatogram of flavonoids from honeysuckle flower extract; Figure S3. Dose–response curves for the AOA using DPPH, ABTS, FRAP, and CUPRAC methods of the samples of honey (H), biomimetic NaDES with honey (GFSw), honey enriched with honeysuckle extract (H_LFL), GFSw enriched with honeysuckle extract (GFSw_LFL) at the concentration 2–200 mg/mL–it can be observed that the AOA of the mixture of honey and GFSw with honeysuckle extract decrease at concentrations greater than 100 mg/mL for DPPH, ABTS methods, and a little bit in the case of CUPRAC, in the case of FRAP methods the AOA cannot be measured at concentrations higher than 100 mg/mL (see main text); Figure S4. Dose–response curves for the AOA using DPPH, ABTS, FRAP, and CUPRAC methods of the samples of honeysuckle extract (LFL) at the concentration of 0.1–10 mg/mL, in the case of FRAP and CUPRAC, the curve was non-linear, and in the case of DPPH and ABTS the value decreases at the concentration 200 mg/mL; Figure S5. Dose–response curves for the AOA using DPPH, ABTS, FRAP, and CUPRAC methods of the samples of honeysuckle extract (LFL) at the concentration of 0.1–5 mg/mL; Figure S6. Dose–response curves for the AOA using DPPH, ABTS, FRAP, and CUPRAC methods of the samples of honey (H), biomimetic NaDES with honey (GFSw), honey enriched with honeysuckle extract (H_LFL), GFSw enriched with honeysuckle extract (GFSw_LFL) at the concentration 2–100 mg/mL of the sample. These curves were used to calculate and evaluate the modulation of LFL. Figure S7. Dose–response curves of the samples generated by SynergyPlot in terms of AOA measured by DPPH: LFL—honeysuckle flower extract, RE—raspberry extract, CA—caffeic acid, EP—epicatechin, AOA—antioxidant activity as a function of polyphenols concentration. Figure S8. Dose–response curves of the samples generated by SynergyPlot in terms of AOA measured by ABTS: LFL—honeysuckle flower extract, RE—raspberry extract, CA—caffeic acid, EP—epicatechin, AOA—antioxidant activity as a function of polyphenols concentration. Table S1. The correlation of AOA behaviour in terms of CI.

Author Contributions: Conceptualization, F.O. and D.C.-A.; methodology, L.D., D.C.-A. and D.P.; validation, D.C.-A.; formal analysis, L.D., D.C.-A. and I.M.; investigation, L.D., D.C.-A. and D.P.; resources, F.O. and I.M.; data curation, L.D. and D.C.-A.; writing—original draft preparation, L.D. and D.C.-A.; writing—review and editing, D.C.-A. and F.O.; visualization, F.O. and N.E.B.; supervision, F.O., D.C.-A. and N.E.B.; project administration, D.C.-A., F.O. and I.M.; funding acquisition, F.O. and I.M. All authors have read and agreed to the published version of the manuscript.

Funding: This work was funded by project POC-A1-A1.2.3-G-2015-P_40_352-SECVENT, Sequential processes to close bioeconomy side stream and innovative bioproducts resulted from these, contract 81/2016, SMIS 105684, funded by cohesion funds of the European Union, subsidiary projects 1391/2022 FructiRan and 1392/2022 NutriCell.

Institutional Review Board Statement: Not applicable.

Informed Consent Statement: Not applicable.

Data Availability Statement: All the data are contained within the article.

Acknowledgments: The authors thank Andra-Lavinia Nichițean, Rom Honey Group Srl, Iași, Romania, for the supply of the mulfloral honey samples, and Rodica Mihaela Frîncu for helpfull discussions on statistical representations.

Conflicts of Interest: The authors declare no conflict of interest.

References

1. Rao, P.V.; Krishnan, K.T.; Salleh, N.; Gan, S.H. Biological and therapeutic effects of honey produced by honey bees and stingless bees: A comparative review. *Rev. Bras. Farmacogn.* **2016**, *26*, 657–664. [CrossRef]
2. Viuda-Martos, M.; Ruiz-Navajas, Y.; Fernández-López, J.; Pérez-Álvarez, J.A. Functional Properties of Honey, Propolis, and Royal Jelly. *J. Food Sci.* **2008**, *73*, R117–R124. [CrossRef]
3. Dai, Y.; Jin, R.; Verpoorte, R.; Lam, W.; Cheng, Y.-C.; Xiao, Y.; Xu, J.; Zhang, L.; Qin, X.-M.; Chen, S. Natural deep eutectic characteristics of honey improve the bioactivity and safety of traditional medicines. *J. Ethnopharmacol.* **2020**, *250*, 112460. [CrossRef]
4. Craveiro, R.; Mano, F.; Paiva, A.; Duarte, A.R.C. Deep Eutectic Solvents: Exploring Their Role in Nature. In *Deep Eutectic Solvents: Synthesis, Properties, and Applications*; John Wiley & Sons: Hoboken, NJ, USA, 2019; pp. 95–110. [CrossRef]
5. Dimitriu, L.; Constantinescu-Aruxandei, D.; Preda, D.; Nichițean, A.L.; Nicolae, C.A.; Faraon, V.A.; Ghiurea, M.; Ganciarov, M.; Băbeanu, N.E.; Oancea, F. Honey and Its Biomimetic Deep Eutectic Solvent Modulate the Antioxidant Activity of Polyphenols. *Antioxidants* **2022**, *11*, 2194. [CrossRef] [PubMed]
6. da Silva, D.T.; Smaniotto, F.A.; Costa, I.F.; Baranzelli, J.; Muller, A.; Somacal, S.; Monteiro, C.S.A.; Vizzotto, M.; Rodrigues, E.; Barcia, M.T.; et al. Natural deep eutectic solvent (NADES): A strategy to improve the bioavailability of blueberry phenolic compounds in a ready-to-use extract. *Food Chem.* **2021**, *364*, 130370. [CrossRef]
7. Fernández, M.; Espino, M.; Gomez, F.J.V.; Silva, M.F. Novel approaches mediated by tailor-made green solvents for the extraction of phenolic compounds from agro-food industrial by-products. *Food Chem.* **2018**, *239*, 671–678. [CrossRef]
8. Morgana, N.M.; Magdalena, E.; Fernandez, M.d.l.A.; Fernanda, S.M. NADES for food industry innovation: Novel bioadditives based on olive oil byproducts. *Food Bioprod. Process.* **2022**, *134*, 193–201. [CrossRef]

9. Mišan, A.; Nađpal, J.; Stupar, A.; Pojić, M.; Mandić, A.; Verpoorte, R.; Choi, Y.H. The perspectives of natural deep eutectic solvents in agri-food sector. *Crit. Rev. Food Sci. Nutr.* **2020**, *60*, 2564–2592. [CrossRef]
10. Mišan, A.; Pojić, M. Chapter Thirteen—Applications of NADES in stabilizing food and protecting food compounds against oxidation. In *Advances in Botanical Research*; Verpoorte, R., Witkamp, G.-J., Choi, Y.H., Eds.; Academic Press: Cambridge, MA, USA, 2021; Volume 97, pp. 333–359.
11. Fuandila, N.N.; Widanarni, W.; Yuhana, M. Growth performance and immune response of prebiotic honey fed pacific white shrimp Litopenaeus vannamei to Vibrio parahaemolyticus infection. *J. Appl. Aquac.* **2020**, *32*, 221–235. [CrossRef]
12. Nagpal, R.; Kaur, A. Synbiotic Effect of Various Prebiotics on In Vitro Activities of Probiotic Lactobacilli. *Ecol. Food Nutr.* **2011**, *50*, 63–68. [CrossRef]
13. Sanz, M.L.; Polemis, N.; Morales, V.; Corzo, N.; Drakoularakou, A.; Gibson, G.R.; Rastall, R.A. In Vitro Investigation into the Potential Prebiotic Activity of Honey Oligosaccharides. *J. Agric. Food Chem.* **2005**, *53*, 2914–2921. [CrossRef]
14. Schell, K.R.; Fernandes, K.E.; Shanahan, E.; Wilson, I.; Blair, S.E.; Carter, D.A.; Cokcetin, N.N. The Potential of Honey as a Prebiotic Food to Re-engineer the Gut Microbiome Toward a Healthy State. *Front. Nutr.* **2022**, *9*, 957932. [CrossRef] [PubMed]
15. Becerril-Sánchez, A.L.; Quintero-Salazar, B.; Dublán-García, O.; Escalona-Buendía, H.B. Phenolic Compounds in Honey and Their Relationship with Antioxidant Activity, Botanical Origin, and Color. *Antioxidants* **2021**, *10*, 1700. [CrossRef]
16. Dżugan, M.; Tomczyk, M.; Sowa, P.; Grabek-Lejko, D. Antioxidant Activity as Biomarker of Honey Variety. *Molecules* **2018**, *23*, 2069. [CrossRef] [PubMed]
17. Alzahrani, H.A.; Boukraa, L.; Bellik, Y.; Abdellah, F.; Bakhotmah, B.A.; Kolayli, S.; Sahin, H. Evaluation of the antioxidant activity of three varieties of honey from different botanical and geographical origins. *Glob. J. Health Sci.* **2012**, *4*, 191–196. [CrossRef]
18. Miguel, M.; Antunes, M.; Faleiro, M. Honey as a Complementary Medicine. *Integr. Med. Insights* **2017**, *12*, 1178633717702869. [CrossRef]
19. Di Lorenzo, C.; Colombo, F.; Biella, S.; Stockley, C.; Restani, P. Polyphenols and Human Health: The Role of Bioavailability. *Nutrients* **2021**, *13*, 273. [CrossRef]
20. Ilies, D.C.; Radulescu, V.; Duţu, L.E. Volatile constituents from the flowers of two species of honeysuckle (Lonicera japonica and Lonicera Caprifolium). *Farmacia* **2013**, *62*, 194–201.
21. Dimitriu, L.; Preda, D.; Constantinescu-Aruxandei, D.; Oancea, F.; Băbeanu, N. Optimization of ultrasound-assisted extraction of polyphenols from honeysuckle (*Lonicera caprifolium*). *AgroLife Sci. J.* **2021**, *10*, 47–55. [CrossRef]
22. Kim, D.-O.; Jeong, S.W.; Lee, C.Y. Antioxidant capacity of phenolic phytochemicals from various cultivars of plums. *Food Chem.* **2003**, *81*, 321–326. [CrossRef]
23. Štefan, M.B.; Vuković Rodríguez, J.; Blažeković, B.; Kindl, M.; Vladimir-Knežević, S. Total Hydroxycinnamic Acids Assay: Prevalidation and Application on Lamiaceae Species. *Food Anal. Methods* **2014**, *7*, 326–336. [CrossRef]
24. Formagio, A.S.N.; Kassuya, C.A.L.; Neto, F.F.; Volobuff, C.R.F.; Iriguchi, E.K.K.; Vieira, M.D.C.; Foglio, M.A. The flavonoid content and antiproliferative, hypoglycaemic, anti-inflammatory and free radical scavenging activities of Annona dioica St. Hill. *BMC Complement. Altern. Med.* **2013**, *13*, 14. [CrossRef] [PubMed]
25. Tõnutare, T.; Moor, U.; Szajdak, L. Strawberry anthocyanin determination by pH differential spectroscopic method—How to get true results? *Acta Sci. Polonorum. Hortorum Cultus* **2014**, *13*, 35–47.
26. Hajimehdipoor, H.; Kondori, B.M.; Amin, G.R.; Adib, N.; Rastegar, H.; Shekarchi, M. Development of a validated HPLC method for the simultaneous determination of flavonoids in Cuscuta chinensis Lam. by ultra-violet detection. *DARU J. Pharm. Sci.* **2012**, *20*, 57. [CrossRef]
27. Dudonné, S.; Vitrac, X.; Coutière, P.; Woillez, M.; Mérillon, J.-M. Comparative Study of Antioxidant Properties and Total Phenolic Content of 30 Plant Extracts of Industrial Interest Using DPPH, ABTS, FRAP, SOD, and ORAC Assays. *J. Agric. Food Chem.* **2009**, *57*, 1768–1774. [CrossRef]
28. Proestos, C.; Lytoudi, K.; Mavromelanidou, O.K.; Zoumpoulakis, P.; Sinanoglou, V.J. Antioxidant capacity of selected plant extracts and their essential oils. *Antioxidants* **2013**, *2*, 11–22. [CrossRef]
29. Benzie, I.F.F.; Strain, J.J. [2] Ferric reducing/antioxidant power assay: Direct measure of total antioxidant activity of biological fluids and modified version for simultaneous measurement of total antioxidant power and ascorbic acid concentration. In *Methods in Enzymology*; Academic Press: Cambridge, MA, USA, 1999; Volume 299, pp. 15–27.
30. Meng, J.; Fang, Y.; Zhang, A.; Chen, S.; Xu, T.; Ren, Z.; Han, G.; Liu, J.; Li, H.; Zhang, Z.; et al. Phenolic content and antioxidant capacity of Chinese raisins produced in Xinjiang Province. *Food Res. Int.* **2011**, *44*, 2830–2836. [CrossRef]
31. Chou, T.C. Theoretical basis, experimental design, and computerized simulation of synergism and antagonism in drug combination studies. *Pharmacol. Rev.* **2006**, *58*, 621–681. [CrossRef]
32. Zheng, S.; Wang, W.; Aldahdooh, J.; Malyutina, A.; Shadbahr, T.; Tanoli, Z.; Pessia, A.; Tang, J. SynergyFinder Plus: Toward Better Interpretation and Annotation of Drug Combination Screening Datasets. *Genom. Proteom. Bioinform.* **2022**, *20*, 587–596. [CrossRef]
33. Brink, M.; Todorov, S.D.; Martin, J.H.; Senekal, M.; Dicks, L.M.T. The effect of prebiotics on production of antimicrobial compounds, resistance to growth at low pH and in the presence of bile, and adhesion of probiotic cells to intestinal mucus. *J. Appl. Microbiol.* **2006**, *100*, 813–820. [CrossRef]
34. Fratianni, F.; Ombra, M.N.; d'Acierno, A.; Caputo, L.; Amato, G.; De Feo, V.; Coppola, R.; Nazzaro, F. Polyphenols Content and In Vitro α-Glycosidase Activity of Different Italian Monofloral Honeys, and Their Effect on Selected Pathogenic and Probiotic Bacteria. *Microorganisms* **2021**, *9*, 1694. [CrossRef]

35. Kong, D.; Li, Y.; Bai, M.; Deng, Y.; Liang, G.; Wu, H. A comparative study of the dynamic accumulation of polyphenol components and the changes in their antioxidant activities in diploid and tetraploid Lonicera japonica. *Plant Physiol. Biochem.* **2017**, *112*, 87–96. [CrossRef] [PubMed]
36. Gao, B.; Zhu, L.; Liu, Z.; Li, Y.; He, X.; Wu, X.; Pehrsson, P.; Sun, J.; Xie, Z.; Slavin, M.; et al. Chemical Composition of Honeysuckle (Lonicerae japonicae) Extracts and Their Potential in Inhibiting the SARS-CoV-2 Spike Protein and ACE2 Binding, Suppressing ACE2, and Scavenging Radicals. *J. Agric. Food Chem.* **2023**, *71*, 6133–6143. [CrossRef] [PubMed]
37. Hsu, H.F.; Hsiao, P.C.; Kuo, T.C.; Chiang, S.T.; Chen, S.L.; Chiou, S.J.; Ling, X.H.; Liang, M.T.; Cheng, W.Y.; Houng, J.Y. Antioxidant and anti-inflammatory activities of Lonicera japonica Thunb. var. sempervillosa Hayata flower bud extracts prepared by water, ethanol and supercritical fluid extraction techniques. *Ind. Crops. Prod.* **2016**, *89*, 543–549. [CrossRef] [PubMed]
38. Zheng, J.; Yu, X.; Maninder, M.; Xu, B. Total phenolics and antioxidants profiles of commonly consumed edible flowers in China. *Int. J. Food Prop.* **2018**, *21*, 1524–1540. [CrossRef]
39. Chaowuttikul, C.; Palanuvej, C.; Ruangrungsi, N. Quantification of chlorogenic acid, rosmarinic acid, and caffeic acid contents in selected Thai medicinal plants using RP-HPLC-DAD. *Braz. J. Pharm. Sci.* **2020**, *56*, e17547. [CrossRef]
40. Mustar, S.; Ibrahim, N. A sweeter pill to swallow: A review of honey bees and honey as a source of probiotic and prebiotic products. *Foods* **2022**, *11*, 2102. [CrossRef]
41. Alves-Santos, A.M.; Sugizaki, C.S.A.; Lima, G.C.; Naves, M.M.V. Prebiotic effect of dietary polyphenols: A systematic review. *J. Funct. Foods* **2020**, *74*, 104169. [CrossRef]
42. Gibson, G.R.; Roberfroid, M.B. Dietary Modulation of the Human Colonic Microbiota: Introducing the Concept of Prebiotics. *J. Nutr.* **1995**, *125*, 1401–1412. [CrossRef]
43. de Melo, F.H.C.; Menezes, F.N.D.D.; de Sousa, J.M.B.; dos Santos Lima, M.; da Silva Campelo Borges, G.; de Souza, E.L.; Magnani, M. Prebiotic activity of monofloral honeys produced by stingless bees in the semi-arid region of Brazilian Northeastern toward Lactobacillus acidophilus LA-05 and Bifidobacterium lactis BB-12. *Food Res. Int.* **2020**, *128*, 108809. [CrossRef]
44. Das, A.; Datta, S.; Mukherjee, S.; Bose, S.; Ghosh, S.; Dhar, P. Evaluation of antioxidative, antibacterial and probiotic growth stimulatory activities of Sesamum indicum honey containing phenolic compounds and lignans. *LWT Food Sci. Technol.* **2015**, *61*, 244–250. [CrossRef]
45. Mohan, A.; Gutierrez-Maddox, N.; Meng, T.; He, N.; Gao, Y.; Shu, Q.; Quek, S.Y. Manuka Honey with Varying Levels of Active Manuka Factor (AMF) Ratings as an Anaerobic Fermentation Substrate for Limosilactobacillus reuteri DPC16. *Fermentation* **2021**, *7*, 128. [CrossRef]
46. Filannino, P.; Gobbetti, M.; De Angelis, M.; Di Cagno, R. Hydroxycinnamic acids used as external acceptors of electrons: An energetic advantage for strictly heterofermentative lactic acid bacteria. *Appl. Environ. Microbiol.* **2014**, *80*, 7574–7582. [CrossRef]
47. Zaunmüller, T.; Eichert, M.; Richter, H.; Unden, G. Variations in the energy metabolism of biotechnologically relevant heterofermentative lactic acid bacteria during growth on sugars and organic acids. *Appl. Microbiol. Biotechnol.* **2006**, *72*, 421–429. [CrossRef] [PubMed]
48. Gaur, G.; Gänzle, M.G. Conversion of (poly)phenolic compounds in food fermentations by lactic acid bacteria: Novel insights into metabolic pathways and functional metabolites. *Curr. Res. Food Sci.* **2023**, *6*, 100448. [CrossRef]

antioxidants

MDPI

Article

Honey and Its Biomimetic Deep Eutectic Solvent Modulate the Antioxidant Activity of Polyphenols

Luminița Dimitriu [1,2], Diana Constantinescu-Aruxandei [1,*], Daniel Preda [1,3], Andra-Lavinia Nichițean [2,4], Cristian-Andi Nicolae [1], Victor Alexandru Faraon [1], Marius Ghiurea [1], Mihaela Ganciarov [1], Narcisa Elena Băbeanu [2] and Florin Oancea [1,2,*]

1 Bioproducts Team, Bioresources Department, National Institute for Research & Development in Chemistry and Petrochemistry—ICECHIM, Splaiul Independenței No. 202, Sector 6, 060021 Bucharest, Romania
2 Faculty of Biotechnologies, University of Agronomic Sciences and Veterinary Medicine of Bucharest, Mărăști Blv., No. 59, Sector 1, 011464 Bucharest, Romania
3 Department of Analytical Chemistry and Environmental Engineering, Faculty of Chemical Engineering and Biotechnologies, University Politehnica Bucharest, Str. Gheorghe Polizu nr/1-7, Sector 1, 011061 Bucharest, Romania
4 Research and Development Department, Rom Honey Group Srl, Str. Grădinari nr. 1, Iași County, 700390 Iași, Romania
* Correspondence: diana.constantinescu@icechim.ro (D.C.-A.); florin.oancea@icechim.ro (F.O.); Tel.: +40-21-316-3071 (D.C.-A. & F.O.)

Abstract: Honey is a highly valued natural product with antioxidant, antimicrobial and anti-inflammatory properties. However, its antioxidant activity (AOA) is not as high as that of other honeybee products, such as propolis. Several polyphenol—honey formulations have been proposed up to now, most of them using maceration of biomass in honey or mixtures with liquid extracts, which either limit polyphenols bioavailability or destroy the characteristics of honey. To improve the health benefits of honey by increasing AOA and keeping its structural and sensory properties, we propose its enrichment in a polyphenol extract of raspberry after solvent evaporation. A honey-biomimetic natural deep eutectic solvent (NaDES) was prepared and compared with honey. The main polyphenols found in the raspberry extract were tested in combination with honey and NaDES, respectively. The AOA was determined by DPPH, ABTS, CUPRAC, and FRAP methods. The AOA behaviour of honey—polyphenol mixtures varied from synergism to antagonism, being influenced by the AOA method, polyphenol type, and/or mixture concentration. The honey-biomimetic NaDES resulted in similar AOA behaviour as with honey mixed with polyphenols. Honey seems to have additional properties that increase synergism or reduce antagonism in some cases. Honey and its biomimetic NaDES modulate AOA of polyphenols extract.

Keywords: honey; natural deep eutectic solvent (NaDES); biomimetic NaDES; antioxidant activity; polyphenols; synergism; antagonism; raspberry extract; polyphenol-enriched honey

check for updates

Citation: Dimitriu, L.; Constantinescu-Aruxandei, D.; Preda, D.; Nichițean, A.-L.; Nicolae, C.-A.; Faraon, V.A.; Ghiurea, M.; Ganciarov, M.; Băbeanu, N.E.; Oancea, F. Honey and Its Biomimetic Deep Eutectic Solvent Modulate the Antioxidant Activity of Polyphenols. *Antioxidants* **2022**, *11*, 2194. https://doi.org/10.3390/antiox11112194

Academic Editors: Josipa Vlainić and Ivana Tlak Gajger

Received: 25 September 2022
Accepted: 2 November 2022
Published: 6 November 2022

1. Introduction

Honey is a natural product produced by honeybees with several biological properties resulting from its multifaceted activities, i.e., antioxidant, antimicrobial, and anti-inflammatory activities. The chemical composition depends on diverse factors such as sugary source, floral nectar or aphid honeydew, environmental conditions, and genetic factors; it consists of more than 80% sugar [1]. Honey is considered to have the characteristics of a natural deep eutectic solvent (NaDES) due to the intermolecular interactions between monosaccharides and disaccharides and hydrogen bonds formed between them [2]. Deep eutectic solvents (DES) are a class of solvents produced by mixing a minimum of two components that at ambient conditions remain together in a liquid state. The DES components have a melting point above that of the eutectic point due to hydrogen bonds that are formed between these components [3].

Polyphenols are secondary metabolites of plants with various biological activities, such as antioxidant, antimicrobial, prebiotic, enzyme inhibition activity, and others [4]. The AOA activity of honey is believed to be mainly the result of the presence of different categories of polyphenols. Compared with other types of honeybee products, such as propolis, honey has a lower AOA per gram of sample. Because honey remains one of the most consumed honeybee products, it is desirable to improve its health benefits, including its antioxidant capacity. One solution is to enrich honey in polyphenols from other sources.

Several previous studies have used different types of approaches, from maceration with unprocessed biomass rich in polyphenols such as propolis, beebread, royal jelly, pollen, plant leaves, and fruits, to mixtures with polyphenol extractions in different ratios [5–10]. In general, the mixtures have been prepared with liquid hydroalcoholic extraction, which does not preserve the characteristics and properties of honey during formulation. On the other hand, macerations in honey could have the disadvantage of limiting the bioavailability of compounds released from the biomass. Separating the unsolubilized residues from polyphenol-rich biomass is difficult due to the high viscosity of honey and restriction on honey heating, which increases the formation of 5-hydroxymethylfurfural (5-HMF) by dehydration of glucose or fructose [11]. Another issue related to honey enrichment with polyphenol-rich biomass (such as propolis) is deterioration of the sensory properties due to the astringency and bitterness of the polyphenols [12].

Raspberry (*Rubus idaeus*) fruits are a good source of polyphenols (phenolic acids, flavonoids, and anthocyanins) for human nutrition, with antioxidant properties and excellent sensory characteristics [13]. Raspberry leaves and fruits added to rape honey in amounts of 0.5% and 1% and, respectively, 1% and 4%, were already demonstrated to enhance the antioxidant properties of honey and to increase its antibacterial and antiviral characteristics [10].

In this study, we enriched honey with dry extracts of raspberry as an example of enriching honey in antioxidant polyphenols after evaporation of the solvent. In this way, the water activity, as well as structural features of honey during formulation, can be preserved and, therefore, its stability and properties. The resulting product is a honeybee product fortified with polyphenols, which has superior sensory characteristics compared with other possible combinations, such as honey and propolis, the latter having strong astringency properties. We also investigated the modulation potential between honey and polyphenols and exploited the contribution of the main sugars in a honey-mimicking DES formulation, sugars that are believed to give the main characteristics of honey as a natural DES.

2. Materials and Methods

2.1. Materials

Fresh raspberries (*R. idaeus*, cv. Remontant, from Domeniul Cerbi, Marginea, Suceava, Romania) and multifloral honey (RomHoney Group, Iași, Romania) were used in this work. The raspberries were dried by lyophilization and were ground to a fine powder using an electrical grinder. The following chemicals were used: pharmaceutical ethanol 96% (Chimopar Srl, Bucharest, Romania), D(+)-Glucose anhydrous extra pure, D(−)-Fructose, extra pure, D(+) Saccharose, reagent grade (Scharlau, Barcelona, Spain) Trolox 97% (Acros Organics, Thermo Fisher Scientific, Pittburghs, PA, USA), Gallic acid, 2,2-Diphenyl-1-picrylhydrazyl (Sigma-Aldrich, Merck Group, Darmstad, Germany), 2,2′-Azino-bis (3-ethylbenzothiazo-line-6-sulfonic acid) diammonium salt, 98%, 2,4,6-tri (2-pyridyl-1,3,5-triazine) 98% (Alfa Aesar, Kandel, Germany), Folin Ciocalteu's phenol reagent, Iron chloride (III) (Merck, Darmstadt, Germany), hydrochloric acid, acetic acid (Chimopar Srl, Bucharest, Romania), sodium acetate (Scharlau, Barcelona, Spain), HPLC standards: ferulic acid, p-coumaric acid, caffeic acid, quercetin dihydrate (Sigma-Aldrich, Merck Group, Darmstad, Germany), syringic acid, luteolin, (+)-rutin trihydrate, (Alfa Aesar, Haverhill, MA, USA), chlorogenic acid, myricetin (Cayman Chemical, Ann Arbor, MI, USA), apigenin, (−) epicatechin (Roth, Karlsruhe, Germany), and kaempferol (Cayman Chemical, Ann Arbor, MI, USA).

2.2. *Hydroalcoholic Extraction of Polyphenols from Raspberry*

The polyphenols were extracted from freeze-dried raspberries by ultrasound-assisted extraction with a 70% ethanol solution, and a ratio of substrate to solvent of 1:10, for 30 min. at room temperature. The extraction was performed in an ultrasonic bath (P = 580 W, frequency = 37 Hz), with the temperature at 20–30 °C by adding ice to the bath. The samples were then centrifuged for 20 min at 8500 rpm, the supernatant was removed, and the same volume of solvent was added over the remaining substrate to repeat the extraction. The two resulting extract fractions were mixed together.

2.3. *Analysis of Polyphenolic Content of Raspberry Extract and Honey*

2.3.1. HPLC Analysis

High-pressure liquid chromatographic (HPLC) analysis of phenolic acids and flavonoids was performed using Dionex Ultimate 3000 equipment (Thermo Fisher Scientific, Waltham, MA, USA) with VWD-3100 detector, and the chromatograms were processed by Chromelleon 7.0 software (Thermo Fisher Scientific, Waltham).

Solid-Phase Extraction

For the extraction of polyphenols from honey, solid-phase extraction was performed based on a previously described method [14] with some modifications. Five grams of honey were dissolved with 10 mL of MilliQ water and passed through a Strata®SDB-L-conditioned cartridge (100 µm styrene-divinylbenzene 500 mg/3 mL, Phenomenex, Torrance, CA, USA) with a mixture of acetonitrile, methanol, and MilliQ water (1:1:1), at a flow rate of 1 mL/min. Elution was effected with a mixture of Methanol-Acetonitrile 2:1 at 1 mL/min.

HPLC Analysis of Phenolic Acids

The analysis of phenolic acids was conducted according to a method described by [15] on a Luna Omega 5 µm Polar C18 100 Å column (250 mm × 4.6 mm) (Phenomenex, Torrance, CA, USA). The method involved using a gradient program with a two-solvent system (A: aqueous solution with 0.1% formic acid and B: methanol), applied as follows: 0–25 min. 5% B, 25–33 min. 30% B, 34–40 min. 5% B. The flow rate was set at 1.25 mL × min^{-1}, and an injection volume of 10 µL was used to detect phenolic acids at 280 nm. The calibration curve consisted of several standard concentrations between 18.125–1000 µg/mL. The coefficients of determination (R^2) were above 0.9996, which indicated good linearity.

HPLC Analysis of Flavonoids

The HPLC analysis of flavonoids from the raspberry extract and honey was performed according to the method described by [16] on an Omega 5 µm Polar C18 100 Å column (250 mm × 4.6 mm) (Phenomenex, Torrance). The compounds were separated with a gradient elution of the mobile phase composed of (A) MeOH and (B) 0.5% H_3PO_4. The gradient elution program was set as follows: 0–10 min 15% A and 85% B, 15–25 min 85% A and 15% B, 25–30 min. 60% A and 40% B. The flow rate of the mobile phase was 1.5 mL/min, and the column temperature was 25 °C. to detect flavonoids at 280 nm.

Flavonoids were identified and quantified by matching the retention time and their spectral characteristics with the standards using a calibration curve.

2.3.2. Total Polyphenol Content

The total polyphenol content (TPC) of the extracts was determined by the Folin-Ciocalteau method described by [17]. Briefly, 10 µL of sample solution or standard solution was mixed with 90 µL double-distilled water (ddH_2O) and 10 µL of Folin Ciocalteu reagent. After 5 min of mixing, 100 µL of 7% Na_2CO_3 and 40 µL ddH_2O were added to the mixture. The absorbance was measured spectrophotometrically using a plate reader (CLARIOstar,

BMG LABTECH, Ortenberg, Germany) at 765 nm after 60 min of incubation at room temperature. The calibration curve was in the range of 5–30 μg/mL of gallic acid in 70% ethanol. The results were expressed as mg gallic acid equivalent/100 g dry weight (DW) of the sample (mg GAE/100 g).

2.3.3. Total Flavonoid Content

The total flavonoid content (TFC) of the extracts was determined using the aluminum chloride/sodium acetate method according to [18] with some modifications. To evaluate the TFC, 0.1 mL of sample/standard was mixed with 0.1 mL of 10% sodium acetate and then 0.12 mL of 2.5% $AlCl_3$ and 0.68 mL of ddH_2O were added to the mixture. The absorbance was read at $\lambda = 430$ nm after 45 min of incubation at room temperature. The results were expressed as quercetin equivalent mg/100 g DW of the sample.

2.3.4. Total Hydroxycinnamic Acid Content

Total hydroxycinnamic acid content (HAT) was determined by a method adapted from the European Pharmacopoeia [19]. Briefly, 0.25 μL of sample/standard was mixed with 50 μL 0.5 M HCl, then 50 μL of solution consisting of 1% (*w*/*v*) $NaNO_2$ and 1% (*w*/*v*) Na_2MoO_4 were added, followed by 50 μL of 8.5% NaOH and 75 μL ddH_2O. The absorbance was read at $\lambda = 524$ nm. A calibration curve with chlorogenic acid at concentrations in the range 0–50 μg/mL in 70% (*v*/*v*) ethanol was performed to quantify hydroxycinnamic acids. The results were expressed as mg chlorogenic acid equivalent/100 g DW of the sample.

2.3.5. Total Anthocyanin Content

Total anthocyanin content (TAC) was determined by the pH differential spectroscopic method [20]. Briefly, 1.5 mL of extracts were diluted in two different buffers: in 0.025 M potassium chloride buffer pH = 1, and in 0.4 M sodium acetate buffer pH = 4.5 respectively. The absorbance (A) was measured at 520 and 700 nm (Ocean Optics UV-VIS-NIR, Orlando, FL, USA) after 30 min of incubation at room temperature. The TAC was calculated using the molar absorptivity coefficient (ε) and molecular weight (MW) of cyaniding 3-glucoside (ε = 26,900 M^{-1} cm^{-1} and MW = 449.2 g/mol). The results were calculated as follows: $A_{sp} = (A_{520}-A_{700})_{pH1.0}-(A_{520}-A_{700})_{pH4.5}$ and $TAC = (A_{sp} \times MW \times DF \times V \times 1000)/(\varepsilon \times L \times m)$, where A_{sp} is the absorbance of sample, DF is dilution factor, L is the cuvette optical pathlength (1 cm), *V*-volume of the extracts (L), and m is the weight of the sample (g). TAC was expressed as mg cyaniding 3-glucoside equivalent/100 g DW of the sample.

2.4. Preparation of Honey with Raspberry Extract/Polyphenolic Standard for AOA Activities

The extracts of raspberry were split equally into two equal fractions that were concentrated to dryness (S_CD) using a semi-automated evaporation system, i.e., a MultiVap54 (Lab tech, Sorisole, Italy) at 40 °C. One of the fraction S_CD was resuspended in honey (H) at a ratio of 1:20 (*w*/*w*), resulting the honey-raspberry mixture sample (H_RE). The other fraction S_CD was resuspended in 70% ethanol solution at the same ratio as in honey (1:20 *w*/*v*), resulting in the RE sample. The sample H_RE was obtained by solubilizing the extract fraction S_CD in honey using an ultrasonic bath, mixing thoroughly, and leaving the polyphenols to diffuse overnight in honey. For AOA, the samples H and H_RE were solubilized in 70% (*v*/*v*) ethanol at a concentration of 0.2 g/mL (*w*/*v*). The AOA of the samples was assayed using four spectrophotometric methods: radical scavenging activity (ABTS and DPPH) and reducing antioxidant power (CUPRAC and FRAP). The AOA was performed at several concentrations, and calibration curves were calculated for each method. The concentration values of RE tested individually were equivalent to the concentrations of RE in mixtures with honey/GFSw. To check if the behaviour of RE held for individual polyphenols dissolved in honey, we prepared the mixture of honey and individual major polyphenols found in the raspberry extract: caffeic acid (CA) and epicatechin (EP). Each polyphenol was solubilized in honey/70% ethanol at 0.5 mg per g (*w*/*w*) of honey or 0.5 mg per mL (*w*/*v*) of 70% ethanol using an ultrasonic bath and the polyphenols were left to diffuse overnight. The AOA was performed at several concentrations, and calibration curves were calculated for each method, as in the case of RE. The concentration values of CA/EP tested individually were equivalent to the concentrations used in mixtures with honey/GFSw. The final concentrations tested were in the range of 5 to 200 mg/mL honey or GFSw and their mixtures, 0.25 to 10 mg/mL RE, and 0.0025 to 0.1 mg/mL CA or EP either individually or at the corresponding mixture concentrations.

2.5. *Antioxidant Activity*

2.5.1. Radical Scavenging Activity by ABTS Assay

The antioxidant method of neutralizing the ABTS radical was determined by the ABTS radical cation discoloration test [21]. $ABTS^+$ was produced by the reaction between 7 mM ABTS in water and 2.45 mM potassium persulfate, incubated in the dark at room temperature for 12–16 h before use. The $ABTS^+$ solution was then diluted with 96% ethanol to have an absorbance of 0.700 ± 0.04 at 734 nm. A volume of 20 µL of sample or standard solution (prepared as described above, 2.4) was mixed with 180 µL of diluted $ABTS^+$ solution, and the absorbance was measured at 734 nm after 30 min of incubation at room temperature.

2.5.2. Radical Scavenging Activity by the DPPH Assay

DPPH (2,2-diphenyl-1-picrylhydrazyl) free radical-scavenging activity of the samples was performed according to [22] with some modification. Briefly, 100 µL of sample/standard solution was mixed with 100 µL of 0.3 mM DPPH solution in 99.6% (*v*/*v*) ethanol. The absorbance was read at λ = 517 nm after 30 min of reaction using a UV-Vis plate reader (CLARIOstar, BMG LABTECH, Ortenberg, Germany).

2.5.3. Cupric-Ion Reducing Antioxidant Capacity (CUPRAC) Assay

The antioxidant method of cupric ion reducing capacity (CUPRAC) was performed according to a method adapted from [23] as follows. Ten microliters of sample/standard solutions were mixed with 30 µL $CuSO_4$ (5 mM), 30 µL neocuproine (3.75 mM) and 280 µL distilled water, reaching a final volume of 350 µL. After 30 min, the absorbance was measured at λ = 450 nm. A calibration curve of Trolox as the standard substance was calculated based on several Trolox concentrations tested. The standard solutions started from a stock solution of 10 mM Trolox in 70% (*v*/*v*) ethanol and were used for the calibration curve within the concentration interval of 0–2 mM Trolox.

2.5.4. Ferric-Ion Reducing Antioxidant Power (FRAP) Assay

The antioxidant method of ferric ion reducing power (FRAP) is based on the ability of antioxidants to reduce the tripyridyltriazine-Fe^{3+} (Fe (III)-TPTZ) complex to the blue-colored tripyridyltriazine-Fe^{2+} (Fe (II)-TPTZ) complex by the action of electron released by the antioxidant.

The determination of the antioxidant power of iron reduction was performed by the method described by [24] with some modifications. The FRAP reagent was prepared by mixing 10 parts of 0.3 M acetate buffer pH 3.6 with one part of 10 mM TPTZ (solubilized in 40 mM HCl) and one part of 20 mM $FeCl_3$ solution (10:1:1). An aliquot of 15 µL of /standard solution was added to the 285 µL FRAP reagent. The absorbance was read at 593 nm after incubation for 30 min at 37 °C in the dark. A calibration curve of Trolox as the standard substance was calculated based on several Trolox concentrations tested. The calibration curve was made from the concentration range of 0–450 µM Trolox/mL in 70% (*v*/*v* ethanol).

2.5.5. Evaluation of Modulation Activity between Honey/GFSw and Polyphenols

In order to establish possible modulations between honey/GFSw and polyphenols, the combination index and isobologram analyses were performed. For DPPH and ABTS, which presented non-linear effect dependence on concentration, the Webb analysis was also performed, in which the theoretical inhibited fraction was calculated by the formula: $100 - ((100 - f_{n,A}) \times (100 - f_{n,B}))$, where $f_{n,A}$ and $f_{n,B}$ represent non-inhibited fractions by *A* and *B* when tested individually, respectively. The combination index (CI) was calculated from the formula:

$$CI = \frac{C_{A,m}}{C_{A,i}} + \frac{C_{B,m}}{C_{B,i}}$$

$C_{A,m}$ and $C_{B,m}$ are the concentrations of *A* and respectively *B* in the mixture that give the same effect as the individual concentrations, $C_{A,i}$ and respectively $C_{B,i}$. *A* and *B* represent the two components that are mixed. *A* was honey/GFSw and *B* was RE/CA/EP. Isobologramic diagrams were produced based on these values. The theory behind the methods is described in reference [25].

The CI and isobologram analysis were determined at IC_{50} and IC_{20} (50% and 20% substrate inhibition, respectively) in the case of DPPH and ABTS. These values were calculated based on the median-effect equation proposed by Chou group that transforms a non-linear dose-effect curve into a linear form:

$$\log(\frac{f_i}{f_n}) = a \times \log(conc) + b$$

where f_i and f_n are the inhibited and non-inhibited fractions, respectively, a is the slope and $b = -a \times IC_{50}$. The non-inhibitory (f_n) and inhibitory (f_i) fractions were expressed as percent and calculated from the formula:

$$f_n = \frac{(A_0 - blank_0) - (A_c - blank_c)}{(A_0 - blank_0)} \times 100$$

and $f_i = 100 - f_n$, respectively, where A_0 and blank 0 are the absorbances of the substrate in the absence of the antioxidant and of the corresponding blank (solvent without substrate), respectively, and A_C and blank C are the absorbances of the substrate in the presence of concentration C of the antioxidant and of the corresponding blank (antioxidant without substrate), respectively. In the case of CUPRAC and FRAP, the Trolox calibration curve was used to express CI at 1 mM Trolox equivalent.

2.6. *Preparation of Honey-Mimetic Natural Deep Eutectic Solvent*

The natural deep eutectic solvent (NaDES), which mimics honey, was prepared based on the content of the main sugars in multifloral honey according to the literature data [26]. The NaDES was synthesized by mixing glucose, fructose, saccharose, and water (1:1.3:0.2:5 by molar ratios). The mixture was heated and stirred at 70 °C until a clear, viscous mixture was formed ($\approx 2^{1/2}$ h). From this point forward, the NaDES formed, abbreviated as GFSw, was cooled to room temperature and kept in a closed bottle until use.

2.7. *Physico-Chemical Characterisation of Honey-Mimetic Natural Deep Eutectic Solvent (GFSw) and Honey*

2.7.1. FTIR Analysis

FTIR-ATR spectroscopy measurements were performed using a Spectrum GX spectrometer (Perkin Elmer, Beaconsfield, UK), applying the Attenuated Total Reflectance (ATR) technique with a diamond crystal, according to the manufacturer's instructions. IR absorption spectra were obtained by the acquisition of 32 scans, with a resolution of 4 cm^{-1} in the region between 4000 and 600 cm^{-1}. The spectra of GFSw were compared with honey.

2.7.2. Thermogravimetric Analysis

Thermogravimetric analysis (TGA) was performed using a TA-Q5000 V3.13 (TA Instruments, Inc., New Castle, DE, USA) device with nitrogen as the purge gas at a 50 mL/min flow rate, according to the manufacturer's instructions. The runs were carried out using a 10–15 mg sample in a platinum pan and a synthetic air atmosphere with 50 mL/min airflow. The temperature range was between 25–700 °C with a heating rate of 10 °C/min.

2.7.3. Differential Scanning Calorimetry Analysis

Differential scanning calorimetry (DSC) analysis was performed using a DSC Q2000 (TA Instruments, Inc., New Castle, DE, USA) under helium flow (25 mL/min), according to the Manufacturer instructions. Samples weighing around 10 mg were packed in aluminum pans, and MDSC analysis was carried out to determine the thermodynamic parameters (transition temperature—Tg, specific heat capacity—ΔCp, enthalpy—ΔH) and the glass transition.

2.7.4. Surface Tension Analysis

The surface tension of DES and honey was measured by optical tensiometer OCA 50EC (DataPhysics Instruments GmbH, Filderstadt, Germany), according to the manufacturer's instructions. The method was based on evaluating the shape of a liquid droplet suspended at the needle end of a syringe. The diameter of the needle had an outer diameter of $\Phi = 1.83$ mm, an inner diameter of $\Phi = 1.36$ mm, and the length of the needle was $l = 38.1$ mm. The shape of the drop represents the result of the interfacial tension of the analyzed liquid (a spherical shape produces a minimum surface area) and the gravity (elongation of the drop due to the mass of the liquid). The Laplace-Young equation was used to determine the surface tension by software calculation.

2.7.5. Measurement of Specific Density

The densities of the DES and honey samples were measured using a density meter Easy D40 (Mettler Toledo, Columbus, OH, USA), according to the manufacturer's instructions. For each sample, three replicates were obtained, and the average was reported.

2.7.6. Measurement of Water Activity

Water activity was measured at 22 °C using LabMaster-aw neo (Novasina AG, Lachen, Switzerland) equipment, according to the manufacturer's instructions. For each determination, four replicates were obtained, and the average was reported.

2.7.7. Measurement of pH

The pH values of samples were measured using a pH-meter SevenCompact 2S10 (Mettler Toledo, Columbus, OH, USA), according to the manufacturer's instructions.

2.7.8. Measurement of Refractive Index and Total Soluble Solids

Refractive index and total soluble solids (TSS) were determined using a digital refractometer (MyBrix, Mettler Toledo, Columbus, OH, USA), according to the manufacturer's instructions. The refractometer was first calibrated with double-distilled water. The total soluble solids of honey and GFSw were represented by total soluble sugar and expressed as Brix degrees (one percent of TSS is considered one ^{0}Brix) [27]. For each determination, four replicates were obtained, and the average values were reported.

2.7.9. Spray-Drying

Honey and GFSw were powdered by a spray-drying method. The honey solution was prepared for spray-drying according to [28] with some modification by mixing with maltodextrin (MD) and ddH_2O to obtain a solution with 75% solids (*w/v*). The ratio between honey and MD was 60:40 (*w:w*). The GFSw solution was prepared in the same way. The spray drying of honey and of the GFSw solutions was performed using a Mini Spray Drier B—290 (Büchi, Flawil, Switzerland). The spray drier was also equipped with a pre-drying air module that worked in parallel during the drying process with the spray drier. During the spray drying process, the pre-drying air module showed a 69–72% dehumidification at 0–1 °C. Honey and GFSw solution was spray dried under the following conditions. The feed solution was introduced, along with the dehumidified drying air through a three-fluid nozzle system mounted on top of the spray drier, the inlet air drying temperature was set at 120 °C, and the debit of the peristaltic pump was set at 10% (3 mL/min). The debit flow meter of drying air was set at 55 mm (670 L/h, with a 1.05 bar pressure drop, meaning that the actual inserted air volume was 1374 L/h at standard temperature and pressure, as recorded in the instructions manual). During the spray-drying process, the outlet temperature was recorded at 50 °C for honey and 74 °C for GFSw. Powders were kept in a desiccator to prevent moisture.

2.7.10. Scanning Electron Microscopy

Scanning electron microscopy (SEM) was performed with TM4000Plus II tabletop electron microscope (Hitachi, Tokyo, Japan) at 5 kV electron acceleration voltage, 200× and 600× magnification, backscattered-electron (BSE) detector, and standard (M) vacuum mode, according to the manufacturer's instructions.

2.7.11. X-ray Diffraction

X-ray diffractograms were obtained with a SmartLab diffractometer (Rigaku, Tokyo, Japan) in "parallel beam" geometry, using Cu-K_α radiation (λ = 1.5406 Å) obtained at an acceleration voltage of 45 kV and emission current of 200 mA, and a scintillator detector, according to the manufacturer's instructions. The diffractograms were recorded in the 2θ range of 5–90° in steps of 0.02° at a speed of 4°/min.

2.8. Preparation of GFSw with Raspberry Extract and with Polyphenolic Standards

The mixtures of GFSw and polyphenols were prepared in a similar way to the mixtures using honey described above. The extracts of raspberry were split equally into two fractions and were concentrated to dryness using a semi-automated evaporation system MultiVap54 (Lab tech, Sorisole, Italy) at 40 °C. One of the samples was resuspended in GFSw at a ratio of 1/20 (*w*/*w*), and the other one was resuspended in 70% ethanol solution at the same ratio of 1/20 (*w*/*v*). The extract was solubilized in GFSw using an ultrasonic bath, mixed thoroughly, and the polyphenols were left to diffuse overnight. The AOA of the samples was assayed using the same spectrophotometric methods: radical scavenging activity (ABTS and DPPH) and reducing antioxidant power (CUPRAC and FRAP).

The individual major polyphenols found to be in raspberry extract (caffeic acid and epicatechin) were solubilized in GFSw at the same concentration and in the same way for honey described above.

2.9. Statistical Analysis

We calculated confidence intervals at 95% confidence for the isobolographic analysis of the AOA activities. The confidence intervals were calculated by subtracting and adding the value 1.96 × SD/sqrt (n), where SD is the standard deviation and n is the number of measurement replicates (n = 3 in all cases).

3. Results

The polyphenolic composition of the raspberry extract was determined based on several assays: total polyphenolic content (TPC), total flavonoid content (TFC), total hydroxycinnamic content (HAT), total anthocyanin content (TAC), and HPLC analysis.

3.1. Screening of Bioactive Compounds in Honey and Raspberry Extract

3.1.1. Total Polyphenols, Flavonoids, and Anthocyanins Content

The results of TPC, TFC, HAT, and TAC of raspberry and honey samples are summarized in Table 1.

Table 1. The results of total polyphenol content (TPC), total flavonoid content (TFC), total anthocyanin content (TAC).

	TPC GAE mg/100 g DW	TFC QE mg/100 g DW	HAT Chae mg/100 g DW	TAC mg cya, 3-Glu equivalent/100 g DW
Raspberry	282 ± 10.72	29.88 ± 1.05	57.92 ± 2.92	62.92 ± 0.64
Honey	4.63 ± 0.30	2.25 ± 0.057	2.89 ± 0.086	-

Values are mean ± SD (n = 3). GAE—gallic acid equivalent, QE—quercetin equivalent, Chae—chlorogenic acid equivalent, Cya3-Glu—cyaniding 3-glucoside, DW—dry weight.

The results indicated that the TPC, TFC, and HAT (282 ± 10.72 GAE mg/100 g DW, 29.88 ± 1.05 QE µg/g DW, and 57.92 ± 2.92 Chae mg/100 g DW) were significantly higher than that of honey (4.63 ± 0.30 GAE mg/100 g DW, 2.25 ± 0.057 QE µg/g DW and 2.89 ± 0.086 Chae mg/100 g DW). According to literature data [29–31], the concentration of phenolic compounds (which also includes polyphenols, flavonoids, hydrocinnamic acids, and anthocyanins) is dependent on numerous factors (species, cultivars, environmental, storage, methods of extractions, and analysis). For these reasons, the concentration of phenolic compounds varies in different scientific articles. Our results correspond to the literature. The value of TPC from raspberry was in the range of reported results by [32], who obtained a total phenolic content in the range 164.54–416.24 mg GAE/100 g, and also was higher than the results reported by [13] −140.31–160 mg/100 g FW. The total flavonoid content was lower than those revealed by [13], who obtained values in the range 88.98–111.14 mg/100 g. The total anthocyanin value was slightly higher than the values reported by [32,33] and lower than the values obtained by [13].

Our results for TPC and TFC were in agreement with literature data [34,35] concerning polyphenols in Romanian honey samples. Our value of TPC was lower than the values reported by [35] and slightly higher than the results reported by other authors [34]. These differences between our data and the data from the literature could be related to the composition of honey, which is affected by various factors, such as the floral and geographical origin, the collection season, the storage, and the harvesting technology.

3.1.2. HPLC Analysis

The identification and quantification of phenolic acids and flavonoids in honey and raspberry were performed by an HPLC technique. In the case of the honey sample, the HPLC analysis of phenolic acids from the honey sample was performed after a preliminary isolation step of the phenolic compound by SPE from the honey matrix. The chromatograms of polyphenolic compounds from raspberry and honey samples are presented in Supplementary Materials (Figures S1–S4). The deconvolution of the peaks was performed in OriginPro 2018 (OriginLab Corporation, USA). The amounts of phenolic acids and flavonoids identified in honey and raspberry are summarized in Table 2.

Phenolic compounds found in the honey analyzed included 4-hydroxybenzoic acid, caffeic acid, p-coumaric acid, protocatechuic acid, ferulic acid, rutin, quercetin, apigenin, and myricetin. Overall, the concentration of phenolic compounds was in agreement with data reported in other scientific articles [34–36]. The content of protocatechuic acid, caffeic acid, myricetin, and 4-hydroxybenzoic acid in our honey sample was higher than the value reported by [34], who obtained 0.15, 0.14, and 0.50 mg/100 g, and 0.08 mg/100 g, respectively. They also reported the concentration of quercetin at 1.23 mg/100 g, which was lower than our value. The hydroalcoholic raspberry extract was analyzed to identify and quantify phenolic compounds. The phenolic compounds found in the raspberry extract were caffeic acid, ferulic acid, p-coumaric acid, epicatechin, rutin, quercetin, kaempferol, apigenin, and myricetin, and corresponded with literature data [13]. The highest content of phenolic acids analyzed was identified as caffeic acid with 770.96 ± 24.06 µg/g, and epicatechin with the highest flavonoid content at 1684.06 ± 77.88 µg/g.

Table 2. Polyphenols (phenolic acids and flavonoids) from raspberry and honey by HPLC analysis.

Polyphenols	Raspberry, µg/g	Honey, µg/g
Phenolic acids:		
Caffeic acid	770.96 ± 24.06	1.88 ± 0.02
Ferulic acid	7.14 ± 0.57	26.60 ± 1.46
p-coumaric acid	2.72 ± 0.39	3.06 ± 0.05
4-hydroxybenzoic acid	-	6.39 ± 0.09
Protocatechuic acid	-	2.66 ± 0.15
Flavonoids:		
Epicatechine	1684.06 ± 77.88	-
Rutin	83.43 ± 1.58	1.60 ± 0.03
Quercetin	16.12 ± 1.31	6.50 ± 0.44
Kaempferol	27.73 ± 2.09	-
Apigenin	22.44 ± 0.28	13.27 ± 0.39
Myricetin	-	8.39 ± 0.27

3.2. Evaluation of the Antioxidant Activity of Honey and Its Formulations with Polyphenols

According to our results, the AOA of honey enriched with raspberry extract was higher than commercial multifloral honey as determined by all methods (DPPH, ABTS, FRAP, and CUPRAC), as can been seen in the Supplementary Material Table S1. The concentration dependence of experimental and theoretical AOA of honey and RE (H_RE versus H + RE) is shown in Supplementary Information Table S2. As can be seen, the AOA varied linearly with concentration in the case of FRAP and CUPRAC and sigmoidal in the case of DPPH and ABTS. A similar trend occurred in the case of individual RE. Based on the individual calibration curves, isobologramic diagrams were built, as seen in Figure 1.

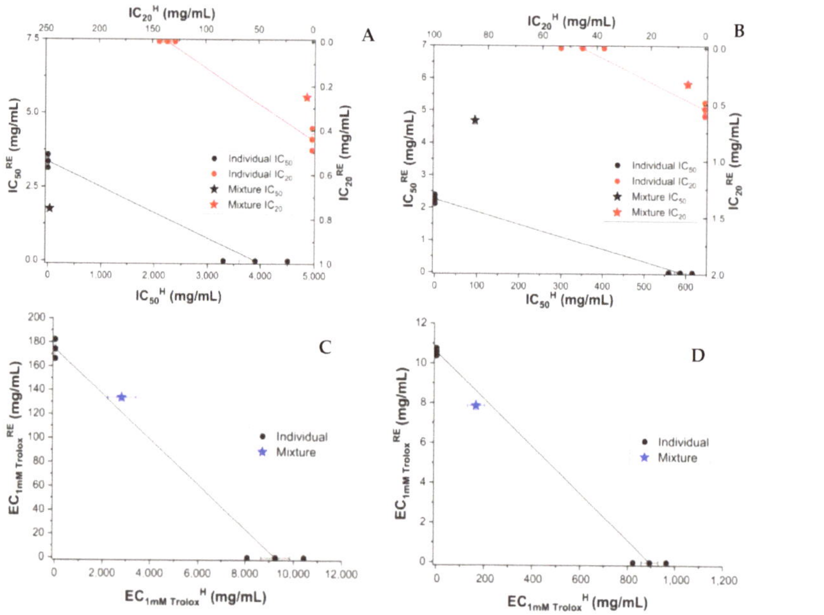

Figure 1. Isobolograms of honey (H) and raspberry extract (RE) based on IC_{50} (half-maximal inhibitory concentration) and IC_{20} (inhibitory concentration at 20% substrate inhibition) for DPPH (**A**), and ABTS (**B**) methods, and based on $EC_{1mM\ Trolox}$ (effective concentration at 1 mM Trolox equivalent of the samples) for CUPRAC (**C**) and FRAP (**D**) methods. The error bars from three measurements are shown for each value. Confidence intervals at 95% confidence are shown by dashed lines.

To compare the behaviour of RE with that of individual polyphenols dissolved in honey, we analyzed the AOA of the mixture of honey and individual major polyphenols from the raspberry extract. As shown above, HPLC analysis showed caffeic acid (CA) and epicatechin to be present in significant amounts. These were chosen to test the antioxidant behaviour induced by polyphenols and honey. For the determination of AOA activity, the polyphenol was resuspended in honey or 70% ethanol at the same polyphenol concentration (0.5 mg/g of honey, 0.5 mg/mL of 70% ethanol, respectively. Caffeic acid and epicatechin enhanced the AOA of honey. (Table S1). The concentration dependence of experimental and theoretical AOA of honey and the polyphenol (CA or EP) is shown in Supplementary Information Table S3. The concentration dependence of AOA was linear in the case of FRAP and CUPRAC and sigmoidal in the case of DPPH and ABTS (Supplementary Information Tables S1 and S3), as seen also for honey and honey with extract mixture (H_RE and H + RE). A similar trend was seen in the case of individual CA and EP. The results of isobolomic representations of honey enriched with CA and epicatechin (EP) are shown in Figures 2 and 3, respectively.

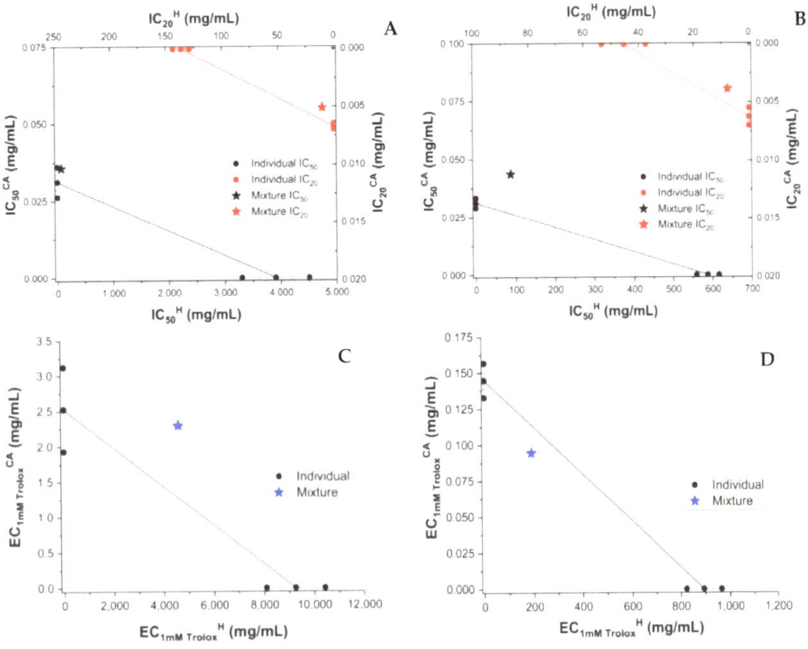

Figure 2. Isobolograms of honey (H) and caffeic acid (CA) based on IC_{50} (half-maximal inhibitory concentration) and IC_{20} (inhibitory concentration at 20% substrate inhibition) for DPPH (**A**) and ABTS (**B**) methods, and based on $EC_{1mM\ Trolox}$ (effective concentration at 1 mM Trolox equivalent of the samples) for CUPRAC (**C**) and FRAP (**D**) methods. The error bars from three measurements are shown for each value. Confidence intervals at 95% confidence are shown by dashed lines.

The CI values were calculated for all combinations (Table 3). The CI values were dependent on method, dose and polyphenol type, and varied between a minimum of 0.532 (DPPH IC50 of H_RE) and 2.885 (DPPH IC50 of H_EP).

Table 3. Combination index (CI) between samples in AOA assays.

AOA Method	H_RE	H_CA	H_EP
FRAP	0.929 ± 0.029	0.866 ±0.021	1.104 ± 0.071
CUPRAC	1.069 ± 0.059	1.426 ± 0.016	1.641 ± 0.086
DPPH IC_{50}	0.532 ± 0.003	1.174 ± 0.083	2.885 ± 0.183
DPPH IC_{20}	0.604 ± 0.019	0.836 ± 0.030	1.436 ± 0.110
ABTS IC_{50}	2.237 ± 0.043	1.552 ±0.081	1.292 ± 0.079
ABTS IC_{20}	0.731 ± 0.032	0.790 ± 0.026	1.438 ± 0.055

H_RE-mixture of honey and raspberry extract, H_CA-mixture of honey-caffeic acid, H_EP—mixture of honey-epicatehin. IC_{50} and IC_{20} represent the analysis for doses at 50% and 20% substrate inhibition, respectively.

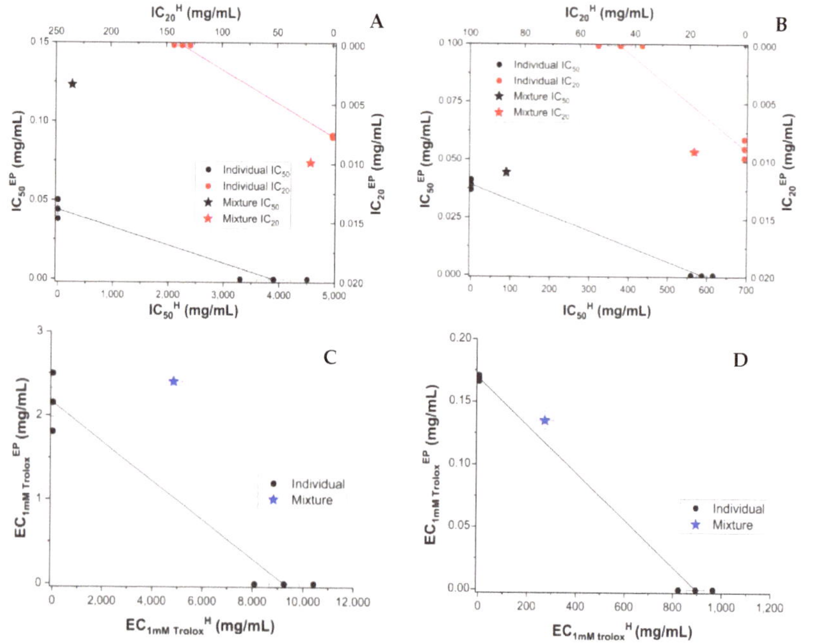

Figure 3. Isobolograms of honey (H) and epicatechin (EP) based on IC_{50} (half-maximal inhibitory concentration) and IC_{20} (inhibitory concentration at 20% substrate inhibition) for DPPH (**A**) and ABTS (**B**) methods, and based on $EC_{1mM\ Trolox}$ (effective concentration at 1 mM Trolox equivalent of the samples) for CUPRAC (**C**) and FRAP (**D**) methods. The error bars from three measurements are shown for each value. Confidence intervals at 95% confidence are shown by dashed lines.

3.3. Comparison between Honey and the Honey-Mimetic NaDES, GFSw

We prepared a honey-mimetic NaDES based on the content of glucose, fructose, sucrose, and water according to the literature data [26]. This NaDES was characterized by different methods and was compared with wildflower honey.

Fourier transform infrared spectroscopy (FTIR) was used to study the interaction between the main components of NaDES, to follow the structural changes induced by the formation of DES, and to compare NaDES with honey (Figure 4).

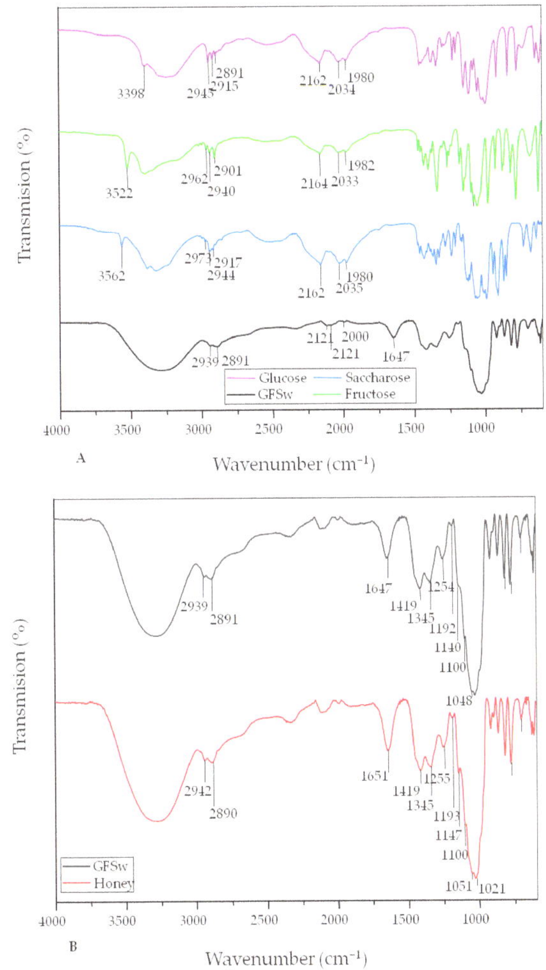

Figure 4. Comparison between the ATR-FTIR spectra of honey and the main components of GFSw—glucose, fructose, and sucrose (**A**) and between the ATR-FTIR spectra of honey and GFSw NaDES (**B**). The absorption bands from 1419 cm^{-1} and 1345 cm^{-1} are characteristic of the bending vibrations of O-CH and C- CH in the structure of carbohydrates or of the bending vibrations coming from OH from the C-OH group. The bands from about 1255–1140 cm^{-1} are characteristic of the stretching vibration of C-H or C-O from carbohydrates. The vibration with a maximum of about 1100 cm^{-1} is a band that can come from the C-O vibration in the C-O-C group. The bands of approximately 1055, 1025, 990, and 777 cm^{-1} can be assigned to C-O stretching from the C-OH group or C-C from the carbohydrate structure. The band at 987–988 cm^{-1} is characteristic of the glycosidic bond C-O-C. The spectral area from 898 to 818 cm^{-1} is characteristic of the anomeric vibrational region of carbohydrates or the C-H deformation group [37,38].

The spectra of the analyzed samples (4000 to 600 cm^{-1}) show the characteristic band of hydrogen bonds at 3280 cm^{-1} (O-H hydrogen bonds), the bands around 2900 cm^{-1}, characteristic of the stretching vibrations of the C-H groups, and the fingerprint region (1500–700 cm^{-1}).

The FTIR spectra showed that there were significant changes upon mixing the three carbohydrates and water compared to individual compounds (Figure 4A). The most significant changes were in the fingerprint region, where the number of bands decreased in GFSw.

The individual carbohydrates were characterized by multiple sharp bands, while in GFSw, these sharp bands disappeared, and the remaining bands became broader. Some bands, such as the sharp band in the region 3400–3600 cm^{-1} and bands in the region 2300–2000 cm^{-1} that were present in the individual carbohydrates disappeared, and a new band appeared at 1647 cm^{-1} in GFSw. This band was present in honey as well (Figure 4B). The FTIR spectrum of GFSw was very similar to the spectrum of honey.

The TG/DTG curves of the samples (Figure 5) showed that the decomposition of honey occurred in consecutive events involving different stages.

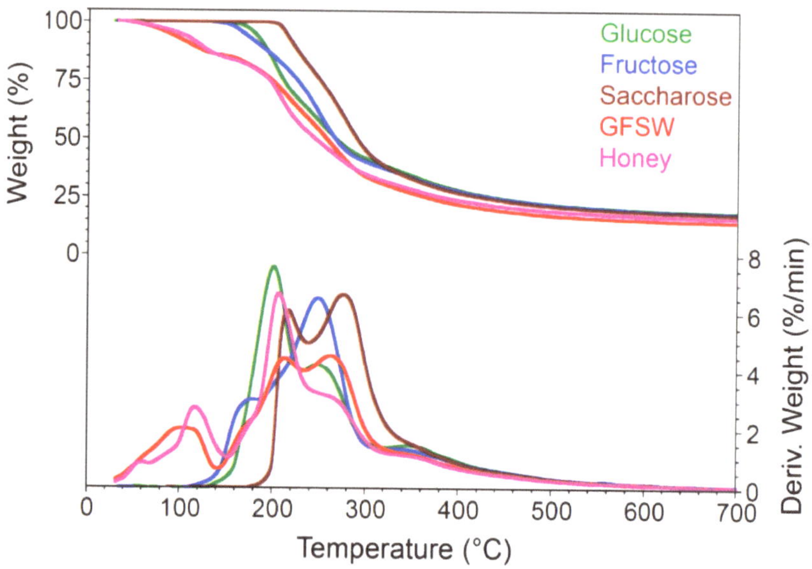

Figure 5. TGA/DTG curves of honey and DES mimetic with honey (GFSw,) and of the main components of GFSw (glucose, saccharose, fructose).

These stages start at room temperature and end close to 600 ° C. The results are summarized in Table 4

Table 4. TGA/DTG analysis of honey and GFS.

Sample	Transition	Temperature Range (°C)		Wt. Loss (%)	Tmax (°C)
Honey	1	32.7	153.7	15.96	117.5
	2	153.7	257.5	37.86	170.2/208.1
	3	257.5	338.0	16.26	273.6
	4	338.0	700.0	13.48	359.7
	Residue			16.44	(700 °C/N_2)
	Ash			0.08	(700 °C/Air)
GFSw	1	32.7	142.6	14.26	104.8
	2	142.6	236.0	26.39	169.1/214.8
	3	236.0	341.7	31.43	264.3
	4	341.7	700.0	13.28	360.2
	Residue			14.63	(700 °C/N_2)
	Ash			0.03	(700 °C/Air)

The first and second transition of thermal decomposition took place between room temperature and about 140–155 °C, and can be attributed to the water loss and volatile constituents and possibly small contributions from protein denaturation; the last two only in the case of honey.

The next thermal events, which occurred between 155 and 700 °C, can be attributed to the thermal decomposition of sugars and materials resulting from caramelization processes.

The percentage of residues at 700 °C corresponded to the content of carbonaceous materials formed as a result of advanced pyrolysis of sugars.

The ash content values resulted from the combustion of carbonaceous materials and represent traces of inorganic materials or graphitized organic materials. Honey and GFSw showed decompositions within similar temperature ranges, but the weight losses presented some differences between the two products at transitions 2 and 3.

Figure 6 shows DSC thermograms obtained after cooling and heating cycles of honey and GFSw samples. The DSC analysis of the first cooling from room temperature to −75 °C showed a glass transition for honey (from −36.0 °C to −53.6 °C, with a mid-point, glass transition temperature (Tg) of −45.0 °C) and GFSw (from −32.7 to −49.4 °C, with Tg of −40.3 °C). The specific heat capacity was similar between honey (0.81 J/(g °C) and GFSw (0.84 J/(g °C).

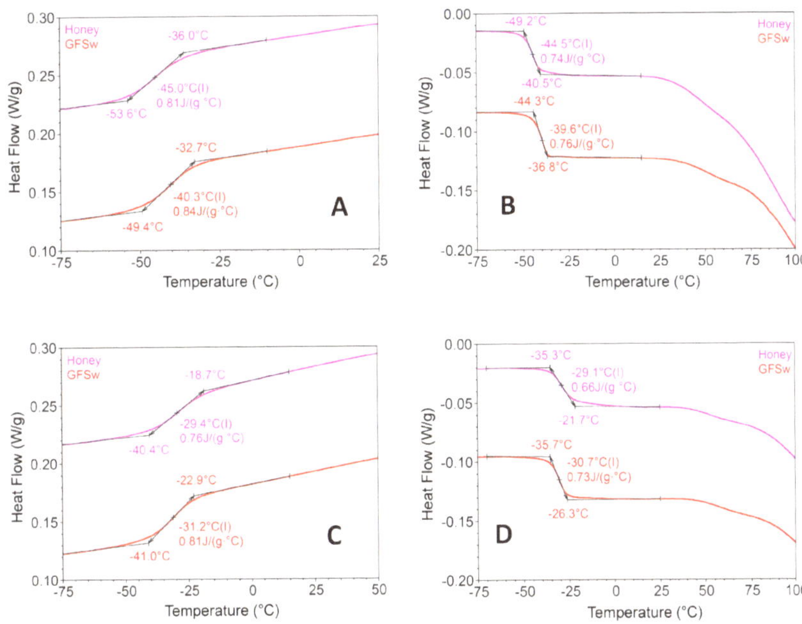

Figure 6. Differential scanning calorimetry (DSC) heating thermograms of honey and GFSw. First cooling cycle in the range of temperatures from 0 °C to −75 °C (**A**), first heating cycle in the range of temperatures from −75°C to 100 °C (**B**), second cooling cycle in the range of temperatures from 0 °C to −75 °C (**C**), second heating cycle in the range of temperatures from −80 °C to 100 °C (**D**).

The corresponding enthalpy was 235.5 J/g for honey and 243.5 J/g for GFSw. This transition occurs when the material changes upon cooling from the rubber-like state into the hard, glassy state [39]. The first heating from −75 °C to 100 °C presented similar reversible transitions, but with less negative temperatures and lower heat capacities, representing transitions from the glassy solid state to the rubbery state upon heating. There was also a change in heat flow at high temperatures (>25 °C), steeper and larger for honey than for GFSw.

In the case of the second cooling and second heating, the mid-point temperatures of both honey and GFSw become less negative compared with the first transitions and much closer between the two samples, but with changing order (mid-point temperatures of GFSw are lower than mid-point temperatures of honey). The temperature differences (ΔT) of Tg values between the first and second transitions were higher with approx. 6.5 °C for honey than for GFSw. The change in heat flow at high temperatures also became similar between the two samples and less pronounced than during the first transitions.

We next wanted to obtain information about the morphological and structural behavior of GFSw compared to honey. Honey and, respectively, GFSw, were spray-dried after mixing with maltodextrin and investigated by SEM and X-ray diffraction techniques. Figure 7 shows several SEM micrographs of the external microstructure of honey and GFSw powders at 200× and 600× magnification. As can been seen, the products show similar morphological features, with smooth surfaces and aggregates of round microparticles with linkages between them, similar to other types of honey reported [28]. There are some small holes present in both honey and GFSw clearly visible at 600× magnification. Although similar, the particles of honey seem to be larger than the particles of GFSw, and we observed additional very small particles on the honey surface, which could be due to the presence of other compounds in honey.

Figure 7. SEM micrographs of spray-dried honey and GFSw at 200× and 600× magnification.

The XRD profiles show two wide main peaks and a much less intense one, similar to honey and GFSw (Figure 8).

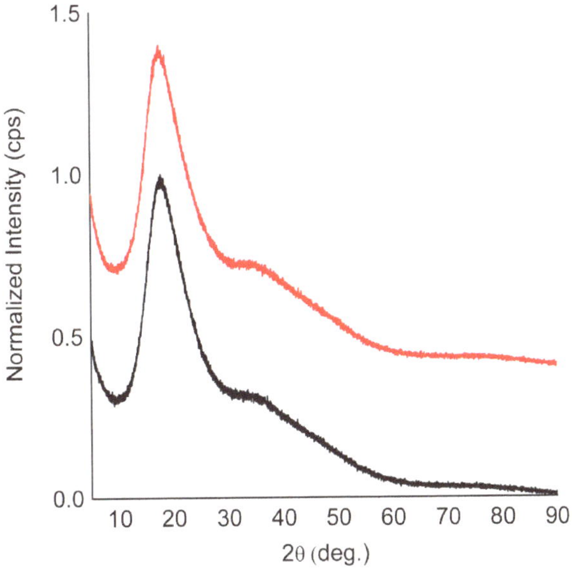

Figure 8. X-ray diffractograms of honey (red) and GFSw (black).

The diffractograms could be deconvoluted by Gaussian decomposition in up to four apparent peaks, at approx. 2θ = 17.58°/17.8°, 35.9°/35.01°, 44.2°/46.2°, and 78.3°/78.1°, for honey/GFSw but with the two intermediate peaks overlapping significantly and the fourth (78°) with a very small amplitude (Supplementary Figures S5 and S6, Table 3).

When fitting with two maxima, the 2θ were approx. 17.59°/17.72° and 36.16°/34.53° for honey/GFSw. The fitting were slightly but not significantly improved from 2 to 4 angles. The corresponding average atomic distances resulting from 2θ were similar between honey and GFSw (Table 5).

Table 5. XRD parameters from the Gaussian deconvolution of diffractograms for honey and GFSw.

Sample	2 Peaks Deconvolution 2θ	Distance (d)
Honey	17.59°; 36.16°	5.04 Å; 2.48 Å
GFSw	17.72°; 34.53°	5.00 Å; 2.59 Å
	4 Peaks Deconvolution	
Honey	17.58°; 35.9°; 44.2°; 78.3°	5.04 Å; 2.50 Å; 2.05 Å; 1.22 Å
GFSw	17.8°; 35.01°; 46.2°; 78.1°	4.99 Å; 2.56 Å; 1.97 Å; 1.22 Å

3.4. Physicochemical Characteristics of Honey and GFSw

The physicochemical characteristics of surface tension, density, pH, water activity, refractive index, and Brix degree of honey (total soluble solids—TSS) and GFSw are summarized in Table 6. Most physicochemical parameters measured (surface tension, density, pH, refractive index, and TSS) had slightly lower values in the case of honey than in the case of GFSw. Water activity was slightly higher in the case of honey compared with GFSw.

We wanted to check if the mixtures GFSw—extract and GFSw—polyphenol standard had the same behaviour as in the case of honey. The NaDES GFSw was enriched with concentrated raspberry extract/standard (caffeic acid or epicatechin) in the same ratio of 1:20 (*w*/*w*) as with honey. The concentration dependence of AOA was similar in the case of GFSw compared to honey (linear for FRAP and CUPRAC and sigmoidal for DPPH and ABTS: Tables S4 and S5), but GFSw had very small AOA activity, as expected due to the lack of polyphenols, except in the case of CUPRAC method, where GFSw activity was similar to the honey activity. The isobologram diagrams are presented in Figures 9–11 for RE, CA and EP, respectively. The CI values are shown in Table 7.

Table 6. Surface tension, density, the pH, water activity refractive index, and total soluble solids (TSS) of honey and GFSw/.

Sample	Surface Tension, mN/m	Density, g/cm^3	pH	Water Activity	Refractive Index	TSS, ^{0}Brix
Honey	80.292 ± 0.167	1.4207 ± 0.00025	3.75 ± 0.045	0.586 ± 0.00059	1.490 ± 0.00075	80.075 ± 0.09
GFSw	82.214 ± 0.015	1.4301± 0.00092	4.31 ± 0.032	0.555 ± 0.00134	1.495 ± 0.000	82.025 ± 0.15

Table 7. Combination index (CI) between samples in AOA assays.

AOA Method	GFSw_RE	GFSw_CA	GFSw_EP
FRAP	0.986 ± 0.021	0.973 ± 0.036	1.149 ± 0.056
CUPRAC	1.409 ± 0.023	1.605 ± 0.065	1.502 ± 0.176
DPPH IC_{50}	0.915 ± 0.043	1.642 ± 0.085	1.884 ± 0.097
DPPH IC_{20}	0.572 ± 0.071	1.077 ± 0.088	1.337 ± 0.108
ABTS IC_{50}	2.646 ± 0.162	1.564 ± 0.078	1.310 ± 0.062
ABTS IC_{20}	1.011 ±0.079	0.897 ± 0.036	1.241 ± 0.047

GFSw_RE-mixture of NaDES and raspberry extract, GFSw_CA-mixture of NaDES-caffeic acid, GFSw_EP—mixture of NaDES-epicatehin. IC_{50} and IC_{20} represent the analysis for doses at 50% and 20% substrate inhibition, respectively.

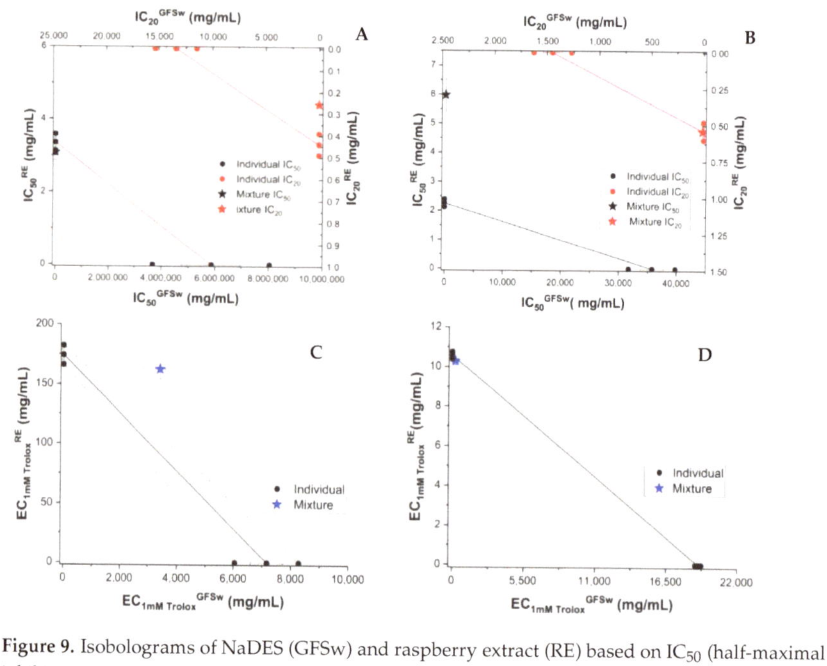

Figure 9. Isobolograms of NaDES (GFSw) and raspberry extract (RE) based on IC_{50} (half-maximal inhibitory concentration) and IC_{20} (inhibitory concentration at 20% substrate inhibition) for DPPH (**A**), and ABTS (**B**) methods, and based on $EC_{1mM\ Trolox}$ (effective concentration at 1 mM Trolox equivalent of the samples) for CUPRAC (**C**) and FRAP (**D**) methods. The error bars from three measurements are shown for each value. Confidence intervals at 95% confidence are shown by dashed lines.

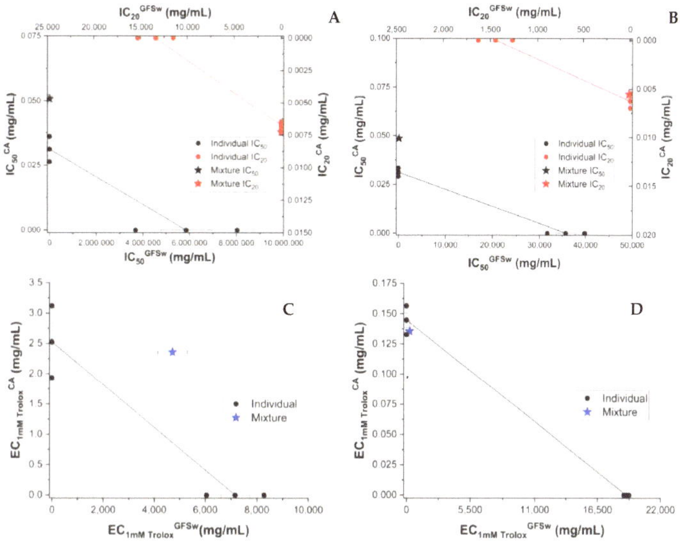

Figure 10. Isobolograms of NaDES (GFSw) and caffeic acid (CA) based on IC_{50} (half-maximal inhibitory concentration) and IC_{20} (inhibitory concentration at 20% substrate inhibition) for DPPH (**A**), and ABTS (**B**) methods, and based on $EC_{1mM\ Trolox}$ (effective concentration at 1 mM Trolox equivalent of the samples) for CUPRAC (**C**) and FRAP (**D**) methods. The error bars from three measurements are shown for each value. Confidence intervals at 95% confidence are shown by dashed lines.

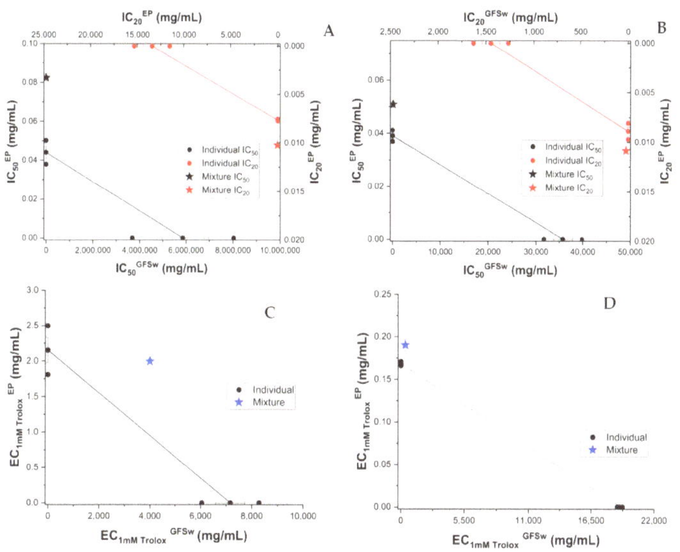

Figure 11. Isobolograms of NaDES (GFSw) and epicatechin (EP) based on IC_{50} (half-maximal inhibitory concentration) and IC_{20} (inhibitory concentration at 20% substrate inhibition) for DPPH (**A**), and ABTS (**B**) methods, and based on $EC_{1mM\ Trolox}$ (effective concentration at 1 mM Trolox equivalent of the samples) for CUPRAC (**C**) and FRAP (**D**) methods. The error bars from three measurements are shown for each value. Confidence intervals at 95% confidence are shown by dashed lines.

The GFSw values that resulted in the same effect as the mixture were extremely high and non-realistic (especially with the DPPH method). This suggests that in this case, NaDES acts more as potentiator or inhibitor rather than synergiser or antagoniser, because of its lack of AOA activity.

The CI values were dependent on method, dose and polyphenols type, and varied between a minimum of 0.572 (DPPH IC_{20} of GFSw_RE) and 2.646 (ABTS IC_{50} of GFSw_EP).

4. Discussion

As mentioned previously, phenolic compounds are those most responsible for the antioxidant activity of honey and plant extracts. The antioxidant properties of phenolic compounds are attributed to their capacity to neutralize free radicals by several mechanisms, such as HAT (hydrogen atom transfer), SET-PT (single electron transfer via proton transfer), sequential proton loss electron transfer or TMC (transition metal chelation) [40].

The AOA of the samples (honey and raspberry extract) could be correlated with the concentrations and profiles of polyphenols analyzed by colourimetric assays (TPC, TFC, HAT) and the HPLC method. As expected, the AOA of honey enriched in raspberry polyphenolic extracts was much higher than the AOA of pure honey at all concentrations tested, irrespective of the AOA method employed.

It has often been claimed that honey constituents, as well as mixtures of honey with other ingredients, act or induce synergistic effects. Still, these claims have not been thoroughly and rigorously investigated and proven. In some studies, synergism was less evident, but we believe this was partially due to the method of evaluation approached [5,6]. In other cases, synergism seemed to depend on the honey type and the AOA method [8].

To the best of our knowledge, we performed here for the first time an investigation on honey—polyphenols and honey biomimetic NaDES—polyphenols modulations as reflected in AOA activities, based on isobolomic and combination index (CI) calculation, which represents a rigorous assessment of synergism or antagonism behaviour. Depending on the values of CI, the activity is theoretically defined as synergic ($CI < 1$), additive ($CI = 1$) and antagonistic ($CI > 1$). In practice, a confidence interval also applies and the values between 0.9 and 1.1 are usually considered as reflecting additive behaviour. The behaviour of RE was highly heterogeneous, varying from relatively strong synergism ($CI = 0.532$ for DPPH IC_{50}) to relatively strong antagonism ($CI = 2.237$ for ABTS IC_{50}), depending on the method and, in the case of DPPH and ABTS on the dose of mixture tested (Table 3). FRAP and CUPRAC showed additive behaviour, which correlated to the experimental and theoretical calibration curves (Table S2). The individual main polyphenols representative of phenolic acids and flavonoids (caffeic acid and epicatechin, respectively) from the extract had heterogeneous behaviour that depended on method and dose in a similar manner to that in the case of CA, and in a different manner to that in the case of EP, compared to RE. CA had a more heterogeneous behaviour than EP, the latter resulting in only different degrees of antagonism and no synergism. The ABTS activity of H_RE and H_CA seem to depend strongly on the dose applied, shifting from synergism at low doses to antagonism at high doses. The CI correlated with the experimental and theoretical calibration curves (Table S3).

A previous study from more than a decade ago applied a comparison approach, using ORAC and EPR techniques and physiologically relevant media [41]. It was found that certain combinations of antioxidant compounds at lower concentrations than in our study showed synergic effects, especially when involving sugar solutions and ascorbic acid together with polyphenols. Some results were contradictory, and further investigations are necessary but have not become available.

More recent work found other contradictory results with respect to sugar's influence on polyphenol activity, with either synergetic or antagonistic behavior or no effect. This depended on compound and sugar types, concentration, and AOA method [42–44]. Based on some studies, a possible explanation for the synergetic effect observed could be related to the stabilization and protection of some polyphenols by sugars and vice versa [45–47]. Still, considering the heterogeneous behavior observed by several groups, this aspect probably has only a partial contribution and prevails only in specific cases, not as a general rule. Sucrose, glucose, and other sugars, for example, were previously shown to be able to quench ·OH radicals [48,49], and it was predicted that sucrose, sucrose radicals, and other sugars could interact with secondary metabolites, such as phenolic compounds, which could determine in some cases sucrose recycling [44]. It is possible that these features play a role in some particular situations, especially involving ·OH radicals. Sugar interaction with aromatic molecules was predicted based on molecular dynamics simulation and NMR, which showed that the interaction is rather hydrophobic in nature (sugar and aromatic rings stacking) than H-bonds driven [50,51].

Honey is considered an example of a natural deep eutectic solvent [2,52]. Most of the physicochemical and structural properties of our honey-mimetic NaDES were similar to those of honey. The FTIR spectral changes observed upon saccharide mixing compared to individual compounds indicate a shift from the crystalline nature characteristic of the saccharide powder to an amorphous structure within NaDES. These differences between crystalline and amorphous carbohydrates/dried melt samples have been reported before [53,54].

Similar ATR-FTIR spectra for honey, as obtained in our study, have been previously reported [28,55–57]. The results prove that the carbohydrate arrangement within GFSw is very similar to that in honey. The appearance of the band at 1640 cm^{-1} can result from the deformation vibrations of the −OH groups of the water present in GFSw. This band is very similar to the one from honey, indicating similar composition and molecular arrangement between GFSw and honey.

Based on XRD analysis, honey was previously characterized as having an amorphous structure. In accordance with the FTIR data, the XRD profile was characterized by broad diffraction maxima, indicating amorphous structures with short-range order (Figure 9). From the Gaussian decomposition of diffractograms, the main average distances between atoms in molecules were found to be similar between honey and GFSw. Few in-depth studies have reported on the diffractogram deconvolution and analysis of honey powders obtained by spray drying. A recent study reported similar results, with a four-peak Gaussian deconvolution profile, but with one difference, i.e., almost half 2θ angle (approx. 23°) compared with our result (approx. 45°). It is unclear at the moment what the cause of this difference is, but we believe that this angle has no physical significance as it did not significantly influence the overall results. As mentioned, there was a significant overlap of peaks 2 and 3, and the second angle (approx. 35°) had similar values between the two-peak fitting and four-peak fitting (Table 3). This aspect needs more in-depth studies and is beyond the purpose of this work. The most important outcome is that honey and GFSw gave very similar diffractograms, showing that both samples have a similar amorphous structure characterized by short-range order and only slightly different in the main distances between atoms [28].

Some differences were expected, such as pH slightly lower in honey than in GFSw due to the presence of organic acids and other molecules, or slightly lower surface tension of honey compared to GFSw due to more complex/slightly different composition. The pH range of honey is 3.5–5.5, and is influenced by various intrinsic and extrinsic factors [58]. The pH of GFSw is within this range. Surface tension is a measure of the interaction strength between the components in a sample. A better understanding of the intermolecular forces that are manifested in the liquid and between the surfaces is obtained from surface tension values [59]. The factors that influence the intermolecular forces within DES are temperature, the nature of HBA/HBD (hydrogen bond acceptor/hydrogen bond donor), and the molar ratio of the components (higher intake of HBA will increase the surface tension of the mixture) [60]. Honey contains about 80% solid components that melt individually above 100 °C (glucose, fructose, sucrose), and about 15–20% of water. The fact that this mixture is liquid at room temperature is due to the optimal combination ratio of the components. Honey has high surface tension due to the hydrogen bonds that are formed between saccharides and water (saccharides have many O atoms with non-participating e^- pairs that participate in the formation of H bonds with H from water), as well as cohesive forces. The higher surface tension in GFSw than in honey could indicate slightly stronger H-bond interactions in GFSw than in honey and/or higher percent of saccharides, which correlate with the higher density and lower water activity, respectively. Considering that TSS was similarly higher in GFSw than in honey, it is possible that a significant contribution comes from the slightly higher percent of saccharides in GFSw than in honey.

The more complex/slightly different composition of honey compared to GFSw is probably responsible also for the differences in the morphological features observed in SEM micrographs, honey resulting in larger spray-dried particles than GFSw.

Other minor differences between honey and GFSw correlated with each other, which shows that the data were consistent. For example, the experimental water content difference between honey and GFSw was similar to that determined by TSS (Δ = 2%) and TGA (Δ = 1.7%). The absolute values were lower (with approx. 4% water content) as determined by TGA compared to TSS for both honey and GFSw, which could represent molecules of water more tightly bound than the rest and which evaporated at higher temperatures. The water activity of honey was approx. 5.6% higher than that of GFSw, implying some water molecules are less tightly bound in honey than in GFSw, besides the contribution of the 2% higher water content in honey than in GFSw. Water activity is a quality parameter that is used to estimate the shelf life and crystallization rate of honey samples. In honey, water activity is influenced by sugar content (glucose, fructose, and other sugars) [61]. Refractive index, density, and TSS correlated with each other for both honey and GFSw.

The TGA profiles of honey and GFSw were similar, with small differences coming from the more complex composition of honey. Below 150 °C, the thermal profile of the tested honey was more complex than that of GFSw, with three apparent transitions compared to one, respectively. It was previously found that the profile in this region depends on the bee species and varieties of honey [62,63]. The differences in this region most probably come from the various volatiles and protein content in honey. The main transitions (between 150 and 340 °C) showed some difference between honey and GFSw, suggesting a higher amount of glucose in honey than in GFSw.

The thermodynamic behavior evidenced by DSC was similar between honey and GFSw, reflecting similar supramolecular structures, as suggested for other honey biomimetic NaDES previously obtained [2]. The water content and water activity correlated with the glass transition temperature (Tg) determined by DSC. The glass transition and specific heat capacity of honey were approx. 5 °C and 0.3 J/(g·°C), respectively, lower than those of GFSw both at first cooling and first heating. Lower Tg means honey freezes harder than GFSw, and this is correlated with the higher water activity, which assures higher plasticity and dynamics [64]. The steeper and larger transition above 25 °C at first heating and the higher ΔT (15.6 °C/15.4 °C versus 9.1 °C/8.9 °C) between the first and second cooling/heating for honey compared to GFSw is also related to water content and water activity. Honey loses water more easily and in higher amount than GFSw, which is reflected in a more significant change in Tg. After water loss, the Tg order was reversed. GFSw than had lower values than honey, which means that in the absence of water, GFSw is more dynamic than honey. The small differences in water content and water activity also correlate with the observation that the temperature difference between the inlet and outlet temperature during the spray-drying process was higher for honey (70 °C) than for GFSw (46 °C). These small variations in some properties and behavior most probably do not have a significant effect on the AOA behavior, so we believe that our GFSw NaDES is a close mimetic of honey, at least in this respect.

The honey-biomimetic NaDES obtained, GFSw, had a similar effect as honey on the AOA of raspberry extract and individual polyphenols (Figures 9–11 versus Figures 1–3 and Table 7 versus Table 3), but in the case of RE and CA honey showed, in general, a higher tendency towards synergism and less antagonism than GFSw. We call the phenomena observed as synergism and antagonism instead of potentiation/inhibition because both

GFSw and especially honey present some AOA activity, although very low in the case of GFSw, except for the CUPRAC method. For convenience of comparison between honey and GFSw, we show the figures combined for each AOA method in Supplementary Material Figures S7–S10. We gathered all CI values within Table 8 and codified in Table 9 the CI value intervals as follows: 0.5–0.7 (+2, strong synergism); 0.7–0.9 (+1, moderate synergism); 0.9–1.1 (0, almost additive); 1.1–1.5 (−1, moderate antagonism); 1.5–2 (−2, moderate—strong antagonism); >2 (−3, strong antagonism). The color code indicates that most combinations presented similar behaviour of honey compared to GFSw.

Table 8. CI values for AOA activity from Tables 3 and 7 together.

H / GFSw	RE (H)	RE (GFSw)	CA (H)	CA (GFSw)	EP (H)	EP (GFSw)
FRAP	0.929 ± 0.029	0.986 ± 0.021	0.866 ± 0.021	0.973 ± 0.036	1.104 ± 0.071	1.149 ± 0.056
CUPRAC	1.069 ± 0.059	1.409 ± 0.023	1.426 ± 0.016	1.605 ± 0.065	1.641 ± 0.086	1.502 ± 0.176
DPPH 50%	0.532 ± 0.003	0.915 ± 0.043	1.174 ± 0.083	1.642 ± 0.085	2.885 ± 0.183	1.884 ± 0.097
ABTS 50%	2.237 ± 0.043	2.646 ± 0.162	1.552 ± 0.081	1.564 ± 0.078	1.292± 0.079	1.310 ± 0.062
DPPH 20%	0.604 ± 0.019	0.572 ± 0.071	0.836 ± 0.030	1.077 ± 0.088	1.436 ± 0.110	1.337 ± 0.108
ABTS 20%	0.731 ± 0.032	1.011 ± 0.079	0.790 ± 0.026	0.897 ± 0.036	1.438 ± 0.055	1.241 ± 0.047

Table 9. CI value intervals and color codification for AOA activity *.

H / GFSw	RE (H)	RE (GFSw)	CA (H)	CA (GFSw)	EP (H)	EP (GFSw)
FRAP	0 (+1)	0 (+1)	+1	0 (+1)	−1	−1
CUPRAC	0 (−1)	−1	−1	−2	−2	−2
DPPH 50%	+2	0 (+1)	−1	−2	−3	−2
ABTS 50%	−3	−3	−2	−2	−1	−1
DPPH 20%	+2	+2	+1	0 (−1)	−1	−1
ABTS 20%	+1	0 (−1)	+1	+1	−1	−1

* The values in brackets indicate the tendency, either towards synergism (+1) or antagonism (−1). The color code indicates highly similar behaviour between honey and GFSw (dark orange) and moderate similar behaviour between honey and GFSw (light orange).

As can been seen in Tables 8 and 9, seven of thirty-six cases (less than 20%) were apparently additive. From the additive ones, half had a tendency towards synergism and half had a tendency towards antagonism. Approximately 20% of cases were clearly synergistic. The majority of cases (70%) from Table 8 are either antagonistic or additive tending to antagonistic. Most synergistic effects were seen in the DPPH and ABTS methods, but only at 20% inhibition (i.e., at lower concentrations of mixtures). At 50% inhibition (higher concentrations of mixtures) DPPH and ABTS present more antagonist cases than FRAP and CUPRAC. This dependence on mixture concentration in the case of DPPH and ABTS is probably related to inhibition caused by honey/GFSw, which could have several explanations, such as higher viscosity, too strong H-bonds and hydrophobic interactions between NaDES and polyphenols, change of redox potential or even increased competition of the weaker antioxidant (honey/GFSw) against the stronger antioxidant (polyphenols). From the cases that are antagonistic, some could probably become additive or even synergistic by modulating the honey: polyphenols ratio.

ABTS and DPPH RE differ in their behaviour with respect to the individual polyphenols. While in the case of ABTS the degree of effect (either synergistic, additive or antagonistic) seems to be cumulative in RE. In DPPH there seems to be an additional synergic interaction between polyphenols in mixture (extract) besides the effect induced by honey/GFSw. In other words, the polyphenols probably synergise each other in the DPPH, but not in the ABTS reaction.

Similar difference can be observed when comparing the two methods that gave linear dependence on concentration (CUPRAC and FRAP), with a relatively cumulative effect in FRAP and a polyphenol—polyphenol synergic effect that compensates for the antagonism between honey/GFSw and individual polyphenols in CUPRAC.

All in all, the data suggest that the polyphenols synergise each other in CUPRAC and DPPH methods, but not in FRAP and ABTS methods.

As mentioned above, for RE and CA, honey induces slightly lower CI than GFSw in general. This could be related, on one hand, to the presence of additional polyphenols that increase the synergism, and on the other hand to other elements present in honey that could synergise/potentiate the reactions more. The polyphenols and/or other elements present in honey seem to synergise CA in most of the cases, but not EP. Further investigations are needed in order to determine the elements in honey responsible for this difference. However, in general, our data show that honey mimetic mixtures of sugars behave similarly to honey, especially when antagonism is present, and in some cases no other compounds are necessary for a certain degree of synergic effect.

A recent study showed that some deep eutectic solvents based on ethylene glycol and choline chloride (ethaline) and, respectively, betaine and citric acid (BCA) can change the redox potential of polyphenols to lower values, and this is influenced by the composition of the solvents [65]. Lower redox potential implies that the polyphenols are more easily oxidized, so they have higher antioxidant capacity. Moreover, BCA was very efficient in stabilizing polyphenols. Honey-biomimetic NaDES was previously shown to improve the bioavailability, bioactivity and heat stability of compounds from *Astragali Radix*, a traditional Chinese medicine and functional food [2]. Taken together, these data imply that NaDES in general could modulate the AOA of polyphenols at different degrees, which could explain the behaviour observed in our study. The exact mechanism for each individual case remains to be established, requiring more in-depth analysis. The effect on redox potential could be one explanation for the heterogeneous behaviour observed, which was dependent on polyphenol type and AOA method.

Taken together, there are several possible mechanisms for the synergism/potentiation observed in some cases: stabilisation of polyphenols, redox potential, crowding space inducing environment by honey/GFSw, presence of enzymes in honey, multiple synergism between polyphenols and other compounds present in honey and extract, among others.

The behaviour seen in our mixtures is characteristic for what are called complex systems, in which unexpected behaviours manifest as a result of multiple interactions.

Some preliminary unpublished results from our group suggest that other extracts such as propolis or sea buckthorn extracts have similar behaviour when mixed with honey, which suggests that there are some general features that manifest independently of the extract type. Multiple functional foods or bioproducts for different biomedical fields based on honey enriched in extracted poplyphenols could be developed. The compositions would need to be optimized in order to reduce antagonism/inhibition and maximize the synergism. The applications will depend on multiple parameters, including the sensory one mentioned in the Introduction, but the synergic AOA will be beneficial in all cases. More work is needed to take into consideration optimizing the extract concentrations, extract—honey ratios, extract composition, honey type and properties and other parameters. Although our developed product does not present synergism in all cases, it still has advantages over simple honey, especially in the cases where hydroalcoholic supplements are forbidden.

5. Conclusions

We obtained honey enriched in polyphenols from raspberry extracts with conserved honey characteristics and enhanced (in some cases) synergic AOA between honey and polyphenols. The honey-biomimetic NaDES with similar properties as honey resulted in similar AOA behavior to honey when mixed with polyphenols, but honey seemed to have additional properties that increase synergism/reduce antagonism in some cases. The AOA behaviour of honey—polyphenols mixtures can be influenced by the AOA method, polyphenol type, mixture concentration and is characteristic for complex systems. The new product can be further optimized to maximize synergism, tested for biological activities and represents a promising functional food with enhanced AOA.

Supplementary Materials: The following supporting information can be downloaded at: https://www.mdpi.com/article/10.3390/antiox11112194/s1.

Author Contributions: Conceptualization, F.O., and D.C.-A.; methodology, L.D., D.C.-A., D.P., V.A.F., C.-A.N., M.G. (Marius Ghiurea) and M.G. (Mihaela Ganciarov); validation, D.C.-A.; formal analysis, L.D. and D.C.-A.; investigation, L.D., D.C.-A., D.P., V.A.F., C.-A.N., M.G. (Marius Ghiurea) and M.G. (Mihaela Ganciarov); resources, F.O. and A.-L.N.; data curation, L.D. and D.C.-A.; writing—original draft preparation, L.D., D.C.-A., V.A.F. and M.G. (Marius Ghiurea); writing—review and editing, D.C.-A. and F.O.; visualization, F.O., and N.E.B.; supervision, F.O., D.C.-A. and N.E.B.; project administration, D.C.-A., F.O. and A.-L.N.; funding acquisition, F.O. and A.-L.N. All authors have read and agreed to the published version of the manuscript.

Funding: This work was funded by project POC-A1-A1.2.3-G-2015-P_40_352-SECVENT, Sequential processes to close bioeconomy side stream and innovative bioproducts resulted from these, contract 81/2016, SMIS 105684, funded by cohesion funds of the European Union, subsidiary projects 2236/2018 and 5544/2020.

Institutional Review Board Statement: Not applicable.

Informed Consent Statement: Not applicable.

Data Availability Statement: All the data is contained within the article.

Acknowledgments: The XRD diffractometer and the spray dryer were purchased with the support of the POS-CCE "Agri-Flux" project, nr. 645/18.03.2014, SMIS-CSNR 48695. The scanning electron microscope was acquired in the frame of the project 15PFE Next-Bexcel. We thank Sanda Maria Doncea for the help with FTIR data interpretation.

Conflicts of Interest: The authors declare no conflict of interest.

References

1. Alvarez-Suarez, J.M.; Gasparrini, M.; Forbes-Hernández, T.Y.; Mazzoni, L.; Giampieri, F. The Composition and Biological Activity of Honey: A Focus on Manuka Honey. *Foods* **2014**, *3*, 420–432. [CrossRef] [PubMed]
2. Dai, Y.; Jin, R.; Verpoorte, R.; Lam, W.; Cheng, Y.-C.; Xiao, Y.; Xu, J.; Zhang, L.; Qin, X.-M.; Chen, S. Natural Deep Eutectic Characteristics of Honey Improve the Bioactivity and Safety of Traditional Medicines. *J. Ethnopharmacol.* **2020**, *250*, 112460. [CrossRef] [PubMed]
3. El Achkar, T.; Greige-Gerges, H.; Fourmentin, S. Understanding the Basics and Properties of Deep Eutectic Solvents. In *Deep Eutectic Solvents for Medicine, Gas Solubilization and Extraction of Natural Substances*; Springer: Berlin, Germany, 2021; pp. 1–40.
4. Cory, H.; Passarelli, S.; Szeto, J.; Tamez, M.; Mattei, J. The Role of Polyphenols in Human Health and Food Systems: A Mini-Review. *Front. Nutr.* **2018**, *5*, 87. [CrossRef] [PubMed]

5. Postali, E.; Peroukidou, P.; Giaouris, E.; Papachristoforou, A. Investigating Possible Synergism in the Antioxidant and Antibacterial Actions of Honey and Propolis from the Greek Island of Samothrace through Their Combined Application. *Foods* **2022**, *11*, 2041. [CrossRef]
6. Freitas, A.S.; Cunha, A.; Oliveira, R.; Almeida-Aguiar, C. Propolis Antibacterial and Antioxidant Synergisms with Gentamicin and Honey. *J. Appl. Microbiol.* **2022**, *132*, 2733–2745. [CrossRef]
7. Juszczak, L.; Gałkowska, D.; Ostrowska, M.; Socha, R. Antioxidant Activity of Honey Supplemented with Bee Products. *Nat. Prod. Res.* **2016**, *30*, 1436–1439. [CrossRef]
8. Bellik, Y.; Selles, S.M.A. In Vitro Synergistic Antioxidant Activity of Honey-*Mentha spicata* Combination. *J. Food Meas. Charact.* **2017**, *11*, 111–118. [CrossRef]
9. Tomczyk, M.; Miłek, M.; Sidor, E.; Kapusta, I.; Litwińczuk, W.; Puchalski, C.; Dżugan, M. The Effect of Adding the Leaves and Fruits of Morus Alba to Rape Honey on Its Antioxidant Properties, Polyphenolic Profile, and Amylase Activity. *Molecules* **2019**, *25*, 84. [CrossRef]
10. Grabek-Lejko, D.; Miłek, M.; Sidor, E.; Puchalski, C.; Dżugan, M. Antiviral and Antibacterial Effect of Honey Enriched with *Rubus* spp. as a Functional Food with Enhanced Antioxidant Properties. *Molecules* **2022**, *27*, 4859. [CrossRef]
11. Yang, W.; Zhang, C.; Li, C.; Huang, Z.Y.; Miao, X. Pathway of 5-hydroxymethyl-2-furaldehyde Formation in Honey. *J. Food Sci. Technol.* **2019**, *56*, 2417–2425. [CrossRef]
12. Habryka, C.; Socha, R.; Juszczak, L. The Effect of Enriching Honey with Propolis on the Antioxidant Activity, Sensory Characteristics, and Quality Parameters. *Molecules* **2020**, *25*, 1176. [CrossRef]
13. Ponder, A.; Hallmann, E. The Effects of Organic and Conventional Farm Management and Harvest Time on the Polyphenol Content in Different Raspberry Cultivars. *Food Chem.* **2019**, *301*, 125295. [CrossRef] [PubMed]
14. Pulcini, P.; Allegrini, F.; Festuccia, N. Fast SPE Extraction and LC-ESI-MS-MS Analysis of Flavonoids and Phenolic Acids in Honey. *Apiacta* **2006**, *41*, 21–27.
15. Robbins, R.J.; Bean, S.R. Development of a Quantitative High-Performance Liquid Chromatography–Photodiode Array Detection Measurement System for Phenolic Acids. *J. Chromatogr. A* **2004**, *1038*, 97–105. [CrossRef] [PubMed]
16. Hajimehdipoor, H.; Kondori, B.M.; Amin, G.R.; Adib, N.; Rastegar, H.; Shekarchi, M. Development of a Validated HPLC Method for the Simultaneous Determination of Flavonoids in Cuscuta chinensis Lam. by Ultra-Violet Detection. *DARU J. Pharm. Sci.* **2012**, *20*, 57. [CrossRef]
17. Kim, D.-O.; Jeong, S.W.; Lee, C.Y. Antioxidant Capacity of Phenolic Phytochemicals from Various Cultivars of Plums. *Food Chem.* **2003**, *81*, 321–326. [CrossRef]
18. Formagio, A.S.N.; Kassuya, C.A.L.; Neto, F.F.; Volobuff, C.R.F.; Iriguchi, E.K.K.; Vieira, M.D.C.; Foglio, M.A. The Flavonoid Content and Antiproliferative, Hypoglycaemic, Anti-inflammatory and Free Radical Scavenging Activities of Annona dioica St. Hill. *BMC Complement. Altern. Med.* **2013**, *13*, 14. [CrossRef]
19. Štefan, M.B.; Vuković Rodríguez, J.; Blažeković, B.; Kindl, M.; Vladimir-Knežević, S. Total Hydroxycinnamic Acids Assay: Prevalidation and Application on Lamiaceae Species. *Food Anal. Methods* **2014**, *7*, 326–336. [CrossRef]
20. Tõnutare, T.; Moor, U.; Szajdak, L. Strawberry Anthocyanin Determination by pH Differential Spectroscopic Method—How to Get True Results? *Acta Sci. Polonorum. Hortorum Cultus Ogrod.* **2014**, *13*, 35–47.
21. Proestos, C.; Lytoudi, K.; Mavromelanidou, O.K.; Zoumpoulakis, P.; Sinanoglou, V.J. Antioxidant Capacity of Selected Plant Extracts and Their Essential Oils. *Antioxidants* **2013**, *2*, 11–22. [CrossRef]
22. Dudonné, S.; Vitrac, X.; Coutière, P.; Woillez, M.; Mérillon, J.-M. Comparative Study of Antioxidant Properties and Total Phenolic Content of 30 Plant Extracts of Industrial Interest Using DPPH, ABTS, FRAP, SOD, and ORAC Assays. *J. Agric. Food Chem.* **2009**, *57*, 1768–1774. [CrossRef] [PubMed]
23. Meng, J.; Fang, Y.; Zhang, A.; Chen, S.; Xu, T.; Ren, Z.; Han, G.; Liu, J.; Li, H.; Zhang, Z.; et al. Phenolic Content and Antioxidant Capacity of Chinese Raisins Produced in Xinjiang Province. *Food Res. Int.* **2011**, *44*, 2830–2836. [CrossRef]
24. Benzie, I.F.F.; Strain, J.J. [2] Ferric Reducing/Antioxidant Power Assay: Direct Measure of Total Antioxidant Activity of Biological Fluids and Modified Version for Simultaneous Measurement of Total Antioxidant Power and Ascorbic Acid Concentration. In *Methods in Enzymology*; Academic Press: Cambridge, MA, USA, 1999; Volume 299, pp. 15–27.
25. Chou, T.-C. Theoretical Basis, Experimental Design, and Computerized Simulation of Synergism and Antagonism in Drug Combination Studies. *Pharmacol. Rev.* **2006**, *58*, 621–681. [CrossRef] [PubMed]
26. Ball, D.W. The Chemical Composition of Honey. *J. Chem. Educ.* **2007**, *84*, 1643. [CrossRef]
27. Albu, A.; Radu-Rusu, C.-G.; Pop, I.M.; Frunza, G.; Nacu, G. Quality Assessment of Raw Honey Issued from Eastern Romania. *Agriculture* **2021**, *11*, 247. [CrossRef]
28. Kozłowicz, K.; Różyło, R.; Gładyszewska, B.; Matwijczuk, A.; Gładyszewski, G.; Chocyk, D.; Samborska, K.; Piekut, J.; Smolewska, M. Identification of Sugars and Phenolic Compounds in Honey Powders with the Use of GC–MS, FTIR Spectroscopy, and X-Ray Diffraction. *Sci. Rep.* **2020**, *10*, 16269. [CrossRef]
29. Balasundram, N.; Sundram, K.; Samman, S. Phenolic Compounds in Plants and Agri-Industrial By-Products: Antioxidant Activity, Occurrence, and Potential Uses. *Food Chem.* **2006**, *99*, 191–203. [CrossRef]
30. Bautista, I.; Boscaiu, M.; Lidón, A.; Llinares, J.V.; Lull, C.; Donat, M.P.; Mayoral, O.; Vicente, O. Environmentally Induced Changes in Antioxidant Phenolic Compounds Levels in Wild Plants. *Acta Physiol. Plant.* **2015**, *38*, 9. [CrossRef]

31. Chen, L.; Xin, X.; Zhang, H.; Yuan, Q. Phytochemical Properties and Antioxidant Capacities of Commercial Raspberry Varieties. *J. Funct. Foods* **2013**, *5*, 508–515. [CrossRef]
32. Alibabic, V.; Skender, A.; Orascanin, M.; Sertovic, E.; Bajric, E. Evaluation of Morphological, Chemical, and Sensory Characteristics of Raspberry Cultivars Grown in Bosnia and Herzegovina. *Turk. J. Agric. For.* **2018**, *42*, 67–74. [CrossRef]
33. Fredes, C.; Montenegro, G.; Zoffoli, J.P.; Santander, F.; Robert, P. Comparison of the Total Phenolic Content, Total Anthocyanin Content and Antioxidant Activity of Polyphenol-Rich Fruits Grown in Chile. *Ciencia e Investigación Agraria Revista Latinoamericana de Ciencias de la Agricultura* **2014**, *41*, 49–60. [CrossRef]
34. Cimpoiu, C.; Hosu, A.; Miclaus, V.; Puscas, A. Determination of the Floral Origin of Some Romanian Honeys on the Basis of Physical and Biochemical Properties. *Spectrochim. Acta Part A Mol. Biomol. Spectrosc.* **2013**, *100*, 149–154. [CrossRef] [PubMed]
35. Scripcă, L.A.; Norocel, L.; Amariei, S. Comparison of Physicochemical, Microbiological Properties and Bioactive Compounds Content of Grassland Honey and other Floral Origin Honeys. *Molecules* **2019**, *24*, 2932. [CrossRef]
36. Pauliuc, D.; Dranca, F.; Oroian, M. Antioxidant activity, total phenolic content, individual phenolics and physicochemical parameters suitability for Romanian honey authentication. *Foods* **2020**, *9*, 306. [CrossRef]
37. Max, J.-J.; Chapados, C. Glucose and Fructose Hydrates in Aqueous Solution by IR Spectroscopy. *J. Phys. Chem. A* **2007**, *111*, 2679–2689. [CrossRef]
38. Kędzierska-Matysek, M.; Matwijczuk, A.; Florek, M.; Barłowska, J.; Wolanciuk, A.; Matwijczuk, A.; Chruściel, E.; Walkowiak, R.; Karcz, D.; Gładyszewska, B. Application of FTIR Spectroscopy for Analysis of the Quality of Honey. *BIO Web Conf.* **2018**, *10*, 02008. [CrossRef]
39. Tomaszewska-Gras, J.; Bakier, T.; Goderska, K.; Mansfeld, K. Differential Scanning Calorimetry for Determining the Thermodynamic Properties of Selected Honeys. *J. Apic. Sci.* **2015**, *59*, 109–118. [CrossRef]
40. Zeb, A. Concept, Mechanism, and Applications of Phenolic Antioxidants in Foods. *J. Food Biochem.* **2020**, *44*, e13394. [CrossRef]
41. Parker, T.L.; Miller, S.A.; Myers, L.E.; Miguez, F.E.; Engeseth, N.J. Evaluation of Synergistic Antioxidant Potential of Complex Mixtures Using Oxygen Radical Absorbance Capacity (ORAC) and Electron Paramagnetic Resonance (EPR). *J. Agric. Food Chem.* **2010**, *58*, 209–217. [CrossRef]
42. Kopjar, M.; Lončarić, A.; Mikulinjak, M.; Šrajbek, Ž.; Šrajbek, M.; Pichler, A. Evaluation of Antioxidant Interactions of Combined Model Systems of Phenolics in the Presence of Sugars. *Nat. Prod. Commun.* **2016**, *11*, 1934578X1601101008. [CrossRef]
43. Lončarić, A.; Pichler, A.; Rašić, N.; Vukoja, I.; Leventić, A.; Kopjar, M. Influence of Phenol and Sugar Interactions on Antioxidant Activity of Pomegranate Juice. *Acta Aliment.* **2018**, *47*, 203–209. [CrossRef]
44. Peshev, D.; Vergauwen, R.; Moglia, A.; Hideg, É.; Van den Ende, W. Towards Understanding Vacuolar Antioxidant Mechanisms: A Role for Fructans? *J. Exp. Bot.* **2013**, *64*, 1025–1038. [CrossRef] [PubMed]
45. Shpigelman, A.; Zisapel, A.; Cohen, Y.; Livney, Y.D. Mechanisms of Saccharide Protection against Epigallocatechin-3-Gallate Deterioration in Aqueous Solutions. *Food Chem.* **2013**, *139*, 1105–1112. [CrossRef]
46. Peinado, J.; López de Lerma, N.; Peinado, R.A. Synergistic Antioxidant Interaction between Sugars and Phenolics from a Sweet Wine. *Eur. Food Res. Technol.* **2010**, *231*, 363–370. [CrossRef]
47. Hanuka Katz, I.; Eran Nagar, E.; Okun, Z.; Shpigelman, A. The Link between Polyphenol Structure, Antioxidant Capacity and Shelf-Life Stability in the Presence of Fructose and Ascorbic Acid. *Molecules* **2020**, *25*, 225. [CrossRef] [PubMed]
48. Baccelli, I.; Luti, S.; Bernardi, R.; Scala, A.; Pazzagli, L. Cerato-Platanin Shows Expansin-like Activity on Cellulosic Materials. *Appl. Microbiol. Biotechnol.* **2014**, *98*, 175–184. [CrossRef]
49. Lo Scalzo, R. EPR Free Radical Scavenging Activity on Superoxide, Hydroxyl and Tert–Butyl Hydroperoxide Radicals by Common Hydrophilic Antioxidants: Effect of Mixing and Influence of Glucose and Citric Acid. *Eur. Food Res. Technol.* **2021**, *247*, 2253–2265. [CrossRef]
50. Wohlert, J.; Schnupf, U.; Brady, J.W. Free Energy Surfaces for the Interaction of D-glucose with Planar Aromatic Groups in Aqueous Solution. *J. Chem. Phys.* **2010**, *133*, 155103. [CrossRef]
51. Tavagnacco, L.; Engström, O.; Schnupf, U.; Saboungi, M.-L.; Himmel, M.; Widmalm, G.; Cesàro, A.; Brady, J.W. Caffeine and Sugars Interact in Aqueous Solutions: A Simulation and NMR Study. *J. Phys. Chem. B* **2012**, *116*, 11701–11711. [CrossRef]
52. Craveiro, R.; Mano, F.; Paiva, A.; Duarte, A.R.C. Deep Eutectic Solvents: Exploring Their Role in Nature. In *Deep Eutectic Solvents*; Wiley Online Library: Hoboken, NJ, USA, 2019; pp. 95–110. [CrossRef]
53. Kačuráková, M.; Wilson, R.H. Developments in Mid-Infrared FT-IR Spectroscopy of Selected Carbohydrates. *Carbohydr. Polym.* **2001**, *44*, 291–303. [CrossRef]
54. Listiohadi, Y.; Hourigan, J.A.; Sleigh, R.W.; Steele, R.J. Thermal Analysis of Amorphous Lactose and α-Lactose Monohydrate. *Dairy Sci. Technol.* **2009**, *89*, 43–67. [CrossRef]
55. Anjos, O.; Campos, M.G.; Ruiz, P.C.; Antunes, P. Application of FTIR-ATR Spectroscopy to the Quantification of Sugar in Honey. *Food Chem.* **2015**, *169*, 218–223. [CrossRef] [PubMed]
56. Gok, S.; Severcan, M.; Goormaghtigh, E.; Kandemir, I.; Severcan, F. Differentiation of Anatolian Honey Samples from Different Botanical Origins by ATR-FTIR Spectroscopy using Multivariate Analysis. *Food Chem.* **2015**, *170*, 234–240. [CrossRef] [PubMed]
57. Svečnjak, L.; Bubalo, D.; Baranović, G.; Novosel, H. Optimization of FTIR-ATR Spectroscopy for Botanical Authentication of Unifloral Honey Types and Melissopalynological Data Prediction. *Eur. Food Res. Technol.* **2015**, *240*, 1101–1115. [CrossRef]
58. Bogdanov, S.; Martin, P.; Lüllmann, C. Harmonised Methods of the European Honey Commission. *Apidologie* extra issue. **1997**, 1–59.

59. Chen, Y.; Chen, W.; Fu, L.; Yang, Y.; Wang, Y.; Hu, X.; Wang, F.; Mu, T. Surface Tension of 50 Deep Eutectic Solvents: Effect of Hydrogen-Bonding Donors, Hydrogen-Bonding Acceptors, Other Solvents, and Temperature. *Ind. Eng. Chem. Res.* **2019**, *58*, 12741–12750. [CrossRef]
60. Haghbakhsh, R.; Taherzadeh, M.; Duarte, A.R.C.; Raeissi, S. A General Model for the Surface Tensions of Deep Eutectic Solvents. *J. Mol. Liq.* **2020**, *307*, 112972. [CrossRef]
61. Subbiah, B.; Blank, U.K.M.; Morison, K.R. A Review, Analysis and Extension of Water Activity Data of Sugars and Model Honey Solutions. *Food Chem.* **2020**, *326*, 126981. [CrossRef]
62. Cardona, Y.; Torres, A.; Hoffmann, W.; Lamprecht, I. Differentiation of Honey from *Melipona* Species Using Differential Scanning Calorimetry. *Food Anal. Methods* **2018**, *11*, 1056–1067. [CrossRef]
63. Pauliuc, D.; Dranca, F.; Ropciuc, S.; Oroian, M. Advanced Characterization of Monofloral Honeys from Romania. *Agriculture* **2022**, *12*, 526. [CrossRef]
64. Suhag, Y.; Nayik, G.A.; Nanda, V. Modelling of Moisture Sorption Isotherms and Glass Transition Temperature of Spray-Dried Honey Powder. *J. Food Meas. Charact.* **2018**, *12*, 2553–2560. [CrossRef]
65. Percevault, L.; Limanton, E.; Nicolas, P.; Paquin, L.; Lagrost, C. Electrochemical Determination and Antioxidant Capacity Modulation of Polyphenols in Deep Eutectic Solvents. *ACS Sustain. Chem. Eng.* **2021**, *9*, 776–784. [CrossRef]

antioxidants

Article

Kelulut Honey Ameliorates Oestrus Cycle, Hormonal Profiles, and Oxidative Stress in Letrozole-Induced Polycystic Ovary Syndrome Rats

Datu Agasi Mohd Kamal [1,2], Siti Fatimah Ibrahim [1], Azizah Ugusman [1] and Mohd Helmy Mokhtar [1,*]

1 Department of Physiology, Faculty of Medicine, Universiti Kebangsaan Malaysia, Kuala Lumpur 56000, Malaysia
2 Department of Biomedical Sciences, Faculty of Medicine and Health Sciences, University Malaysia Sabah, Kota Kinabalu 88400, Malaysia
* Correspondence: helmy@ukm.edu.my; Tel.: +60-3-9145-8617

Abstract: Kelulut honey (KH) has been proven to have excellent antioxidative and anti-inflammatory properties with unique physicochemical characteristics. Therefore, we investigated the isolated and combined effects of KH, metformin, or clomiphene in alleviating oxidative stress and reproductive and metabolic abnormalities in polycystic ovary syndrome (PCOS). Female Sprague-Dawley (SD) rats were given 1 mg/kg/day of letrozole for 21 days to induce PCOS. PCOS rats were then divided into six treatment groups: untreated, metformin (500 mg/kg/day), clomiphene (2 mg/kg/day), KH (1 g/kg/day), combined KH (1 g/kg/day) and metformin (500 mg/kg/day), and combined KH (1 g/kg/day) and clomiphene (2 mg/kg/day). All treatments were administered orally for 35 days. The physicochemical characteristics of KH were assessed through hydroxymethylfurfural, free acidity, diastase number, moisture content, sugar profile, metals, and mineral compounds. Additionally, we determined the semivolatile organic compounds present in KH through gas chromatography-mass spectrometry (GC/MS) analysis. KH and its combination with metformin or clomiphene were shown to improve the oestrus cycle, hormonal profile, and oxidative stress in PCOS rats. However, KH did not reduce the fasting blood glucose, insulin, and body weight gain in PCOS rats. These findings may provide a basis for future studies to discover the potential use of KH as a complementary treatment for women with PCOS.

Keywords: kelulut honey; antioxidative; honey physicochemical properties; PCOS

Citation: Kamal, D.A.M.; Ibrahim, S.F.; Ugusman, A.; Mokhtar, M.H. Kelulut Honey Ameliorates Oestrus Cycle, Hormonal Profiles, and Oxidative Stress in Letrozole-Induced Polycystic Ovary Syndrome Rats. *Antioxidants* **2022**, *11*, 1879. https://doi.org/10.3390/antiox11101879

Academic Editors: Josipa Vlainić and Ivana Tlak Gajger

Received: 23 August 2022
Accepted: 18 September 2022
Published: 22 September 2022

1. Introduction

Kelulut honey (KH) is multi-floral stingless bee honey from *Trigona* spp. Over 500 species of stingless bees and 64 distinct genera have been found worldwide [1]. Stingless bees habituate the warm and humid forests and are often found in tropical countries. KH is native to Southeast Asia and can be found in Malaysia, Indonesia, and Thailand. This honey has been used traditionally as a remedy for anti-ageing and improving the immune system in these countries. The physicochemical properties of KH are unique. It differs from the Apis honeybees; hence, its quality should not be assessed according to the standards of the Apis honeybees, such as Codex Alimentarius and the International Honey Commission (2009). In addition, numerous studies have shown that stingless bee honey does not meet the quality standards for Apis bee honey [2], and the International Standard for stingless bee honey has still not been established. However, in 2017, Malaysia published a quality standard for Malaysian stingless bees as quality control for marketed stingless bee honey [3].

KH has demonstrated beneficial health effects as evidenced by numerous studies. For example, KH is reported to have high antibacterial potency [4], chemopreventive properties in the colorectal cancer rat model [5], and anti-inflammatory properties by protecting against lipopolysaccharide (LPS)-induced chronic subclinical systemic inflammation in rats [6]. KH also prevents keloid scar formation by attenuating TGFβ-induced epithelial to mesenchymal transition in primary human keratinocytes and improves ethanol-induced gastric ulcers in rats [7]. Additionally, KH shows anti-obesity properties by improving lipid and cholesterol profiles [8,9].

The most prominent property of KH is its antioxidative actions. In many settings, KH has demonstrated excellent antioxidative activities. A previous study reported that treatment with KH improved the oxidative damage in streptozotocin-induced diabetic rats by increasing superoxide dismutase (SOD) activity and glutathione (GSH) levels and decreasing the protein carbonyl (PC) and malondialdehyde (MDA) levels in the sperm and testis of the diabetic rats [10]. Another study showed that KH could ameliorate glucocorticoid-induced osteoporosis by reducing lipid peroxidation [11]. In LPS-induced chronic subclinical systemic inflammation in rats, KH treatment significantly improved the levels of oxidative stress markers such as MDA, 8-hydroxy-2′-deoxyguanosine (8-OHdG), GSH, glutathione peroxidase (GPx), and glutathione S-transferase (GST) [6]. In fact, several studies have found that KH possesses a higher antioxidant content than common bee honey such as Acacia and Tualang honey [12–14]. Trehalulose, a highly active antioxidant, is one of the discovered content in KH that is responsible for its antioxidative actions [15]. Taken together, KH has been shown to have potential therapeutic benefits for a variety of diseases related to oxidative stress.

Polycystic ovary syndrome (PCOS) is a complex disorder that affects the reproductive, endocrine, metabolic, and psychological systems [16]. The anomalies of ovarian steroidogenesis, insulin resistance, hyperinsulinism, abnormalities of gonadotrophin production, and follicular arrest are some of the aetiologies of PCOS [17,18]. Additionally, it has been demonstrated that low-grade inflammation and oxidative stress play a role in the aetiology of PCOS [19–21]. Symptoms of PCOS include hyperandrogenism, anovulation, infertility, obesity, menstrual cycle irregularities, dyslipidaemia, and hirsutism [22]. Treatment for patients with PCOS typically focuses on the symptoms, as there is yet no definitive treatment for PCOS [23].

Medications used to treat PCOS include clomiphene citrate and metformin [24]. However, these medications are linked to a number of side effects, such as diarrhoea, vaginal and uterine bleeding, breast discomfort, hot flashes, and abdominal pain [25]. Therefore, discovering a natural supplement that could be used as a complementary treatment for PCOS with minimal side effects is of great interest. Previously, inositol, a naturally occurring substance found primarily in fruits, grains, and beans, was shown to be effective in treating metabolic and reproductive disorders in PCOS women [26,27]. This finding strengthens the potential use of nutraceuticals in treating typical symptoms of PCOS and the need for more research to be conducted in this field. Recently, we found that KH could improve the regulation of the oestrus cycle and ovarian histomorphological alterations in rats with letrozole-induced PCOS [28]. In view of these promising effects, the current study aimed to focus on the antioxidative effect of KH on letrozole-induced PCOS rats. Additionally, we demonstrated the physicochemical properties of KH, including the semivolatile component, to verify the quality and discover the potential beneficial compound in KH.

2. Materials and Methods

2.1. Honey Sample

Kelulut honey (KH) was gathered in Negeri Sembilan, Malaysia, by an experienced local beekeeper. The nearby herbal plant provides nectar for the bees to collect. The KH was kept raw at 4 °C in amber bottles away from heat sources and sunlight until further analysis.

2.2. *Physicochemical Profiling of KH*

2.2.1. Moisture Content

The Petri dish was weighed empty before adding 1 g of KH. The uncovered dish was then placed in an oven at 60 °C for two hours. Subsequently, the dish was transferred into a desiccator to cool it before reweighting. The heating and weighing steps continued for another two hours in the oven until a weight change of less than 2 mg was achieved. The moisture content was determined using the following formula:

$$\text{Moisture content}: \frac{W_2 - W_1}{W} \times 100 \quad (1)$$

Moisture content: $\frac{W_2-W_1}{W} \times 100$
W_2 = Weight of empty dish + sample (g)
W_1 = Weight of dried dish + dried sample (g)
W = Weight of the sample (g)

2.2.2. Ash

The ash content was calculated by weighing 5 g of KH samples in a platinum dish and heating them to a consistent weight in a laboratory furnace at 600 °C following the AOAC Method 920.181 [29]. Ash content was measured in triplicate. Total ash content was measured in triplicate and expressed as the percentage of residue left after dry oxidation by weight (g/100 g honey), which was calculated using the following equation:

$$\text{Ash}(\%) = [(m_1 - m_2)/m_0] \times 100 \quad (2)$$

Note that m_0 is the mass of the KH taken, m_1 is the mass of the dish plus ash, and m_2 is the mass of the platinum dish before it was calcined.

2.2.3. Free Acidity

The titrimetric method was used to determine free acidity following the AOAC Official Method 962.19 [29]. In a 250 mL beaker, 10 g of KH sample was dissolved in 75 mL CO_2-free water. The solution was swirled with a magnetic stirrer while the electrode of the pH meter was submerged and titrated to pH 8.5 by adding 0.05 M NaOH solution.

2.2.4. Diastase

Diastase activity was calculated according to the AOAC Official Method 958.09 [29]. A buffered mixture of soluble starch and KH was incubated in a thermostatic bath at 40 °C. Subsequently, 1 mL aliquot was taken at 5 min intervals, and a Perkin Elmer Luminescence Spectrophotometer was used to measure the sample's absorption at 660 nm (Norwalk, CT, USA). The diastase number was determined using the same length of time it took for the absorbance to reach 0.235. The results were reported in Schade units per gram of honey as the volume (mL) of 1% starch hydrolysed by an enzyme in 1 g of honey in 1 h.

2.2.5. Minerals and Metals

Determination of minerals and heavy metals was performed using inductively coupled plasma–optical emission spectrophotometry (ICP-OES) described by Aghamirlou et al. [30]. Briefly, samples were weighed at approximately 1 g. The sample was then mixed with 10 mL of HNO_3 and heated for 1 h and 30 min at 95 °C in a microwave. After cooling to room temperature, 50 mL of distilled water was added, and the sample filtrate was analysed by ICP-OES (Agilent 7500ce, Agilent Technologies Inc. Palo Alto, CA, USA). Table 1 shows the ICP-OES operating conditions.

Table 1. The ICP-OES operating conditions.

Nebulizer	Micromist
RF generator (W)	1550
Argon flow rate (L min^{-1})	0.85
Nebuliser pump (rps)	0.10
Scanning condition	Number of replicates 3, dwelling time 1 s
H2 flow (L min^{-1})	3.5
He flows (L min^{-1})	4.0

2.2.6. Determination of Hydroxymethylfurfural (HMF)

The method of [31] was used to determine the HMF content. A 50 mL volumetric flask containing 0.5 mL of Carrez solution I and 0.5 mL of Carrez solution II was filled with precisely 5 g of KH dissolved in 25 mL of water. Water was added to the flask to make it 50 mL larger, and the resultant solution was then filtered. After discarding the first 10 mL of filtrate, two aliquots of 5 mL each were added to the test tubes. The reference tube received 5 mL of a 0.2% sodium bisulphite solution, whereas the sample tube received 5 mL of pure water. A UV-visible spectrometer was used to measure the solution's absorbance at two wavelengths, 284 and 336 nm. The equation below was used to determine the HMF content:

$$\text{HMF (mg/100 g)} = (A_{284} - A_{336}) \times 14.97 \times 5 \quad (3)$$

Note that 14.97 was the factor, and 5 was the mass of the sample. A_{284} was the absorbance at 284 nm, and A_{336} was the absorbance at 336 nm.

2.2.7. Sugar Profiling

The high-performance ion chromatography (HPIC)-based approach, as described in [32], was used to profile the sugar levels in the KH samples. The amount of 200 mg of KH was dissolved in a few millilitres of water. This solution was transferred quantitatively into a volumetric flask, filled to the 100 mL mark with water, thoroughly mixed, and then filtered through a 0.02 m nylon membrane filter (Whatman). The sample was then injected into the Thermo Scientific™ Dionex™ ICS-5000+ machine and detected by an electrochemical detector. HPIC-grade solvents were applied to a Dionex CarboPac PA20, Analytical (3 × 150 mm) column from Thermo Fisher Scientific in Waltham, MA, USA.

2.2.8. Semivolatile Organic Compound (SVOC) Determination

Gas chromatography-mass spectrometry (GC/MS) was used for SVOC determination on the KH sample based on modifications to the U.S. E.P.A. 8270 protocol [32]. The KH samples were extracted with 99.8% dichloromethane (GC grade, Merck, Darmstadt, Germany) before concentrating on the minimum injection volume. The sample was injected into the splitless inlet of an Agilent 7890B, 5977B MSD GC-MS system (Agilent Technologies, Santa Clara, CA, USA). The carrier gas, helium, flowed at a 1.0 mL/min rate. The peaks collected were compared with those in the NIST collection to identify the SVOCs (Gaithersburg, MD, USA). Table 2 lists the GC and MS running circumstances.

2.3. Animal Preparation

The ethical review and approval of the study protocol were granted by the National University of Malaysia Animal Ethics Committee (Ethical Approval Code FISIO/FP/2020/MOHD HELMY/14-MAY/1104-JUNE-2020-MAY-2023). Female Sprague–Dawley (SD) rats weighing 120–150 g and exhibiting at least two consecutive regular oestrus cycles were used in this study. The experimental animals were supplied by Laboratory Animal Research Unit (LARU), Faculty of Medicine, Universiti Kebangsaan Malaysia. Rats were kept in individual cages, allowed to acclimate for a week, kept in an air-conditioned room at 24 ± 2 °C with a 12 h light/12 h dark cycle, and provided with regular food pellets and water ad libitum. Animals were weighed twice a week, and the oestrus phase was determined daily by observing vaginal smears under an Olympus BX40 light microscope (Olympus Corporation, Tokyo, Japan).

Table 2. GC/MS parameters for the determination of SVOCs in KH.

GC Parameters	
Inlet mode	Splitless
Splitless time (min)	16
Carrier gas, flow, flow rate	Helium, constant pressure 10 psi, 1.7 mL/min
Oven	50 °C, 0.1 min
Chromatographic column	30 m × 0.25 mm internal diameter × 0.5 µm film thickness DB-UI-8270D ULTRA INERT (Agilent Technologies, Santa Clara, CA, USA)
MS Parameters	
Transfer line temperature (°C)	300
Source temperature (°C)	230
Ionisation mode	Electron ionisation (EI)
Electron energy (eV)	70
Acquisition mode	Full scan 40–650 m/z
MS Library	NIST MS Search 2.2 (Gaithersburg, MD, USA)

Animal Treatment

The rats were divided into two main groups (Figure 1). The first group (normal control group, $n = 6$) received distilled water throughout the study (56 days). The second group ($n = 36$) was administered with 1 mg/kg/day of letrozole orally, once daily for 21 days, to induce PCOS as reported in a previous study [33]. As validated in our previous study, PCOS was induced in all rats administered with 1 mg/kg/day of letrozole for 21 days, determined by examining the dioestrous days, blood glucose levels, and ovarian histology [28].

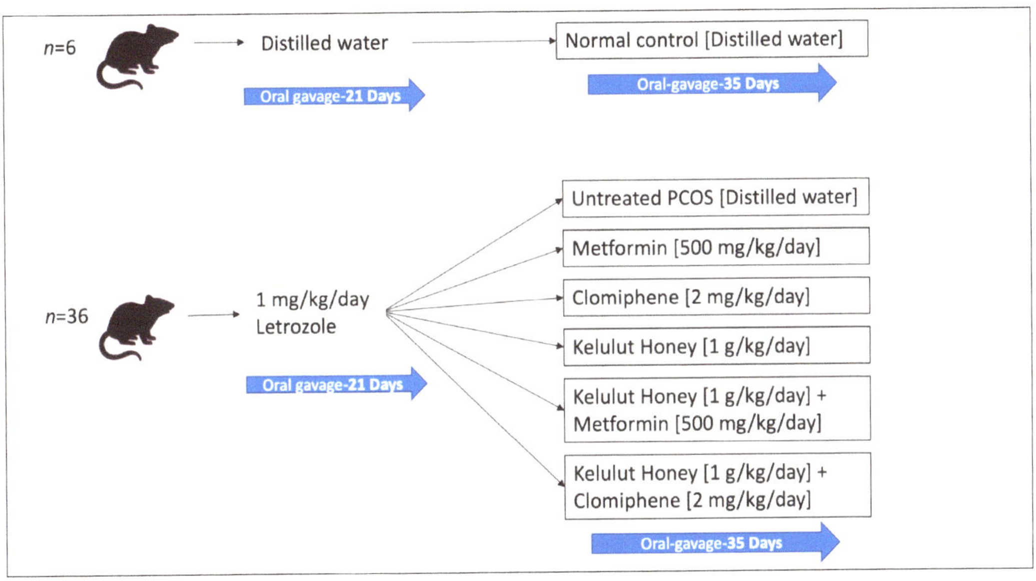

Figure 1. A schematic diagram of the animal grouping and treatments.

The PCOS rats ($n = 36$) were then randomly distributed into six experimental groups ($n = 6$ per group): untreated PCOS rats that received distilled water, treatment with metformin (500 mg/kg/day), treatment with clomiphene (2 mg/kg/day), treatment with KH

(1 g/kg/day), combined treatment with KH (1 g/kg/day) and metformin (500 mg/kg/day), and combined treatment with KH (1 g/kg/day) and clomiphene (2 mg/kg/day). All treatments were administered orally for 35 days. The doses of metformin (500 mg/kg/day) and clomiphene (2 mg/kg/day) were based on the study by Ndeingang et al. [34]. KH dose (1 g/kg/day) and treatment duration were determined from our pilot study [28]. Metformin is used for insulin resistance in women with PCOS, whereas clomiphene is used to induce ovulation [24]. We designed the groups to receive metformin and clomiphene or a combination of these drugs with KH to assess the effect of KH treatment and to investigate any synergistic effect. All animals were euthanised by ketamine-xylazine overdose (0.3 mL/100 g body weight) at the end of the 35 days [35].

2.4. Determination of Oestrous Cycle

Every day at 9:00 am, all the rats had vaginal smears taken using cotton buds dipped in 0.9% saline. The vaginal fluid was then collected by rolling a cotton bud onto a glass slide. Oestrous cycles were tracked until the end of the study. The cells were stained with methylene blue and viewed under a light microscope. The smears were categorised as one of the four oestrous cycle phases previously described [36]. A smear of the proestrous phase has a significant number of rounded and nucleated epithelial cells. The oestrous phase, in contrast, is characterised by smears that primarily contain irregular cornified cells without a nucleus. Leukocytes were mainly found in the dioestrous phase and were tiny, rounded cells without a nucleus. The metoestrous phase is characterised by having the same ratio of leukocytes, cornified cells, and nucleated epithelial cells in a smear [36].

2.5. Determination of Fasting Blood Glucose

This test was performed before the rats were sacrificed. Rats were fasted for eight hours, and a sample of tail blood was used to measure blood glucose using a handheld glucometer (Accucheck performa, Roche Diagnostics, Basel, Switzerland).

2.6. Determination of Serum Hormone Levels and Insulin by Enzyme-Linked Immunosorbent Assay (ELISA) Technique

Prior to rat sacrifice, samples of retro-orbital blood were collected in serum separator tubes (BD Vacutainer SSTTM, Becton Dickinson, Franklin Lakes, NJ, USA). Samples were allowed to clot at 4 °C before centrifugation at 3000× *g* for 15 min and stored at −80 °C until hormonal analysis was performed.

The serum hormone levels were analysed using a competitive ELISA for oestradiol, progesterone, and testosterone and a sandwich ELISA for follicle-stimulating hormone (FSH), luteinising hormone (LH), and insulin. All samples were tested in duplicate according to the manufacturer's guidelines (Elabscience, Houston, TX, USA). In brief, reference wells were prepared by mixing biotinylated antibody solution and reference standards. Biotinylated antibody solution and serum samples were added to the test wells. The plate was then sealed and incubated for 45 min at 37 °C. The plate washing was done with a wash buffer and repeated thrice. Horseradish peroxidase (HRP) conjugate was added to each well, and the plate was incubated for 30 min at 37 °C. Then, the plate washing process was repeated five times. A substrate reagent was then added to each well, and the plate was incubated for 15 min at 37 °C. Finally, a stop solution was added to each well to stop the reaction, and the absorbance of the solution in the wells was immediately measured at 450 nm using a microplate reader. Standard dilutions of known hormone concentrations were used to create the standard curve. The produced standard curve was used to determine the serum hormone concentrations.

2.7. Oxidative Stress Status Evaluation

Ovarian tissue lysates were prepared based on previous methods [37]. The tissues were weighed and powdered in liquid nitrogen using a mortar and pestle. The tissue powder was then mixed with phosphate-buffered saline (PBS; 0.1 M, pH 7.4) in a ratio of 1:9 (w/v), and centrifuged for 10 min at 4 °C. The protein concentration in the ovarian tissue lysates was measured using the Bradford assay [38].

Levels of catalase (CAT, as U/mg protein), total superoxide dismutase (SOD, as U/mg protein), glutathione peroxidase (GSH, as mg U/mg protein), and malondialdehyde (MDA, as µmol/g protein) in ovarian tissues were determined using calorimetric diagnostic kits (Elabscience, Houston, TX, USA) according to the manufacturer's guidelines.

2.8. Statistical Analysis

The data were reported as mean ± SEM. GraphPad Software (GraphPad Inc., San Diego, CA, USA) was used to determine the differences between the groups using one-way ANOVA and Tukey's multiple comparison tests. Statistical significance was defined as $p < 0.05$.

3. Results

3.1. Physicochemical Profile of KH for Quality Determination

KH is a dark amber with low viscosity and has a sweet aroma with a mildly acidic taste. The physicochemical analysis (Table 3) and sugar profile (Table 4) of KH were found to comply with the Malaysian KH Standard [3]. Figure 2 shows the HPIC chromatogram of the KH sugar profile. Meanwhile, the metal and mineral analyses reported here are in the range of other studies, as shown in Table 5.

Table 3. Comparison of the physicochemical analysis of our KH and Malaysian KH Standard (Raw).

Analysis	KH	Malaysian KH Standard (Raw) [3]
Moisture Content	14.5 g/100 g	Not more than 35.0%
Ash Content	0.1 g/100 g	1.0 g/100 g
Hydroxyl Methyl Fulfural	<0.1 mg/100 g	Not more than 30.0 mg/kg
Free Acidity	269 meq/kg	2.5 to 3.8 pH
Diastase Number (DN)	<3 Schade Unit	Not stated

Table 4. Comparison of sugar contents in KH and Malaysian Standard KH (Raw).

Sugar Analysis	Value	Malaysian Standard KH (Raw) [3]
Fructose	9.6 g/100 g	Fructose and glucose (sum), not more than 85.0 g/100 g
Glucose	7.9 g/100 g	Fructose and glucose (sum), not more than 85.0 g/100 g
Sucrose	<0.100 g/100 g	Not more than 7.5 g/100 g
Maltose	0.845 g/100 g	Not more than 9.5 g/100 g
Lactose	<0.100 g/100 g	Not stated
Galactose	<0.100 g/100 g	Not stated
Total sugars	18.3 g/100 g	Not stated

3.2. GC-MS Semivolatile Organic Compound Analysis

A mass spectral analysis of KH identified six semivolatile compounds (Table 6) including 2,4-Dimethylhept-1-ene-(retention time (RT): 2.515, 2.598, 2.671 min), Tetradecane-(RT: 6.790 min), 2,4-Di-tert-butylphenol (RT: 7.369 min), n-Hexadecanoic acid-(RT: 9.405 min) and Octadecanoic acid (RT: 10.184 min), and z-10-Octdecen-1-ol acetate-(RT: 10.972 min). All compounds were identified by comparing their mass spectra with those found in the NIST mass spectral library. Figure 3 shows the GC-MS spectrum of the semivolatile organic compounds present in KH.

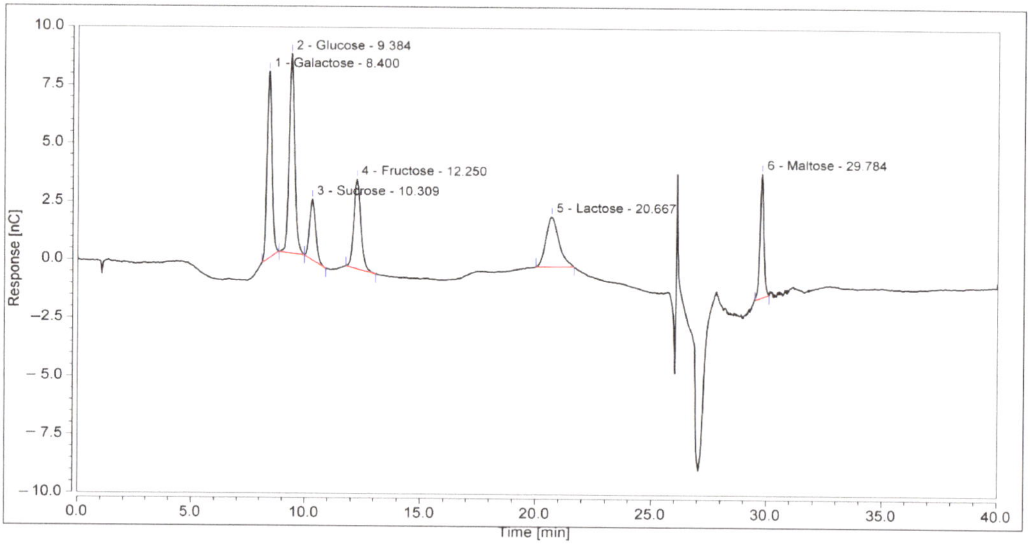

Figure 2. HPIC chromatograms of the sugar profile of KH. The retention times of the identified sugars were galactose, 8.4 min; glucose, 9.384 min; sucrose, 10.309 min; fructose, 12.250 min; lactose, 20.667 min; and maltose, 29.784 min.

Table 5. Comparison of metal and mineral contents in our KH and other analyses.

Metal and Mineral Analysis	Result (mg/kg)	Other Studies (mg/kg)
Calcium	361	59.513–191.9 [12,39]
Antimony	<0.100	No report found
Iron	0.9	6.57–10.90 [12,39]
Arsenic	<0.100	0.019 [39]
Potassium	894	370.65–732.2 [12,39]
Magnesium	40.7	10.09–33.81 [12,39]
Cadmium	<0.100	0.002–0.03 [12,39]
Sodium	36.4	108.78–589.7 [12,39]
Phosphorus	15.4	0.206 [39]
Sulphur	32.8	No report found
Lead	<0.100	0.154 [39]
Tin	2.89	No report found
Mercury	<0.050	0.022 [39]

Table 6. Semivolatile organic compounds identified in KH.

Retention Time (min)	GC-MS Semivolatile Organic Compounds	Molecular Weight (Da)	Cas. No.
2.515	2,4-Dimethylhept-1-ene	126.24	19549-87-2
2.598	2,4-Dimethylhept-1-ene	126.24	19549-87-2
2.671	2,4-Dimethylhept-1-ene	126.24	19549-87-2
6.790	Tetradecane	198.39	629-59-4
7.369	2,4-Di-tert-butylphenol	206.32	96-76-4
9.405	n-Hexadecanoic acid	256.42	57-10-3
10.184	Octadecanoic acid	284.5	57-11-4
10.972	z-10-Octadecen-1-ol acetate	310.5	Not found

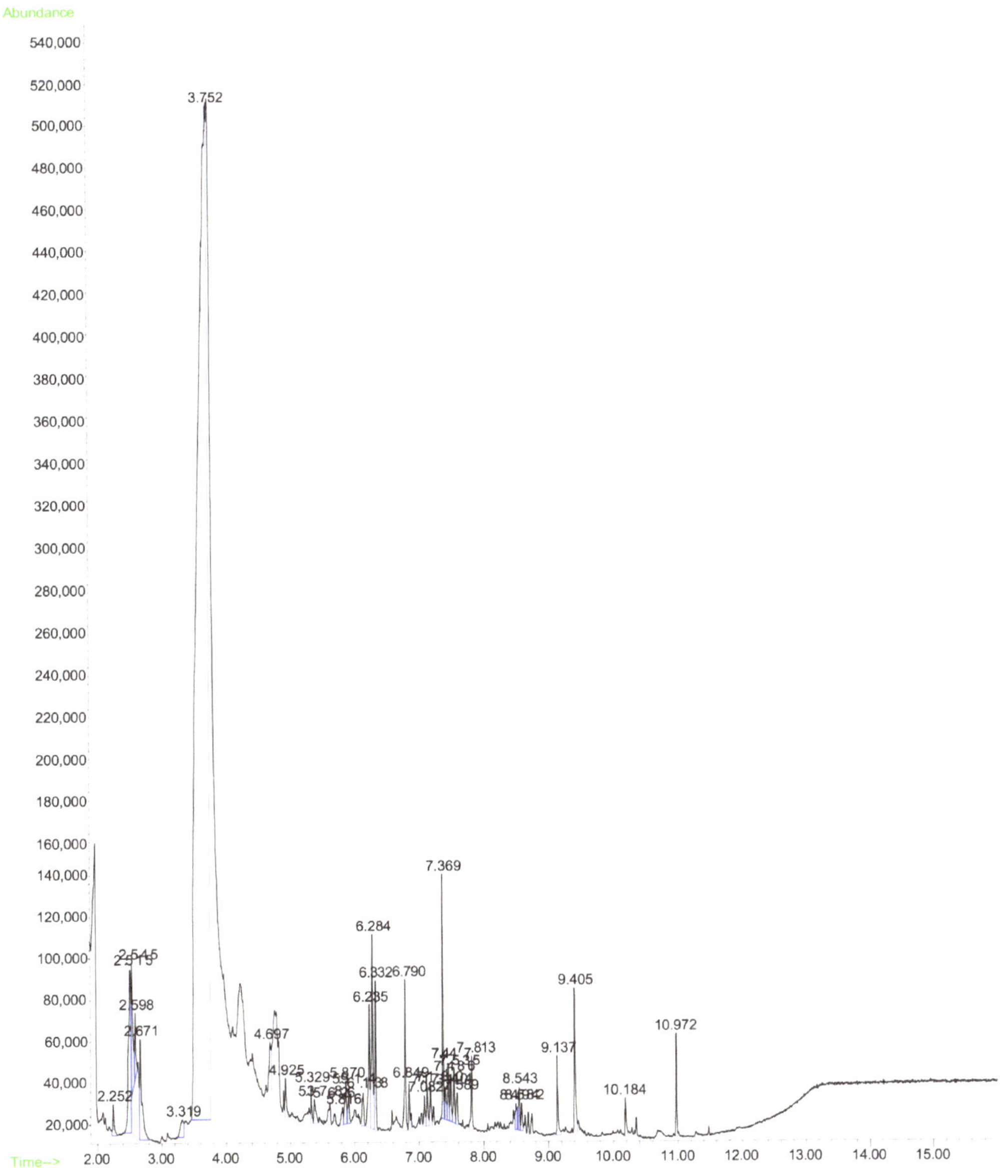

Figure 3. GC-MS spectrum of the semivolatile organic compounds present in KH.

3.3. *Effects of KH on Oestrus Cycle and Body Weight Gain*

The percentage of dioestrus days (Figure 4a) was the lowest in the normal control group (44.94 ± 1.68%). Letrozole induction caused dioestrus days to increase significantly ($p < 0.05$) in untreated PCOS rats compared with normal control rats (83.59 ± 1.95% vs. 44.94 ± 1.68%, $p < 0.05$). Treatment with clomiphene (65.37 ± 2.29%), KH only (67.75 ± 2.55%), KH with metformin (71.42 ± 1.53%), and KH with clomiphene (63.18 ± 1.78%) significantly reduced the percentage of dioestrus days ($p < 0.05$) compared with the untreated PCOS group (83.59 ± 1.95%). Meanwhile, no significant difference in the percentage of dioestrus days was recorded among the clomiphene group, KH group, combined KH + metformin group, and combined KH + clomiphene group. Additionally, treatment with metformin alone (78.67 ± 1.62%) did not change the percentage of dioestrus days compared with the untreated PCOS group.

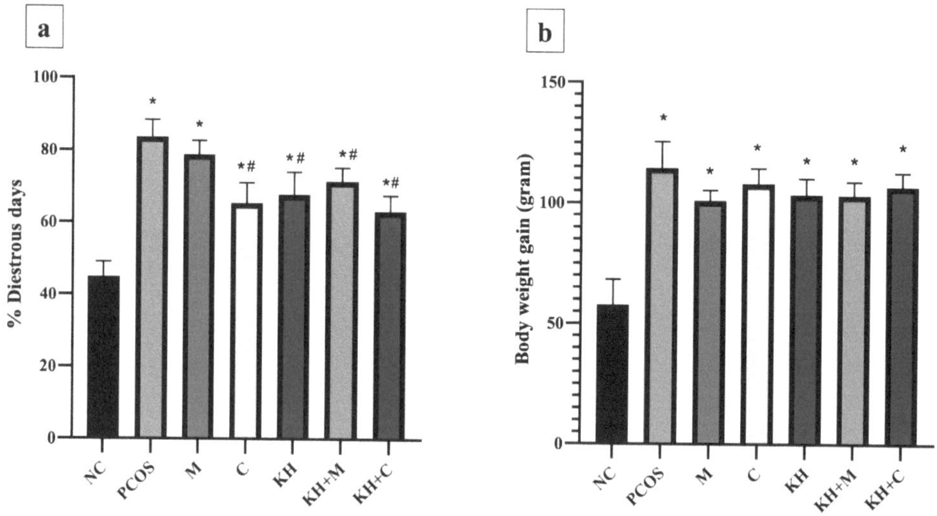

Figure 4. Effects of KH on (**a**) the percentage of dioestrus days and (**b**) rat body weight gain. NC: normal control; PCOS: Untreated PCOS; M: PCOS + Metformin; C: PCOS + Clomiphene; KH: PCOS + Kelulut honey; KH + M: PCOS + Kelulut honey + Metformin; KH + C: PCOS + Kelulut honey + Clomiphene. * $p < 0.05$ significance against the normal control group, # $p < 0.05$ significance against the untreated PCOS group. $n = 6$ per treatment group.

Meanwhile, Figure 4b illustrates the effects of KH on body weight gain. Letrozole induction significantly increased the body weight gain in all PCOS rats compared with the normal control group. However, no significant difference was recorded between the untreated PCOS rats and any other treatment groups.

3.4. *Effect of KH on Fasting Blood Glucose and Insulin Levels*

The effect of KH on fasting blood glucose levels is shown in Figure 5a. Blood glucose levels were significantly elevated ($p < 0.05$) in all PCOS rats compared with normal control rats (6.45 ± 0.18 mmol/L). The untreated PCOS group had the highest blood glucose (10.77 ± 0.17 mmol/L). Treatment with metformin or combined KH + metformin significantly reduced the fasting blood glucose compared with the untreated PCOS rats (8.42 ± 0.15 mmol/L, 8.92 ± 0.06 mmol/L vs. 10.77 ± 0.17 mmol/L, $p < 0.05$). We found that treatment with KH (10.23 ± 0.08 mmol/L), clomiphene (10.28 ± 0.09 mmol/L), or combined KH + clomiphene (10.28 ± 0.07 mmol/L) did not result in a significant reduction in blood glucose levels compared with the untreated PCOS rats (10.77 ± 0.17 mmol/L).

Figure 5b illustrates the effects of KH on insulin levels. Insulin levels were significantly increased in all PCOS rats compared with normal control rats ($p < 0.05$). Treatment with metformin significantly reduced the insulin level compared with the untreated PCOS rats (53.83 ± 1.04 pg/mL vs. 71.60 ± 0.49 pg/mL, $p < 0.05$). Treatment with clomiphene (68.14 ± 0.92 pg/mL), KH (68.41 ± 0.45 pg/mL), combined KH + metformin (67.99 ± 0.43 pg/mL), or combined KH + clomiphene (68.75 ± 1.36 pg/mL) did not reduce the insulin levels compared with the untreated PCOS rats (71.60 ± 0.49 pg/mL).

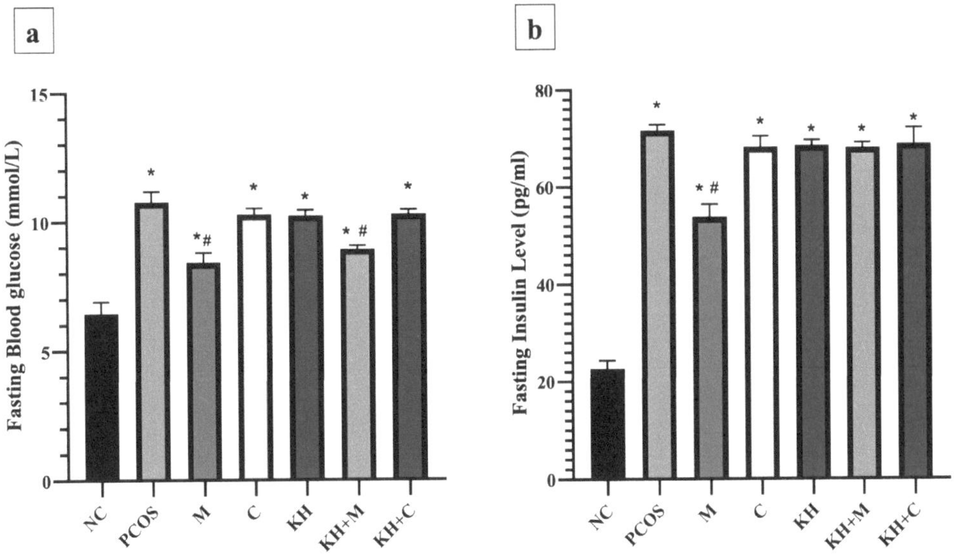

Figure 5. Effects of KH on the (**a**) fasting blood glucose and (**b**) insulin levels. NC: normal control; PCOS: Untreated PCOS; M: PCOS + Metformin; C: PCOS + Clomiphene; KH: PCOS + Kelulut honey; KH + M: PCOS + Kelulut honey + Metformin; KH + C: PCOS + Kelulut honey + Clomiphene. * $p < 0.05$ significance against the normal control group, # $p < 0.05$ significance against the untreated PCOS group. $n = 6$ per treatment group.

3.5. Effect of KH on Serum Testosterone, Oestradiol, Progesterone, LH, and FSH

Figure 6a demonstrates the effect of KH on testosterone levels. Testosterone levels were significantly elevated in untreated PCOS rats as compared with normal control rats (2.65 ± 0.19 ng/mL vs. 1.16 ± 0.02 ng/mL, $p < 0.05$). The elevated testosterone level was significantly reversed ($p < 0.05$) by treating the rats with clomiphene (1.12 ± 0.12 ng/mL), combined KH + clomiphene (1.25 ± 0.08 ng/mL), metformin (1.38 ± 0.07 ng/mL), combined KH + metformin (1.47 ± 0.09 ng/mL), and KH (1.54 ± 0.09 ng/mL). However, no significant differences were recorded among the treatment groups. Meanwhile, as shown in Figure 6b, oestradiol levels were significantly reduced in the untreated PCOS rats compared with the normal control group (1.74 ± 0.05 pg/mL vs. 2.94 ± 0.11 pg/mL, $p < 0.05$). Treatment with metformin (2.86 ± 0.19 pg/mL), combined KH + metformin (2.96 ± 0.22 pg/mL), clomiphene (3.08 ± 0.12 pg/mL), or combined KH + clomiphene (3.18 ± 0.07 pg/mL) significantly increased the oestradiol levels ($p < 0.05$) as compared with the untreated PCOS rats (1.74 ± 0.05 pg/mL). On the other hand, treatment with KH alone did not increase the oestradiol levels compared with the untreated PCOS rats.

As shown in Figure 6c, progesterone levels were significantly reduced in untreated PCOS rats compared with the normal control group (2.05 ± 0.04 ng/mL vs. 4.45 ± 0.15 ng/mL, $p < 0.05$). Only treatment with KH + metformin (4.40 ± 0.09 ng/mL), metformin (4.22 ± 0.07 ng/mL), and combined KH + clomiphene (4.02 ± 0.06 ng/mL) significantly increased the progesterone levels ($p < 0.05$) compared with the untreated PCOS rats (2.05 ± 0.04 ng/mL). Treatment with clomiphene (2.41 ± 0.13 ng/mL) and KH alone (2.44 ± 0.08 ng/mL) did not increase the progesterone levels compared with the untreated PCOS rats (2.05 ± 0.04 ng/mL). Meanwhile, Figure 6d illustrates that letrozole induction caused LH levels to increase significantly in untreated PCOS rats compared with the normal control rats (22.40 ± 0.32 mIU/mL vs. 11.01 ± 0.3 mIU/mL, $p < 0.05$). Treatment with KH alone (17.45 ± 0.19 mIU/mL), combined KH + clomiphene (18.48 ± 0.27 mIU/mL), and combined KH + metformin (18.75 ± 0.43 mIU/mL) significantly reduced the levels of LH compared with untreated PCOS rats ($p < 0.05$). No reduction in LH levels was recorded when the PCOS rats were treated with metformin (21.64 ± 0.27 mIU/mL) or clomiphene alone (21.52 ± 0.26 mIU/mL). However, no changes in FSH levels were found in any groups (Figure 6e).

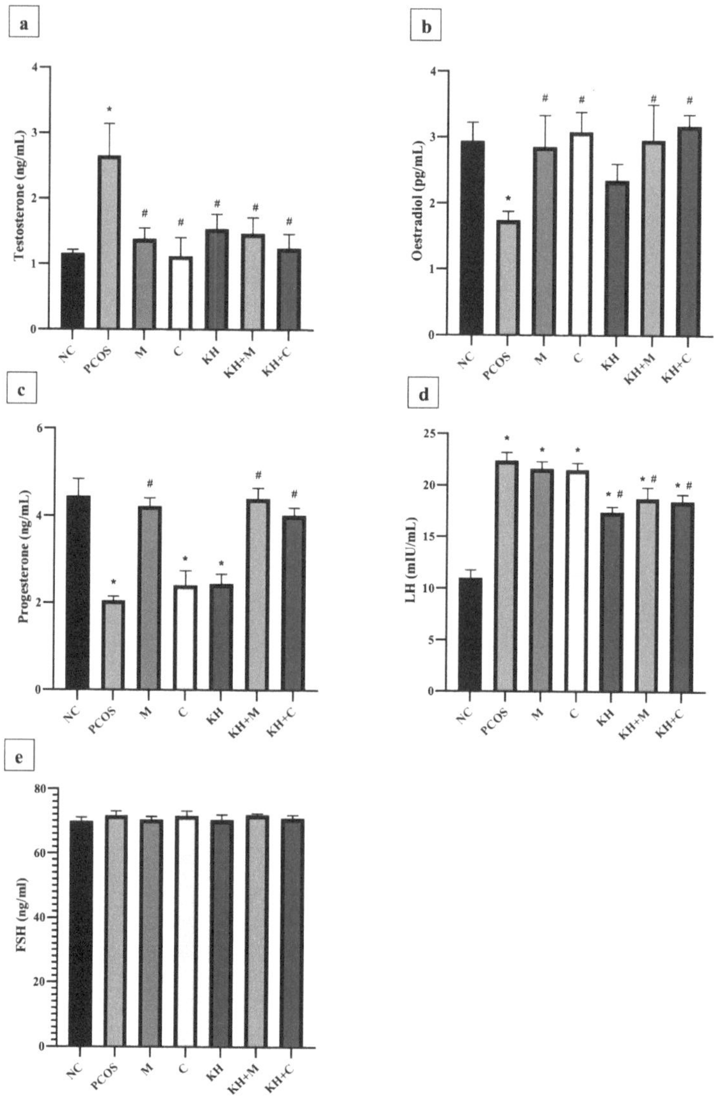

Figure 6. Effects of KH on serum (**a**) testosterone, (**b**) oestradiol, (**c**) progesterone, (**d**) LH, and (**e**) FSH levels. NC: normal control; PCOS: Untreated PCOS; M: PCOS + Metformin; C: PCOS + Clomiphene; KH: PCOS + Kelulut honey; KH + M: PCOS + Kelulut honey + Metformin; KH + C: PCOS + Kelulut honey + Clomiphene. * $p < 0.05$ significance against the normal control group, # $p < 0.05$ significance against the untreated PCOS group. $n = 6$ per treatment group.

3.6. Effect of KH on Ovarian Oxidative Stress

Figure 7a illustrates the effect of KH on catalase activity in ovarian tissues. Letrozole induction decreased the catalase activity in untreated PCOS rats compared with the normal control group (1.78 ± 0.16 U/mg prot vs. 3.57 ± 0.09 U/mg prot, $p < 0.05$). This reduction was significantly reversed ($p < 0.05$) by treatment with KH + metformin (3.43 ± 0.04 U/mg prot), KH + Clomiphene (3.17 ± 0.05 U/mg prot), and KH alone (3.09 ± 0.03 U/mg prot), whereas treatment with metformin (2.12 ± 0.03 U/mg prot) or clomiphene alone (2.06 ± 0.03 U/mg prot) did not increase the catalase activity compared with the untreated PCOS rats.

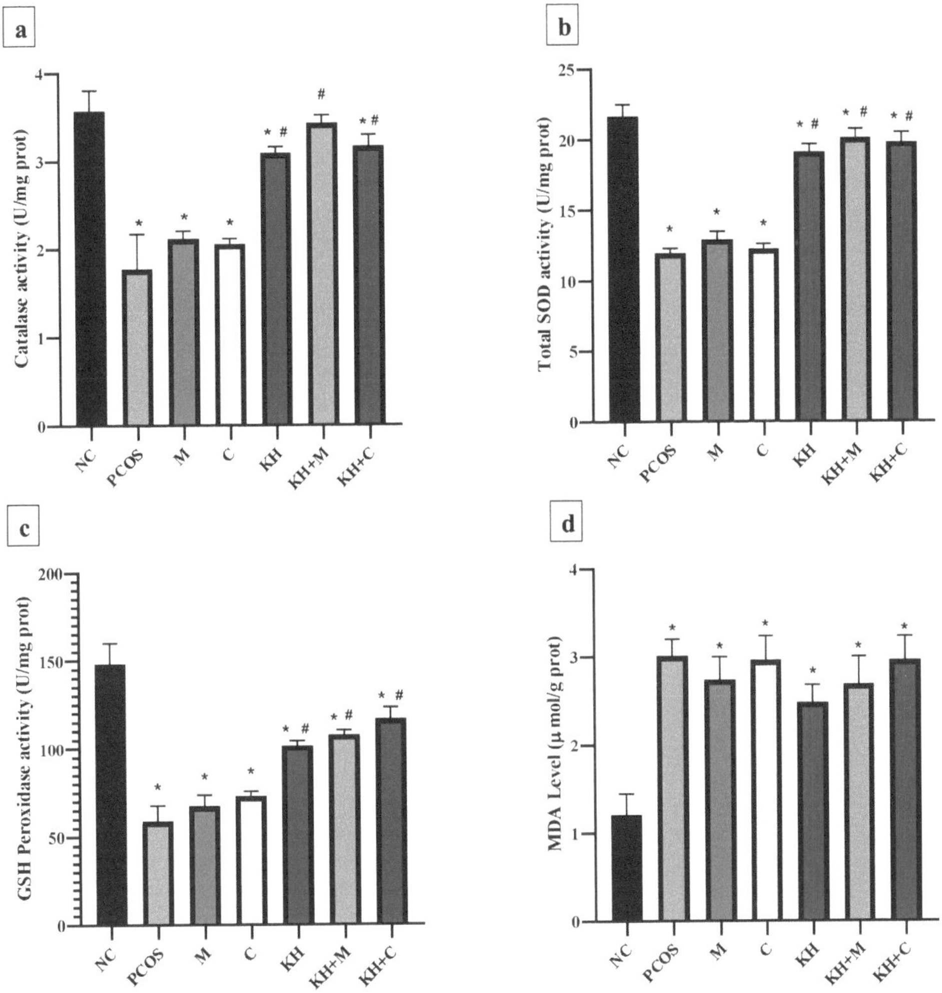

Figure 7. Effects of KH on ovarian (**a**) catalase activity, (**b**) total SOD activity, (**c**) GSH peroxidase activity, and (**d**) MDA levels. NC: normal control; PCOS: Untreated PCOS; M: PCOS + Metformin; C: PCOS + Clomiphene; KH: PCOS + Kelulut honey; KH + M: PCOS + Kelulut honey + Metformin; KH + C: PCOS + Kelulut honey + Clomiphene. * $p < 0.05$ significance against the normal control group, # $p < 0.05$ significance against the untreated PCOS group. $n = 6$ per treatment group.

Total SOD activity in the ovarian tissues is shown in Figure 7b. SOD activity was found to be decreased in untreated PCOS rats as compared with the normal control group (11.97 ± 0.12 U/mg prot vs. 21.68 ± 0.33 U/mg prot, $p < 0.05$). Treatment with KH (19.11 ± 0.23 U/mg prot), combined KH + metformin (20.12 ± 0.26 U/mg prot), and combined KH + clomiphene (19.79 ± 0.28 U/mg prot) significantly increased the total

SOD activities ($p < 0.05$) compared with the untreated PCOS rats (11.97 ± 0.12 U/mg prot). However, treatment with metformin (12.90 ± 0.23 U/mg prot) or clomiphene alone (12.24 ± 0.14 U/mg prot) did not increase the total SOD activity compared with the untreated PCOS rats.

Figure 7c demonstrates the effect of different treatments on GSH peroxidase activity in ovarian tissues. GSH peroxidase level was reduced in the untreated PCOS rats compared with the normal control rats (59.09 ± 4.29 U/mg prot vs. 148.22 ± 5.88 U/mg prot, $p < 0.05$). As in SOD and catalase analyses, treatment with KH (101.72 ± 1.39 U/mg prot), combined KH + clomiphene (116.66 ± 3.27 U/mg prot), and combined KH + metformin (107.76 ± 1.26 U/mg prot) significantly increased the GSH peroxidase activity ($p < 0.05$) as compared with the untreated PCOS rats (148.22 ± 5.88 U/mg prot). Treatment with metformin (67.33 ± 3.19 U/mg prot) or clomiphene alone (73.14 ± 1.33 U/mg prot) did not increase the GSH peroxidase activity compared with the untreated PCOS group.

MDA level in the ovarian tissues (Figure 7d) was found to be significantly increased in untreated PCOS rats (3.01 ± 0.09 µmol/g prot) and all the other treatment groups compared with the normal control rats (1.21 ± 0.12 µmol/g prot). However, no differences in MDA levels were recorded in any treatment groups.

4. Discussion

The physicochemical analysis of honey is essential for evaluating its quality and nutritional content. In this study, the physicochemical findings of KH complied with the range provided by the Malaysian Standard for stingless bee honey [3]. Our sample's moisture content was lower than other KH analyses [40–42], which may suggest a better shelf life. It has been noted that the environmental conditions during harvesting and storage have an impact on the moisture content of KH. Honey with a high water content has a greater potential for fermentation, making its preservation and storage more challenging [43]. In addition, the value of HMF is a commonly accepted indicator of honey freshness. In a freshly obtained honey samples, HMF is typically absent and increases over time. HMF is a by-product of the breakdown of simple carbohydrates, particularly fructose. The HMF content has been reported to be affected by a number of variables, including heating, storage conditions, honey pH, and honey adulterated with sugars [44]. We recorded a very low value of HMF (<0.1 mg/100 g), which implies a good quality of the honey sample.

This study also revealed that KH has lower glucose and fructose values than other analyses of KH [39,42]. However, a review by Nordin et al. that analysed stingless bee honey physicochemical characteristics around the globe found that the range of glucose and fructose values was between 12.5 g/100 g and 75.7 g/100 g, with which our KH sample complies [2]. KH contains a lower sugar value than Apis sp bee honey [2]. The Malaysian Standard for stingless bee honey has determined a value of fructose and glucose (sum) to be not more than 85.0 g/100 g [3]. On the other hand, the mineral content of the honey is reported to relate to its nutritional benefit and depends on its botanical and geographical origin [39,45]. The mineral elements we reported were within the permissible values set by the World Health Organization (WHO) [46,47]. In minimal amounts, some heavy metals are nutritionally essential for human health. In contrast, a high concentration of heavy metals leads to toxicity which has the potential to cause disease in humans. The poisoning is due to these heavy metals' inability to be metabolised by the body, thus leading to the heavy metals being accumulated to toxic levels within the human tissues without being degraded. Cobalt, molybdenum, and nickel are heavy metals of no biological importance and were not detected in our sample [48].

We demonstrated the semivolatile organic compounds identified in KH, which revealed six main chemical compounds, namely, 2,4-Dimethylhept-1-ene, Tetradecane, 2,4-Di-tert-butylphenol, n-Hexadecanoic acid, Octadecanoic acid, and z-10-Octadecen-1-ol acetate. The compound 2,4-Dimethylhept-1-ene is reported to play a role in human metabolites [49]. In contrast, tetradecane is noted as a phytochemical found in *Prosopis farcta* plants [50] and almonds [51]. Furthermore, a study reported that 2,4-Di-tert-butyl phenol is an antifungal and antioxidant bioactive compound purified from a newly isolated *Lactococcus* sp. [52]. In comparison, n-Hexadecanoic acid was reported to have anti-inflammatory [53] and cytotoxic properties [54]. Interestingly, octadecanoic acid or its synonym, stearic acid, was found to inhibit tumour development in a rat mammary carcinoma model induced by nitroso-methyl urea (NMU) [55]. Furthermore, a study comparing stearic acid with other saturated fatty acids in human studies has confirmed that stearic acid possesses cholesterol-lowering properties [56–58]. Another study showed that stearic acid (19 g/day) taken by healthy males could enhance their thrombogenic and atherogenic risk factor profiles [59]. Thus, the various health properties of semivolatile organic compounds found in KH enhanced its nutritional and antioxidative value.

As demonstrated in our previous study [28], KH did not affect the blood glucose levels in PCOS rats. This agrees with previous observations which found that daily intake of KH for 30 days caused no changes in fasting blood glucose in patients who have impaired fasting blood glucose [60]. Furthermore, we demonstrated that metformin treatment or combining KH with metformin reduced fasting blood glucose. This was expected due to metformin's validated fasting glucose-lowering effect [61]. In this study, clomiphene demonstrated no effect on fasting blood glucose and insulin levels. Meanwhile, another study revealed that clomiphene does not improve glucose tolerance tests in PCOS rats [34]. In contrast, Emam et al. found an improvement in the fasting glucose levels of PCOS rats with clomiphene treatment [62]. A study reported that clomiphene increases insulin-like growth factor binding protein-1 and reduces insulin-like growth factor-I but does not correct insulin resistance associated with women with polycystic ovarian syndrome [63].

The oestrus cycle is an expression of the female reproductive system since it reflects the status of the ovary, uterus, and hormonal physiology. We demonstrated that KH treatment could normalise the oestrus cycle in PCOS rats, and this effect is comparable to clomiphene. Furthermore, the combination of KH with metformin and clomiphene showed significant normalisation of the oestrus cycle. These results strengthen our previous finding which revealed oestrus cycle improvement in PCOS-induced rats treated with KH [28]. Previously, Tualang honey has been demonstrated to improve the bisphenol A-induced disruption of the oestrus cycle [64]. Hence, KH may be explored more for its potential to enhance the female menstrual cycle.

According to Sohaei et al., women with PCOS are overweight or obese in 38–88% of cases [65]. Clinically significant improvements in PCOS symptoms are seen with a modest weight loss of 5–10% [66]. In accordance with previous findings [28], our current study showed that KH treatment for 35 days did not affect the rat body weight gain. Previously, a study showed that six-week KH supplementation to high-fat diet-induced obese rat models could reduce the rat body weight and BMI [8]. The difference in the duration of honey treatment may contribute to the disparity of the findings. According to Atangwho et al., the duration of honey supplementation determines the body weight alteration [67]. Therefore, longer treatment duration may be needed to explore the KH effect on body weight.

Disturbance in sex steroid hormones was among the main findings in women with PCOS and PCOS animal models [68,69]. As validated in numerous animal PCOS induction studies [68], we demonstrated increased testosterone and LH levels but decreased progesterone and oestradiol levels with PCOS induction. However, we could not find any changes in the FSH level. With KH treatment, the testosterone level was significantly reversed to near normal. A similar trend was recorded in the clomiphene and metformin groups or in the combination of KH with both drugs. In agreement with previous studies [62,70,71], metformin and clomiphene were reported to reverse the elevated testosterone level in letrozole-induced PCOS rats. In women with PCOS, metformin causes a rapid decrease in LH-stimulated testosterone secretion [72]. Meanwhile, another study reported that clomiphene treatment does not affect testosterone levels in PCOS-induced rats [34]. Clomiphene possesses oestrogenic and anti-oestrogenic properties, but its exact mode of action is yet to be known. Clomiphene stimulates the release of the gonadotropins, FSH and LH, which leads to the development and maturation of the ovarian follicle, ovulation, and subsequent development and function of the corpus luteum, thus resulting in pregnancy [73].

As for the LH levels, KH significantly reversed the increment induced by PCOS induction. Interestingly, a combination of KH with clomiphene and metformin reduced LH levels but not in clomiphene or metformin-only groups. This suggests a synergistic effect of KH with metformin and clomiphene. Similarly, Ndeingang et al. also found that clomiphene treatment does not affect the LH levels in PCOS rats [34]. Meanwhile, Ibrahim et al. discovered that metformin treatment reduced LH increment in PCOS-induced rats [71]. In a previous study, six months of metformin treatment successfully reduced the LH levels in women with PCOS [74]. The improvement in testosterone and LH levels in this study could be explained by our previous finding in which KH ameliorated the altered cystic follicles, antral follicles, and corpus luteum in PCOS-induced rat [28]. This improvement of the folliculogenesis process by KH can hinder the follicular hyperandrogenism and restore the sex steroid-related mechanism alteration in PCOS. In addition, a study proposed that local Nigerian honey may regulate the pituitary gland by modulating the feedback mechanism to alter the sex steroid hormonal level [75]. Meanwhile, another study also demonstrated honey could alter the hypothalamic-pituitary-adrenal axis [76].

In this study, we demonstrated that KH did not affect oestradiol, progesterone, or FSH levels. Serum oestradiol and FSH levels in letrozole-induced PCOS rats showed variations in different studies, as revealed by Ryu et al. [68]. Furthermore, a study that used the same model of letrozole-induced PCOS rats found that treatment with letrozole, metformin, and clomiphene did not affect the FSH levels [34]. According to a report, honey could be oestrogenic or anti-oestrogenic depending on its concentration [32,77,78]. In another study, female rats supplemented with Tualang honey for eight weeks demonstrated no changes in their oestradiol levels [79]. Meanwhile, other researchers have proved that Tualang honey reduces oestradiol and progesterone levels in ovariectomised rats [80]. However, Ismail et al. found no difference in testosterone, progesterone, and oestradiol levels with Gelam honey supplementation in ovariectomised rats [32].This may suggest further study on the effect of honey on the sex steroid hormone. In this study, metformin and clomiphene treatment caused oestradiol levels to increase, as similar findings have recorded previously [34,81]. Interestingly, while KH alone did not affect the oestradiol and progesterone levels in this study, the combination of KH with metformin or clomiphene increased both hormones significantly. This again suggests that KH has a synergistic effect with metformin and clomiphene.

Supplementation with KH in PCOS-induced rats significantly improved their oxidative stress status. We found that KH significantly increased catalase, total SOD, and GSH peroxidase activities compared with the untreated PCOS rats. In fact, higher values of catalase, total SOD, and GSH peroxidase activities were recorded when combining KH with clomiphene or metformin, in which treatment with the drugs alone did not improve the oxidative stress status in PCOS rats. Meanwhile, we found that the MDA level was not affected by KH. Our findings strengthen the findings regarding the antioxidative properties of KH, which are recorded in two other studies demonstrating that KH increases SOD and GSH levels in testicular oxidative damage [10] and increases SOD in an osteoporosis rat model [11]. Another study recorded findings similar to ours, in which letrozole induction did not alter the MDA level in rat ovaries [34]. The KH antioxidant effect is attributed to its phenolic content as suggested in previous studies [82,83]. The positive results shown by the KH and combination of KH with metformin or clomiphene in improving catalase, SOD, and GSH peroxidase activities were concomitant with its effect on improving testosterone and LH levels. This implies an interplay between the reduction of oxidative stress and the enhancement of physiological hormonal processes manifested by the oestrus cycle improvement seen in these groups.

Reactive oxygen species and oxidative stress markers such as SOD, glutathione peroxidase, and catalase have been reported to have a regulatory role in both physiological and pathological processes of the oestrous cycle, folliculogenesis, oocyte maturation, ovarian steroidogenesis, and sex steroid hormone level [84–86]. Furthermore, as in our PCOS-induced rats, GSH peroxidase activity was found to be significantly reduced in the follicular fluid of tobacco-smoking women. In addition, GSH peroxidase activities were higher in fertilised oocytes than in non-fertilised oocytes [87]. In another study, a mitochondrial malfunction with reduced GSH levels and O_2 consumption was found in PCOS patients with insulin resistance [88]. Park et al. proved that the inhibition of catalase causes DNA damage and chromosome misalignment during meiotic maturation in mouse oocytes [89]. Meanwhile, in vitro studies demonstrated that LH could increase the mRNA and protein levels of SOD and catalase in the bovine corpus luteum [90]. This may explain the normalisation of the oestrus cycle, testosterone, and LH level with improvement in SOD, glutathione peroxidase, and catalase seen with KH treatment on PCOS rats.

5. Conclusions

Taken together, our results show that KH and its combination with metformin or clomiphene improve oxidative stress, hormonal profile, and oestrus cycle in rats with PCOS. These results may provide a basis for future studies to discover the potential use of KH as a complementary treatment for women with PCOS.

Author Contributions: Conceptualization, M.H.M., S.F.I. and D.A.M.K.; formal analysis, M.H.M., S.F.I. and D.A.M.K.; methodology, M.H.M. and D.A.M.K.; investigation, M.H.M. and D.A.M.K.; writing—original draft preparation, M.H.M. and D.A.M.K.; writing—review and editing, M.H.M., S.F.I. and A.U.; supervision, M.H.M., S.F.I. and A.U.; funding acquisition, M.H.M. All authors have read and agreed to the published version of the manuscript.

Funding: This research was funded by an institutional grant from Universiti Kebangsaan Malaysia (UKM; FF-2020-241).

Institutional Review Board Statement: The study was conducted according to the guidelines of the Declaration of Helsinki and approved by the Animal Ethics Committee of Universiti Kebangsaan Malaysia (Ethical Approval Code FISIO/FP/2020/MOHD HELMY/14-MAY/1104-JUNE-2020-MAY-2023).

Informed Consent Statement: Not applicable.

Data Availability Statement: The data is contained within the article.

Acknowledgments: The authors thank Aini Farzana Zulkefli from the Department of Physiology, Faculty of Medicine, UKM, for providing technical support during the animal treatment.

Conflicts of Interest: The authors declare no conflict of interest.

References

1. Michener, C.D. The Meliponini. In *Pot-Honey: A Legacy of Stingless Bees*; Vit, P., Pedro, S.R.M., Roubik, D., Eds.; Springer: New York, NY, USA, 2013; pp. 3–17. [CrossRef]
2. Nordin, A.; Sainik, N.Q.A.V.; Chowdhury, S.R.; Saim, A.B.; Idrus, R.B.H. Physicochemical properties of stingless bee honey from around the globe: A comprehensive review. *J. Food Compos. Anal.* **2018**, *73*, 91–102. [CrossRef]
3. Standard, M. *Kelulut (Stingless Bee) Hone—Specification*; Department of Standards Malaysia: Selangor, Malaysia, 2017; p. 25.
4. Zainol, M.I.; Mohd Yusoff, K.; Mohd Yusof, M.Y. Antibacterial activity of selected Malaysian honey. *BMC Complementary Altern. Med.* **2013**, *13*, 129. [CrossRef]
5. Saiful Yazan, L.; Muhamad Zali, M.F.S.; Mohd Ali, R.; Zainal, N.A.; Esa, N.; Sapuan, S.; Ong, Y.S.; Tor, Y.S.; Gopalsamy, B.; Voon, F.L.; et al. Chemopreventive Properties and Toxicity of Kelulut Honey in Sprague Dawley Rats Induced with Azoxymethane. *BioMed Res. Int.* **2016**, *2016*, 4036926. [CrossRef]
6. Ranneh, Y.; Akim, A.M.; Hamid, H.A.; Khazaai, H.; Fadel, A.; Mahmoud, A.M. Stingless bee honey protects against lipopolysaccharide induced-chronic subclinical systemic inflammation and oxidative stress by modulating Nrf2, NF-κB and p38 MAPK. *Nutr. Metab.* **2019**, *16*, 15. [CrossRef] [PubMed]
7. Yazan, L.S.; Zainal, N.A.; Ali, R.M.; Zali, M.; Shyfiq, M.F.; Sze, O.Y.; Sim, T.Y.; Gopalsamy, B.; Ling, V.F.; Sapuan, S. Antiulcer Properties of Kelulut Honey against Ethanol-Induced Gastric Ulcer. *Pertanika J. Sci. Technol.* **2018**, *26*, 121–132.
8. Mohd Rafie, A.Z.; Syahir, A.; Wan Ahmad, W.A.N.; Mustafa, M.Z.; Mariatulqabtiah, A.R. Supplementation of Stingless Bee Honey from *Heterotrigona itama* Improves Antiobesity Parameters in High-Fat Diet Induced Obese Rat Model. *Evid.-Based Complementary Altern. Med.* **2018**, *2018*, 6371582. [CrossRef]

9. Ramli, N.Z.; Chin, K.-Y.; Zarkasi, K.A.; Ahmad, F. The Beneficial Effects of Stingless Bee Honey from Heterotrigona itama against Metabolic Changes in Rats Fed with High-Carbohydrate and High-Fat Diet. *Int. J. Environ. Res. Public Health* **2019**, *16*, 4987. [CrossRef]
10. Budin, S.; Jubaidi, F.; Azam, S.; Mohammed Yusof, N.; Taib, I.S.; Mohamed, J. Kelulut honey supplementation prevents sperm and testicular oxidative damage in streptozotocin-induced diabetic rats. *J. Teknol.* **2017**, *79*, 89–95. [CrossRef]
11. Kamaruzzaman, M.; Thanu, A.; Yusof, M.; Soelaiman, I.; Ramli, E. Kelulut honey ameliorates glucocorticoid induced osteoporosis via its antioxidant activity in rats. *Asian Pac. J. Trop. Biomed.* **2019**, *9*, 493. [CrossRef]
12. Muhammad, N.I.I.; Sarbon, N.M. Physicochemical profile, antioxidant activity and mineral contents of honey from stingless bee and honey bee species. *J. Apic. Res.* **2021**, *50*, 61–69. [CrossRef]
13. Ranneh, Y.; Ali, F.; Zarei, M.; Akim, A.M.; Hamid, H.A.; Khazaai, H. Malaysian stingless bee and Tualang honeys: A comparative characterization of total antioxidant capacity and phenolic profile using liquid chromatography-mass spectrometry. *LWT* **2018**, *89*, 1–9. [CrossRef]
14. Jalil, M.A.A.; Kasmuri, A.R.; Hadi, H. Stingless Bee Honey, the Natural Wound Healer: A Review. *Ski. Pharmacol. Physiol.* **2017**, *30*, 66–75. [CrossRef] [PubMed]
15. Fletcher, M.T.; Hungerford, N.L.; Webber, D.; Carpinelli de Jesus, M.; Zhang, J.; Stone, I.S.J.; Blanchfield, J.T.; Zawawi, N. Stingless bee honey, a novel source of trehalulose: A biologically active disaccharide with health benefits. *Sci. Rep.* **2020**, *10*, 12128. [CrossRef] [PubMed]
16. Teede, H.; Deeks, A.; Moran, L. Polycystic ovary syndrome: A complex condition with psychological, reproductive and metabolic manifestations that impacts on health across the lifespan. *BMC Med.* **2010**, *8*, 41. [CrossRef]
17. Dumesic, D.A.; Oberfield, S.E.; Stener-Victorin, E.; Marshall, J.C.; Laven, J.S.; Legro, R.S. Scientific Statement on the Diagnostic Criteria, Epidemiology, Pathophysiology, and Molecular Genetics of Polycystic Ovary Syndrome. *Endocr. Rev.* **2015**, *36*, 487–525. [CrossRef]
18. Rosenfield, R.L. Current concepts of polycystic ovary syndrome pathogenesis. *Curr. Opin. Pediatrics* **2020**, *32*, 698–706. [CrossRef]
19. Mancini, A.; Bruno, C.; Vergani, E.; d'Abate, C.; Giacchi, E.; Silvestrini, A. Oxidative Stress and Low-Grade Inflammation in Polycystic Ovary Syndrome: Controversies and New Insights. *Int. J. Mol. Sci.* **2021**, *22*, 1667. [CrossRef] [PubMed]
20. Khashchenko, E.; Vysokikh, M.; Uvarova, E.; Krechetova, L.; Vtorushina, V.; Ivanets, T.; Volodina, M.; Tarasova, N.; Sukhanova, I.; Sukhikh, G. Activation of Systemic Inflammation and Oxidative Stress in Adolescent Girls with Polycystic Ovary Syndrome in Combination with Metabolic Disorders and Excessive Body Weight. *J. Clin. Med.* **2020**, *9*, 1399. [CrossRef] [PubMed]
21. Murri, M.; Luque-Ramírez, M.; Insenser, M.; Ojeda-Ojeda, M.; Escobar-Morreale, H.F. Circulating markers of oxidative stress and polycystic ovary syndrome (PCOS): A systematic review and meta-analysis. *Hum. Reprod. Update* **2013**, *19*, 268–288. [CrossRef] [PubMed]
22. McCartney, C.R.; Marshall, J.C. Polycystic Ovary Syndrome. *N. Engl. J. Med.* **2016**, *375*, 54–64. [CrossRef] [PubMed]
23. Lee, Y.H.; Yang, H.; Lee, S.R.; Kwon, S.W.; Hong, E.J.; Lee, H.W. Welsh Onion Root (*Allium fistulosum*) Restores Ovarian Functions from Letrozole Induced-Polycystic Ovary Syndrome. *Nutrients* **2018**, *10*, 1430. [CrossRef] [PubMed]
24. Morgante, G.; Massaro, M.G.; Di Sabatino, A.; Cappelli, V.; De Leo, V. Therapeutic approach for metabolic disorders and infertility in women with PCOS. *Gynecol. Endocrinol.* **2018**, *34*, 4–9. [CrossRef]
25. Domecq, J.P.; Prutsky, G.; Mullan, R.J.; Sundaresh, V.; Wang, A.T.; Erwin, P.J.; Welt, C.; Ehrmann, D.; Montori, V.M.; Murad, M.H. Adverse effects of the common treatments for polycystic ovary syndrome: A systematic review and meta-analysis. *J. Clin. Endocrinol. Metab.* **2013**, *98*, 4646–4654. [CrossRef]
26. Kamenov, Z.; Gateva, A. Inositols in PCOS. *Molecular* **2020**, *25*, 5566. [CrossRef]
27. Kalra, B.; Kalra, S.; Sharma, J.B. The inositols and polycystic ovary syndrome. *Indian J. Endocrinol. Metab.* **2016**, *20*, 720–724. [CrossRef]
28. Kamal, D.A.M.; Ibrahim, S.F.; Ugusman, A.; Mokhtar, M.H. Effects of Kelulut Honey on Oestrus Cycle Regulation and Histomorphological Changes in Letrozole-Induced Polycystic Ovary Syndrome Rats: A Preliminary Study. *Life* **2022**, *12*, 890. [CrossRef]
29. Horwitz, W. *Official Methods of Analysis of AOAC International*, 17th ed.; AOAC International: Gaithersburg, MD, USA, 2000.
30. Aghamirlou, H.M.; Khadem, M.; Rahmani, A.; Sadeghian, M.; Mahvi, A.H.; Akbarzadeh, A.; Nazmara, S. Heavy metals determination in honey samples using inductively coupled plasma-optical emission spectrometry. *J. Environ. Health Sci. Eng.* **2015**, *13*, 39. [CrossRef] [PubMed]
31. White, J.W., Jr. Spectrophotometric Method for Hydroxymethylfurfural in Honey. *J. Assoc. Off. Anal. Chem.* **2020**, *62*, 509–514. [CrossRef]
32. Ismail, N.H.; Osman, K.; Zulkefli, A.F.; Mokhtar, M.H.; Ibrahim, S.F. The Physicochemical Characteristics of Gelam Honey and Its Outcome on the Female Reproductive Tissue of Sprague-Dawley Rats: A Preliminary Study. *Molecular* **2021**, *26*, 3346. [CrossRef]
33. Kafali, H.; Iriadam, M.; Ozardali, I.; Demir, N. Letrozole-induced polycystic ovaries in the rat: A new model for cystic ovarian disease. *Arch. Med. Res.* **2004**, *35*, 103–108. [CrossRef]
34. Ndeingang, E.C.; Defo Deeh, P.B.; Watcho, P.; Kamanyi, A. *Phyllanthus muellerianus* (Euphorbiaceae) Restores Ovarian Functions in Letrozole-Induced Polycystic Ovarian Syndrome in Rats. *Evid.-Based Complementary Altern. Med.* **2019**, *2019*, 2965821. [CrossRef] [PubMed]

35. Kamal, D.A.M.; Ibrahim, S.F.; Mokhtar, M.H. Effects of Testosterone on the Expression of Connexin 26 and Connexin 43 in the Uterus of Rats During Early Pregnancy. *In Vivo* **2020**, *34*, 1863–1870. [CrossRef] [PubMed]
36. Ajayi, A.F.; Akhigbe, R.E. Staging of the estrous cycle and induction of estrus in experimental rodents: An update. *Fertil. Res. Pract.* **2020**, *6*, 5. [CrossRef] [PubMed]
37. Md. Salleh, M.F.R.R.; Aminuddin, A.; Hamid, A.A.; Salamt, N.; Japar Sidik, F.Z.; Ugusman, A. Piper sarmentosum Roxb. Attenuates Vascular Endothelial Dysfunction in Nicotine-Induced Rats. *Front. Pharmacol.* **2021**, *12*, 1495. [CrossRef]
38. Bradford, M.M. A rapid and sensitive method for the quantitation of microgram quantities of protein utilizing the principle of protein-dye binding. *Anal. Biochem.* **1976**, *72*, 248–254. [CrossRef]
39. Kek, S.P.; Chin, N.L.; Tan, S.W.; Yusof, Y.A.; Chua, L.S. Classification of Honey from Its Bee Origin via Chemical Profiles and Mineral Content. *Food Anal. Methods* **2017**, *10*, 19–30. [CrossRef]
40. Lim, D.C.C.; Abu Bakar, M.F.; Majid, M. Nutritional composition of stingless bee honey from different botanical origins. *IOP Conf. Ser. Earth Environ. Sci.* **2019**, *269*, 012025. [CrossRef]
41. Shamsudin, S.; Selamat, J.; Sanny, M.; Razak, S.-B.A.; Jambari, N.N.; Mian, Z.; Khatib, A. Influence of origins and bee species on physicochemical, antioxidant properties and botanical discrimination of stingless bee honey. *Int. J. Food Prop.* **2019**, *22*, 239–264. [CrossRef]
42. Shamsudin, S.; Selamat, J.; Sanny, M.; Shamsul Bahari, A.R.; Jambari, N.N.; Khatib, A. A Comparative Characterization of Physicochemical and Antioxidants Properties of Processed Heterotrigona itama Honey from Different Origins and Classification by Chemometrics Analysis. *Molecular* **2019**, *24*, 3898. [CrossRef]
43. Babarinde, G.O.; Babarinde, S.A.; Adegbola, D.O.; Ajayeoba, S.I. Effects of harvesting methods on physicochemical and microbial qualities of honey. *J. Food Sci. Technol.* **2011**, *48*, 628–634. [CrossRef]
44. Pasias, I.N.; Kiriakou, I.K.; Proestos, C. HMF and diastase activity in honeys: A fully validated approach and a chemometric analysis for identification of honey freshness and adulteration. *Food Chem.* **2017**, *229*, 425–431. [CrossRef]
45. Solayman, M.; Islam, M.A.; Paul, S.; Ali, Y.; Khalil, M.I.; Alam, N.; Gan, S.H. Physicochemical Properties, Minerals, Trace Elements, and Heavy Metals in Honey of Different Origins: A Comprehensive Review. *Compr. Rev. Food Sci. Food Saf.* **2016**, *15*, 219–233. [CrossRef] [PubMed]
46. Moniruzzaman, M.; Chowdhury, M.A.; Rahman, M.A.; Sulaiman, S.A.; Gan, S.H. Determination of mineral, trace element, and pesticide levels in honey samples originating from different regions of Malaysia compared to manuka honey. *BioMed Res. Int.* **2014**, *2014*, 359890. [CrossRef] [PubMed]
47. WHO. *Evaluation of Certain Food Additives and Contaminants: Sixty-Eighth Report of the Joint FAO/WHO Expert Committee on Food Additives*; World Health Organization: Geneva, Switzerland, 2007.
48. Ajibola, A.; Chamunorwa, J.P.; Erlwanger, K.H. Nutraceutical values of natural honey and its contribution to human health and wealth. *Nutr. Metab.* **2012**, *9*, 61. [CrossRef] [PubMed]
49. CHEBI. CHEBI:77746—Human Metabolite. Available online: https://www.ebi.ac.uk/chebi/searchId.do;jsessionid=E40442A18CEEBAC8040EF1DB197C2005?chebiId=CHEBI%3A77746 (accessed on 15 July 2022).
50. Amal, M.S.; Mosad, A.G.; Mohamed, S.A.-A.; Hassan, M.F.M.; Omar, M.K.; Ahmed, K.E.-Z.; Mamdouh, A.-M. Chemical constituents and biological activities of different solvent extracts of Prosopis farcta growing in Egypt. *J. Pharmacogn. Phytother.* **2017**, *9*, 67–76. [CrossRef]
51. Xu, Y.; Liao, M.; Wang, D.; Jiao, S. Physicochemical quality and volatile flavor compounds of hot air-assisted radio frequency roasted almonds. *J. Food Processing Preserv.* **2020**, *44*, e14376. [CrossRef]
52. Varsha, K.K.; Devendra, L.; Shilpa, G.; Priya, S.; Pandey, A.; Nampoothiri, K.M. 2,4-Di-tert-butyl phenol as the antifungal, antioxidant bioactive purified from a newly isolated Lactococcus sp. *Int. J. Food Microbiol.* **2015**, *211*, 44–50. [CrossRef] [PubMed]
53. Aparna, V.; Dileep, K.V.; Mandal, P.K.; Karthe, P.; Sadasivan, C.; Haridas, M. Anti-Inflammatory Property of n-Hexadecanoic Acid: Structural Evidence and Kinetic Assessment. *Chem. Biol. Drug Des.* **2012**, *80*, 434–439. [CrossRef] [PubMed]
54. Ravi, L.; Krishnan, K. Cytotoxic Potential of N-hexadecanoic Acid Extracted from Kigelia pinnata Leaves. *Asian J. Cell Biol.* **2016**, *12*, 20–27. [CrossRef]
55. Habib, N.A.; Wood, C.B.; Apostolov, K.; Barker, W.; Hershman, M.J.; Aslam, M.; Heinemann, D.; Fermor, B.; Williamson, R.C.N.; Jenkins, W.E. Stearic acid and carcinogenesis. *Br. J. Cancer* **1987**, *56*, 455–458. [CrossRef] [PubMed]
56. Grundy, S.M. Influence of stearic acid on cholesterol metabolism relative to other long-chain fatty acids. *Am. J. Clin. Nutr.* **1994**, *60*, 986S–990S. [CrossRef] [PubMed]
57. Bonanome, A.; Grundy, S.M. Effect of Dietary Stearic Acid on Plasma Cholesterol and Lipoprotein Levels. *N. Engl. J. Med.* **1988**, *318*, 1244–1248. [CrossRef]
58. Hunter, J.E.; Zhang, J.; Kris-Etherton, P.M. Cardiovascular disease risk of dietary stearic acid compared with trans, other saturated, and unsaturated fatty acids: A systematic review. *Am. J. Clin. Nutr.* **2009**, *91*, 46–63. [CrossRef] [PubMed]
59. Kelly, F.D.; Sinclair, A.J.; Mann, N.J.; Turner, A.H.; Abedin, L.; Li, D. A stearic acid-rich diet improves thrombogenic and atherogenic risk factor profiles in healthy males. *Eur. J. Clin. Nutr.* **2001**, *55*, 88–96. [CrossRef] [PubMed]
60. Rashid, M.R.; Nor Aripin, K.N.; Syed Mohideen, F.B.; Baharom, N.; Omar, K.; Md Taujuddin, N.M.S.; Mohd Yusof, H.H.; Addnan, F.H. The Effect of Kelulut Honey on Fasting Blood Glucose and Metabolic Parameters in Patients with Impaired Fasting Glucose. *J. Nutr. Metab.* **2019**, *2019*, 3176018. [CrossRef]

61. Freemark, M.; Bursey, D. The effects of metformin on body mass index and glucose tolerance in obese adolescents with fasting hyperinsulinemia and a family history of type 2 diabetes. *Pediatrics* **2001**, *107*, E55. [CrossRef]
62. Emam, S.R.; Abd-Elsalam, R.M.; Azouz, A.A.; Ali, S.E.; El Badawy, S.A.; Ibrahim, M.A.; Hassan, B.B.; Issa, M.Y.; Elmosalamy, S.H. Linum usitatissimum seeds oil down-regulates mRNA expression for the steroidogenic acute regulatory protein and Cyp11A1 genes, ameliorating letrezole-induced polycystic ovarian syndrome in a rat model. *J. Physiol. Pharm.* **2021**, *72*, 55–67. [CrossRef]
63. de Leo, V.; la Marca, A.; Morgante, G.; Ciotta, L.; Mencaglia, L.; Cianci, A.; Petraglia, F. Clomiphene citrate increases insulin-like growth factor binding protein-1 and reduces insulin-like growth factor-I without correcting insulin resistance associated with polycystic ovarian syndrome. *Hum. Reprod.* **2000**, *15*, 2302–2305. [CrossRef]
64. Zaid, S.S.; Othman, S.; Kassim, N.M. Potential protective effect of Tualang honey on BPA-induced ovarian toxicity in prepubertal rat. *BMC Complementary Altern. Med.* **2014**, *14*, 509. [CrossRef]
65. Sohaei, S.; Amani, R.; Tarrahi, M.J.; Ghasemi-Tehrani, H. The effects of curcumin supplementation on glycemic status, lipid profile and hs-CRP levels in overweight/obese women with polycystic ovary syndrome: A randomized, double-blind, placebo-controlled clinical trial. *Complement Med.* **2019**, *47*, 102201. [CrossRef]
66. Barber, T.M.; Hanson, P.; Weickert, M.O.; Franks, S. Obesity and Polycystic Ovary Syndrome: Implications for Pathogenesis and Novel Management Strategies. *Clin. Med. Insights Reprod. Health* **2019**, *13*, 1179558119874042. [CrossRef]
67. Atangwho, I.J.; Ibeneme, C.E.; Egbung, G.E.; Ibeneme, E.; Eno, M.A.; Nwankpa, P. Effect of long-term feeding of the Obudu natural honey and table sugar-sweetened diets on obesity and pro-inflammatory biomarkers in rats. *BMC Nutr.* **2020**, *6*, 3. [CrossRef]
68. Ryu, Y.; Kim, S.W.; Kim, Y.Y.; Ku, S.Y. Animal Models for Human Polycystic Ovary Syndrome (PCOS) Focused on the Use of Indirect Hormonal Perturbations: A Review of the Literature. *Int. J. Mol. Sci.* **2019**, *20*, 2720. [CrossRef] [PubMed]
69. Abdul Hamid, F.; Abu, M.A.; Abdul Karim, A.K.; Ahmad, M.F.; Aziz, N.H.A.; Mohd Kamal, D.A.; Mokhtar, M.H. Sex Steroid Receptors in Polycystic Ovary Syndrome and Endometriosis: Insights from Laboratory Studies to Clinical Trials. *Biomedicines* **2022**, *10*, 1705. [CrossRef] [PubMed]
70. Rudic, J.; Jakovljevic, V.; Jovic, N.; Nikolic, M.; Sretenovic, J.; Mitrovic, S.; Bolevich, S.; Bolevich, S.; Mitrovic, M.; Raicevic, S.; et al. Antioxidative Effects of Standardized Aronia melanocarpa Extract on Reproductive and Metabolic Disturbances in a Rat Model of Polycystic Ovary Syndrome. *Antioxidants* **2022**, *11*, 1099. [CrossRef]
71. Ibrahim, Y.F.; Alorabi, M.; Abdelzaher, W.Y.; Toni, N.D.M.; Thabet, K.; Hegazy, A.; Bahaa, H.A.; Batiha, G.E.-S.; Welson, N.N.; Morsy, M.A.; et al. Diacerein ameliorates letrozole-induced polycystic ovarian syndrome in rats. *Biomed. Pharmacother.* **2022**, *149*, 112870. [CrossRef] [PubMed]
72. Kurzthaler, D.; Hadziomerovic-Pekic, D.; Wildt, L.; Seeber, B.E. Metformin induces a prompt decrease in LH-stimulated testosterone response in women with PCOS independent of its insulin-sensitizing effects. *Reprod. Biol. Endocrinol.* **2014**, *12*, 98. [CrossRef]
73. Wallach, E.E.; Adashi, E.Y. Clomiphene citrate: Mechanism(s) and site(s) of action—A hypothesis revisited*. *Fertil. Steril.* **1984**, *42*, 331–344. [CrossRef]
74. Genazzani, A.D.; Battaglia, C.; Malavasi, B.; Strucchi, C.; Tortolani, F.; Gamba, O. Metformin administration modulates and restores luteinizing hormone spontaneous episodic secretion and ovarian function in nonobese patients with polycystic ovary syndrome. *Fertil Steril* **2004**, *81*, 114–119. [CrossRef] [PubMed]
75. Nervey, D.; Igbigbi, P.; Avwioro, G. The Influence of Excessive and Prolonged Ingestion of Honey on Sex Hormones and Prostate Specific Antigen in Adult Male Wistar Rats. *Med. Sci. Int. Med. J.* **2012**, *1*, 161–170. [CrossRef]
76. Al-Rahbi, B.; Zakaria, R.; Othman, Z.; Hassan, A.; Ahmad, A.H. Enhancement of BDNF Concentration and Restoration of the Hypothalamic-Pituitary-Adrenal Axis Accompany Reduced Depressive-Like Behaviour in Stressed Ovariectomised Rats Treated with Either Tualang Honey or Estrogen. *Sci. World J.* **2014**, *2014*, 310821. [CrossRef] [PubMed]
77. Erejuwa, O.O.; Sulaiman, S.A.; Wahab, M.S.A. Effects of Honey and Its Mechanisms of Action on the Development and Progression of Cancer. *Molecular* **2014**, *19*, 2497–2522. [CrossRef]
78. Tsiapara, A.V.; Jaakkola, M.; Chinou, I.; Graikou, K.; Tolonen, T.; Virtanen, V.; Moutsatsou, P. Bioactivity of Greek honey extracts on breast cancer (MCF-7), prostate cancer (PC-3) and endometrial cancer (Ishikawa) cells: Profile analysis of extracts. *Food Chem.* **2009**, *116*, 702–708. [CrossRef]
79. Mosavat, M.; Ooi, F.K.; Mohamed, M. Stress hormone and reproductive system in response to honey supplementation combined with different jumping exercise intensities in female rats. *BioMed. Res. Int.* **2014**, *2014*, 123640. [CrossRef] [PubMed]
80. Zaid, S.S.M.; Sulaiman, S.A.; Sirajudeen, K.N.M.; Othman, N.H. The effects of tualang honey on female reproductive organs, tibia bone and hormonal profile in ovariectomised rats—Animal model for menopause. *BMC Complementary Altern. Med.* **2010**, *10*, 82. [CrossRef] [PubMed]
81. Jahan, S.; Abid, A.; Khalid, S.; Afsar, T.; Qurat Ul, A.; Shaheen, G.; Almajwal, A.; Razak, S. Therapeutic potentials of Quercetin in management of polycystic ovarian syndrome using Letrozole induced rat model: A histological and a biochemical study. *J. Ovarian Res.* **2018**, *11*, 26. [CrossRef] [PubMed]
82. Karim, N. Antioxidant Properties of Stingless Bee Honey and Its Effect on the Viability of Lymphoblastoid Cell Line. *Med. Health* **2019**, *14*, 91–105. [CrossRef]
83. Haron, H.; Talib, R.A.; Subramaniam, P.; Arifen, Z.N.Z.; Ibrahim, M. A Comparison Of Chemical Compositions In Kelulut Honey From Different Regions. *Malays. J. Anal. Sci.* **2022**, *26*, 447–456.

84. Agarwal, A.; Gupta, S.; Sharma, R.K. Role of oxidative stress in female reproduction. *Reprod. Biol. Endocrinol.* **2005**, *3*, 28. [CrossRef] [PubMed]
85. Wang, S.; He, G.; Chen, M.; Zuo, T.; Xu, W.; Liu, X. The Role of Antioxidant Enzymes in the Ovaries. *Oxidative Med. Cell. Longev.* **2017**, *2017*, 4371714. [CrossRef] [PubMed]
86. Suzuki, T.; Sugino, N.; Fukaya, T.; Sugiyama, S.; Uda, T.; Takaya, R.; Yajima, A.; Sasano, H. Superoxide dismutase in normal cycling human ovaries: Immunohistochemical localization and characterization. *Fertil Steril* **1999**, *72*, 720–726. [CrossRef]
87. Paszkowski, T.; Traub, A.I.; Robinson, S.Y.; McMaster, D. Selenium dependent glutathione peroxidase activity in human follicular fluid. *Clin. Chim. Acta* **1995**, *236*, 173–180. [CrossRef]
88. Victor, V.M.; Rocha, M.; Bañuls, C.; Alvarez, A.; de Pablo, C.; Sanchez-Serrano, M.; Gomez, M.; Hernandez-Mijares, A. Induction of oxidative stress and human leukocyte/endothelial cell interactions in polycystic ovary syndrome patients with insulin resistance. *J. Clin. Endocrinol. Metab.* **2011**, *96*, 3115–3122. [CrossRef] [PubMed]
89. Park, Y.S.; You, S.Y.; Cho, S.; Jeon, H.J.; Lee, S.; Cho, D.H.; Kim, J.S.; Oh, J.S. Eccentric localization of catalase to protect chromosomes from oxidative damages during meiotic maturation in mouse oocytes. *Histochem. Cell Biol.* **2016**, *146*, 281–288. [CrossRef] [PubMed]
90. Kawaguchi, S.; Sakumoto, R.; Okuda, K. Induction of the expressions of antioxidant enzymes by luteinizing hormone in the bovine corpus luteum. *J. Reprod. Dev.* **2013**, *59*, 219–224. [CrossRef] [PubMed]

antioxidants

Article

^{1}H-NMR Metabolic Profiling, Antioxidant Activity, and Docking Study of Common Medicinal Plant-Derived Honey

Maha Montaser [1], Asmaa T. Ali [2], Ahmed M. Sayed [1], Usama Ramadan Abdelmohsen [3,4], Ehab W. Zidan [5], Raha Orfali [6], Mostafa E. Rateb [7], Mohamed A. Zaki [8], Hossam M. Hassan [8,*], Rabab Mohammed [8,*] and Mohamed S. Hifnawy [9]

1 Department of Pharmacognosy, Faculty of Pharmacy, Nahda University, Beni-Suef 62513, Egypt
2 Department of Biochemistry, Faculty of Pharmacy, Nahda University, Beni-Suef 62511, Egypt
3 Department of Pharmacognosy, Faculty of Pharmacy, Minia University, Minia 61519, Egypt
4 Department of Pharmacognosy, Faculty of Pharmacy, Deraya University, New Minia City 61111, Egypt
5 Agricultural Research Centre, Department of Bee Research, Plant Protection Research Institute, Giza 12618, Egypt
6 Department of Pharmacognosy, College of Pharmacy, King Saud University, Riyadh 11451, Saudi Arabia
7 School of Computing, Engineering & Physical Sciences, University of the West of Scotland, Paisley PA1 2BE, UK
8 Department of Pharmacognosy, Faculty of Pharmacy, Beni-Suef University, Beni-Suef 62514, Egypt
9 Department of Pharmacognosy, Faculty of Pharmacy, Cairo University, Cairo 11787, Egypt
* Correspondence: hossam.mokhtar@nub.edu.eg (H.M.H.); rababmohammed@pharm.bsu.edu.eg (R.M.)

Citation: Montaser, M.; Ali, A.T.; Sayed, A.M.; Abdelmohsen, U.R.; Zidan, E.W.; Orfali, R.; Rateb, M.E.; Zaki, M.A.; Hassan, H.M.; Mohammed, R.; et al. ^{1}H-NMR Metabolic Profiling, Antioxidant Activity, and Docking Study of Common Medicinal Plant-Derived Honey. *Antioxidants* **2022**, *11*, 1880. https://doi.org/10.3390/antiox11101880

Academic Editors: Josipa Vlainić and Ivana Tlak Gajger

Received: 13 August 2022
Accepted: 19 September 2022
Published: 22 September 2022

Abstract: The purpose of this investigation was to determine ^{1}H-NMR profiling and antioxidant activity of the most common types of honey, namely, citrus honey (HC1) (*Morcott tangerine* L. and *Jaffa orange* L.), marjoram honey (HM1) (*Origanum majorana* L.), and clover honey (HT1) (*Trifolium alexandrinum* L.), compared to their secondary metabolites (HC2, HM2, HT2, respectively). By using a ^{1}H-NMR-based metabolomic technique, PCA, and PLS-DA multivariate analysis, we found that HC2, HM2, HC1, and HM1 were clustered together. However, HT1 and HT2 were quite far from these and each other. This indicated that HC1, HM1, HC2, and HM2 have similar chemical compositions, while HT1 and HT2 were unique in their chemical profiles. Antioxidation potentials were determined colorimetrically for scavenging activities against DPPH, ABTS, ORAC, 5-LOX, and metal chelating activity in all honey extract samples and their secondary metabolites. Our results revealed that HC2 and HM2 possessed more antioxidant activities than HT2 in vitro. HC2 demonstrated the highest antioxidant effect in all assays, followed by HM2 (DPPH assay: IC_{50} 2.91, 10.7 μg/mL; ABTS assay: 431.2, 210.24 at 50 ug/mL Trolox equivalent; ORAC assay: 259.5, 234.8 at 50 ug/mL Trolox equivalent; 5-LOX screening assay/IC_{50}: 2.293, 6.136 ug/mL; and metal chelating activity at 50 ug/mL: 73.34526%, 63.75881% inhibition). We suggest that the presence of some secondary metabolites in HC and HM, such as hesperetin, linalool, and caffeic acid, increased the antioxidant activity in citrus and marjoram compared to clover honey.

Keywords: honey; antioxidant; 5-LOX; metabolomics; NMR; docking

1. Introduction

Floral honey shows compositional variety, especially in aroma and flavor, because different plants contribute their own bioactive constituents. The fragrant white flowers of *Murcott tangarins*, which are a hybrid of *Citrus reticulate* and *Citrus sinensis* [1], are produced singly or in a cluster of up to six flowers. The flower's oil consists mainly of linalool, limonene, sabinene, and trans-nerolidol and is used as stomachic carminative, antimicrobial agent, and flavoring agent [2]. The flower of marjoram (*Origanum mjorana*), commonly known as "sweet marjoram", is a perennial herb native to eastern Mediterranean countries. Marjoram is used worldwide as a spice product. Essential oils from aerial parts of the plants are used in the flavor, perfumery, and pharmaceutical industries. Marjoram is well known for its insecticidal and medicinal value with antioxidant, anticancer, and antimicrobial activities [3,4]. Egyptian clover, *Trifolium alexandrinum* L., is a winter crop widely grown in Egypt, with white or yellow flowers. Its biological activities include antioxidant, anti-inflammatory, anticestodal, and cytotoxic activities, and it is used as a chemoprotective agent against cancers and cardiovascular diseases [4].

Honey is a naturally sweet substance made by honey bees from floral nectar, plant secretions, or plant-sucking bee excretions. After nectar or honeydew collection, transformation by interacting with certain substances in the bee, and maturation, it is then deposited inside the beehive. Honey is produced under different climatic conditions, but the main ingredients in most types of honey are similar.

The process by which nectar saccharides are converted into honey in honeybees comprises regurgitation, evaporation, and enzymatic conversion [5]. One of nature's most complex foods, honey is the only sweetener that can be ingested without going through a human digestive process [6]. Honey's energy content is primarily determined by sugars, which account for 95% of its dry weight and are mainly composed of the monosaccharides fructose and glucose [7]. Honey contains roughly 25 oligosaccharides in addition to the two primary sugar components (tri- and tetra-saccharides). The characterization of the honey's carbohydrate profile was published in different studies [7,8]. Water makes up 12–22% of the composition of honey. Honey's organoleptic and nutritional qualities are characterized by additional minor components such as organic acids, vitamins, minerals, proteins, amino acids, enzymes, volatiles, and phenolic compounds [9].

Honey's beneficial health effects, such as its antioxidant, anti-inflammatory, antibacterial, and immune system-stimulating qualities, are due to minor components in addition to its high nutritional value [6,10].

In addition to being a supersaturated solution of glucose and fructose, honey contains 200 other minor metabolites typically present between 0.01 and 10 ppm [11]. Using Nuclear Magnetic Resonance (NMR) spectroscopy to analyze honey has some benefits over other traditional analytical techniques such as GC and GC-MS [12,13]. These include the simultaneous detection of multiple components, the availability of a wealth of information in a single measurement, the high reproducibility and comparability of the data with a high statistical confidence level, and the minimal needs for sample and pre-processing [13,14]. In particular, the metabolomics approach based on NMR spectroscopy, in conjunction with multivariate statistical analysis, is a potent fingerprinting tool that has been effectively utilized for biomarker identification, origin discrimination, and food quality control [15–17]. This approach examines metabolite profiles and finds the main discriminating components that differentiate honey varieties. Additionally, numerous studies demonstrated that ^{1}H-NMR-based screening techniques are effective tools for the quick examination of honey's authenticity [13].

The natural antioxidant properties of honey are well recognized; it contains flavonoids, aromatic acids, and polyphenols derived from plants. Other bioactive components such as organic acids, amino acids, vitamins, and proteins are also present [18–20]. Esters are essential for honey's antibacterial and antioxidant activities [21,22]. In addition to phenolics, honey includes enzymes with antibacterial properties, such as glucose oxidase, diastase, invertase, catalase, and peroxidase [18,19,23]. Phenolic acids, flavonoids, vitamins, enzymes, and a trace amount of minerals, mainly copper and iron, are thought to be responsible for honey's redox properties [24,25]. However, little is understood about the antioxidant properties of honey and the metabolic processes that underlie each component, whether through reducing power or radical scavenging activity, due to their synergistic interactions or the additive combined action of these minor components [10].

The secondary plant metabolites that honey bees acquire with flower nectars are thought to be responsible for honey's health-promoting qualities. The variety of honey's secondary metabolites is correlated with its biological activities [26].

Generally, several elements, including botanical, geographic, climatic, and seasonal [27], influence the chemical composition of honey and its quality. Other variables may be external, such as the environment, beekeeper honey treatment practices, storage conditions, and intentional producer adulteration [28]. Several studies have demonstrated that most chronic diseases, including cancer, coronary artery disease, and neurological deterioration, are caused by oxidative damage. Additionally, it has been established that honey's medicinal efficacy is invariably linked to its antioxidant activity against reactive oxygen species [29]. As a result, current research has concentrated on the composition of the three types of honey and their biological capabilities, including antioxidants, as honey includes a high concentration of free radical scavengers, which support a balance between the levels of antioxidants and free radical production [30]. The high concentration of reducing sugars, more than 65% in honey, such as glucose and fructose, may result in increased reducing antioxidant power in the DPPH method, resulting in a positive error in determining antioxidant activity [29].

Oxidative stress builds up in our bodies as time passes, leading to various illnesses. Oxidative stress may be caused by many metabolic activities within the body and outside stimuli, such as exposure to ultraviolet radiation (UV) and pollutants in the environment [31]. Scientific investigations have shown that free radicals, DNA damage, and cell malignancy are directly linked. Moreover, oxidative stress is involved in the formation of type II diabetes. Because of the significant healthcare costs incurred from these disorders, practical solutions are required to relieve the burdens on people and society. Moreover, the extensive secondary metabolites in honey extracts give them enormous preventive and therapeutic capabilities [32].

More research on phytochemicals has revealed many modern medications, including those now being researched. Natural product-derived bioactive chemicals are more effective therapeutic agents with fewer side effects than synthetics [33]. Polyphenolic natural compounds are a focus of research in both medical supplies and nutrition. In addition to scavenging free radicals, polyphenols may also have potent immunological modulatory and hormone action-inhibitory properties [34]. Polyphenols are also thought to be effective peroxyl radical scavengers, owing to the hydrogen mobility in their molecular structures [21]. Among polyphenols, phenolic acids are perhaps the most abundant in honey. Additionally, they have been observed to affect honey's flavor and physical appearance, most notably in the color [35].

In this study, we compared the antioxidation potentials of the three most famous regularly used honeys in Egypt (citrus honey, marjoram honey, and trifolium honey—HC1, HM1, and HT1) and their secondary metabolites (HC2, HM2, and HT2). Their radical scavenging potentials were also evaluated using DPPH, APTS, ORAC, and 5-LOX. Additionally, metal chelating activities were also determined, since many free metals have been linked to the production of free radicals [36,37]. The most active compounds suggested in each honey were investigated by the ^{1}H-NMR fingerprint technique using mathematical models correlating their presence to antioxidant activity.

In addition, compounds that might contribute to the 5-LOX inhibitory activity of the tested honey samples were predicted depending on a series of in silico and modeling experiments. This work is one of the few studies [38,39] in metabolomics that has attempted to correlate the antioxidant activity of the three most famous types of honey in Egypt to their ^{1}H-NMR profiles.

2. Materials and Methods

2.1. Honey Preparation and Collection

2.1.1. Honey Samples

Honey samples were collected in the 2019 season from private apiaries as follows: citrus honey (*Morcott tangerine* L. and *Jaffa orange* L.) from Wadi Almollak, Ismailia Governorate, in April; marjoram honey (*Origanum majorana* L.) from Sawiris Al-Gali Tamiya, Fayoum Governorate, in May; clover honey (*Trifolium alexandrinum* L.) in Mansoura, Dakahlia Governorate, at the end of June.

2.1.2. Preparation of Reference Slides

A pollen library of all the common plant species found in the honey-producing regions was assembled as a reference library for identifying the pollen extracted from the honey samples. The direct method was used to create reference slides of plant pollen [40]. Fresh plants' flower buds under study were stripped off their anthers, then washed in an ether-filled watch glass. The ether was decanted, and the pollen was rinsed with fresh ether and left to dry once a ring of pollen had formed at the edge of the ether solution. After being transferred to a microscope slide, the pollen grains were mounted in Kaiser's glycerin jelly and sealed with paraffin after being warmed to 40 °C.

2.1.3. Qualitative Analysis of Pollen in Honey Samples

The investigation was based on the idea that microscopic elements were concentrated by centrifuging the honey that had been dissolved in water, examining the sediments, and examining them under a microscope. The method for pollen analysis was followed as previously described [41]. Shortly after being dissolved in 20 mL of warm distilled water (about 40 °C), a sub-sample of honey (10 g) was centrifuged twice (at 2000 rpm) for 10 min. After drying with slight heating at 40 °C, the entire sediment was placed on a slide and spread over an area of 20 mm × 20 mm. Glycerin/gelatin was used to mount the sediment, and an alcoholic solution of fuchsin was used to stain it mildly. Slides were examined under a microscope and identified using the reference.

2.2. Multivariate and Statistical Analysis

2.2.1. Metabolites Extraction

The secondary metabolite content of honey was extracted using the solid phase extraction (SPE) technique [42]. In brief, 200 g of the available honey samples was thoroughly mixed with 400 mL of deionized H_2O and 400 mL of MeOH until completely fluid, centrifuged for 10 min at 400 rpm to remove solid particles, and then dried at 40 °C under reduced pressure. Additionally, 100 g of the honey samples was dissolved in 500 mL of acid water (adjusted to pH 2.0). To enable metabolite adsorption, the solution was treated with 100 g of amberlite XAD-LH20 resin (100 m) and gently swirled for 30 min. After stirring, 250 mL of acidic H_2O and 250 mL of deionized H_2O were used to wash the resin. To extract the adsorbed metabolites, MeOH (750 mL) was used to wash the resin. As soon as the resin gained its previous white appearance, this step was repeated three to four times. The methanol extract was concentrated at 40 °C in a rotatory evaporator (Buchi, G. Switzerland) before being used for chemical profiling and antioxidant assays [26,43].

2.2.2. H-NMR Analysis

^{1}H NMR (400 MHz) spectra were carried out (Bruker, Munich, Germany) using tetramethylsilane TMS as internal standard and dimethyl sulfoxide (DMSO-d_6) as a solvent. Chemical shift values are reported in ppm. Elemental analyses were performed at the Microanalytical Unit, Faculty of Science Ain Shams University.

2.2.3. Multivariate and Statistical Analysis

MetaboAnalyst is a web-based statistical analysis platform that considers ^{1}H-NMR data. A single zip file comprising the sample name, peak list (ppm), and peak intensities was required for this investigation. The initial step in data normalization was to normalize the raw data using Pareto scaling and the median. After that, multivariate analysis was carried out statistically using unsupervised principal component analysis (PCA) and supervised partial least squares–discriminant analysis (PLS-DA).

2.3. Antioxidant Activity

2.3.1. DPPH (Diphenyl-1-picrylhydrazyl) Assay

The 1,1-diphenyl-2-picrylhydrazyl (DPPH) assay, as outlined in [44], was used to measure the honey samples' efficiency in scavenging free radicals. By dissolving 2 mg in 100 mL of MeOH, the solution (20 mg/L) was created. Then, 0.75 mL of methanolic honey solution was added to 1.5 mL of solution in various concentrations ranging from 20 to 40 mg/mL. The absorbance was measured at 517 nm after 15 min of incubation at 25 °C. The use of ascorbic acid served as a positive control. The ascorbic acid calibration curve (10 and 50 mg/L) was used to calculate the concentration of honey sample needed to scavenge 50% of the ascorbic acid (IC_{50}) [45]. The experiment was conducted in triplicate, and the following formula was used to determine the DPPH radical scavenging activity:

$$(\text{Acontrol} - \text{Asample})/(\text{Acontrol} \times 100) = \text{DPPH radical scavenging activity (percent)}$$

where Asample is the absorbance when a honey extract is present; Acontrol is the absorbance of the control reaction when a honey extract is not present.

2.3.2. ABTS Antioxidant Assay

With a few alterations noted by [46], the method of [47] was applied to assess the free radical scavenging activity. To create the stable ABTS radical cation, the ABTS free radical solution was created (final concentration: 7 mM/L) and incubated for 16 h with potassium persulphate (final concentration: 2.45 mM/L). Five times more ABTS solution was diluted to achieve an absorbance of 2.0–2.4 at 645 nm. In the concentration range of 0 to 0.125 mmol/L, a typical Trolox solution was created. Honey samples were made in water at 1 g/mL concentration. Using a FLUOstar Omega microtiter plate reader (BMG LabTech, Australia) set to 25 °C, 100 mL of ABTS solution was injected into each well of a 96-well flat-bottomed plate. To obtain corrected values, the absorbance of sample and Trolox were measured before and after injection of the ABTS solution. The Trolox-equivalent antioxidant capacity (TEAC) was calculated as mol of Trolox per gram of honey using the following equation, and used to express the antioxidant activity against free radicals.

$$\text{Scavenging activity (percent) of ABTS} = [1 - \text{Ax}/\text{A0}] \times 100$$

where Ax is the absorbance of the leftover ABTS following the reaction with Trolox and honey solution and A0 is the absorbance obtained using pure water. Trolox percent inhibition was calibrated using a calibration curve. Micromoles of Trolox equivalents per gram of honey (mol TE/g of honey) were used to express the results.

2.3.3. 5-Lipoxygenase Inhibitor Screening Assay

It is known that the oxidation of unsaturated fatty acids containing 1–4 diene groups is catalyzed by lipoxygenase. The appearance of a conjugate diene at 234 nm was used to track the transformation of linoleic acid into 13-hydroperoxy linoleic acid using a UV/visible spectrophotometer. Rutin and nordihydroguaiaretic acid (NDGA), known to inhibit soybean lipoxygenase, were employed as controls. The reaction was started by mixing 2.0 mL of sodium linoleate (100 M) in phosphate buffer with aliquots (50 µL) of daily-prepared lipoxygenase solution at a concentration sufficient to produce an easily quantifiable initial rate of reaction. The enzymatic reactions were carried out in the absence or presence of an inhibitor, and their kinetics were determined. The inhibitors were dissolved in DMSO to the extent that an aliquot (30 µL) produced a final concentration of no more than 100 ppm in each assay. A 30 µL aliquot of the inhibitors yielded a final concentration of no more than 100 ppm in each assay after being thoroughly dissolved in DMSO. The initial reaction rate was calculated using the slope of the straight-line portion of the curve and compared to the control (30 µL of phosphate buffer (pH 9.0) instead of 30 µL of the inhibitor solution) to determine the percentage inhibition of the enzyme activity. The concentration that inhibited 50% of the enzyme (IC_{50}) was established by charting the inhibition percentages as a function of the inhibitor concentration [48]. Each inhibitor concentration was tested in triplicate, and the results were averaged (IC_{50} 100 g/mL). Aqueous extracts were not used in this study. A negative lipoxygenase assay result did not always imply that a plant was incapable of acting as an anti-inflammatory agent. Throughout the intricate process of inflammation, the active molecules may have impacts at additional sites [49].

2.3.4. Oxygen Radical Absorbance Capacity (ORAC) Assay

The ORAC test was created as previously reported [50]. AAPH produced free radicals in this experiment, which caused fluorescein to oxidize and lose its fluorescence. Trolox (5 g/mL, final concentration of 20 M) was employed as a reference, and all reagents were made in phosphate buffer (pH 7.0). A final volume of 200 µL test solutions comprising fluorescein (16.7 nM), honey at concentrations ranging from 1 to 10 mg/mL, and AAPH at a concentration of 2.2 mg/mL were placed in each well of the plate reader (Ultimate Concentration). After adding the AAPH, the plate was shaken for 5 s, and fluorescence was observed every 60 s for 110 cycles at wavelengths of 535 and 485 nm for emission and excitation, respectively. ORAC values were calculated using the area under the curve (AUC) method and expressed in mol Trolox equivalent (TE)/g for all fluorescence experiments carried out at 37 °C. A blank containing AAPH, fluorescein, and phosphate buffer was provided (pH 7).

2.3.5. Determination of Metal Chelating Activity

According to previous instructions [51], metal chelating activity was assessed by adding 0.1 mM FeSO4 (0.2 mL) and 0.25 mM ferrozine (0.4 mL) to 0.2 mL of honey extract. The mixture's absorbance at 562 nm was measured after 10 min of room temperature incubation.

The metal chelating activity is calculated as (Acontrol − Asample)/Acontrol × 100, where Asample is the absorbance in the presence of the extract, and Acontrol is the absorbance of the control reaction (without extract).

2.3.6. Statistical Evaluations of In Vitro Experiments

Each experiment was run three times to ensure accuracy and validity. The mean and standard deviation of three different trials were shown in this example of data presentation. GraphPad 5.0 was used to conduct statistical analysis (GraphPad Software, Inc., La Jolla, CA, USA). Data comparison was performed using the ANOVA, where statistical significance was found to exist when the p-value was <0.05.

2.4. In Silico and Modeling Investigation

As previously described, binding free energy estimation (ΔG binding) and molecular dynamic simulations were performed [52,53]. The Supplementary Materials file has a detailed description of these procedures.

2.5. Identification of Isolated Compounds

The methanolic extract was fractionated by column chromatography with silica gel eluted with chloroform/methanol gradient elution 99/1 to 1/1. The highest active fraction (HC2) was subjected to an isolation process; three compounds were isolated and purified by preparative TLC (8:2, *v/v*) chloroform/methanol.

^{1}H-NMR spectra of isolated compounds dissolved in DMSO-d_6 were determined with 400 MH_Z spectrometers.

3. Results

3.1. NMR Analysis

A range of metabolites have been identified or suggested as being responsible for the antioxidant activity of these types of honey from the literature. ^{1}H-NMR analysis of the six honey samples using MestreNova revealed their metabolite profiles. We found that the honey samples showed more antioxidant activity than the honey extract samples. The ^{1}H-NMR analysis indicated the presence of some minor metabolites in all three types of honey samples (HC2-HM2-HT2), which might be responsible for antioxidant activity (gallic acid, *p*-coumaric acid, quercetin, cinnamic acid, and chlorogenic acid) [54–56], and their chemical structures are presented in Figure 1.

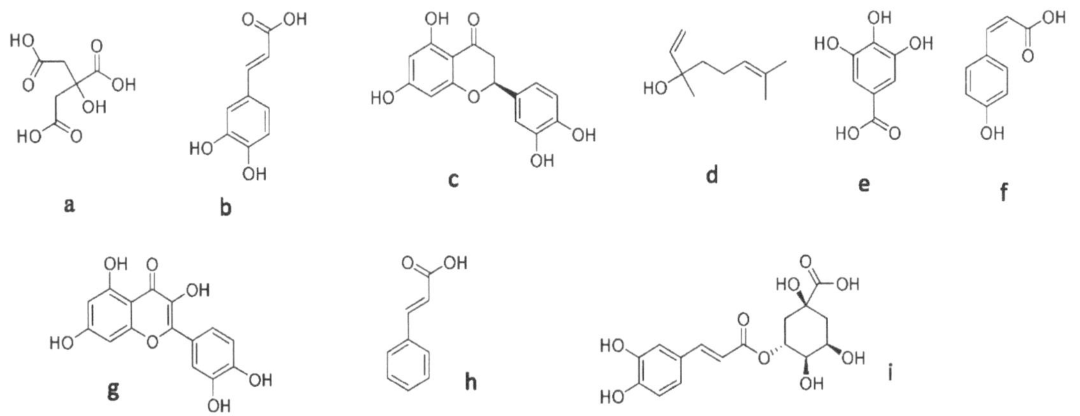

Figure 1. Structure of the secondary metabolites: (**a**) citric acid, (**b**) caffeic acid, (**c**) hesperetin, (**d**) linalool, (**e**) gallic acid, (**f**) *p*-coumaric acid, (**g**) quercetin, (**h**) cinnamic acid, and (**i**) chlorogenic acid.

Linalool, hesperetin, and caffeic acid were found only in citrus and marjoram honey, not clover honey. This could be why citrus and marjoram honey samples were more active as antioxidants than trifolium honey [55,57–61].

On the other hand, three compounds were isolated using the column chromatography of HC2 fraction; they were identified comparing their NMR data to the previously reported ones. Compound **1** was characterized as caffeic acid; its ^{1}H-NMR spectral data were in good agreement with published data [62,63].

Since ^{1}H-NMR spectral data of compound **2** revealed aromatic protons between δ_H 6.18 and 7.66 and phenolic OH groups between δ_H 9.36 and 12.48, respectively, and agreed with the literature [64,65], it was determined to be quercetin.

Compound **3** was characterized as hesperetin, as its ^{1}H-NMR data were in agreement with previous data [66,67].

From these data, we found that citrus honey was more active as an antioxidant then marjoram honey, and the least antioxidant honey was clover or trifolium honey.

3.2. Multivariate Data Analysis

According to the PCA and PLS-DA multivariate analysis (Figure 2), we found that HC2, HM2, HC1, and HM1 were clustered together at PC1 = −10,000, PC2 = −4500 (Figure 2A) and Component 1 = −1000, Component 2 = −4500 (Figure 2B), respectively. However, HT1 and HT2 were plotted far from HC1, HM1, HC2, and HM2. There was a disparity between HC1, HM1, HC2, HM2, and HT1 (10.7% in PCA, 13.7% in PLS-DA) and HT2 (81.2% in PCA, 78.1% in PLS-DA). This finding indicates that HC1, HM1, HC2, and HM2 have similar chemical compositions, while HT1 and HT2 were unique in their chemical profiles.

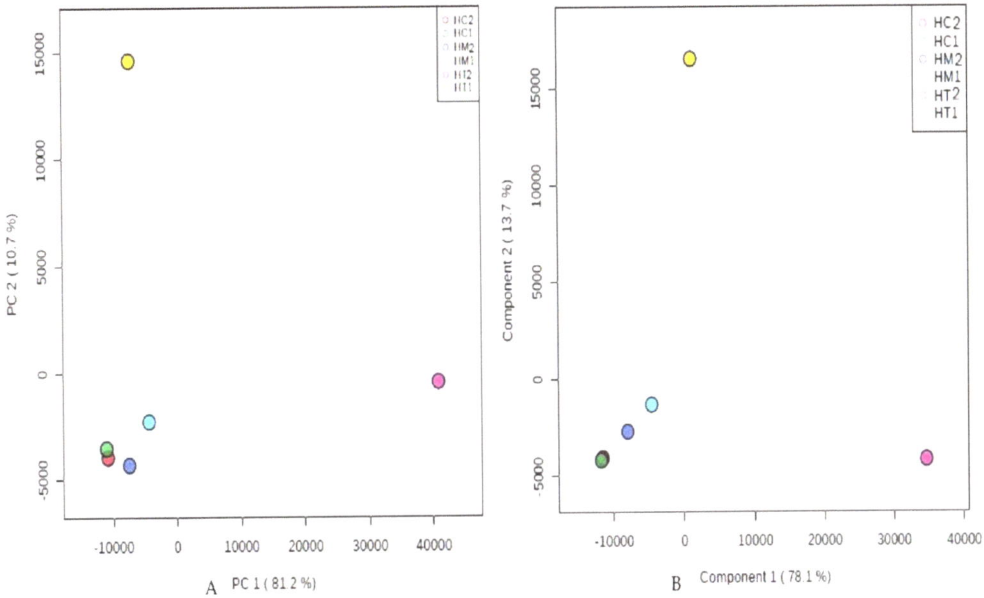

Figure 2. PCA and PLS-DA score plots of the ^{1}H-NMR-derived data of the studied honey bee products ((**A**) and (**B**), respectively).

3.3. Antioxidant Capacity

DPPH, ABTS, metal chelating activity, ORAC, and 5-LOX assays were used to evaluate the antioxidant capabilities of honey extracts and their secondary metabolite samples. Trolox, a substitute for vitamin E, resveratrol, and ascorbic acid were employed as standards. For each gram of honey, scavenging capacity was calculated and given in micromoles of reference standard equivalent.

3.3.1. DPPH Radical Scavenging Activity and Metal Chelating Activity

The DPPH assay was utilized to evaluate the scavenging capacity of honey samples, with ascorbic acid serving as the positive control. The unpaired electron in DPPH reacts with a hydrogen atom provided by honey's free radical scavenging antioxidant, converting the purple-colored odd electron DPPH to its reduced yellow form. To determine the scavenging ability of honey, the degree of decolorization would be determined using a UV/visible spectrophotometer. The lower the IC_{50} value, the greater the capacity of honey to scavenge radicals, as lowering DPPH requires less radical scavenging capacity from honey. According to Figure 3, the most active scavenging agent is HC2 (citrus honey secondary metabolites) with IC_{50} value of 2.91 µg/mL, followed by HM2 (marjoram honey secondary metabolites), HT2 (trifolium or clover honey secondary metabolites), HT1 (trifolium honey extract), HC1 (citrus honey extract), and HM1 (marjoram honey extract) samples with IC_{50} values of 10.7 µg/mL, 20.5 µg/mL, 220.43 µg/mL, 350.32 µg/mL, and 470.42 µg/mL, respectively.

The antioxidant capacities of the secondary metabolites and honey sample extracts were assessed in relation to various radicals (Figure 3, Table 1). The activity against the $ABTS^{\bullet+}$ radical varied between 66.96 and 185.36 µmol TE µM/10 g Trolox and ranged between 120.48 and 431.2 µmol TE µM/50 g Trolox. For HC2 honey, higher values were reported. Additionally, a metal chelating test using honey extracts was evaluated because excess free irons have been linked to the production and generation of free radicals in biological systems. The six extracts showed substantial chelation activities in concentration-dependent manners, with each sample tested with a concentration of 10 µg/mL and 50 µg/mL (Figure 3). Both concentrations revealed that HC2 and HM2 had the strongest activity with 35.62% and 31.98% inhibition, respectively, while HT1 and HC1 had the least.

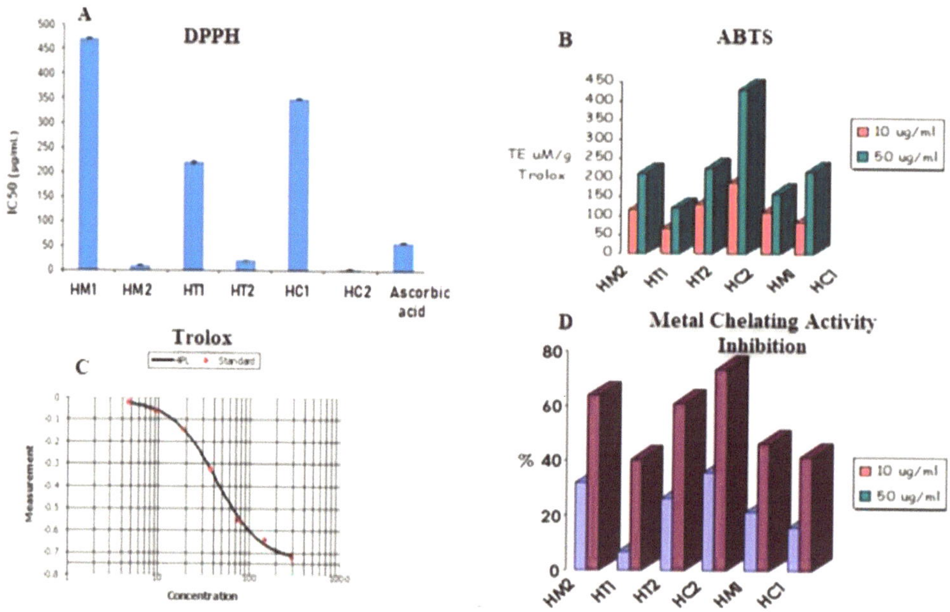

Figure 3. (**A**) DPPH and (**B**) ABTS radical scavenging activities. (**C**) Trolox was used as positive control for SBTS assay. (**D**) Metal chelating activities of different honey extracts. Data are reported as mean ± SE values (n = 3).

Table 1. Free radical scavenging capacities of honey extracts measured with ABTS assay and metal chelating activities at different concentrations on a micro-well plate.

	ABTS		Metal Chelating Activity	
	TE µM/g Trolox TE µM/g Trolox = 36.56/250 × 1000 = 146.23		% Inhibition	
Sample	**10 µg/mL**	**50 µg/mL**	**10 µg/mL**	**50 µg/mL**
HM2	113.88 ± 0.432 c	210.24 ± 1.68 c	31.98978 ± 1.475 c	63.75881 ± 0.7625 b
HT1	66.96± 1.25 a	120.48 ± 1.33 b	6.953549 ± 2.0365 a	40.21628 ± 1.2905 a
HT2	128.96± 0.458 c	224 ± o.655 c	26.50544 ± 1.598 c	60.74911 ± 0.83 b
HC2	185.36± 1.34 b	431.2 ± 2.15 a	35.62371 ± 1.3935 c	73.34526 ± 0.5475 c
HM1	111.0 ± 0.857 c	158.36 ± 0.442 b	21.51157 ± 1.71 b	46.32486 ± 1.1535 a
HC1	83.04 ± 0.612 a	214.76 ± 2.1 c	15.82658 ± 1.835 b	41.68769 ± 1.2575 a
Control	113.88 ± 0.035 c	210.24 ± 0.023 c		

Means with different superscripts (a, b, c) between treatments in the same column are significantly different at $p < 0.05$. Data are represented as mean ± SE values (n = 3).

3.3.2. ORAC Antioxidant Capacity and Lipoxygenase Inhibition Activity

The Oxygen Radical Absorbance Capacity (ORAC) experiment revealed that honey samples have the following antioxidant capacities: HC2 > HM2 > HT2 > HM1 > HC1 > HT1. As shown in Table 2, honey extracts showed strong antioxidant activity due to the secondary metabolites in HC2 and HM2, which had respective values of 259.5, 0.448 and 235.8, 1.03 molTE/g, while HC1 and HT1 had ORAC values of 209.7, 0.198 and 180, 0.672 molTE/g, respectively. As shown in Figure 4A,B, HC2 had the greatest ORAC value of all the extracts tested, demonstrating its potential as a free radical scavenger. The investigated honey extracts significantly inhibited 5-lipoxygenase (5-LOX) activity in a similar pattern; among the honey extracts, HC2 (IC_{50} 2.293 g/mL) displayed the highest antioxidant activity (Table 2), while HC1 (IC_{50} 31.87 g/mL) displayed the lowest inhibition of 5-LOX activity (Figure 4C).

Table 2. Free radical scavenging capacities of honey extracts measured with ORAC assay and inhibition (%) of 5-lipoxygenase activity obtained. IC_{50} values represent the mean ± SD of three determinations.

	ORAC TE μM/L		5-LOX
Sample	**10 μg/mL**	**50 μg/mL**	**IC_{50} μg/mL ± SD**
HM2	168.3 ± 0.839 [c]	235.8 ± 1.03 [c]	6.136 ± 0.4 [a]
HT1	101 ± 0.606 [a]	180 ± 0.672 [a]	23.36 ± 1.4 [b]
HT2	147.9 ± 0.0776 [c]	235.4 ± 0.0776 [c]	10.34 ± 0.6 [a]
HC2	150.1 ± 0.616 [c]	259.5 ± 0.448 [c]	2.293 ± 0.1 [a]
HM1	115.4 ± 0.175 [a]	226.8 ± 0.286 [b]	77.59 ± 4.6 [c]
HC1	134.7 ± 0.69 [b]	209.7 ± 0.198 [b]	31.87 ± 1.9 [b]
NDGA			2.696 ± 0.2 [a]

Means with different superscripts (a, b, c) between treatments in the same column are significantly different at $p < 0.05$. Data are represented as mean ± SD values ($n = 3$).

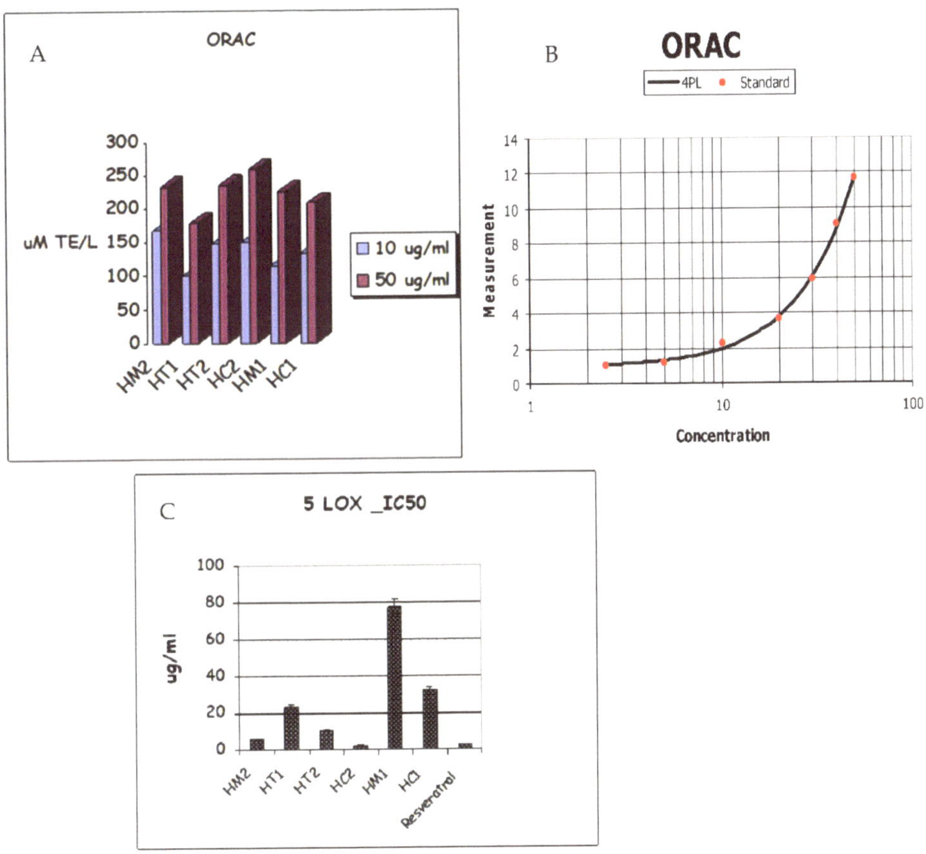

Figure 4. (**A**) ORAC antioxidative activities. (**B**) TE/Trolox standard equivalent. (**C**) Lipoxygenase (5-LOX) inhibition of honey extracts and their metabolites.

3.4. Molecular Modeling Study

Several modeling and molecular simulation-based experiments were conducted to determine which compounds might be responsible for the observed inhibitory activity against 5-LOX. First, the structures of all identified compounds (Figure 1) were prepared and docked inside the 5-LOX active site (PDB code: 6N2W). The resulted docking poses for each structure were almost identical; hence, we selected the top-scoring pose for each structure for the subsequent in silico experiment (Table 3). The purpose of the docking step was to putatively generate the static binding mode of each structure inside the enzyme active site. To validate the docking protocol used for the first docking step, the structure of the reported 5-LOX inhibitor, nordihydroguaiaretic acid (NDGA) [68], was re-docked inside the enzyme active site. The produced binding pose was almost identical to that of the co-crystalized one with RMSD of 0.47Å.

Table 3. Docking and ΔG binding scores of the identified compounds inside the active site of 5-LOX, along with their H-bonding and hydrophobic interactions.

Structure	Docking Score (kcal/mol)	ΔG Binding (kcal/mol)	Average RMSD (Å)	H-Bonding	Hydrophobic Interaction
Caffeic acid	−7.1	−7.5	1.7	HIS-372	LEU-607
Hesperetin	−7.5	−7.9	2.8	HIS-372	TRP-599, LEU-607
Quercetin	−7.6	−8.4	4.0	HIS-372	TRP-599, LEU-607
Chlorogenic acid	−6.1	−5.3	>5	ARG-596	TRP-599
Cinnamic acid	−6.9	−4.7	>5	-	TRP-599, LEU-607
p-coumaric acid	−6.2	−4.1	>5	ARG-596	TRP-599
Citric acid	−5.3	−3.6	>5	ARG-596, TRP-599	-
Gallic acid	−4.7	−2.3	>5	TRP-599	-
Linalool	−3.5	−1.1	>5	-	TRP-599
NDGA *	−7.9	−8.6	4.7	HIS-372, ARG-596	TRP-599, LEU-607

* NDGA is the previously reported 5-LOX co-crystallized inhibitor [68].

Second, generated binding poses from the previous step were used to estimate each structure's absolute binding free energy (ΔG binding) inside the 5-LOX active site. This step was carried out by conducting a series of molecular dynamics simulations (MDS) according to the free energy perturbation (FEP) protocol [69]. The main purpose of this step was to estimate the relative affinity of each structure towards the enzyme active site. The top three structures (Table 3) with the lowest ΔG binding (<−7 kcal/mol) were then chosen along with NDGA for subsequent 50 ns MDS runs to investigate their stability inside the 5-LOX active site.

Third, caffeic acid, hesperetin, and quercetin showed the highest affinities toward the 5-LOX active site (ΔG binding < −7 kcal/mol); they were then subjected to 50 ns long MDS experiments to explore their dynamic binding stability and mode.

The most populated poses were extracted from each MDS run and are depicted in Figure 5. The binding mode of each structure was aligned with that of NDGA to show their degree of similarity, particularly with Fe^{+2} ion. Interestingly, the catechol moiety of the three structures (i.e., caffeic acid, hesperetin, and quercetin) was aligned perfectly with NDGA, establishing H-bonds with HIS-372 and coordinate interactions with Fe^{+2} ion. In addition, only hesperetin and quercetin were able to establish hydrophobic interactions with both TRP-599 and LEU-607.

RMSDs of caffeic acid, hesperetin, quercetin, and NDGA inside the 5-LOX active site ranged from 1.7 Å to 4.1 Å, indicating stability over the simulation. In addition, quercetin showed the lowest fluctuation during the MDS run, while caffeic acid showed the highest. Overall, we could conclude from this modeling and simulation-based experiments that caffeic acid, hesperetin, and quercetin were the potential 5-LOX inhibitors inside the honey samples, particularly HC2, the most potent 5-LOX inhibitor, which was rich in caffeic acid and hesperetin.

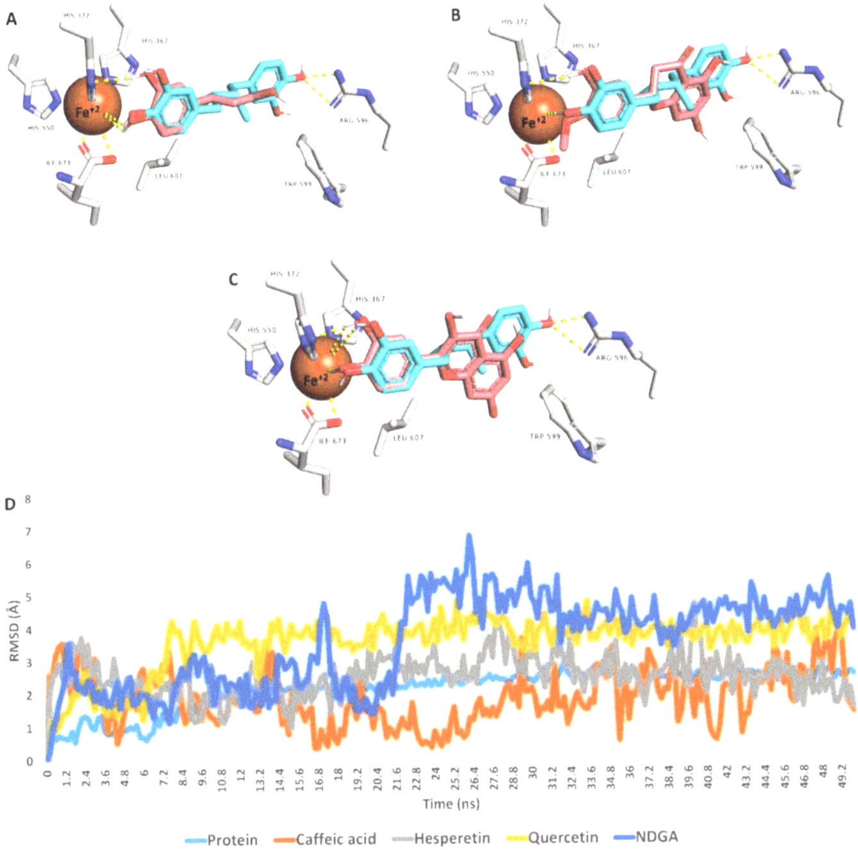

Figure 5. Binding modes of caffeic acid, hesperetin, and quercetin inside the active site of 5-LOX (brick red-colored structures; (**A–C**), respectively). Each structure with found to be aligned with NDGA (cyan-colored structure), the reported 5-LOX co-crystallized inhibitor, at their catechol moiety. These binding modes were extracted from the MDS runs as the most populated poses. D is the RMSD of each structure inside the 5-LOX active site over 50 ns long MDS.

4. Discussion

From the literature, a range of metabolites have been suggested to be responsible for the antioxidant activity of honey [26]. However, this work is one of the few studies in metabolomics that have attempted to correlate the antioxidant activity of the three most famous types of honey in Egypt—citrus honey (HC1), marjoram honey (HM1), and clover honey (HT1)—compared to their secondary metabolites (HC2, HM2, HT2) by ^{1}H-NMR profiles, PCA, and PLS-DA multivariate analysis.

Honey's antioxidant activity is influenced by several parameters, including concentration, temperature, light, substrate type, physical system state, and the existence of micro-components that function as pro-oxidants or synergists [70]. Additionally, it has been proposed that honey's organic acids, such as gluconic, malic, and citric acids, contribute to antioxidant activity by chelating metals, thus enhancing the activity of flavonoids through synergistic effects [71]. Additionally, the enzymes glucose oxidase and catalase contribute to antioxidant action by their capacity to extract oxygen from the medium [72]. The essential elements of honey responsible for its antioxidant activity include phenolic, flavonoid, and carotenoid concentration, along with ascorbic acid and enzymes related to floral capacity [29,73,74].

In this study, we focused on the secondary metabolites that may present in minor concentrations, such as gallic acid, *p*-coumaric acid, quercetin, cinnamic acid, and chlorogenic acid, which might be responsible for antioxidant activity [54–56]. To exclude the effect of organic acids and focus on secondary metabolites by using ^{1}H-NMR metabolomics, six honey samples were used from crude and prepared honey, and the analysis revealed that their metabolite profiles have more antioxidant activities due to the presence of secondary metabolites compared to the honey extract samples. The ^{1}H-NMR analysis indicated the presence of these secondary metabolites in all three types of honey samples (HC2, HM2, HT2), whereas linalool, hesperetin, and caffeic acid were found only in citrus and marjoram honey, not clover honey. This could be why citrus and marjoram honey samples were more active as antioxidants than trifolium honey [55,57–61]. Moreover, using PCA and PLS-DA multivariate analysis confirmed our result that HC1, HM1, HC2, and HM2 have similar secondary metabolites, while HT1 and HT2 were unique in their chemical profiles, as we found that HC2, HM2, HC1, and HM1 were clustered together; however, HT1 and HT2 were plotted far from them and each other.

The antioxidant activities of three different types of Egyptian honey (citrus, clover, and marjoram) and their metabolites were evaluated and tested. Clover honey's antioxidant activity values were generally lower than those of marjoram and citrus honey [29,47,75] based on colorimetrically scavenging activities against DPPH, ABTS, ORAC, 5-LOX, and metal chelating activity in all honey extract samples in concentration-dependent ways. This ideal concentration must be determined to accurately estimate the antioxidant activity of honey types from various floral origins. On the other hand, the relative quantities of minor chemicals, which may be essential to the antioxidant effect, may partially account for the variations in the honey. Numerous authors showed a linear relationship between the amount of all phenolic components and the antioxidant power of plant extracts [47,76,77]. Although their mode of action is unknown, phenolic chemicals and flavonoids are principally responsible for antioxidant activity, with minor chemical compounds [78].

It has been established that 5-LOX contributes to general cellular oxidative stress [79,80]. Accordingly, several previous reports have shown the potential of 5-LOX inhibitors in reducing the 5-LOX-mediated elevated cellular oxidative stress, particularly in inflammatory conditions [81,82], which can lead to cardiovascular, neuronal, and kidney dysfunctions [79,83]. Several theories have been proposed; for example, cinnamon extract activity was indicated as it is linked to free radical sequestration, hydrogen donation, metallic ion chelation, or even has a role as a superoxide or hydroxyl radical substrate. The antioxidant characteristics of these bioactive substances also interfere with propagation processes [84].

Honey's total phenolic content is essential to correlate to its antioxidant properties. Our current study showed a lower IC_{50} value by HC2 and HM2, suggesting that the sample had more potent antioxidant properties, in agreement with several previous works [21,85]. This showed that flavonoids, along with other honey constituents such as glucose and fructose, could also contribute to the reducing power as one of the primary factors influencing the honey samples' reduction capacity. More research is needed to determine which phenolic components are responsible for honey's antioxidant action. Because honey is a complex mix of many different compounds with diverse activity, the involvement of non-phenolic chemicals, which are significant for antioxidant properties, must be examined. Honey also contains amino acids, which have antioxidant properties. Histidine, taurine, glycine, and alanine are a few free amino acids that have antioxidant potential [86]. The relationship between radical scavenging activity and total phenolic content was stronger than between radical scavenging activity and proline content [25]. Honey's antioxidant action appears complicated because it is connected to various substances, including enzymes, sugars, and plant substrates.

5. Conclusions

Our study results reveal that HC2 and HM2 possess the most potential in vitro antioxidant activities. The citrus honey extract (HC2) demonstrated the highest antioxidant activity in all assays (DPPH assay: IC_{50} 2.91 μg/mL; ABTS assay: 431.2 at 50 μg/mL Trolox equivalent; ORAC assay: 259.5 at 50 μg/mL Trolox equivalent; 5-LOX screening assay/IC_{50}: 2.293 μg/mL; metal chelating activity at 50 μg/mL: 73.34526% inhibition), followed by HM2 extract (DPPH assay: IC_{50} 10.7 μg/mL; ABTS assay: 210.24 at 50 μg/mL Trolox equivalent; ORAC assay: 234.8 at 50 μg/mL Trolox equivalent; 5-LOX screening assay/IC_{50}: 6.136 μg/mL; metal chelating activity at 50 μg/mL: 63.75881% inhibition).

These results reveal that the secondary metabolites in HC and HM, which were identified as hesperetin, linalool, and caffeic acid, are responsible for increasing the antioxidant activities in citrus and marjoram honey, more than in clover honey. Chromatographing of the HC2 fraction resulted in the isolation and identification of three compounds (caffeic acid, quercetin, and hesperetin) by ^{1}H-NMR.

Supplementary Materials: The following supporting information can be downloaded at: https://www.mdpi.com/article/10.3390/antiox11101880/s1, Figure S1. ^{1}HNMR chart of Citrus honey extract and its 2nd metabolites (H1 and Hc); Figure S2. ^{1}HNMR chart of Marjoram honey extract and its 2nd metabolites (H2 and HM); Figure S3. ^{1}HNMR chart of Clover honey extract and its 2nd metabolites (H3 and HT); Figure S4. ^{1}HNMR chart of compound no 2: Qurcetin; Figure S5. ^{1}HNMR chart of compound no 3: Hesperetin; Figure S6. ^{1}HNMR chart of compound no 1: Caffeic acid. Refs. [87–91] in Supplementary Materials.

Author Contributions: Conceptualization, M.M., A.T.A., E.W.Z., H.M.H. and R.M.; methodology, M.M., A.T.A., A.M.S., R.O. and M.E.R.; software, A.M.S., U.R.A. and M.E.R.; validation, R.O., M.E.R., H.M.H., R.M. and M.S.H.; formal analysis, M.M., A.T.A., A.M.S., U.R.A. and E.W.Z.; investigation, M.M., A.T.A., A.M.S., U.R.A. and E.W.Z.; resources, A.M.S., R.O., M.E.R., H.M.H., R.M. and M.S.H.; data curation, M.M., A.T.A., A.M.S., U.R.A. and E.W.Z.; writing—original draft preparation, M.M., A.T.A. and A.M.S.; writing—review and editing, M.A.Z., U.R.A., R.O., M.E.R., H.M.H., R.M. and M.S.H.; supervision, H.M.H., R.M. and M.S.H.; project administration, R.O. and M.E.R.; funding acquisition, R.O. and M.E.R. All authors have read and agreed to the published version of the manuscript.

Funding: This research was funded by the Researchers Supporting Project (number RSP2022R431), King Saud University, Riyadh, Saudi Arabia.

Institutional Review Board Statement: Not applicable.

Informed Consent Statement: Not applicable.

Data Availability Statement: Data is contained within the article and Supplementary Materials.

Acknowledgments: The authors extend their appreciation to the Researchers Supporting Project (number RSP2022R431), King Saud University, Riyadh, Saudi Arabia for funding this research work.

Conflicts of Interest: The authors declare no conflict of interest.

References

1. Feger, W.; Brandauer, H.; Ziegler, H. Analytical Investigation of Murcott (Honey) Tangerine Peel Oil. *J. Essent. Oil Res.* **2003**, *15*, 143–147. [CrossRef]
2. Tan, F.C.; Swain, S.M. Functional Characterization of AP3, SOC1 and WUS Homologues from Citrus (*Citrus sinensis*). *Physiol. Plant.* **2007**, *131*, 481–495. [CrossRef] [PubMed]
3. Gharib, F.A.; Badr, S.E.A.; Al-Ghazali, B.A.S.; Zahran, M.K. Chemical Composition, Antioxidant and Antibacterial Activities of Lavender and Marjoram Essential Oils. *Egypt. J. Chem.* **2013**, *56*, 1–24.
4. Abdalla, M.M.F.; Abd El-Naby Zeinab, M. Inbreeding and Fertility in Egyptian Clover, Trifolium Alexandrinum. *J. Pharmacogn. Phyther.* **2012**, *4*, 16–25. [CrossRef]
5. Jamróz, M.K.; Paradowska, K.; Zawada, K.; Makarova, K.; Kaźmierski, S.; Wawer, I. ^{1}H and ^{13}C NMR-Based Sugar Profiling with Chemometric Analysis and Antioxidant Activity of Herbhoneys and Honeys. *J. Sci. Food Agric.* **2014**, *94*, 246–255. [CrossRef] [PubMed]

6. Kazalaki, A.; Misiak, M.; Spyros, A.; Dais, P. Identification and Quantitative Determination of Carbohydrate Molecules in Greek Honey by Employing ^{13}C NMR Spectroscopy. *Anal. Methods* **2015**, *7*, 5962–5972. [CrossRef]
7. Consonni, R.; Cagliani, L.R.; Cogliati, C. NMR Characterization of Saccharides in Italian Honeys of Different Floral Sources. *J. Agric. Food Chem.* **2012**, *60*, 4526–4534. [CrossRef]
8. Pascual-Maté, A.; Osés, S.M.; Marcazzan, G.L.; Gardini, S.; Muiño, M.A.F.; Sancho, M.T. Sugar Composition and Sugar-Related Parameters of Honeys from the Northern Iberian Plateau. *J. Food Compos. Anal.* **2018**, *74*, 34–43. [CrossRef]
9. Bogdanov, S.; Jurendic, T.; Sieber, R.; Gallmann, P. Honey for Nutrition and Health: A Review. *J. Am. Coll. Nutr.* **2008**, *27*, 677–689. [CrossRef]
10. Gheldof, N.; Wang, X.-H.H.; Engeseth, N.J. Identification and Quantification of Antioxidant Components of Honeys from Various Floral Sources. *J. Agric. Food Chem.* **2002**, *50*, 5870–5877. [CrossRef]
11. Ciulu, M.; Spano, N.; Pilo, M.I.; Sanna, G. Recent Advances in the Analysis of Phenolic Compounds in Unifloral Honeys. *Molecules* **2016**, *21*, 451. [CrossRef]
12. Karabagias, I.K.; Vlasiou, M.; Kontakos, S.; Drouza, C.; Kontominas, M.G.; Keramidas, A.D. Geographical Discrimination of Pine and Fir Honeys Using Multivariate Analyses of Major and Minor Honey Components Identified by ^{1}H NMR and HPLC along with Physicochemical Data. *Eur. Food Res. Technol.* **2018**, *244*, 1249–1259. [CrossRef]
13. Spiteri, M.; Jamin, E.; Thomas, F.; Rebours, A.; Lees, M.; Rogers, K.M.; Rutledge, D.N. Fast and Global Authenticity Screening of Honey Using ^{1}H-NMR Profiling. *Food Chem.* **2015**, *189*, 60–66. [CrossRef]
14. Schievano, E.; Stocchero, M.; Morelato, E.; Facchin, C.; Mammi, S. An NMR-Based Metabolomic Approach to Identify the Botanical Origin of Honey. *Metabolomics* **2012**, *8*, 679–690. [CrossRef]
15. Consonni, R.; Cagliani, L.R. Geographical Characterization of Polyfloral and Acacia Honeys by Nuclear Magnetic Resonance and Chemometrics. *J. Agric. Food Chem.* **2008**, *56*, 6873–6880. [CrossRef]
16. Beretta, G.; Caneva, E.; Regazzoni, L.; Bakhtyari, N.G.; Facino, R.M. A Solid-Phase Extraction Procedure Coupled to ^{1}H NMR, with Chemometric Analysis, to Seek Reliable Markers of the Botanical Origin of Honey. *Anal. Chim. Acta* **2008**, *620*, 176–182. [CrossRef]
17. Girelli, C.R.; Del Coco, L.; Fanizzi, F.P. ^{1}H NMR Spectroscopy and Multivariate Analysis as Possible Tool to Assess Cultivars, from Specific Geographical Areas, in EVOOs. *Eur. J. Lipid Sci. Technol.* **2016**, *118*, 1380–1388. [CrossRef]
18. Miguel, M.G.; Antunes, M.D.; Faleiro, M.L. Honey as a Complementary Medicine. *Integr. Med. Insights* **2017**, *12*, 1–15. [CrossRef]
19. Combarros-Fuertes, P.; Fresno, J.M.; Estevinho, M.M.; Sousa-Pimenta, M.; Tornadijo, M.E.; Estevinho, L.M. Honey: Another Alternative in the Fight against Antibiotic-Resistant Bacteria? *Antibiotics* **2020**, *9*, 774. [CrossRef]
20. Kuś, P.M. Honey as Source of Nitrogen Compounds: Aromatic Amino Acids, Free Nucleosides and Their Derivatives. *Molecules* **2020**, *25*, 847. [CrossRef]
21. Al-Mamary, M.; Al-Meeri, A.; Al-Habori, M. Antioxidant Activities and Total Phenolics of Different Types of Honey. *Nutr. Res.* **2002**, *22*, 1041–1047. [CrossRef]
22. Sateriale, D.; Facchiano, S.; Colicchio, R.; Pagliuca, C.; Varricchio, E.; Paolucci, M.; Volpe, M.G.; Salvatore, P.; Pagliarulo, C. In Vitro Synergy of Polyphenolic Extracts from Honey, Myrtle and Pomegranate against Oral Pathogens, *S. mutans* and *R. dentocariosa*. *Front. Microbiol.* **2020**, *11*, 1–11. [CrossRef] [PubMed]
23. Ahmed, S.; Sulaiman, S.A.; Baig, A.A.; Ibrahim, M.; Liaqat, S.; Fatima, S.; Jabeen, S.; Shamim, N.; Othman, N.H. Honey as a Potential Natural Antioxidant Medicine: An Insight into Its Molecular Mechanisms of Action. *Oxid. Med. Cell. Longev.* **2018**, *2018*, 8367846. [CrossRef] [PubMed]
24. Erlund, I. Review of the Flavonoids Quercetin, Hesperetin, and Naringenin. Dietary Sources, Bioactivities, Bioavailability, and Epidemiology. *Nutr. Res.* **2004**, *24*, 851–874. [CrossRef]
25. Meda, A.; Lamien, C.E.; Romito, M.; Millogo, J.; Nacoulma, O.G. Determination of the Total Phenolic, Flavonoid and Proline Contents in Burkina Fasan Honey, as Well as Their Radical Scavenging Activity. *Food Chem.* **2005**, *91*, 571–577. [CrossRef]
26. Suarez, A.F.L.; Tirador, A.D.G.; Villorente, Z.M.; Bagarinao, C.F.; Sollesta, J.V.N.; Dumancas, G.G.; Sun, Z.; Zhan, Z.Q.; Saludes, J.P.; Dalisay, D.S. The Isorhamnetin-Containing Fraction of Philippine Honey Produced by the Stingless Bee *Tetragonula biroi* Is an Antibiotic against Multidrug-Resistant *Staphylococcus aureus*. *Molecules* **2021**, *26*, 688. [CrossRef]
27. Nicolson, S.W.; Nepi, M.; Pacini, E. *Nectaries and Nectar*; Springer: Berlin/Heidelberg, Germany, 2007; Volume 4.
28. Soria, A.C.; González, M.; De Lorenzo, C.; Martınez-Castro, I.; Sanz, J. Characterization of Artisanal Honeys from Madrid (Central Spain) on the Basis of Their Melissopalynological, Physicochemical and Volatile Composition Data. *Food Chem.* **2004**, *85*, 121–130. [CrossRef]
29. Ferreira, I.C.F.R.; Aires, E.; Barreira, J.C.M.; Estevinho, L.M. Antioxidant Activity of Portuguese Honey Samples: Different Contributions of the Entire Honey and Phenolic Extract. *Food Chem.* **2009**, *114*, 1438–1443. [CrossRef]
30. Kishore, R.K.; Halim, A.S.; Syazana, M.S.N.; Sirajudeen, K.N.S. Tualang Honey Has Higher Phenolic Content and Greater Radical Scavenging Activity Compared with Other Honey Sources. *Nutr. Res.* **2011**, *31*, 322–325. [CrossRef]
31. Murapa, P.; Dai, J.; Chung, M.; Mumper, R.J.; D'Orazio, J. Anthocyanin-rich Fractions of Blackberry Extracts Reduce UV-induced Free Radicals and Oxidative Damage in Keratinocytes. *Phyther. Res.* **2012**, *26*, 106–112. [CrossRef]
32. Ibrahim, M.A.; Koorbanally, N.A.; Kiplimo, J.J.; Islam, M.S. Anti-Oxidative Activities of the Various Extracts of Stem Bark, Root and Leaves of *Ziziphus Mucronata* (Rhamnaceae) in Vitro. *J. Med. Plants Res.* **2012**, *6*, 4176–4184.

33. Facchini, P.J.; De Luca, V. Opium Poppy and Madagascar Periwinkle: Model Non-model Systems to Investigate Alkaloid Biosynthesis in Plants. *Plant J.* **2008**, *54*, 763–784. [CrossRef]
34. Havsteen, B.H. The Biochemistry and Medical Significance of the Flavonoids. *Pharmacol. Ther.* **2002**, *96*, 67–202. [CrossRef]
35. Steeg, E.; Montag, A. Minor Ingredients of Honey with Flavor Relevancy. 2. Sensorially Active Decomposition Products of Carboxylic-Acids and Glycosidically Bonded Aromates. *Dtsch. Leb.* **1988**, *84*, 147–150.
36. Martemucci, G.; Costagliola, C.; Mariano, M.; D'andrea, L.; Napolitano, P.; D'Alessandro, A.G. Free Radical Properties, Source and Targets, Antioxidant Consumption and Health. *Oxygen* **2022**, *2*, 48–78. [CrossRef]
37. Phaniendra, A.; Jestadi, D.B.; Periyasamy, L. Free Radicals: Properties, Sources, Targets, and Their Implication in Various Diseases. *Indian J. Clin. Biochem.* **2015**, *30*, 11–26. [CrossRef]
38. Alhadrami, H.A.; Sayed, A.M.; El-Gendy, A.O.; Shamikh, Y.I.; Gaber, Y.; Bakeer, W.; Sheirf, N.H.; Attia, E.Z.; Shaban, G.M.; Khalifa, B.A. A Metabolomic Approach to Target Antimalarial Metabolites in the Artemisia Annua Fungal Endophytes. *Sci. Rep.* **2021**, *11*, 1–11. [CrossRef]
39. More, G.K.; Meddows-Taylor, S.; Prinsloo, G. Metabolomic Profiling of Antioxidant Compounds in Five Vachellia Species. *Molecules* **2021**, *26*, 6214. [CrossRef]
40. Sawyer, R. *Pollen Identification for Beekeepers*; Pickard, R.S., Ed.; University College Cardiff Press: Cardiff, UK, 1981.
41. Louveaux, J.; Maurizio, A.; Vorwohl, G. Methods of Melissopalynology. *Bee World* **1978**, *59*, 139–157. [CrossRef]
42. Ferreres, F.; Tomáas-Barberáan, F.A.; Gil, M.I.; Tomáas-Lorente, F. An HPLc Technique for Flavonoid Analysis in Honey. *J. Sci. Food Agric.* **1991**, *56*, 49–56. [CrossRef]
43. Ferreres, F.; Andrade, P.; Tomfis-Barberfin, F.A. *Flavonoids from Portuguese Heather Honey*; Zeitschrift für Lebensmittel-Untersuchung und -Forschung: Berlin, Germany, 1994; Volume 199, pp. 32–37. [CrossRef]
44. Isla, M.I.; Craig, A.; Ordoñez, R.; Zampini, C.; Sayago, J.; Bedascarrasbure, E.; Alvarez, A.; Salomón, V.; Maldonado, L. Physico Chemical and Bioactive Properties of Honeys from Northwestern Argentina. *LWT Food Sci. Technol.* **2011**, *44*, 1922–1930. [CrossRef]
45. Barbosa-Pereira, L.; Angulo, I.; Paseiro-Losada, P.; Cruz, J.M. Phenolic Profile and Antioxidant Properties of a Crude Extract Obtained from a Brewery Waste Stream. *Food Res. Int.* **2013**, *51*, 663–669. [CrossRef]
46. Brangoulo, H.L.; Molan, P.C. Assay of the Antioxidant Capacity of Foods Using an Iron (II)-Catalysed Lipid Peroxidation Model for Greater Nutritional Relevance. *Food Chem.* **2011**, *125*, 1126–1130. [CrossRef]
47. Baltrušaitytė, V.; Venskutonis, P.R.; Čeksterytė, V. Radical Scavenging Activity of Different Floral Origin Honey and Beebread Phenolic Extracts. *Food Chem.* **2007**, *101*, 502–514. [CrossRef]
48. Njenga, E.W.; Viljoen, A.M. In Vitro 5-Lipoxygenase Inhibition and Anti-Oxidant Activity of *Eriocephalus* L. (Asteraceae) Species. *S. Afr. J. Bot.* **2006**, *72*, 637–641. [CrossRef]
49. Akula, U.S.; Odhav, B. In Vitro 5-Lipoxygenase Inhibition of Polyphenolic Antioxidants from Undomesticated Plants of South Africa. *J. Med. Plants Res.* **2013**, *2*, 207–212.
50. Gillespie, K.M.; Chae, J.M.; Ainsworth, E.A. Rapid Measurement of Total Antioxidant Capacity in Plants. *Nat. Protoc.* **2007**, *2*, 867–870. [CrossRef]
51. Chew, Y.-L.; Goh, J.-K.; Lim, Y.-Y. Assessment of in Vitro Antioxidant Capacity and Polyphenolic Composition of Selected Medicinal Herbs from Leguminosae Family in Peninsular Malaysia. *Food Chem.* **2009**, *116*, 13–18. [CrossRef]
52. Alhadrami, H.A.; Alkhatabi, H.; Abduljabbar, F.H.; Abdelmohsen, U.R.; Sayed, A.M. Anticancer Potential of Green Synthesized Silver Nanoparticles of the Soft Coral Cladiella Pachyclados Supported by Network Pharmacology and In Silico Analyses. *Pharmaceutics* **2021**, *13*, 1846. [CrossRef]
53. Alhadrami, H.A.; Burgio, G.; Thissera, B.; Orfali, R.; Jiffri, S.E.; Yaseen, M.; Sayed, A.M.; Rateb, M.E. Neoechinulin A as a Promising SARS-CoV-2 Mpro Inhibitor: In Vitro and in Silico Study Showing the Ability of Simulations in Discerning Active from Inactive Enzyme Inhibitors. *Mar. Drugs* **2022**, *20*, 163. [CrossRef] [PubMed]
54. Roby, M.H.H.; Abdelaliem, Y.F.; Esmail, A.H.M.; Mohdaly, A.A.A.; Ramadan, M.F. Evaluation of Egyptian Honeys and Their Floral Origins: Phenolic Compounds, Antioxidant Activities, and Antimicrobial Characteristics. *Environ. Sci. Pollut. Res.* **2020**, *27*, 20748–20756. [CrossRef] [PubMed]
55. Hamdy, A.A.; Ismail, H.M.; Al-Ahwal, A.E.-M.A.; Gomaa, N.F. Determination of Flavonoid and Phenolic Acid Contents of Clover, Cotton and Citrus Floral Honeys. *J. Egypt. Public Health Assoc.* **2009**, *84*, 245–24559. [PubMed]
56. Roby, M.H.H.; Sarhan, M.A.; Selim, K.A.H.; Khalel, K.I. Evaluation of Antioxidant Activity, Total Phenols and Phenolic Compounds in Thyme (*Thymus vulgaris* L.), Sage (*Salvia officinalis* L.), and Marjoram (*Origanum majorana* L.) Extracts. *Ind. Crops Prod.* **2013**, *43*, 827–831. [CrossRef]
57. Ahmad Khera, R.; Nadeem, F.; Idrees Jilani, M. Essential Chemical Constituents and Medicinal Uses of Marjoram (*Origanum majorana* L.)—A Comprehensive Review. *Ijcbs* **2016**, *9*, 56.
58. Abou-Shaara, H.F. Potential Honey Bee Plants of Egypt. *Cercet. Agron. Mold.* **2015**, *48*, 99–108. [CrossRef]
59. Castro-Vázquez, L.; Díaz-Maroto, M.C.; González-Viñas, M.A.; Pérez-Coello, M.S. Differentiation of Monofloral Citrus, Rosemary, Eucalyptus, Lavender, Thyme and Heather Honeys Based on Volatile Composition and Sensory Descriptive Analysis. *Food Chem.* **2009**, *112*, 1022–1030. [CrossRef]
60. Castro-Vázquez, L.; Díaz-Maroto, M.C.; Pérez-Coello, M.S. Aroma Composition and New Chemical Markers of Spanish Citrus Honeys. *Food Chem.* **2007**, *103*, 601–606. [CrossRef]

61. Alissandrakis, E.; Daferera, D.; Tarantilis, P.A.; Polissiou, M.; Harizanis, P.C. Ultrasound-Assisted Extraction of Volatile Compounds from Citrus Flowers and Citrus Honey. *Food Chem.* **2003**, *82*, 575–582. [CrossRef]
62. Jeong, C.H.; Jeong, H.R.; Choi, G.N.; Kim, D.O.; Lee, U.; Heo, H.J. Neuroprotective and Anti-Oxidant Effects of Caffeic Acid Isolated from Erigeron Annuus Leaf. *Chin. Med.* **2011**, *6*, 1–9. [CrossRef]
63. Lim, E.-K.; Higgins, G.S.; Li, Y.; Bowles, D.J. Regioselectivity of Glucosylation of Caffeic Acid by a UDP-Glucose: Glucosyltransferase Is Maintained in Planta. *Biochem. J.* **2003**, *373*, 987–992. [CrossRef]
64. Sambandam, B.; Thiyagarajan, D.; Ayyaswamy, A.; Raman, P. Extraction and Isolation of Flavonoid Quercetin from the Leaves of Trigonella Foenum-Graecum and Their Anti-Oxidant Activity. *Int. J. Pharm. Pharm. Sci.* **2016**, *8*, 120–124.
65. Selvaraj, K.; Chowdhury, R.; Bhattacharjee, C. Isolation and Structural Elucidation of Flavonoids from Aquatic Fern Azolla Microphylla and Evaluation of Free Radical Scavenging Activity. *Int. J Pharm. Pharm. Sci.* **2013**, *5*, 743–749.
66. Al-Ashaal, H.A.; El-Sheltawy, S.T. Antioxidant Capacity of Hesperidin from Citrus Peel Using Electron Spin Resonance and Cytotoxic Activity against Human Carcinoma Cell Lines. *Pharm. Biol.* **2011**, *49*, 276–282. [CrossRef] [PubMed]
67. Choi, J.S.; Woo, W.S.; Young, H.S.; Park, J.H. Phytochemical Study on Prunus Davidiana. *Arch. Pharmacal Res.* **1990**, *13*, 374–378. [CrossRef]
68. Gilbert, N.C.; Gerstmeier, J.; Schexnaydre, E.E.; Börner, F.; Garscha, U.; Neau, D.B.; Werz, O.; Newcomer, M.E. Structural and Mechanistic Insights into 5-Lipoxygenase Inhibition by Natural Products. *Nat. Chem. Biol.* **2020**, *16*, 783–790. [CrossRef]
69. Kim, S.; Oshima, H.; Zhang, H.; Kern, N.R.; Re, S.; Lee, J.; Roux, B.; Sugita, Y.; Jiang, W.; Im, W. CHARMM-GUI Free Energy Calculator for Absolute and Relative Ligand Solvation and Binding Free Energy Simulations. *J. Chem. Theory Comput.* **2020**, *16*, 7207–7218. [CrossRef]
70. Russo, A.; Acquaviva, R.; Campisi, A.; Sorrenti, V.; Di Giacomo, C.; Virgata, G.; Barcellona, M.L.; Vanella, A. Bioflavonoids as Antiradicals, Antioxidants and DNA Cleavage Protectors. *Cell Biol. Toxicol.* **2000**, *16*, 91–98. [CrossRef]
71. Pokorny, J.; Yanishlieva, N.; Gordon, M.H. *Antioxidants in Food: Practical Applications*; CRC Press: Cambridge, UK; Woodhead Publishing Ltd.: Boca Raton, FL, USA, 2001; ISBN 0849312221.
72. Rajalakshmi, D.; Narasimhan, S. *Food Antioxidants: Technological, Toxicological, Health Perspective*; Marcel Dekker, Inc.: New York, NY, USA, 1996; 512p.
73. Buratti, S.; Benedetti, S.; Cosio, M.S. Evaluation of the Antioxidant Power of Honey, Propolis and Royal Jelly by Amperometric Flow Injection Analysis. *Talanta* **2007**, *71*, 1387–1392. [CrossRef]
74. Blasa, M.; Candiracci, M.; Accorsi, A.; Piacentini, M.P.; Albertini, M.C.; Piatti, E. Raw Millefiori Honey Is Packed Full of Antioxidants. *Food Chem.* **2006**, *97*, 217–222. [CrossRef]
75. Al, M.L.; Daniel, D.; Moise, A.; Bobis, O.; Laslo, L.; Bogdanov, S. Physico-Chemical and Bioactive Properties of Different Floral Origin Honeys from Romania. *Food Chem.* **2009**, *112*, 863–867. [CrossRef]
76. Katalinic, V.; Milos, M.; Kulisic, T.; Jukic, M. Screening of 70 Medicinal Plant Extracts for Antioxidant Capacity and Total Phenols. *Food Chem.* **2006**, *94*, 550–557. [CrossRef]
77. Zalibera, M.; Staško, A.; Šlebodová, A.; Jančovičová, V.; Čermáková, T.; Brezová, V. Antioxidant and Radical-Scavenging Activities of Slovak Honeys–An Electron Paramagnetic Resonance Study. *Food Chem.* **2008**, *110*, 512–521. [CrossRef] [PubMed]
78. Viuda-Martos, M.; Ruiz-Navajas, Y.; Fernández-López, J.; Pérez-Álvarez, J.A. Functional Properties of Honey, Propolis, and Royal Jelly. *J. Food Sci.* **2008**, *73*, R117–R124. [CrossRef] [PubMed]
79. Osher, E.; Weisinger, G.; Limor, R.; Tordjman, K.; Stern, N. The 5 Lipoxygenase System in the Vasculature: Emerging Role in Health and Disease. *Mol. Cell. Endocrinol.* **2006**, *252*, 201–206. [CrossRef]
80. Joshi, Y.B.; Praticò, D. The 5-Lipoxygenase Pathway: Oxidative and Inflammatory Contributions to the Alzheimer's Disease Phenotype. *Front. Cell. Neurosci.* **2015**, *8*, 436. [CrossRef]
81. Li, C.; Zhang, W.; Fang, S.; Lu, Y.; Zhang, L.; Qi, L.; Huang, X.; Huang, X.; Wei, E. Baicalin Attenuates Oxygen-Glucose Deprivation-Induced Injury by Inhibiting Oxidative Stress-Mediated 5-Lipoxygenase Activation in PC12 Cells. *Acta Pharmacol. Sin.* **2010**, *31*, 137–144. [CrossRef]
82. Liu, Y.; Wang, W.; Li, Y.; Xiao, Y.; Cheng, J.; Jia, J. The 5-Lipoxygenase Inhibitor Zileuton Confers Neuroprotection against Glutamate Oxidative Damage by Inhibiting Ferroptosis. *Biol. Pharm. Bull.* **2015**, *38*, 1234–1239. [CrossRef]
83. Maccarrone, M.; Meloni, C.; Manca-di-Villahermosa, S.; Cococcetta, N.; Casciani, C.U.; Finazzi-Agrò, A.; Taccone-Gallucci, M. Vitamin E Suppresses 5-Lipoxygenase-Mediated Oxidative Stress in Peripheral Blood Mononuclear Cells of Hemodialysis Patients Regardless of Administration Route. *Am. J. Kidney Dis.* **2001**, *37*, 964–969. [CrossRef]
84. Mathew, S.; Abraham, T.E. Studies on the Antioxidant Activities of Cinnamon (*Cinnamomum verum*) Bark Extracts, through Various in Vitro Models. *Food Chem.* **2006**, *94*, 520–528. [CrossRef]
85. Sancho, M.T.; Pascual-Maté, A.; Rodríguez-Morales, E.G.; Osés, S.M.; Escriche, I.; Periche, Á.; Fernández-Muiño, M.A. Critical Assessment of Antioxidant-related Parameters of Honey. *Int. J. Food Sci. Technol.* **2016**, *51*, 30–36. [CrossRef]
86. Wu, H.-C.; Shiau, C.-Y.; Chen, H.-M.; Chiou, T.-K. Antioxidant Activities of Carnosine, Anserine, Some Free Amino Acids and Their Combination. *J. Food Drug Anal.* **2003**, *11*, 148–153. [CrossRef]
87. Seeliger, D.; de Groot, B.L. Ligand docking and binding site analysis with PyMOL and Autodock/Vina. *J. Comput. Aided Mol. Des.* **2010**, *24*, 417–422. [CrossRef]
88. Bowers, K.J.; Chow, D.E.; Xu, H.; Dror, R.O.; Eastwood, M.P.; Gregersen, B.A.; Klepeis, J.L.; Kolossvary, I.; Moraes, M.A.; Sacerdoti, F.D.; et al. Scalable algorithms for molecular dynamics simulations on commodity clusters. In Proceedings of the SC'06:

Proceedings of the 2006 ACM/IEEE Conference on Supercomputing, Tampa, FL, USA, 11–17 November 2006; IEEE: New York, NY, USA, 2006; p. 43.
89. Release, S. 3: *Desmond Molecular Dynamics System, DE Shaw Research, New York, NY, 2017*; Maestro-Desmond Interoperability Tools, Schrödinger: New York, NY, USA, 2017.
90. Phillips, J.C.; Braun, R.; Wang, W.; Gumbart, J.; Tajkhorshid, E.; Villa, E.; Chipot, C.; Skeel, R.D.; Kalé, L.; Schulten, K. Scalable molecular dynamics with NAMD. *J. Comput. Chem.* **2005**, *26*, 1781–1802. [CrossRef]
91. Ngo, S.T.; Tam, N.M.; Quan, P.M.; Nguyen, T.H. Benchmark of Popular Free Energy Approaches Revealing the Inhibitors Binding to SARS-CoV-2 Mpro. *J. Chem. Inf. Model.* **2021**, *61*, 2302–2312. [CrossRef]

antioxidants

MDPI

Article

Content of Phenolic Acids as a Marker of Polish Honey Varieties and Relationship with Selected Honey-Quality-Influencing Variables

Anna Puścion-Jakubik *, Elżbieta Karpińska, Justyna Moskwa and Katarzyna Socha

Department of Bromatology, Faculty of Pharmacy with the Division of Laboratory Medicine, Medical University of Białystok, Mickiewicza 2D Street, 15-222 Białystok, Poland; elzbieta.karpinska@umb.edu.pl (E.K.); justyna.moskwa@umb.edu.pl (J.M.); katarzyna.socha@umb.edu.pl (K.S.)

* Correspondence: anna.puscion-jakubik@umb.edu.pl; Tel.: +48-8574-854-69

Abstract: Phenolic acids are an important component of honey. Literature data indicate their pro-health properties and diversified content in different varieties. Therefore, the aim of our study was to evaluate the content of phenolic acids in bee honey. The material for the research was 49 samples of honey obtained from beekeepers from Poland. Selected phenolic acids were determined by HPLC with PDA detection. Additionally, total phenolic content (TPC), color intensity, color on the Pfund scale, water content, electrical conductivity, and FRAP were assessed. A higher trans-ferulic acid content is accompanied by a stronger free radical scavenging ability. It was shown that buckwheat honeys are characterized by a high TPC value (196.59 mg GAE/100 g), color intensity (2109.2 mAU), color on the Pfund scale (159.8 mm Pfund), and high activity in the FRAP assay (0.403 equivalent of μmol Fe^{2+}/mL). The median obtained in the DPPH test for this honey variety was 41.1%. Moreover, the highest median of 4-hydroxybenzoic acid (3.129 mg/100 g) in buckwheat honey was shown. Buckwheat honeys have promising antioxidant properties and should be included in diets low in antioxidants.

Keywords: honeybee; buckwheat honey; Poland; markers; phenolic acids

check for updates

Citation: Puścion-Jakubik, A.; Karpińska, E.; Moskwa, J.; Socha, K. Content of Phenolic Acids as a Marker of Polish Honey Varieties and Relationship with Selected Honey-Quality-Influencing Variables. *Antioxidants* **2022**, *11*, 1312. https://doi.org/10.3390/antiox11071312

Academic Editors: Josipa Vlainić, Ivana Tlak Gajger and Alessandra Napolitano

Received: 19 April 2022
Accepted: 28 June 2022
Published: 30 June 2022

1. Introduction

Bee honey is a product of very diverse composition. It includes, among others, sugar compounds, water, proteins, organic acids, vitamins, minerals, phenolic compounds, enzymes, and many other ingredients [1]. Honey available for sale should be properly labeled, including, inter alia, the name of the variety. Beekeepers define a variety based on the color, consistency, smell, taste or on the basis of observation of the plants from which the bees collect nectar or honeydew.

Earlier publications indicate that a large percentage of honey is incorrectly labeled [2]. The classic method for determining the type of honey is the melissopalinological method, which consists of counting pollen grains under a microscope and classifying them into botanical species. This is a time-consuming method that requires detailed observation of the grains. Sometimes, it is emphasized that its results are ambiguous and difficult to interpret. Therefore, other methods of identifying honey varieties are being sought. For example, an electronic potentiometric tongue has been developed to help identify the honey variety [3]. The nuclear magnetic resonance (NMR) method was used to distinguish between nectar and honeydew honey [4]. Another method that can be used to identify honey varieties is the method of fluorescence spectroscopy. It was used to distinguish, among others, acacia, linden, and sunflower honey [5].

Other methods of honey classification are based on searching for characteristic markers or identifying fingerprints. For example, high-performance liquid chromatography with diode array detection and tandem mass spectrometry (HPLC-DAD-MS/MS) was used to distinguish between chaste honey and rape honey. The following markers were considered: ferulic acid, kaempferol, and morin. Additionally, chromatographic fingerprints at 270 nm and 360 nm were identified. The above methods were used in conjunction with chemometric techniques [6]. An attempt to identify the honey variety on the basis of its antioxidant properties was also undertaken by Dżugan et al. (2018). Buckwheat honey had the strongest antioxidant properties, and rape honey had the weakest [7].

In addition, the health-promoting properties of bee honey may be conditioned by the presence of compounds with antioxidant properties, including phenolic acids. The literature describes many beneficial properties of bee honey, including its use in the treatment of burns and ulcers [8], rosacea [9], acute cough [10], and bedsores [11]. HPLC is one of the most popular methods used to determine the content of individual compounds with antioxidant properties [12].

Phenolic substances, which are phenol derivatives, are synthesized by plants. They are divided into simple phenols and polyphenols. Polyphenols contain more than one hydroxyl in their molecule structure. Polyphenols can exist in free form or in combination with other substances, such as glycosides (made of aglycone and sugar residue). Phenolic acids include compounds derived from cinnamic and benzoic acids, including caffeic acid, ferulic acid, *p*-hydroxybenzoic acid, *p*-coumaric acid, protocatechuic acid, syringic acid, and vanillic acid. They can interact with biologically active molecules and protect them against damage [13]. In addition to phenolic acids, the antioxidant properties of honey are due to, among others, flavonoids, vitamins (such as vitamins C and E), and minerals (including zinc and manganese) [1]. For example, the literature data indicate that the most common flavonoids in acacia honeys are: apigenin, chrysin, galangin, genistein, kaempferol, luteolin, myricetin, pinobanksin, pinocembrin, and quercetin. Those characteristic of manuka honey are: chrysin, galangin, isorhamnetin, kaempherol, luteolin, pinobanksin, pinocembrin, and quercetin [14].

Therefore, the aim of the research was to assess whether selected phenolic acids can be a marker of individual varieties of honey from Poland, as well as to correlate the content of these acids with selected parameters determining the quality of the honey, such as color scale, color intensity, total phenolic content, water content, electrical conductivity, and % free radical scavenging in DPPH assay.

2. Materials and Methods

2.1. Materials

The research material consisted of 49 samples of natural bee honeys: buckwheat ($n = 15$), linden ($n = 9$), multi-flower light ($n = 3$), dandelion ($n = 4$), nectar–honeydew ($n = 4$), rape ($n = 8$), honeydew ($n = 3$), and heather ($n = 3$). Honey was purchased in Poland; each sample was made by a different beekeeper. Until analysis, the honeys were stored at 4 °C.

All solvents were HPLC grade, and all chemicals were analytical and reagent grade. Formic acid (min. 98%) was obtained from Merck (Darmstadt, Germany). Methanol was purchased from J.T. Baker (Avantor, Gliwice, Poland).

Ultrapure water was obtained from Simplicity™185 Water Purification System (Merck Millipore, Darmstadt, Germany).

HPLC standards of polyphenols such as: 3,4-dihydroxybenzoic acid (3,4-DHBA), 4-hydroxybenzoic acid (4-HBA), caffeic acid (CA), *p*-coumaric acid (p-CA), syringic acid (SA), trans-ferulic acid (*t*-FA), vanillic acid (VA), and reagents for determining the total content of phenolic compounds (Folin–Ciocalteu reagent, Na_2CO_3) were obtained from Sigma–Aldrich (St. Louis, MO, USA). Individual stock solutions of each analyte and a mixture of them were prepared in methanol.

2.2. *Methods*

2.2.1. Identification of the Varieties of Honey

The classification of variety was made on the basis of melissopalinological analysis, in accordance with the Regulation of the Minister of Agriculture and Rural Development [15]. From each honey, 10 g was weighed in a centrifuge tube, supplemented with 50 °C water to 20 mL, mixed, and centrifuged for 10 min at 3000 rpm. The precipitate was decanted, water was added again, and centrifuged. When the sediment was about 0.1 cm, a layer of water of about 0.5 ml was left above it, and when it was about 0.3 cm - a layer of about 1 ml of water was left, a homogeneous suspension was obtained and applied to a microscope slide. At least two microscopic preparations were made of each honey, in which pollen grains were classified to botanical varieties. On the basis of the grains present in the greatest predominance in given bee honey, each was given a variety name.

2.2.2. Determination of Water Content

Honey in an amount of 5 g was weighed into a test tube, closed with a stopper, and placed in a water bath from 45 °C until brought to a liquid state. Then, a few drops of honey were placed in the refractometer, and the refractive index was read. In the case of temperatures above 20 °C, the factor was increased by 0.00023/1 °C, and in the case of temperatures below 20 °C, it was reduced in a similar manner. Then, the water content was read from the table in the Regulation. For each honey, at least 2 determinations were made. The results are expressed in % [15].

2.2.3. Determination of Electrical Conductivity

Based on the water content of each honey, the amount to be weighed was calculated according to the following formula:

$$M = \frac{20\ \text{g} \times 100}{MS},$$

where:

M—the mass of honey to be weighed (g),

MS—dry matter content, calculated as the difference between 100% and the water content, expressed as %.

The honey was weighed out and made up to 100 mL with distilled water. The conductivity cell was rinsed with the sample, and a honey solution (40 mL) was placed in a water bath at a temperature of 20 °C; when the temperature of the solution was 20 °C, the electrical conductivity was measured. The electrical conductivity of honey was calculated according to the formula:

$$S = K \times G, \text{ where :}$$

SH—specific conductivity of honey ($\text{mS} \times \text{cm}^{-1}$),

K—constant of the conductivity cell (cm^{-1}),

G—conductivity (mS).

2.2.4. Determination of Color Intensity

In order to determine the color intensity, 5 ± 0.001 g of honey was weighed, and water at 45 °C temperature was added at a volume of 10 g and mixed thoroughly. The solution was then sonicated and filtered through a 0.45 µm filter. The absorbance of the solutions was measured at 450 and 720 nm. The final result was the difference in absorbance at the two wavelengths, expressed in mAU. For each sample, three determinations were performed, and the final result was the mean result [16].

2.2.5. Determination of Color on the Pfund Scale

To determine the color of natural bee honey using the Pfund scale, 5 ± 0.001 g of sample was weighed, the samples were each dissolved into 10 mL of distilled water, and they were mixed well. The samples were then placed in a water bath at 50 °C to dissolve the sugar crystals. After obtaining a clear solution, absorbance was measured at 635 nm against distilled water. The Pfund color scale was calculated using the formula:

$$\text{mm Pfund} = -38.70 + 371.39 \times \text{ Absorbance}.$$

The final result is the average of three measurements [17].

2.2.6. Determination of Total Phenolic Content (TPC)

The total content of phenolic compounds was determined by reaction with the Folin–Ciocalteu reagent [18]. A calibration curve was prepared using a gallic acid working solution with a concentration of 2 g/L. A 1 ± 0.001 g sample was taken from each honey. Honey was dissolved in distilled water to a volume of 10 mL and then centrifuged at 2000 rpm for 5 min. Next, 0.25 mL of supernatant was collected; then, 1.25 mL of 0.2 N Folin–Ciocalteu was added, and the sample was stirred for 5 min. Next, 1 mL of Na_2CO_3 solution was added, mixed, and incubated in the dark at room temperature for 2 h. The contents of each tube were then mixed, and the absorbance at 760 nm against water was measured using a Hitachi U-2001 spectrophotometer. The results are presented as the mean of 3 determinations, in mg gallic acid/100 g honey.

2.2.7. Determination of Radical Scavenging Activity by DPPH Assay

The ability of bee honeys to scavenge radicals was performed on the basis of the method described by Sánchez-Moreno et al. [19]. Bee honeys were dissolved in distilled water to obtain a concentration of 1 g/mL. In total, 200 µL was taken, and 1800 µL of a DPPH solution with a concentration of 0.04 mg/mL was added. The absorbance at 517 nm was measured with a spectrophotometer U-2001 (Hitachi, Tokyo, Japan). The samples were then incubated at room temperature, protected from light, for 30 min. After the incubation period, the absorbance was measured again. The % of free radical scavenging was calculated:

$$\text{DPPH } [\%] = \left[\frac{A0 - A30}{A30}\right] \times 100\%,$$

where $A0$ is the absorbance at time 0, and $A30$ is the absorbance over 30 min.

2.2.8. Determination of FRAP

To perform the FRAP test, the FRAP reagent was prepared (2.5 mL of a 10 mM TPTZ solution in 40 mM HCl, 2.5 mL of 20 mM FeC13, and 25 mL of 0.3 M acetate buffer pH 3.6) [20].

To 20 µL of honey solution, 180 µL of FRAP reagent was added, and the mixture was incubated at 37 °C for 10 min. The absorbance of the mixture was then measured at 593 nm with a plate reader (UVM 340, Biogenet, Józefów, Poland). The results are presented as the equivalent of µmol Fe^{2+}/mL of the sample [20].

2.2.9. Preparation of Samples for HPLC Analysis–Isolation of Phenolic Compounds

Honey samples (5 g) were dissolved in 50 mL of water (adjusted to pH 2 with HCl) until completely fluid. This solution (50 mL) was then filtered through a Sep-Pak C18 cartridge (tube type SPE, Supelclean LC-18 SPE Tubes 3 mL/500 mg, Supelco Analytical, Bellefonte, PA, USA), which was previously activated with methanol (10 mL) followed by water (10 mL). The phenolic compounds were retained on the column, whilst all sugars and other polar compounds were eluted with water, and then polyphenols were eluted with 2.5 mL of a methanol–water mixture (70%, v/v) in order to validate the efficiency of extraction SPE and similar activities dealing with standards.

Phenolic fractions in methanol evaporated under reduced pressure (22 °C). The residue was redissolved in a mixture of distilled water and HPLC-grade methanol, in proportions such as phase (22.5 MeOH parts: 76.5 H_2O parts: 1 CH_3COOH parts). The prepared sample was analyzed by HPLC with photodiode array (PDA) detection. The applied extraction method enabled recovery values for analyzed compounds of higher than 85%.

2.2.10. HPLC Analysis

HPLC analyses of honey extracts were performed using an Flexar HPLC system (Perkin Elmer, Waltham, MA, USA) with a photodiode array detector (PDA) and using Chromera LC-PDA software (Perkin Elmer, Waltham, MA, USA). Separations were carried out with reversed-phase column Synergi 4 μm C-18 (Merck, Darmstadt, Germany; 250 × 4.60 mm, particle size 4 micron, 80A), SecurityGuard Cartridges Fusion-RP 4 × 3.0 mm ID. A mobile phase of 22.5 MeOH:76.5 H_2O:1 CH_3COOH was used; a constant solvent flow rate (1 mL/min) was applied. The total analysis time was 50 min. An isocratic separation method was used using the mobile phase (22.5 MeOH:76.5 H_2O:1 CH_3COOH). The temperature of the column oven was set at 25 °C. The phenolic acids were detected at 254, 265, and 326 nm, since the most honey phenolic compounds show their UV absorption maxima around these three wavelengths. The comparison of UV spectra and retention times with standard compounds enabled the identification of phenolic acids presented in the analyzed honey extracts. These compounds were quantified against their external standards. The injection volume was 20 μL.

Each sample was analyzed three times, and the method was proved by repeatability test by determining peak area and retention reproducibility for different classes of compounds.

Table 1 presents data on the optimization of the method, including LOQ (limit of quantitation) and LOD (limit of detection).

Table 1. Characteristics of the developed method.

Compounds	RT	LOD (mg/100 g)	LOQ (mg/100 g)
3,4–DHBA	9.183	0.099	0.300
4–HBA	16.570	0.092	0.278
VA	20.284	0.089	0.271
CA	21.756	0.106	0.322
SA	23.886	0.147	0.445
p–CA	40.572	0.138	0.418
t–FA	50.040	0.084	0.255

3,4–DHBA—3,4-dihydroxybenzoic acid, 4–HBA—4-hydroxybenzoic acid, CA—caffeic acid, LOD—limit of detection, LOQ—limit of quantitation, *p*–CA—*p*-coumaric acid, RT—retention time, SA—syringic acid, *t*–PA—*trans*-ferulic acid, VA—vanillic acid.

The concentrations of 4-HBA, VA, and *t*-FA were read at 254 nm and 3,4-DHBA at 265 nm. However, the 326 nm wavelength was the best to read for CA, *p*-CA, and SA. During the optimization of the chromatographic conditions, the necessary quality parameters of the method were taken into account, including retention factors, relative retention factors, and resolution. The resolution of the compounds was 1.5 and above, with the exception of 3,4-DHBA, where the average resolution was 1.0–1.3.

2.2.11. Statistical Analysis

Statistical analyses were performed using Statistica v.13.3 software. Values of $p < 0.05$ were considered significantly different. The correlation between all the measured parameters was evaluated using Spearman's correlation coefficient.

In order to compare the values for several independent groups the Kuskal–Wallis ANOVA tests were performed.

Chemometric analyzes were also performed, including cluster analysis (CA) and principal component analysis (PCA). In the CA, agglomeration was chosen as the method of grouping. The agglomeration method is single bond, and the distance measure is Euclidean distance.

3. Results

3.1. Varieties of Bee Honey

The first analytical step was to assess whether the marking of honey by beekeepers was correct in order to correctly identify the compounds present in the tested honey at a later stage. We have shown that three of the honeys labeled as 'buckwheat' were of a different type. None of the dandelion honeys were of this variety. Among linden honeys, an incorrect declaration of variety was found in over 56% of the honey samples. Among nectar–honeydew honeys, one out of four samples should be marked differently (Table 2).

Table 2. The percentage of honey with the correct and incorrect definitions of the variety.

Variety–Declarations of Beekeepers	The Number of Attempts Correctly Classified	The Number of Attempts Is Classified Incorrectly
buckwheat (n = 15)	12	3
dandelion (n = 4)	0	4
heather (n = 3)	3	0
honeydew (n = 3)	3	0
linden (n = 9)	4	5
multi-flower light (n = 3)	3	0
nectar–honeydew (n = 4)	3	1
rape (n = 8)	8	0

Figure 1 shows pictures of pollen grains characteristic of buckwheat honey (Figure 1a), for heather honey (Figure 1b), for linden honey (Figure 1c), and for rapeseed honey (Figure 1d).

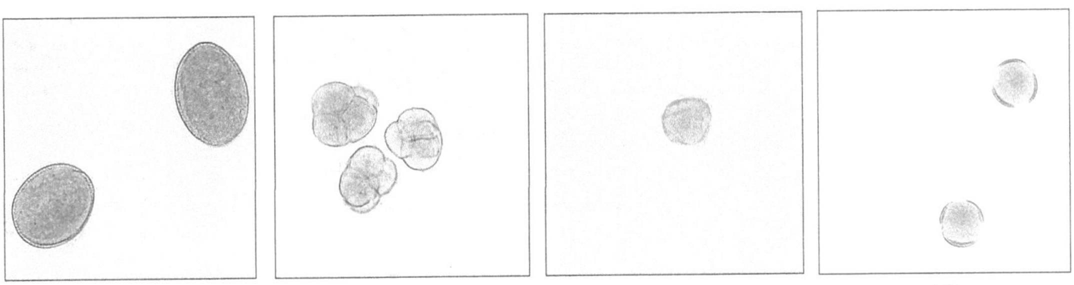

(a) (b) (c) (d)

Figure 1. Pollen grains from honey plants: (**a**) *Fagopyrum esculentum* Moench, (**b**) *Calluna vulgaris* (L.) Hull, (**c**) *Tilia* L., (**d**) *Brassica napus* L. *var. napus*.

3.2. Selected Quality Parameters

Selected quality parameters examined as part of the quality assessment and the search for markers of honey from Poland are determination of the color of honey on the Pfund scale, determination of the color intensity, total phenolic compounds (TPC), water content, and electrical conductivity (Table 3).

We showed that buckwheat honeys were characterized by the highest color value on the Pfund scale (median: 159.8 mm Pfund)—this value was significantly higher compared to the color of linden (44.9 mm Pfund), multifloral light (37.4 mm Pfund), and rape honey (84.8 mm Pfund). A similar tendency was observed for the determination of color intensity: buckwheat honey (2109.2 mAU) had the highest median. Honey of this variety was also characterized by the highest TPC value (196.59 mg GAE/100 g), as well as the highest activity in the FRAP test (0.403 equivalent of μmol Fe^{2+}/mL). Interestingly, honeys of this variety have the ability to scavenge free radicals by about 41.1%. Honeydew honeys, on the other hand, showed the highest specific electrical conductivity (1.181 mS × cm^{-1}), significantly higher than that of rape honey (0.242 mS × cm^{-1}).

Table 3. The value of the parameters for individual varieties of honey.

Variety (Sign)		Colour Scale (mm Pfund)	Colour Intensity (mAU)	TPC (mg GAE/100 g)	Water Content (%)	Electrical Conductivity (mS × cm^{-1})	DPPH (% Free Radical Scavenging)	FRAP (Equivalent of µmol Fe^{2+}/mL)
Buckwheat (B)	Av. ± SD	166.4 ± 29.4	1816.0 ± 688.0	182.60 ± 61.08	18.9 ± 0.5	0.400 ± 0.043	42.0 ± 4.5	0.402 ± 0.010
	Min-Max	125.8–218.5	711.0–2634.7	44.95–241.87	18.1–19.9	0.326–0.507	34.9–52.7	0.379–0.417
	Med	159.8	2109.2	196.59	18.9	0.391	41.1	0.403
	Q1-Q3	147.9–189.0	1229.0–2291.8	142.29–236.63	18.4–19.3	0.380–0.416	39.8–44.0	0.398–0.409
Heather (He)	Av. ± SD	125.2 ± 14.8	575.8 ± 179.5	91.78 ± 4.25	19.2 ± 0.7	0.552 ± 0.027	46.4 ± 3.7	0.141 ± 0.002
	Min-Max	111.1–140.7	468.0–783.0	87.72–96.20	18.6–19.9	0.534–0.583	42.3–49.5	0.139–0.143
	Med	124.0	476.3	91.42	19.0	0.538	47.5	0.140
	Q1-Q3	111.1–140.7	468.0–783.0	87.72–96-20	18.6–19.9	0.5334–0.583	42.3–49	0.139–0.143
Honeydew (Ho)	Av. ± SD	109.9 ± 95.9	587.1 ± 327.0	86.0 ± 55.3	16.3 ± 0.6	1.728 ± 1.072	58.6 ± 4.0	0.323 ± 0.017
	Min-Max	49.8–220.5	215.3–830.0	42.8–148.3	15.7–16.8	1.041–2.963	55.9–63.2	0.312–0.343
	Med	59.5	716.0	67.07	16.4	1.181	56.7 * B	0.315
	Q1-Q3	49.8–220.5	215.3–830.0	42.78–148.30	15.7–16.8	1.041–1.922	55.9–63.2	0.312–0.343
Linden (L)	Av. ± SD	43.5 ± 19.6	84.0 ± 44.0	29.23 ± 10.60	16.7 ± 0.7	0.502 ± 0.104	58.6 ± 1.4	0.083 ± 0.012
	Min-Max	20.0–64.2	49.0–148.3	18.24–43.69	15.7–17.1	0.396–0.597	56.6–59.7	0.071–0.099
	Med	44.9 ** B	69.3 *** B	27.50 ** B	16.9	0.508	59.0 ** B	0.081
	Q1-Q3	27.8–59.2	57.5–110.5	22.44–36.03	16.2–17.1	0.413–0.592	57.5–59.7	0.075–0.091
Multifloral dark (Md)	Av. ± SD	124.4 ± 25.6	1424.7 ± 803.1	187.6 ± 194.3	19.2 ± 0.8	0.416 ± 0.026	56.7 ± 2.8	0.218 ± 0.015
	Min-Max	91.9–154.1	280.0–2160.0	55.60–467.83	18.1–20.0	0.380–0.437	53.2–59.3	0.198–0.232
	Med	125.8	1629.3	113.50	19.3	0.423	57.2 * B	0.221
	Q1-Q3	107.3–141.5	953.7–1895.7	56.37–318.85	18.7–19.7	0.397–0.435	54.4–59.0	0.206–0.230
Multifloral light (Ml)	Av. ± SD	33.3 ± 24.3	155.6 ± 90.9	32.04 ± 3.80	18.6 ± 0.7	0.431 ± 0.109	45.3 ± 6.5	0.052 ± 0.032
	Min-Max	1.0–64.7	64.0–272.0	28.86–38.26	18.0–19.3	0.308–0.584	37.4–52.8	0.014–0.090
	Med	37.4 *** B	128.0 * B	30.85 ** B	18.1	0.452	46.8	0.062 ** B
	Q1-Q3	18.9–44.2	85.0–229.0	29.44–32.79	18.1–19.3	0.344–0.466	39.8–49.6	0.023–0.070

Table 3. *Cont.*

Variety (Sign)	Colour Scale (mm Pfund)	Colour Intensity (mAU)	TPC (mg GAE/100 g)	Water Content (%)	Electrical Conductivity (mS × cm^{-1})	DPPH (% Free Radical Scavenging)	FRAP (Equivalent of µmol Fe^{2+}/mL)
Nectar–honeydew (Nh)	115.2 ± 55.2	322.6 ± 231.7	57.08 ± 11.54	17.8 ± 1.7	0.641 ± 0.031	57.4 ± 4.2	0.205 ± 0.010
	75.4–192.6	93.7–623.0	47.20–70.74	16.5–20.3	0.609–0.670	52.4–61.8	0.197–0.219
	96.5	286.8	55.19	17.2	0.642	57.8 * B	0.203
	75.5–155.0	145.8–499.3	47.51–66.65	16.8–18.8	0.614–0.667	54.1–60.8	0.198–0.213
Rape (R)	81.47 ± 31.88	127.7 ± 48.69	33.18 ± 6.28	18.7 ± 0.8	0.284 ± 0.092	47.9 ± 5.7	0.030 ± 0.012
	17.5–125.6	62.0–231.0	20.40–43.94	17.7–20.6	0.169–0.449	37.8–58.7	0.012–0.058
	84.8 ** B	126.0 *** B	35.10 ** B	18.6	0.242 ** Ho, ** Nh	48.0	0.030 *** B
	66.4–98.8	86.0–150.3	30.17–36.74	18.1–19.1	0.215–0.352	45.6–52.3	0.022–0.035

* $p < 0.05$, ** $p < 0.01$, *** $p < 0.001$.

3.3. *Profile of Phenolic Acids and the Variety of Honey*

HPLC analysis showed the presence of seven phenolic compounds in honey from Poland: CA, *p*-CA, 3,4-DHEA, *t*-FA, SA, VA, and 4-HBA.

Figure 2 shows the chromatograms for standard substances at three wavelengths: 254 nm, 265 nm, and 326 nm.

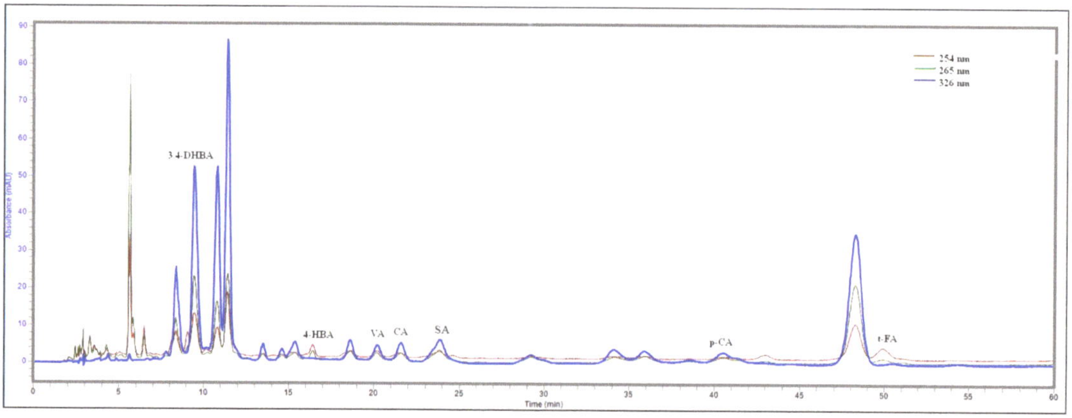

Figure 2. Chromatogram of the analyzed phenolic acid standards.

The calculated levels of individual identified phenolic compounds in analyzed honey are shown in Table 4. The ANOVA analysis of variance showed differences in the content of individual phenolic acids between the groups. Each of the varieties of honey is characterized by a high or low content of a specific phenolic compound.

It has been shown that the content of individual phenolic compounds for varieties of honey is characteristic. 3,4-DHBA was the highest median in linden (1.993 mg/100 g) and buckwheat (1.421 mg/100 g) honey. The next compound, 4-HBA, was characteristic for buckwheat (3.129 mg/100 g) and mulitfloral dark (1.934 mg/100 g) honey. The other determined phenolic acids such as CA, VA, SA, and *t*-FA were of highest value in linden honey (1.746, 0.304, 1.107, 1.954 mg/100 g, respectively). Moreover, p-CA was of a similar level to buckwheat (0.804 mg/100 g) and mulitfloral dark (0.789 mg/100 g) honey (Table 4).

In buckwheat honey, the highest median of 4-HBA was found—this value was significantly higher than that of the content in linden, multifloral light, nectar–honeydew, and rape. This indicates that the above compound can be considered a marker of the authenticity of buckwheat honey.

Another analyzed compound was 3,4-DHBA. Our study showed that linden honey had a significantly higher content of this phenolic acid than rape honey and CA compared to buckwheat honey.

Among the determined compounds, no characteristic concentrations were found for heather, honeydew, multifloral, nectar–honeydew, and rape honeys.

Table 4. The value of the phenolic acids for individual varieties of honey (mg/100 g).

Variety (Sign)		3,4-DHBA	4-HBA	CA	*p*-CA	VA	SA	*t*-FA
Buckwheat (B)	Av. ± SD Min-Max Med Q1-Q3	1.403 ± 0.419 0.784–2.233 1.421 1.101–1.558	3.203 ± 0.736 1.699–4.432 3.129 2.869–3.515	0.194 ± 0.073 <LOD-0.325 0.207 0.177–0.219	0.784 ± 0.129 0.558–1.004 0.804 0.678–0.870	0.151 ± 0.043 <LOD-0.193 0.165 <LOD-0.180	0.186 ± 0.127 <LOD-0.329 <LOD <LOD-0.198	0.095 ± 0.050 <LOD-0.175 <LOD <LOD-0.152
Heather (He)		0.539 ± 0.056 0.505–0.604 0.509 0.505–0.604	0.895 ± 0.172 0.736–1.078 0.873 0.736–1.078	0.215 ± 0.025 0.189–0.239 0.216 0.189–0.239	0.386 ± 0.059 0.340–0.452 0.367 0.340–0.452	0.162 ± 0.143 <LOD-0.273 0.211 <LOD-0.273	0.860 ± 0.159 0.705–1.023 0.852 0.705–1.023	0.106 ± 0.092 <LOD-0.166 0.152 <LOD-0.166
Honeydew (Ho)		7.646 ± 12.383 0.354–21.944 0.639 0.354–21.944	0.287 ± 0.090 0.184–0.348 0.329 0.184–0.348	0.252 ± 0.024 0.225–0.268 0.264 0.225–0.268	0.249 ± 0.089 0.171–0.346 0.230 0.171–0.346	0.368 ± 0.472 0.133–0.913 0.107 0.133–0.913	0.166 ± 0.183 <LOD-0.317 <LOD <LOD-0.317	0.475 ± 0.493 <LOD-0.985 0.442 <LOD-0.985
Linden (L)		2.064 ± 0.278 1.818–2.454 1.993 1.872–2.257	0.200 ± 0.051 0.152–0.271 0.188 * B 0.165–0.235	1.679 ± 0.338 1.227–1.998 1.746 ** B 1.427–1.931	0.212 ± 0.211 <LOD-0.403 0.221 * B 0.151–0.392	0.312 ± 0.080 0.240–0.402 0.304 * B 0.245–0.380	1.085 ± 0.276 0.726–1.399 1.107 * L 0.910–1.259	1.973 ± 2.142 <LOD-3.982 1.954 0.123–3.822
Multifloral dark (Md)		0.680 ± 0.258 0.443–1.045 0.615 0.517–0.842	1.572 ± 0.964 0.185–2.235 1.934 0.917–2.227	0.261 ± 0.075 0.190–0.334 0.260 0.197–0.325	0.631 ± 0.325 0.144–0.804 0.789 0.463–0.800	0.108 ± 0.047 <LOD-0.146 0.128 <LOD-0.136	<LOD <LOD <LOD <LOD	0.633 ± 1.164 <LOD-2.376 0.087 <LOD-1.267
Multifloral light (Ml)		0.646 ± 0.551 0.193–1.596 0.541 0.354–0.545	0.191 ± 0.061 0.108–0.251 0.179 ** B 0.167–0.251	0.332 ± 0.333 <LOD-0.885 0.225 0.203–0.346	0.357 ± 0.185 0.158–0.235 0.373 0.235–0.377	0.108 ± 0.070 <LOD-0.165 0.118 <LOD-0.145	0.178 ± 0.396 <LOD-0.885 <LOD <LOD	1.741 ± 2.479 0.094–5.654 0.155 <LOD -2.782

Table 4. *Cont.*

Variety (Sign)	3,4-DHBA	4-HBA	CA	*p*-CA	VA	SA	*t*-FA
Nectar–honeydew (Nh)	1.019 ± 0.849 0.233–2.222 0.810 0.488–1.550	0.235 ± 0.068 0.140–0.298 0.251 * B 0.193–0.277	0.512 ± 0.667 <LOD-1.493 0.278 0.134–0.891	0.299 ± 0.280 0.154–0.717 0.178 0.174–0.457	0.116 ± 0.033 <LOD-0.189 0.107 <LOD-0.152	0.410 ± 0.541 <LOD-1.141 0.249 <LOD-0.819	2.773 ± 4.607 0.090–9.656 0.673 * B 0.193–5.353
Rape (R)	0.455 ± 0.379 <LOD-1.482 0.350 ** L, ** B 0.265–0.385	0.256 ± 0.139 0.109–0.484 0.174 *** B 0.153–0.411	0.334 ± 0.245 0.177–0.870 0.227 0.199–0.237	0.262 ± 0.145 0.155–0.581 0.242 *** B 0.168–0.325	0.102 ± 0.068 <LOD-0.246 0.114 <LOD-0.127	0.267 ± 0.313 <LOD-0.847 0.165 <LOD-0.419	0.124 ± 0.060 <LOD-0.237 0.128 0.096–0.155

3,4-DHEA—3,4-dihydroxybenzoic acid, 4-HBA—4-hydroxybenzoic acid, <LOD—below the detection limit, CA—caffeic acid, *p*-CA—*p*-coumaric acid, SA—syringic acid, *t*-FA—*trans*-ferulic acid, VA—vanillic acid. * $p< 0.05$, ** $p < 0.01$, *** $p < 0.001$.

3.4. Correlations

The analysis of the correlation (Table 5) between the content of phenolic compounds in honey showed a strong relationship between the content of 4-HBA and *p*-CA (r = 0.82, $p < 0.000$), between VA and SA (r = 0.60, $p < 0.001$), and between SA and CA (r = 0.51, $p < 0.000$). Among the remaining parameters, the correlation between color intensity and TPC (r = 0.90, $p < 0.001$), and color in Pfund scale and color intensity (r = 0.82, $p < 0.001$) should be emphasized. Additionally, it is worth noting the correlation between color intensity and 4-HBA (r = 0.84, $p < 0.000$).

Table 5. Correlations between individual parameters (p <0.05).

Parameter 1	Parameter 2	r	*p*
Color in Pfund scale	Colour intensity	0.82	0.001
Color in Pfund scale	TPC	0.77	0.001
Color in Pfund scale	Diastase number	0.51	0.001
Color in Pfund scale	3,4-DHBA	0.75	0.001
Color in Pfund scale	SA	−0.33	0.021
Color in Pfund scale	*p*-CA	0.51	0.001
Color in Pfund scale	*t*-FA	−0.57	0.001
Color in Pfund scale	CA	−0.45	0.001
Colour intensity	TPC	0.90	0.001
Colour intensity	Diastase number	0.51	0.001
Colour intensity	Water	0.33	0.020
Colour intensity	4-HBA	0.84	0.001
Colour intensity	VA	−0.39	0.005
Colour intensity	SA	−0.45	0.001
Colour intensity	*p*-CA	0.68	0.001
Colour intensity	*t*-FA	−0.52	0.001
Colour intensity	CA	−0.46	0.001
DPPH	Water	−0.37	0.008
DPPH	*p*-CA	−0.35	0.01
DPPH	*t*-FA	0.45	0.001
TPC	Diastase number	0.58	0.001
TPC	3,4-DHBA	0.33	0.020
TPC	4-HBA	0.79	0.001
TPC	VA	−0.30	0.038
TPC	*p*-CA	0.60	0.001
TPC	*t*-FA	−0.57	0.001
TPC	CA	−0.31	0.001
Diastase number	4-HBA	0.56	0.001
Diastase number	*p*-CA	0.55	0.001
Water	Electrical conductivity	−0.37	0.009
Water	4-HBA	0.31	0.026
Water	VA	−0.37	0.009
Water	*p*-CA	0.32	0.023
Water	CA	−0.36	0.011
Electrical conductivity	3,4-DHBA	0.40	0.005
Electrical conductivity	VA	0.29	0.040
Electrical conductivity	CA	0.42	0.002
FRAP	Colour in Pfund scale	0.68	0.001
FRAP	Colour intensity	0.82	0.001
FRAP	TPC	0.82	0.001
FRAP	Diastase number	0.50	0.001
FRAP	Electrical conductivity	0.38	0.008
FRAP	3,4-DHBA	0.53	0.001
FRAP	4-HBA	0.73	0.001

Table 5. *Cont.*

Parameter 1	Parameter 2	r	*p*
FRAP	*p*-CA	0.58	0.001
FRAP	*t*-FA	−0.38	0.006
3,4-DHBA	SA	0.30	0.034
3,4-DHBA	CA	0.37	0.009
4-HBA	SA	−0.29	0.040
4-HBA	*p*-CA	0.82	0.001
4-HBA	*t*-FA	−0.46	0.001
4-HBA	CA	−0.36	0.011
VA	SA	0.60	0.000
VA	*p*-CA	−0.32	0.024
VA	CA	0.60	0.001
SA	*p*-CA	−0.31	0.028
SA	CA	0.51	0.001
p-CA	*t*-FA	−0.30	0.038
p-CA	CA	−0.29	0.040
t-FA	CA	0.47	0.001

3,4-DHEA—3,4-dihydroxybenzoic acid, 4-HBA—4-hydroxybenzoic acid, CA—caffeic acid, *p*-CA—*p*-coumaric acid, SA—syringic acid, *t*-FA—trans-ferulic acid, TPC—total phenolic content, VA—vanillic acid.

It should be emphasized that we noted a positive correlation between the % of free radical scavenging in DPPH assay and the t-FA content (r = 0.45, $p < 0.001$).

3.5. Chemometric Analyzes

Cluster analysis performed for variables showed groups that are similar. One group was p-CA and 4-HBA, while the other group was *t*-FA, CA, SA, VA, and 3,4-DHBA (Figure 3).

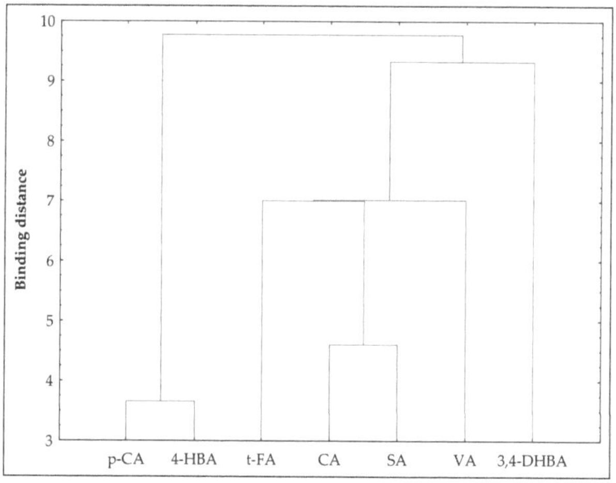

Figure 3. Cluster analysis for variables.

The analysis carried out for the cases, based on the contents of phenolic acids, mainly distinguished honeydew honey. The focus on linden honey is also worth emphasizing. Multiflower dark honeys have also qualified for the group containing buckwheat honey (Figure 4).

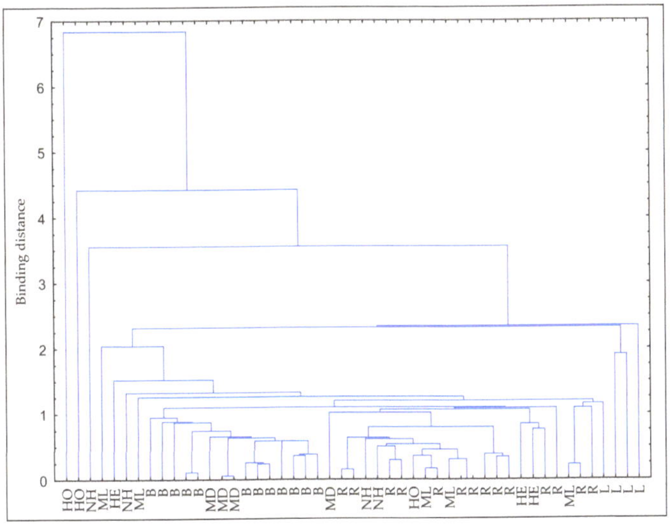

Figure 4. Cluster analysis for cases. B—buckwheat honey, HE—heather honey, HO—honeydew honey, L—linden honey, MD—multifloral dark honey, ML—multifloral light honey, NH—nectar–honeydew honey, R—rape honey.

Then, principal components analysis (PCA) was carried out, the purpose of which was to reduce the variables and classify the honey varieties. The first main component accounted for 43.84% of the variance; the second, 17.31%; the third, 14.45% (total 75.60%); and the subsequent components less than 10% of the variance.

On the basis of the eigenvectors, it can be assessed that factor 1 is related to the following components: *p*-CA (0.44), 4-HBA (0.41), SA (−0.46), and CA (−0.47). The second component is related to 4-HBA (0.54) and *p*-CA (0.50), and the third component 3,4-DHBA (−0.79). Figure 5 shows 2W plots of factor coordinates of the variables. Points are significant factor loadings for individual components. The farther a given load is from the center of the circle, the greater the correlation of the variable with the factor axis. Figure 6 present 2W plots of cases depending on the phenolic acids.

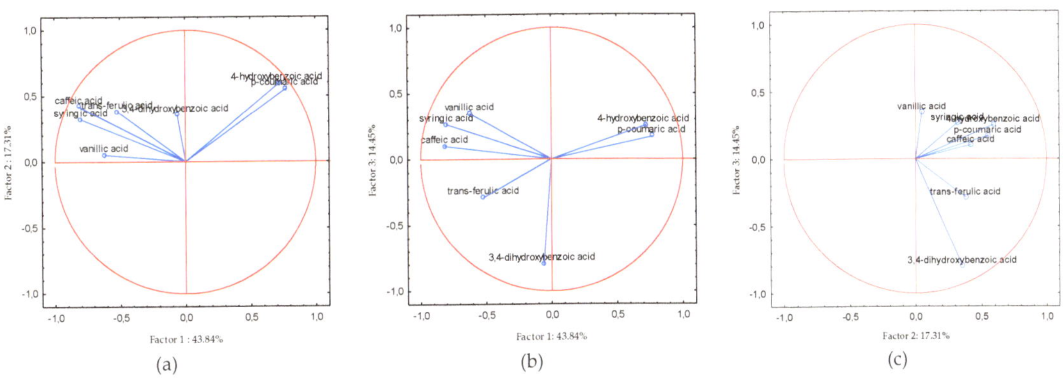

Figure 5. Projection of variables depending on the phenolic acids in a two – factor plane: factor 1 × factor 2 (**a**), factor 1 × factor 3 (**b**), factor 2 × factor 3 (**c**).

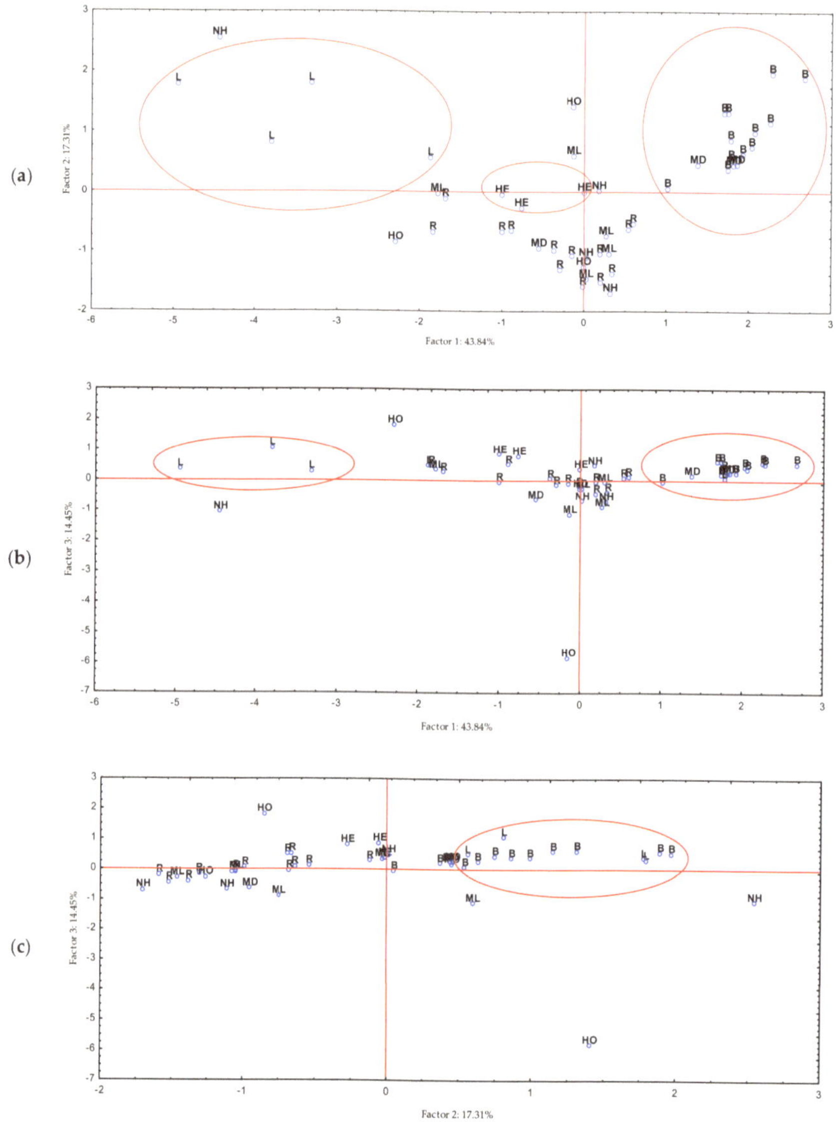

Figure 6. Projection of cases depending on the phenolic acids in a two – factor plane: factor 1 × factor 2 (**a**), factor 1 × factor 3 (**b**), factor 2 × factor 3 (**c**).

4. Discussion

Natural bee honeys are characterized by a very rich composition, which determines their health-promoting properties. We have made an attempt to search for compounds that are characteristic of honey obtained in Poland. For rape, multifloral, nectar–honeydew, and honeydew honey, no characteristic phenolic compounds were found that could be considered determinants of the authenticity of these varieties.

Taking into account other quality criteria, honeydew honeys are distinguished by having the highest median of electrical conductivity (1.181 mS × cm^{-1}).

Multifloral honeys are characterized by a complex composition, without the dominant presence of one plant, which may result in the lack of advantage of a specific phenolic acid.

In addition, for rape honey, it may be necessary to establish a method with other acids. Moreover, rape honeys show the lowest electrical conductivity—the median was 0.242 mS $\times$ cm^{-1}.

Buckwheat honeys from Poland show the darkest color, which results in the highest color value on the Pfund scale (median: 159.8 mm Pfund), the highest color intensity (2109.2 mAU), and the highest total phenolic content (196.59 mg GAE/100 g). Moreover, in these honeys, we showed the highest medians of 4-HBA (3.129 mg/100 g) and *p*-CA (0.804 mg/100 g). The studies conducted by Starowicz et al. (2021) [21] showed a lower value of TPC in honey of this variety (average: 141.1 mg GAE/100 g), while our research allowed us to conclude that the average value of this parameter is 182.60 $\pm$ 61.08 mg GAE/100 g.

Heather honeys were characterized by the highest median of one of the determined phenolic acids: SA (0.852 mg/100 g). Research by Ecem Bayram et al. (2020) [22] indicated that 3,4-DHBA is a characteristic compound for this variety of honey. SA was a relatively abundant compound (193.77 and 242.33 mg/L). TPC in this variety, in line with our results, was 91.78 $\pm$ 4.25 mg GAE/100 g, while the results published by Starowicz et al. (2021) [21] indicated a much higher content, at the level of 159.2 mg GAE/100 g.

Linden honey, despite the low content of phenolic compounds (27.50 mg GAE/100 g), was surprisingly characterized by high contents of 3,4-DHBA (1.993 mg/100 g), CA (1.746 mg/100 g), SA (1.107 mg/100 g), VA (0.304 mg/100 g), and *t*-FA (1.954 mg/100 g). Dimitrova et al. (2007) [23] determined the content of, inter alia, phenolic acids in 49 honey samples. In the case of linden honey (n = 4), the authors provided only the maximum value of CA—this was 1.57 mg/kg. Our analysis showed about a 10 times higher content of this ingredient, at the level of 1.679 $\pm$ 0.338 mg/100 g, and the maximum value was 1.998 mg/100 g. The research carried out on linden honey from Turkey showed a characteristically high CA content (642.94 mg/L) [22], which was consistent with our observations (1.679 $\pm$ 0.338 mg/100 g). The content of SA was indicated only as of the maximum value (0.29 mg/kg)—in our study, the average content was 1.085 $\pm$ 0.276 mg/kg. The average VA content in honey of this variety was indicated by the authors at the level of 1.19 mg/kg, while our research indicated a value almost two times higher (0.312 mg/100 g). These results are slightly divergent due to the fact that the apiaries from which the honey was obtained differed in geographical location—in the case of Dimitrova et al. (2007) [23], these were honeys from producers from Denmark, France, Germany, Italy, the Netherlands, Portugal, United Kingdom, and Spain, while in our study, all honeys were from Poland. Our analyses also show the existence of many dependencies between the measured phenolic acids, as well as other quality parameters. A high positive correlation between the contents of VA and SA and between SA and CA may indicate the common presence of individual phenolic acids in nectar, which is particularly visible in the case of linden honey.

The content of phenolic acids such as 3,4-DHBA, 4-HBA, VA, SA, *p*-CA, FA, CA in buckwheat and heather honey from Poland can be compared with the results obtained by Jasicka-Misiak et al. in 2012 [24], including heather (n = 15) and buckwheat honey (n = 7). In this study, similar contents of 3,4-HBA were obtained: our research showed the content of this compound at the level of 1.403 $\pm$ 0.419 mg/100 g, while Jasicka-Misiak et al. [24] showed a level of about 1.228 mg/100 g (average content based on the determination of seven samples). The average VA content found in our study was approximately 10 times lower than in the study published by Jasicka-Misiak et al. [24]. The content of CA and SA in heather and buckwheat honey is low, in some cases below the detection limit, which is confirmed by both our research and the above-mentioned team of authors. According to our determinations, the p-CA content in buckwheat honey ranged from 0.558 to 1.004 mg/100 g, while the results obtained by Jasicka-Misiak et al. [24] were more divergent and indicated contents of 0.026 to 4.551 mg/100 g, and their average was almost three times higher.

Buckwheat honey show the highest value of the TPC parameter. Numerous scientific publications indicate their rich composition; they are characterized, among others, by the presence of many volatile compounds, such as the occurrence of i.a. isovaleric acid in honey of this variety [25].

Searching for biomarkers of honey varieties is a task that has been carried out for over a dozen years [26]. For example, Cabras et al. (1999) [27] showed that the marker for strawberry honey is 2,5-dihydroxyphenylacetic acid, called homogentisic acid. Its content is around 378 ± 92 mg/kg. On the other hand, studies characterizing heather honey from Poland showed the presence of a less common compound: 4-hydroxy-3-(1-methylethyl) benzaldehyde [28]. Lumichrome is indicated as a honey marker for polish yellow sweet clover [29].

Literature data show that the phenolic acids contained in honey can penetrate lymphocytes and protect DNA from oxidative damage by scavenging hydrogen peroxide and chelating ferrous ions, as shown in studies on mice [30].

Single phenolic acids show very promising activities. For example, ferulic acid has been shown to have anti-inflammatory properties [31] and potential anti-cancer properties [32], protocatechuic acid has anti-viral properties [33], and *p*-coumaric antidiabetic and antihyperlipidemic properties [34]. The above examples show that bee honey, being a mixture of many compounds with antioxidant properties, may show multidirectional activity.

The research conducted by Wilczyńska et al. (2010) showed that buckwheat honeys can be characterized by up to 100.00% of free radical scavenging capacity. Heather honeys turned out to be even more effective—all tested samples were characterized by a result of 100%. The lowest capacity was recorded for acacia honeys—from 25.58 to 35.90%. In our study, the median for buckwheat honeys was 41.1%. Moreover, Wilczyńska et al. showed that, in buckwheat honeys, the highest value of TPC was recorded (180.07 mg GAE/100 g). Our research showed a median of 196.59, with the highest value being 241.87 mg GAE/100 g [35].

Another study published by Pentoś et al. (2020) aimed to compare selected antioxidant properties of honey from Poland with Manuka honey. It was shown that Manuka honey has a TPC value of 492.65 ± 1.32 mg GAE/100 g, while the honey from Poland with the highest value of this parameter was buckwheat honey (334.04 ± 1.26 mg GAE/100 g). The honey with the second-highest TPC value was heather honey (183.85 ± 1.27 mg GAE/100 g) [36].

Dżugan et al. (2017) assessed, inter alia, results obtained in the FRAP test by Polish honeys. The highest result was obtained by buckwheat honeys (3635.49 ± 1328.22 μmol TE/kg), followed by honeydew (2153.37 ± 663.92 μmol TE/kg), while the lowest result was found for rapeseed honeys (656.73 ± 119.40 TE/kg). Our results were presented in a different way, but the trend was similar—we obtained the following results: 0.402 ± 0.010, 0.323 ± 0.017, and 0.030 ± 0.012 μmol Fe^{2+}/mL, respectively [7]. The studies by Beretta et al. (2005) also confirm that buckwheat and honeydew honeys are characterized by one of the highest results (800.7–23.8 and 772.0–21.5 μM) [16].

Our study has some limitations. We tested different amounts of honey samples belonging to a particular variety. This was due to the improper labeling of honey by beekeepers. Future research should be based on an even selection of the number of samples. It seems necessary to develop a method that will allow the determination of all phenolic acids present in honey—this will allow for the creation of detailed characteristics and the development of characteristic ranges of variability. Additionally, it seems necessary to characterize the varieties in terms of the content of individual flavonoids.

5. Conclusions

Phenolic acids can be considered markers of the authenticity of Polish honeybee varieties, in particular, syringic acid, vanillic acid, and coffee acid for linden honeys, *p*-coumaric acid and 4-hydroxybenzoic acid for buckwheat honeys, and vanillic acid for honeydew honeys. Moreover, buckwheat honeys show the highest median of the TPC parameter, which indicates a high content of phenolic compounds in the honeys of this variety. This variety of honey can be recommended to enrich the diet with antioxidant ingredients.

Author Contributions: Conceptualization, A.P.-J. and E.K.; methodology, E.K., A.P.-J., and J.M.; software, A.P.-J. and E.K.; validation, A.P.-J., K.S., and E.K.; formal analysis, K.S.; investigation, A.P.-J. and K.S.; resources, A.P.-J., E.K., and K.S.; data curation, A.P.-J., E.K., J.M., and K.S.; writing—original draft preparation, A.P.-J. and E.K.; writing—review and editing, K.S.; visualization, A.P.-J.; supervision, K.S.; project administration, A.P.-J.; funding acquisition, A.P.-J. All authors have read and agreed to the published version of the manuscript.

Funding: This research was funded by MEDICAL UNIVERSITY OF BIAŁYSTOK, grant number SUB/2/DN/21/005/2216.

Institutional Review Board Statement: Not applicable.

Informed Consent Statement: Not applicable.

Data Availability Statement: Data is contained within the article.

Conflicts of Interest: The authors declare no conflict of interest.

References

1. Da Silva, P.M.; Gauche, C.; Gonzaga, L.V.; Costa, A.C.; Fett, R. Honey: Chemical composition, stability and authenticity. *Food Chem.* **2016**, *196*, 309–323. [CrossRef] [PubMed]
2. Puścion-Jakubik, A.; Socha, K.; Borawska, M.H. Comparative study of labelled bee honey from Poland and the result of the melissopalynological analysis. *J. Apic. Res.* **2020**, *59*, 928–938. [CrossRef]
3. Major, N.; Marković, K.; Krpan, M.; Sarić, G.; Hruškar, M.; Vahčić, N. Rapid honey characterization and botanical classification by an electronic tongue. *Talanta* **2011**, *85*, 569–574. [CrossRef] [PubMed]
4. Simova, S.; Atanassov, A.; Shishiniova, M.; Bankova, V. A rapid differentiation between oak honeydew honey and nectar and other honeydew honeys by NMR spectroscopy. *Food Chem.* **2012**, *134*, 1706–1710. [CrossRef]
5. Lenhardt, L.; Bro, R.; Zeković, I.; Dramićanin, T.; Dramićanin, M.D. Fluorescence spectroscopy coupled with PARAFAC and PLS DA for characterization and classification of honey. *Food Chem.* **2015**, *175*, 284–291. [CrossRef]
6. Zhou, J.; Yao, L.; Li, Y.; Chen, L.; Wu, L.; Zhao, J. Floral classification of honey using liquid chromatography-diode array detection-tandem mass spectrometry and chemometric analysis. *Food Chem.* **2014**, *145*, 941–949. [CrossRef]
7. Dżugan, M.; Tomczyk, M.; Sowa, P.; Grabek-Lejko, D. Antioxidant Activity as Biomarker of Honey Variety. *Molecules* **2018**, *23*, 2069. [CrossRef]
8. Al-Waili, N.; Salom, K.; Al-Ghamdi, A.A. Honey for wound healing, ulcers, and burns; data supporting its use in clinical practice. *Sci. World J.* **2011**, *11*, 766–787. [CrossRef]
9. Braithwaite, I.; Hunt, A.; Riley, J.; Fingleton, J.; Kocks, J.; Corin, A.; Helm, C.; Sheahan, D.; Tofield, C.; Montgomery, B.; et al. Randomised controlled trial of topical kanuka honey for the treatment of rosacea. *BMJ Open* **2015**, *5*, e007651. [CrossRef]
10. Oduwole, O.; Udoh, E.E.; Oyo-Ita, A.; Meremikwu, M.M. Honey for acute cough in children. *Cochrane Database Syst. Rev.* **2018**, *4*, CD007094. [CrossRef]
11. Saha, A.; Chattopadhyay, S.; Azam, M.; Sur, P.K. The role of honey in healing of bedsores in cancer patients. *South Asian J. Cancer* **2012**, *1*, 66–71. [CrossRef] [PubMed]
12. Martinello, M.; Mutinelli, F. Antioxidant Activity in Bee Products: A Review. *Antioxidants* **2021**, *10*, 71. [CrossRef]
13. Vacek, J.; Ulrichová, J.; Klejdus, B.; Šimánek, V. Analytical methods and strategies in the study of plant polyphenolics in clinical samples. *Anal. Methods* **2010**, *6*, 604–613. [CrossRef]
14. Olas, B. Honey and Its Phenolic Compounds as an Effective Natural Medicine for Cardiovascular Diseases in Humans? *Nutrients* **2020**, *12*, 283. [CrossRef] [PubMed]
15. Regulation of the Minister of Agriculture and Rural Development dated 14 January 2009. On Methods of Analysis Related to the Performance Evaluation of Honey [In Polish, Rozporządzenie Ministra Rolnictwa I Rozwoju Wsi Z Dnia 14 Stycznia 2009 R. W Sprawie Metod Analiz Związanych Z Dokonywaniem Oceny Miodu]. 2009; Dz.U. 2009 nr 17 poz. 94. Available online: https://isap.sejm.gov.pl/isap.nsf/DocDetails.xsp?id=WDU20090170094 (accessed on 2 March 2022).
16. Beretta, G.; Granata, P.; Ferrero, M.; Orioli, M.; Facino, R.M. Standarization of antioxidant properties of honey by a combination of spectrophotometric/fluorimetric assays and chemometrics. *Anal. Chim. Acta* **2005**, *533*, 185–191. [CrossRef]
17. Ferreira, I.C.; Aires, E.; Barreira, J.C.; Estevinho, L.M. Antioxidant activity of Portuguese honey samples: Different contributions of the entire honey and phenolic extract. *Food Chem.* **2009**, *114*, 1438–1443. [CrossRef]
18. Djeridane, A.; Yousfi, M.; Nadjemi, B.; Maamri, S.; Djireb, F.; Stocker, P. Phenolic extracts from various Algerian plants as strong inhibitors of porcine liver carboxylesterase. *J. Enzym. Inhib. Med. Chem.* **2006**, *21*, 719–726. [CrossRef]
19. Sánchez-Moreno, C.; Larrauri, J.A.; Saura-Calixto, F. A procedure to measure the antiradical efficiency of polyphenols. *J. Sci. Food Agric.* **1998**, *76*, 270–276. [CrossRef]

20. Benzie, I.F.; Strain, J.J. The ferric reducing ability of plasma (FRAP) as a measure of "antioxidant power": The FRAP assay. *Anal. Biochem.* **1996**, *239*, 70–76. [CrossRef]
21. Starowicz, M.; Ostaszyk, A.; Zieliński, H. The Relationship between the Browning Index, Total Phenolics, Color, and Antioxidant Activity of Polish-Originated Honey Samples. *Foods* **2021**, *10*, 967. [CrossRef]
22. Ecem Bayram, N.; Canli, D.; Gercek, Y.C.; Bayram, S.; Çelik, S.; Güzel, F.; Morgil, H.; Oz, G.C. Macronutrient ans micronutrient levels and phenolic compound characteristics of monofloral honey samples. *J. Food Nutr. Res.* **2020**, *59*, 311–322.
23. Dimitrova, B.; Gevrenova, R.; Anklam, E. Analysis of phenolic acids in honeys of different floral origin by solid-phase extraction and high-performance liquid chromatography. *Phytochem. Anal.* **2007**, *18*, 24–32. [CrossRef] [PubMed]
24. Jasicka-Misiak, I.; Poliwoda, A.; Dereń, M.; Kafarski, P. Phenolic compounds and abscisic acid as potential markers for the floral origin of two Polish unifloral honeys. *Food Chem.* **2012**, *131*, 1149–1156. [CrossRef]
25. Pasini, F.; Gardini, S.; Marcazzan, G.L.; Caboni, M.F. Buckwheat honeys: Screening of composition and properties. *Food Chem.* **2013**, *141*, 2802–2811. [CrossRef]
26. Wang, X.; Chen, Y.; Hu, Y.; Zhou, J.; Chen, L.; Lu, X. Systematic review of the characteristic markers in honey of various botanical, geographic, and entomological origins. *ASC Food Sci. Technol.* **2022**, *2*, 206–220. [CrossRef]
27. Cabras, P.; Angioni, A.; Tuberoso, C.; Floris, I.; Reniero, F.; Guillou, C.; Ghelli, S. Homogentisic acid: A phenolic acid as a marker of strawberry-tree (Arbutus unedo) honey. *J. Agric. Food Chem.* **1999**, *47*, 4064–4067. [CrossRef]
28. Jerković, I.; Kuś, P.M.; Tuberoso, C.I.; Šarolić, M. Phytochemical and physical-chemical analysis of Polish willow (*Salix* spp.) honey: Identification of the marker compounds. *Food Chem.* **2014**, *145*, 8–14. [CrossRef]
29. Jasicka-Misiak, I.; Makowicz, E.; Stanek, N. Polish Yellow Sweet Clover (*Melilotus officinalis* L.) Honey, Chromatographic Fingerprints, and Chemical Markers. *Molecules* **2017**, *22*, 138. [CrossRef]
30. Cheng, N.; Wang, Y.; Cao, W. The Protective Effect of Whole Honey and Phenolic Extract on Oxidative DNA Damage in Mice Lymphocytes Using Comet Assay. *Plant Foods Hum. Nutr.* **2017**, *72*, 388–395. [CrossRef]
31. Liu, Y.M.; Shen, J.D.; Xu, L.P.; Li, H.B.; Li, Y.C.; Yi, L.T. Ferulic acid inhibits neuro-inflammation in mice exposed to chronic unpredictable mild stress. *Int. Immunopharmacol.* **2017**, *45*, 128–134. [CrossRef]
32. Wang, T.; Gong, X.; Jiang, R.; Li, H.; Du, W.; Kuang, G. Ferulic acid inhibits proliferation and promotes apoptosis via blockage of PI3K/Akt pathway in osteosarcoma cell. *Am. J. Transl. Res.* **2016**, *8*, 968–980. [PubMed]
33. Ou, C.; Shi, N.; Yang, Q.; Zhang, Y.; Wu, Z.; Wang, B.; Compans, R.W.; He, C. Protocatechuic acid, a novel active substance against avian influenza virus H9N2 infection. *PLoS ONE* **2014**, *9*, e111004. [CrossRef]
34. Amalan, V.; Vijayakumar, N.; Indumathi, D.; Ramakrishnan, A. Antidiabetic and antihyperlipidemic activity of p-coumaric acid in diabetic rats, role of pancreatic GLUT 2: In vivo approach. *Biomed. Pharm.* **2016**, *84*, 230–236. [CrossRef]
35. Wilczyńska, A. Phenolic content and antioxidant activity of different types of Polish honey—A short report. *Pol. J. Food Nutr. Sci.* **2010**, *60*, 309–313.
36. Pentoś, K.; Łuczycka, D.; Oszmiański, J.; Lachowicz, S.; Pasternak, G. Polish honey as a source of antioxidants—A comparison with Manuka honey. *J. Apic. Res.* **2020**, *59*, 939–945. [CrossRef]

antioxidants

Article

Therapeutic Effects of *Heterotrigona itama* (Stingless Bee) Bee Bread in Improving Hepatic Lipid Metabolism through the Activation of the Keap1/Nrf2 Signaling Pathway in an Obese Rat Model

Zaida Zakaria [1], Zaidatul Akmal Othman [2], Joseph Bagi Suleiman [3], Khairul Mohd Fadzli Mustaffa [4], Nur Asyilla Che Jalil [5], Wan Syaheedah Wan Ghazali [1], Ninie Nadia Zulkipli [1], Mahaneem Mohamed [1,6,*] and Khaidatul Akmar Kamaruzaman [1]

1 Department of Physiology, School of Medical Sciences, Universiti Sains Malaysia, Kubang Kerian 16150, Kelantan, Malaysia
2 Unit of Physiology, Universiti Sultan Zainal Abidin, Kuala Terengganu 20400, Terengganu, Malaysia
3 Department of Science Laboratory Technology, Akanu Ibiam Federal Polytechnic, Unwana P.O. Box 1007, Ebonyi State, Nigeria
4 Institute for Research in Molecular Medicine, Universiti Sains Malaysia, Kubang Kerian 16150, Kelantan, Malaysia
5 Department of Pathology, School of Medical Sciences, Universiti Sains Malaysia, Kubang Kerian 16150, Kelantan, Malaysia
6 Unit of Integrative Medicine, School of Medical Sciences, Universiti Sains Malaysia, Kubang Kerian 16150, Kelantan, Malaysia
* Correspondence: mahaneem@usm.my; Tel.: +60-97676158

Citation: Zakaria, Z.; Othman, Z.A.; Suleiman, J.B.; Mustaffa, K.M.F.; Jalil, N.A.C.; Ghazali, W.S.W.; Zulkipli, N.N.; Mohamed, M.; Kamaruzaman, K.A. Therapeutic Effects of *Heterotrigona itama* (Stingless Bee) Bee Bread in Improving Hepatic Lipid Metabolism through the Activation of the Keap1/Nrf2 Signaling Pathway in an Obese Rat Model. *Antioxidants* **2022**, *11*, 2190. https://doi.org/10.3390/antiox11112190

Academic Editors: Josipa Vlainić and Ivana Tlak Gajger

Received: 6 October 2022
Accepted: 2 November 2022
Published: 5 November 2022

Abstract: Bee bread (BB) has traditionally been used as a dietary supplement to treat liver problems. This study evaluated the therapeutic effects of *Heterotrigona itama* BB from Malaysia on obesity-induced hepatic lipid metabolism disorder via the regulation of the Keap1/Nrf2 pathway. Male *Sprague Dawley* rats were fed with either a normal diet or high-fat diet (HFD) for 6 weeks to induce obesity. Following 6 weeks, obese rats were treated either with distilled water (OB group), BB (0.5 g/kg body weight/day) (OB + BB group) or orlistat (10 mg/kg body weight/day) (OB + OR group) concurrent with HFD for another 6 weeks. BB treatment suppressed Keap1 and promoted Nrf2 cytoplasmic and nuclear translocations, leading to a reduction in oxidative stress, and promoted antioxidant enzyme activities in the liver. Furthermore, BB down-regulated lipid synthesis and its regulator levels (SIRT1, AMPK), and up-regulated fatty acid β-oxidation in the liver of obese rats, being consistent with alleviated lipid levels, improved hepatic histopathological changes (steatosis, hepatocellular hypertrophy, inflammation and glycogen expression) and prevented progression to non-alcoholic steatohepatitis. These results showed the therapeutic potentials of *H. itama* BB against oxidative stress and improved lipid metabolism in the liver of obese rats possibly by targeting the Keap1/Nrf2 pathway, hence proposing its role as a natural supplement capable of treating obesity-induced fatty liver disease.

Keywords: *Heterotrigona itama*; bee bread; high-fat diet; obesity; MAFLD; oxidative stress; lipid metabolism; Keap1/Nrf2 pathway; SIRT1; AMPK

1. Introduction

The worldwide prevalence of overweight and obesity has increased by two-folds since 1980 to an extent that almost a third of the global population is now considered to be overweight or obese [1]. Obesity adversely affects most of the body's physiological functions and is regarded as "the biggest public health threat for this century". It is caused by various factors, including a high-fat diet (HFD), and increases the risk of developing numerous co-morbidities, including type 2 diabetes mellitus, insulin resistance, hyperlipidemia, several types of cancers, cardiovascular diseases and liver diseases [2].

Metabolic dysfunction-associated fatty liver disease (MAFLD), previously known as non-alcoholic fatty liver disease (NAFLD) [3,4] is strongly related to obesity and is recognized as a complex metabolic syndrome of abnormal liver metabolism. There is growing evidence that supports a key role of oxidative stress resulting from the generation of reactive oxygen species (ROS) in the progression of MAFLD and its progressive form non-alcoholic steatohepatitis (NASH) [5]. Mitochondria are the primary intracellular sites of oxygen consumption; hence, they are the main source of ROS generation in MAFLD. It is believed that excessive hepatic lipid buildup leads to structural abnormalities of hepatic mitochondria and impairment of the electron chain, as well as augmented lipid peroxidation and ROS generation, which further stimulates oxidative stress and inflammation in the liver [6–8]. As oxidative stress is classified as the disequilibrium between ROS and antioxidants, a counterbalance by a complex antioxidant defense system such as the key antioxidant enzymes superoxide dismutase (SOD), catalase (CAT), glutathione peroxidase (GPx), glutathione-S-transferase (GST) and glutathione reductase (GR) is necessary to avoid the generation of ROS and henceforth oxidative stress in the liver. Previous studies have demonstrated the correlation between the reduction in antioxidant enzyme activities and the severity of MAFLD in clinical and animal studies [9–12]. Additionally, it has also been reported that the activation of nuclear factor erythroid 2-related factor 2 (Nrf2), which is the key transcription factor regulating the expression of antioxidant enzyme genes [13], results in the inhibition of oxidative stress and a reduction in hepatic inflammation and fibrosis [14]. Under quiescent cellular environments, Nrf2 remains inactive by forming a complex with Kelch-like ECH-associated protein 1 (Keap1), a Nrf2 repressor protein in the cytoplasm. However, during exposure to oxidative stress and electrophilic substances, Keap1 becomes oxidized and releases Nrf2 to be translocated to the nucleus where it stimulates the transcription of the gene containing the antioxidant response element (ARE) which activates the translation of antioxidant genes [15]. Numerous studies have indicated the effects of sterol regulatory element binding protein-1c (SREBP-1c) or peroxisome proliferator-activated receptor alpha (PPARα)-dependent pathways on lipogenesis and fatty acid oxidation, respectively, in hepatic lipid metabolism [16]. SREBP-1c controls the activities of lipogenic enzymes including fatty acid synthase (FAS) and its expression is up-regulated by the increased insulin level in the circulation and liver [17]. Meanwhile, PPARα has been demonstrated to be involved in the oxidation of fatty acids and mediates the activity of carnitine palmitoyltransferase 1α (CPT1α), which catalyzes the rate-limiting step in fatty acid β-oxidation. Consequently, a reduced level of PPARα leads to hyperlipidemia and lipid deposition in the hepatocytes [18]. In addition, both AMPK-activated protein kinase (AMPK) and sirtuin 1 (SIRT1) signaling also play a role in maintaining an energy balance and regulating lipid metabolism in the liver, muscle and adipose tissue [19,20]. Activated AMPK can suppress the levels of lipogenic-related protein including SREBP1 and FAS, as well as up-regulate the activities of enzymes responsible for fatty acid β-oxidation (PPARα and CPT1); hence, it can reduce lipid accumulation [21]. Likewise, an increased level of lipid accumulation in the liver of obese mice has been linked with the inhibition of AMPK and SIRT1 activities, hence aggravating the development of MAFLD [22]. Of note, it has also been reported that the activation of the Keap1/Nrf2 signaling pathway to maintain a redox status mediates lipid metabolism-related genes (SREBP-1c and PPARα) to reduce hepatic steatosis in an HFD-induced obese animal model [23]. Thus, the hepatotherapeutic agents with anti-oxidative stress may have a potential therapeutic modality on lipid accumulation.

Generally, stingless bees are found mainly in tropical and subtropical regions of the world, including Africa, Southeast Asia, Australia and South America. In Malaysia, over 30 species of stingless bee (*Trigona* spp.), which is also known as 'kelulut', have been documented in Peninsular Malaysia and out of this, 17 species were identified to inhabit the virgin forest [24]. The most common species of *Trigona* spp. in Malaysia is *Heterotrigona itama*, which is reared commercially due to high demand [25]. Bee bread is formed from the fermentation of mixtures of nectar, pollen and digestive enzymes secreted from the bee's salivary glands [26,27]. The chemical compositions of bee bread are mainly comprised of proteins, free amino acids, carbohydrates, fatty acids, a wide range of sugars and high reducing sugars [28,29] and vitamins [30]. Moreover, each bee bread has a different composition which varies from different regions, climatic and seasonal variations, floral origins [31,32] and soil type [27]. Numerous data have shown that bee bread produced by *H. itama* possesses beneficial therapeutic effects on metabolic diseases including cardiovascular disease [33–35], renal disorder [36] and male infertility [37–39]. Moreover, our prior study revealed that concurrent administration of HFD and *H. itama* bee bread at 0.5 g/kg b.w./day for 12 weeks exhibited protection against obesity, liver oxidative damage and inflammation, as well as prevented the liver from NASH and fibrosis in rats [40]. Bee bread was also previously demonstrated to have an anti-lipogenic effect by reducing the expressions of FAS and acetyl-CoA carboxylase (ACC) in the obesity-induced fatty liver disease rat model [41]. Furthermore, numerous findings have been reported on the anti-obesity effects of flavonoids and phenolic acids, hence, increasing the liver antioxidant and detoxification system as well as improving lipid metabolism [42–45]. In this current study, we investigated whether the intake of *H. itama* bee bread would have beneficial therapeutic effects on liver oxidative stress and lipid accumulation after the induction of obesity in rats. Moreover, the underlying molecular mechanism by which bee bread alleviates redox imbalance and lipid metabolism disorder via the activation of the Keap1/Nrf2 pathway was explicated.

2. Materials and Methods

2.1. Bee Bread Collection and Preparation

Bee bread sample from the stingless bee *H. itama* was self-harvested and collected from a local stingless bee farm (Mentari Technobee PLT, Kelantan, Malaysia). The sample initial weight was measured, dried using a food dehydrator (Domani Industries Ltd., Foshan, China) (35 °C) and ground into powder form before being stored in −20 °C until further use [33,38].

2.2. High-Performance Liquid Chromatography (HPLC) Detection and Quantification of Polyphenolic Compounds

HPLC was performed based on the method published by Suleiman et al. (2021) [46]. Bee bread powder was suspended in water and methanol to produce aqueous and methanol solutions, respectively, to achieve final concentrations of 100 mg/mL. The solutions were then vortexed, sonicated and followed with centrifugation at 20,111× g for 5 min prior to HPLC analysis. The samples were analyzed using a Dionex RS3000 system (Thermo Scientific, Waltham, MA, USA). The chromatic separation was achieved at 25 °C on a Zorbax SB-C18 column (3.5 μm, 4.6 mm I.D × 150 mm) (Agilent Technologies, Santa Clara, CA, USA). A binary solvent system was employed consisting of 0.1% formic acid in water as solvent A and 0.1% formic acid in methanol (40:60, *v*/*v*) as solvent B. The chromatographic analyses were conducted at a run time of 0, 20, 25, 25.1 and 30 min. The flow rate was 1.0 mL/min, and the injection volume was 20 μL. The eluted components were monitored at 340 nm. The standard substances of gallic acid, caffeic acid, mangiferin, trans-ferulic acid, 2-hydroxycinnamic acid, trans-3-hydroxycinnamic acid, quercetin, kaempferol and apigenin were purchased from Sigma (St. Louis, MO, USA) and used as reference compounds.

2.3. Animals and Diet

Thirty-two male *Sprague Dawley* rats aged from 8 to 10 weeks (200–230 g) were purchased from the Laboratory Animal Research and Service Centre (ARASC), Universiti Sains Malaysia (USM). All the animal experiments complied with the National Institute of Health Guide for the Care and Use of Laboratory Animals and were approved by the USM

Institutional Animal Care and Use Committee (IACUC) (USM/IACUC/2018/(113)(933)). Each rat was individually housed in a polypropylene cage with sterilized husk bedding in a room with a 12 h light–dark cycle, a controlled temperature at 22–24 °C and a relative humidity of 55–70%. All the rats had free access to a normal rat chow pellet and clean drinking water during one week of the acclimatization period.

All animals were fed with either a normal diet or an HFD. The normal diet was a standard Altromin pellet (Altromin Spezialfutter GmbH & Co. KG, Lage, Germany) composed of soy, wheat and corn with approximately 24% protein, 64% carbohydrates and 12% fat in terms of caloric content. Meanwhile, the HFD was prepared based on a prior study with slight modifications, composed of 32 g of animal ghee, 68 g of powdered normal diet, 12% of cholesterol powder, as well as 300 mg calcium and 100 IU vitamin D3, which contained approximately 12% protein, 46% carbohydrate and 31% fat in terms of the caloric content [47].

2.4. Experimental Design

After adaptation, all the rats were indiscriminately divided into two experimental groups and fed ad libitum as follows: (1) Control (CON) group (n = 8); rats fed a normal diet and 1 mL distilled water once daily for 12 weeks, (2) HFD group (n = 24): rats fed HFD and oral gavage of 1 mL distilled water for 6 weeks to induce obesity. The Lee obesity index was used to confirm the obesity using a previously reported formula [48]: $\frac{\sqrt[3]{bodyweight\ (g)}}{naso-anal\ length\ (cm)} \times 1000$ and a value of Lee obesity of more than 315 was considered as obese [49]. Following 6 weeks, the confirmed obese rats in the HFD group were then separated into 3 random experimental groups (n = 8/group), and treated as follows for 6 weeks: (1) Obese group (OB): rats fed HFD and oral gavage of 1 mL distilled water, (2) Bee bread group (OB + BB): rats fed HFD and oral gavage of bee bread (0.5 g/kg b.w./day), and (3) Orlistat group (OB + OR): rats fed with HFD and oral gavage of orlistat (10 mg/kg b.w./day). A pilot study was conducted to identify the best dose of bee bread (0.5, 1.0 and 1.5 g/kg b.w./day) for improving some hepatic parameters in obese male rats. After 6 weeks, the findings demonstrated that the best dose of bee bread 0.5 g/kg b.w./day was selected for this animal study as this dose exerted the most improvement on the Lee obesity index, liver function and hepatic steatosis in obese male rats (Table S1 and Figure S1). Meanwhile, the dose of orlistat (Xepa-Soul Pattinson Sdn. Bhd. Melaka, Malaysia) at 10 mg/kg b.w./day was selected based on a prior reported study on obese rats [50]. Both bee bread and orlistat were weighed before being dissolved in distilled water to a final volume of 1 mL before being dispensed to the animals.

2.5. Measurements of Obesity Parameters and Nutritional Composition

Body weights were monitored weekly throughout the experiment and the changes in the body weight between week 12 and week 0 were recorded as body weight gain. The Lee obesity index was calculated at the end of the experimental period. The food intake was monitored daily and the averages of food and calorie intakes for each rat were determined.

2.6. Blood and Tissue Collection

At the end of the 12th week, all the animals were anaesthetized intraperitoneally using 90 mg/kg ketamine and 5 mg/kg xylazine following a 12 h fast. Then, blood was collected from the rat's posterior vena cava in tubes containing a gel clot activator. The blood samples were centrifuged (3000× g, 15 min) to obtain the serum and stored at −80 °C for biochemical assays. Meanwhile, the livers from each experimental group were dissected, rinsed in ice-cold normal saline solution, blotted dry and cut into three portions. The liver tissue was stored in the RNAlater (Sigma-Aldrich, St. Louis, MO, USA) at −80 °C for qRT-PCR analysis. The second portion of tissue was homogenized using a tissue homogenizer (IKA Labortechnik Co., Ltd., Wilmington, NC, USA) in 10 volumes of ice-cold phosphate buffer saline (pH 7.4) and centrifuged (3000× g, 20 min). The separated supernatant was stored at −80 °C until further use for liver biochemical analyses. The last portion of the liver tissue was rapidly fixed in 10% formalin for at least 48–72 h for immunohistochemical and histopathological analyses.

2.7. Determination of Serum Glucose, Insulin and HOMA-IR

Serum glucose (Qayee-Bio Life Science Co., Ltd., Shanghai, China) and insulin (Elabscience Biotechnology Inc. Co., Ltd. Wuhan, Hubei, China) levels were tested with commercially available kits according to the manufacturers' instructions. The homeostatic model of assessment-insulin resistance (HOMA-IR) was calculated as referred to in the prior study [51].

2.8. Evaluations of Lipid Profiles

The levels of both triglyceride (TG) and total cholesterol (TC) in the serum were assessed using a commercialized kit (ARCHITECT c kit, Abbott, IL, USA) according to an enzymatic colorimetric method. Meanwhile, the level of low-density lipoprotein (LDL) was measured according to a formula as described in the previous study [52], i.e., LDL (mmol/L) = (TC − HDL − (TG/5). Furthermore, the level of high-density lipoprotein (HDL) was determined using Biosino Direct HDL-Cholesterol reagent kit (Biosino Bio-Technology and Science Inc., Beijing, China) by the eliminations of LDL-Cholesterol, chylomicron and VDLD-Cholesterol by cholesterol oxidase, cholesterol esterase and catalase.

2.9. Liver Biochemical Analyses

The hepatic TG, TC, SIRT1 and AMPK concentrations were determined using the commercial kits obtained from Qayee-Bio Life Science Co., Ltd. (Shanghai, China), referring to the manufacturers' protocols. Meanwhile, nitric oxide (NO) concentration was assayed with common commercially available biochemical kits obtained from Elabscience Biotechnology Inc. Co., Ltd. (Wuhan, China) according to the manufacturer's instruction. The level of lipid peroxidation in the liver was determined as a thiobarbituric acid reactive substance (TBARs) according to Chatterjee et al. (2000) [53]. The absorbance of a colored complex produced from the reaction of thiobarbituric acid and malondialdehyde was measured at 532 nm. The SOD activity was calculated according to a method by Al Batran et al. (2013) [54]. The enzyme activity was assessed by the measurement of diformazone, a final product formed from a reduced superoxide ion by tetrazolium blue nitro (NBT) at wavelength 560 nm. The CAT activity was evaluated according to Góth, (1991) [55] based on the enzyme-catalyzed decomposition of hydrogen peroxide and an assay of the remaining hydrogen peroxide with molybdate ions. The enzyme activity was assessed by measurement of the yellowish complex formed from the reaction spectrophotometrically (BioTek Instruments, Winooski, VT, USA) at 405 nm. The activity of GPx was assessed according to Doğan et al. (1994) [56]. The enzyme activity was evaluated by the measurement of the change in the concentration of NADPH at wavelength 340 nm. Estimation of GST activity in the liver was measured according to Habig et al. (1974) [57] which was based on glutathione conjugation to 1-chloro-2,4-dinitrobenzene (CDNB) as a substrate. The enzyme activity was measured spectrophotometrically at wavelength 340 nm. The activity of GR was estimated according to Luchese et al. (2009) [58] based on the reduction of GSSG catalyzed by GR in the presence of NADPH to form GSH and $NADP^+$. The decrease in absorbance due to the decreased concentration of NADPH was determined spectrophotometrically at 340 nm. The level of total glutathione (GSH) was evaluated according to Annuk et al. (2001) [59] with some modifications. The rate of yellow complex (5-thio-2-nitrobenzoic acid) formed from the reaction between the sulfhydryl group of GSH and 5,5′-dithiobis-2-nitrobenzoic acid (DTNB) was measured at wavelength 405 nm. The liver total antioxidant capacity (TAC) was evaluated by referring to a former described method by Koracevic et al. (2001) [60]. In this assay, the antioxidants from the liver homogenate inhibited the formation of TBARS and the reaction was calculated by measuring spectrophotometric absorbance at 532 nm. The protein contents of these samples were quantified using a bicinchoninic acid (BCA) protein assay kit according to the manufacturer's instructions (Thermo Fisher Scientific, Inc., Waltham, MA, USA) and normalized to the data of liver tissue biochemical parameters.

2.10. RNA Extraction and RT-qPCR Analysis

Total RNA from the liver tissue was extracted using Innu Prep RNA mini kit (Analytik Jena, Jena, Germany) and treated with PureLink^{TM} DNase (Invitrogen, Thermo Fisher Scientific Inc. USA) to remove any contaminating genomic DNA according to the manufacturers' instructions. The concentration and purity of the RNA preparation were quantified by measuring the absorbances at 260 and 280 nm on a μDrop^{TM} Plate (Thermo Fisher Scientific Inc., Waltham, MA, USA) and only samples with OD260/280 of 1.8–2.0 were included in this study. Then, the quality and integrity of RNA were assessed using agarose gel electrophoresis stained with fluorescent dye in 1x LB buffer (Faster Better Media LLC, Hunt Valley, MD, USA). RT-qPCR was carried out in a 20 μL volume reaction containing 10 μL of SYBR Lo-ROX One-step Mix, 10 μM of each primer, 0.2 μL of reverse transcriptase, 0.4 μL of RNase inhibitor, 16 μL of DEPC-treated water and 4 μL of RNA (SensiFAST^{TM} SYBR Lo-ROX One-step kit, Bioline USA Inc., Taunton, MA, USA) using Stratagene Mx3000P qPCR system (Agilent Technologies, Santa Clara, CA, USA). Thermal cycling conditions included cDNA synthesis at 45 °C for 10 min, polymerase activation at 95 °C for 2 min, PCR amplification for 40 cycles each one consisting of denaturation at 95 °C for 5 s, annealing at 60 °C for 10 s and extension at 72 °C for 5 s, followed by melt curve stage. All primers were selected from GenBank and synthesized by Integrated DNA Technologies (IDT, Malaysia). The GAPDH gene was used as an internal control to normalize target gene expressions. Three replicates of each reaction were conducted and the relative mRNA transcription levels were calculated according to the method of the $2^{-\Delta\Delta Ct}$ method [61]. The primer sequences used for this study are shown in Table 1.

Table 1. Primers used for PCR amplification.

Gene Name	Accession Number	Sequence	Reference
SREBP-1c	NM_001276707.1	Forward: 5′-GCCTGCTTGGCTCTTCTCT-3′ Reverse: 5′-GCTTGTTTGCGATGTCTCC-3′	[62]
FAS	NM_017332.1	Forward: 5′-TCGACTTCAAAGGACCCAGC-3′ Reverse: 5′-ACTGCACAGAGGTGTTAGGC-3′	[63]
PPARα	NM_013196.1	Forward: 5′-ATTCGGCTAAAGCTGGCGTA-3′ Reverse: 5′-TGCATTGTGTGACATCCCGA-3′	[63]
CPT1α	BC072522.1	Forward: 5′-GGACATTCCTCTCTCAGGTTTC-3′ Reverse: 5′-ACCTCCTCCTTTGAACACATAC-3′	[63]
GAPDH	NM_017008	Forward: 5′-TCACCACCATGGAGAAGGC-3′ Reverse: 5′-GCTAAGCAGTTGGTGGTGCA-3′	[39]

SREBP-1c, sterol regulatory element-binding protein-1c; FAS, fatty acid synthase; PPARα, peroxisome proliferator activated receptor α; CPT1α, carnitine palmitoyl transferase 1α; GAPDH, glyceraldehyde 3-phosphate dehydrogenase.

2.11. Immunohistochemical Detections of Keap1 and Nrf2 Expressions

Serial sections of formalin-fixed, paraffin-embedded liver samples were de-waxed, rehydrated and subjected to antigen retrieval treatment in a Tris-EDTA buffer solution (0.1 M, pH 9.0) with 0.05% Tween-20. The endogenous peroxide activity was blocked for 5 min using 3% hydrogen peroxide. Then, the liver sections were incubated using rabbit polyclonal primary antibodies against Keap1 and Nrf2 (Cloud- Clone Corp, Katy, TX, USA) (1:100) at 4 °C overnight. The primary antibodies were detected using a secondary antibody containing goat anti-rabbit (Dako EnVision^{TM} + System/HRP labelled polymer) (Agilent Technologies, Inc., Santa Clara, CA, USA). Immunospecific reactivity was visualized by Dako 3,3′-diaminobenzidine (DAB) chromogen substrate reagent (1:1) mixed solution (Agilent Technologies, Inc., Santa Clara, CA, USA), counterstained with hematoxylin (Merck, Darmstadt, Germany), then dehydrated in alcohol and xylene before being mounted. The protein expressions of Keap1 and Nrf2 were analyzed by two independent pathologists (blinded to the treatment the rats received) according to a previous study [64]. The immunoreactive score was assessed by multiplying the staining intensity (0, colorless; 1, light yellow; 2, brownish yellow; 3, brown) with a percentage of positively stained cells (0, negative; 1, 10%; 2, 11–50%; 3, 51–75%; 4, 75–100%).

2.12. Liver Histopathological Examination

The formalin-fixed liver tissues were dehydrated in a series of ethanol and embedded in paraffin wax. Sections (3-μm thick) were cut and stained with hematoxylin-eosin (H&E) (both Merck, Darmstadt, Germany) and analyzed under an Olympus BX41 (Olympus Co., Tokyo, Japan). The histopathological changes such as the hepatocellular vesicular steatosis (i.e., macro- or micro-vesicular steatosis and hypertrophy) and inflammatory cell infiltration were assessed and followed by grading and scoring referring to the NAFLD activity score (NAS) [65,66] as follows: NASH; $\geq$5 and non-NASH; $\leq$3. A periodic acid-Sciff (PAS) staining was performed to detect glycogen accumulation (magenta color staining) in liver tissue. The glycogen score was evaluated following the percentage of positively stained cells as follows: 0 (0–15%), 1 (16–25%), 2 (26–50%) or 3 (76–100%) [67]. All the histopathological assessments were carried out by two liver pathologists blinded to the diet.

2.13. Statistical Analysis

Statistical analysis was carried out using GraphPad Prism, 8th Version Software (GraphPad Software Inc., Maryland, USA). All data were checked for normality and variance of the data sets using the Shapiro–Wilk and D'Agostino–Pearson Omnibus normality test, respectively. All data are expressed as means $\pm$ standard error of the means (SEM). One-way ANOVA was used followed by Tukey post-hoc test to determine the differences between the groups. $p < 0.05$ was considered to be statistically significant.

3. Results

3.1. Phenolic Compound Analysis of *H. itama* Bee Bread Using High-Performance Liquid Chromatography (HPLC)

Figure 1 shows the HPLC profiles of (a) aqueous and (b) methanol bee bread extracts, respectively. The retention time point of the standards was compared to the HPLC profiles of both bee bread extracts, and the quantities for each of the chemical markers within the bee bread extracts were calculated and are summarized in Table 2. Aqueous bee bread extract had a relatively higher concentration of trans-3-hydroxycinnamic acid, followed by 2-hydroxycinnamic acid, gallic acid and mangiferin. Similarly, the methanol bee bread extract possessed a relatively higher concentration of trans-3-hydroxycinnamic acid, followed by quercetin, apigenin, kaempferol, 2-hydroxycinnamic acid, mangiferin, caffeic acid and trans-ferulic acid (Table 2).

Table 2. Contents of phenolic compounds in the aqueous and methanol extracts of bee bread.

Phenolic Compound	Molecular Formula	Retention Time (min)	Area (mAU $\times$ min)	Relative Area (%)
Aqueous extract:				
Gallic acid	$C_7H_6O_5$	5.93	1.106	1.59
Mangiferin	$C_{19}H_{18}O_{11}$	13.09	1.093	1.57
Trans 3-hydroxycinnamic acid	$C_9H_8O_3$	15.05	40.235	57.77
2-hydroxycinnamic acid	$C_9H_8O_3$	15.77	1.379	1.98
Methanol extract:				
Caffeic acid	$C_9H_8O_4$	12.53	2.153	0.50
Mangiferin	$C_{19}H_{18}O_{11}$	13.09	4.170	0.97
Trans ferulic acid	$C_{10}H_{10}O_4$	14.62	1.035	0.24
Trans 3-hydroxycinnamic acid	$C_9H_8O_3$	15.05	74.221	17.31
2-hydroxycinnamic acid	$C_9H_8O_3$	15.77	4.909	1.15
Quercetin	$C_{15}H_{10}O_7$	17.64	18.878	4.40
Kaempferol	$C_{15}H_{10}O_6$	18.73	7.999	1.87
Apigenin	$C_{15}H_{10}O_5$	19.01	9.499	2.22

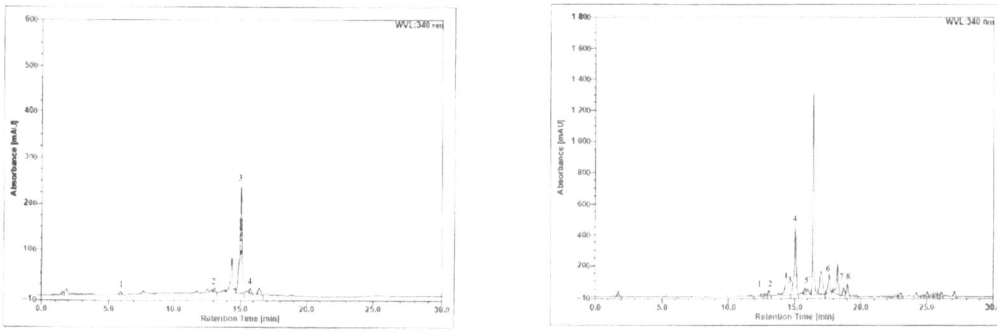

(a) (b)

Figure 1. HPLC profiles of (**a**) aqueous and (**b**) methanol extracts of bee bread samples at 340 nm. (**a**) Peak 1, gallic acid; peak 2, mangiferin; peak 3, trans 3-hydroxycinnamic acid; peak 4, 2-hydroxycinnamic acid. (**b**) Peak 1, caffeic acid; peak 2, mangiferin; peak 3, trans ferulic acid; peak 4, trans 3-hydroxycinnamic acid; peak 5, 2-hydroxycinnamic acid; peak 6, quercetin; peak 7, kaempferol; peak 8, apigenin.

3.2. Effects of H. itama Bee Bread on Obesity Parameters and Nutritional Composition

To determine the anti-obesity property of bee bread supplementation in obese rats, the Lee obesity index and body weight gain were recorded in the present study. As demonstrated in Table 3, 12-week HFD feeding significantly increased ($p < 0.05$) the Lee obesity index and body weight gain in the OB group. Meanwhile, the administration of bee bread significantly reduced ($p < 0.05$) these parameters. Similar patterns of results were also present in the OB + OR group.

Table 3. Obesity parameters and nutritional composition of rats in the experimental groups.

	CON	OB	OB + BB	OB + OR
Lee obesity index	306.4 ± 1.60	331.5 ± 2.69 ^{a}	311.4 ± 2.45 ^{b}	316.9 ± 2.64 ^{a,b}
Body weight gain (g)	103.0 ± 9.26	204.7 ± 10.98 ^{a}	154.9 ± 14.47 ^{a,b}	157.0 ± 6.98 ^{a,b}
Average daily food intake (g/day)	20.97 ± 0.50	20.65 ± 0.71	18.58 ± 0.70	19.08 ± 0.72
Average daily calorie intake (kJ/day)	282.7 ± 6.47	446.2 ± 15.35 ^{a}	401.5 ± 15.17 ^{a}	412.3 ± 15.46 ^{a}

Data are presented as mean ± SEM, n = 8/group. CON, control; OB, obese; OB + BB, obese + bee bread 0.5 g/kg body weight/day; OB + OR, obese + orlistat 10 mg/kg body weight/day. One-way ANOVA, followed by Tukey post-hoc test. ^{a} $p < 0.05$ vs. CON group, ^{b} $p < 0.05$ vs. OB group.

Furthermore, no significant differences ($p > 0.05$) were found in the average food intake among all the experimental groups. However, all animals fed with the HFD demonstrated significant increases ($p < 0.05$) in the average of calorie intake compared to the CON group. Nonetheless, no significant differences ($p > 0.05$) were found in the average calorie intake between all the HFD-fed groups (Table 3).

3.3. Effects of H. itama Bee Bread on Serum Glucose, Insulin Resistance and Lipid Profile

Indeed, hyperglycemia, hyperinsulinemia and insulin resistance are closely associated with obesity. To investigate the effects of bee bread on these parameters, the levels of glucose, insulin and HOMA-IR were evaluated in this current experiment. As shown in Table 4, the OB group showed significantly increased ($p < 0.05$) blood glucose and insulin levels after 12 weeks of HFD intervention compared to the CON group. In addition, compared with the CON group, the rats in the OB group also demonstrated a markedly higher ($p < 0.05$) insulin resistance index, HOMA-IR. However, bee bread supplementation significantly decreased ($p < 0.05$) blood glucose and insulin levels as well as improved the insulin resistance as shown by the reduced ($p < 0.05$) HOMA-IR index and those effects were also present in the OB + OR group, except for the HOMA-IR index.

Table 4. Serum fasting glucose, insulin resistance and lipid profile of rats in the experimental groups.

	CON	OB	OB + BB	OB + OR
Fasting glucose (mg/dL)	70.50 ± 1.45	83.00 ± 3.54 ^{a}	73.22 ± 1.36 ^{b}	74.50 ± 1.93 ^{b}
Fasting insulin (ng/mL)	0.64 ± 0.09	3.71 ± 1.33 ^{a}	1.22 ± 0.22 ^{b}	1.80 ±0.09 ^{b}
HOMA-IR	0.12 ± 0.01	0.38 ± 0.07 ^{a}	0.22 ± 0.03 ^{b}	0.26 ± 0.05 ^{a}
TG (mmol/L)	0.49 ± 0.03	0.92 ± 0.03 ^{a}	0.60 ± 0.06 ^{b}	0.77 ± 0.08 ^{a}
TC (mmol/L)	1.66 ± 0.10	2.79 ± 0.40 ^{a}	1.97 ± 0.06 ^{b}	2.01 ± 0.09 ^{b}
LDL-C (mmol/L)	0.80 ± 0.04	1.80 ± 0.24 ^{a}	1.12 ± 0.10 ^{b}	1.03 ± 0.08 ^{b}
HDL-C (mmol/L)	0.53 ± 0.02	0.40 ± 0.02 ^{a}	0.49 ± 0.04	0.66 ± 0.03 ^{a,b,c}

Data are presented as mean ± SEM, n = 8/group. CON, control; OB, obese; OB + BB, obese + bee bread 0.5 g/kg body weight/day; OB + OR, obese + orlistat 10 mg/kg body weight/day; HOMA-IR, homeostatic model of assessment-insulin resistance; TG, triglyceride; TC, total cholesterol; LDL-C, low-density lipoprotein cholesterol; HDL-C, high-density lipoprotein cholesterol. One-way ANOVA, followed by Tukey post-hoc test. ^{a} $p < 0.05$ vs. CON group, ^{b} $p < 0.05$ vs. OB group, ^{c} $p < 0.05$ vs. BB group.

In addition, biochemical analysis of the lipid profile was also evaluated in the present study to determine the effect of bee bread on hyperlipidemia. After 12 weeks of HFD administration, compared to the CON group, serum TG, TC and LDL-C were increased significantly ($p < 0.05$), meanwhile serum HDL was reduced significantly ($p < 0.05$) in the OB group. Bee bread treatment significantly reduced ($p < 0.05$) these serum lipid levels, except for HDL-C, compared to those in the OB group, whereas, orlistat administration markedly reduced ($p < 0.05$) serum TC and LDL-C and significantly increased ($p < 0.05$) serum HDL-C compared to those in the OB group (Table 4).

3.4. *Effects of* H. itama *Bee Bread on Accumulations of Hepatic Lipid, NASH Activity and Glycogen*

The liver is a vital organ for controlling lipid metabolism and prolonged excessive lipid accumulation commonly leads to hepatic steatosis. Figure 2A,B demonstrates the significantly increased ($p < 0.05$) hepatic TG and TC levels in the OB group after 12 weeks of HFD intake, compared to the CON group, whereas, treatments with bee bread and orlistat markedly alleviated ($p < 0.05$) these hepatic lipid contents.

Photomicrographs of the liver samples stained with H&E are shown in Figure 2C. In the CON group, the liver structure was normal without any pathological symptoms. In this specimen, there was high preservation of hepatocytes and the lining cells of both sinusoids and postinusoidal venules, as well as the structural integrity of the hepatic lobule. In the hepatocytes of obese rats, a large number of lipid droplets (micro- and macro-vesicular steatosis) were observed. The liver tissue from this group also had more dilated sinusoids, portal triads and inflammatory cells infiltrated largely deposited at the portal region. In contrast, these histopathological changes were clearly reduced in the bee bread and orlistat groups although they were not completely reversed. Moreover, the NAS score of the liver samples is shown in Figure 2E. Clearly, the liver from the CON group had no presence of MAFLD. In contrast, the liver of the OB group demonstrated an average NAS of 7, which markedly demonstrated the presence of NASH in this group. Treatments with bee bread and orlistat demonstrated an average NAS of $\leq$4 which revealed the presence of simple steatosis in these groups.

Increased hepatic gluconeogenesis, reduced glycogenesis and elevation in lipogenesis are common features of impaired hepatic insulin sensitivity. Therefore, we next evaluated the effect of bee bread on glycogen storage in the liver tissue using PAS staining (Figure 2D). The livers of the CON group displayed a high PAS staining intensity, which reflected the number of glycogen particles in the cytoplasm of hepatocytes, whereas, the staining intensity in the OB group was dramatically alleviated. Moreover, in the area of severe fat accumulation, positive staining was barely observed. The staining intensity of glycogen in the liver of the bee bread-treated group was much stronger than that in the OB group and this finding was similarly observed in the orlistat group. In addition, Figure 2F shows the grading for glycogen in all the liver sections. The results demonstrated a substantially decreased ($p < 0.05$) glycogen score in the OB group in comparison with the CON group, meanwhile, treatments with bee bread and orlistat significantly improved ($p < 0.05$) these changes.

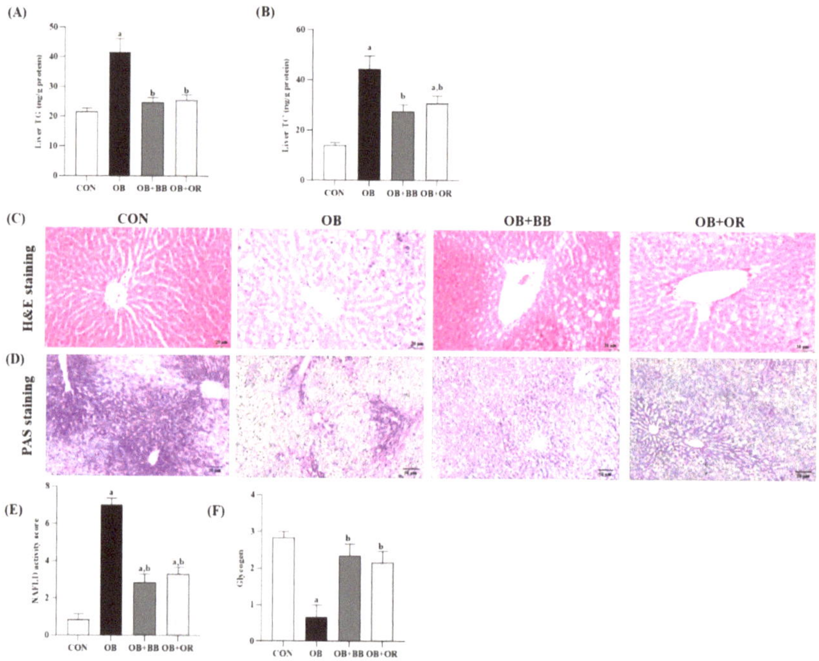

Figure 2. Effect of bee bread on liver lipid levels and hepatic steatosis. (**A**) Liver TG level, (**B**) liver TC level, (**C**) H&E staining (magnification ×400, scale bars represent 20 μm) and (**D**) PAS staining (magnification ×100, scale bars represent 50 μm) analyses of liver, (**E**) NASH scoring, and (**F**) glycogen scoring of rats in all experimental groups. Values are presented as mean ± SEM, n = 8/group. CON, control; OB, obese; OB + BB, obese + bee bread 0.5 g/kg body weight/day; OB + OR, obese + orlistat 10 mg/kg body weight/day; TG, triglyceride; TC, total cholesterol. One-way ANOVA, followed by Tukey post-hoc test. ^{a} $p < 0.05$ vs. CON group, ^{b} $p < 0.05$ vs. OB group.

3.5. Effect of *H. itama* Bee Bread on Liver Oxidant–Antioxidant Parameters

Oxidative stress is linked with obesity-related fatty liver disease. We explored the effects of bee bread supplementation on several oxidant-antioxidant markers, as well as on the Keap1 and Nrf2 pathways as reported in Figures 3 and 4, respectively. Our results showed that, compared to the CON group, there were significantly increased ($p < 0.05$) concentrations of oxidative stress markers including TBARS and NO in the OB group. Furthermore, there were also significant reductions ($p < 0.05$) in the activities of SOD, CAT, GPx, GST and GR enzymes, as well as in the levels of GSH and TAC in the OB group compared to the CON group. Treatment with bee bread notably reversed ($p < 0.05$) all these effects and equivalent outcomes were also demonstrated by the OB + OR group following orlistat treatment (Figure 3A–I).

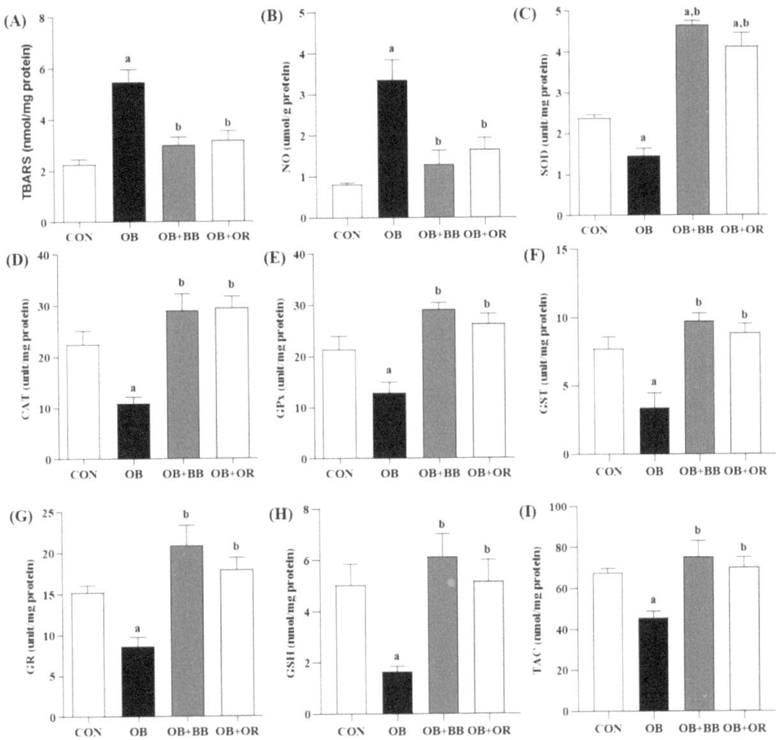

Figure 3. Effect of bee bread on liver oxidant–antioxidant parameters of rats in all experimental groups. (**A**) TBARS, (**B**) NO, (**C**) SOD, (**D**) CAT, (**E**) GPx, (**F**) GST, (**G**) GR, (**H**) GSH and (**I**) TAC in the liver of obese rats. Values are presented as mean ± SEM, n = 8/group. CON, control; OB, obese; OB + BB, obese + bee bread 0.5 g/kg body weight/day; OB + OR, obese + orlistat 10 mg/kg body weight/day; TBARS, thiobarbituric acid reactive substances; NO, nitric oxide; SOD; superoxide dismutase; CAT; catalase; GPx, glutathione peroxidase; GST, glutathione S-transferase; GR; glutathione reductase; GSH, total glutathione; TAC; total antioxidant capacity. One-way ANOVA, followed by Tukey post-hoc test. ^{a} $p < 0.05$ vs. CON group, ^{b} $p < 0.05$ vs. OB group.

Next, we used immunohistochemical analysis to further discuss the antioxidative effect of bee bread in activating the Keap1/Nrf2 pathway in the liver of obese rats, as shown in Figure 4. There was a significantly increased ($p < 0.05$) liver Keap1 expression level in the cytoplasm of obese rats, whereas, 6-week bee bread treatment significantly suppressed ($p < 0.05$) this Keap1 expression level (Figure 4A,C). Furthermore, the results also showed that, compared to the control group, the expressions of Nrf2 both in the cytoplasm and nucleus decreased ($p < 0.05$) after overfeeding of HFD for 12 weeks in the obese group, which indicated a lower rate of Nrf2 translocation from the cytoplasm into the nucleus. At the same time, the immunoexpression results clearly showed that the expression of cytoplasmic Nrf2 was significantly decreased ($p < 0.05$) and the expression of nuclear Nrf2 was significantly increased ($p < 0.05$) after bee bread treatment, and the results were also comparable in the orlistat group when compared to the obese group. These results demonstrated a higher translocation of cytoplasmic Nrf2 into the nucleus following bee bread treatment and these findings were similarly observed after orlistat administration in the OB + OR group (Figure 4B,D).

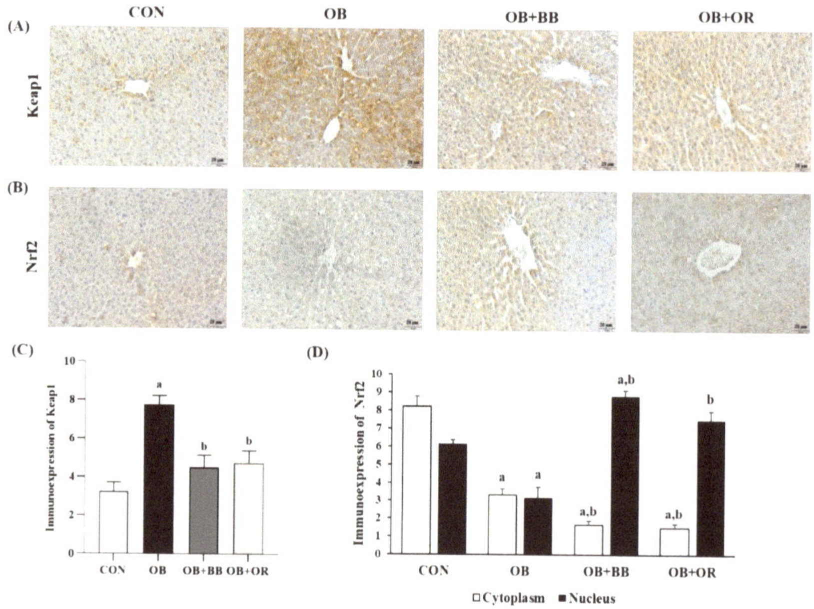

Figure 4. Effect of bee bread on (**A**) Keap1 and (**B**) Nrf2 immunohistochemical expressions in liver sections of rats in all experimental groups. (**C**) Immunohistochemical analyses of Keap1 and (**D**) cytoplasmic and nuclear Nrf2 in liver sections. Magnification ×400, scale bars represent 20 μm. Values are presented as mean ± SEM, $n = 6$/group. CON, control; OB, obese; OB + BB, obese + bee bread 0.5 g/kg body weight/day; OB + OR, obese + orlistat 10 mg/kg body weight/day; Keap1, Kelch-like ECH-associated protein 1; Nrf2, nuclear factor erythroid 2-related factor 2. One-way ANOVA, followed by Tukey post-hoc test. ^{a} $p < 0.05$ vs. CON group, ^{b} $p < 0.05$ vs. OB group.

3.6. Effects of *H. itama* Bee Bread on the Expression of Hepatic Lipid Metabolism-Related Genes, and SIRT1 and AMPK Protein Levels

To further investigate the underlying mechanisms by which bee bread improves hepatic lipid metabolism, we analyzed the effects of bee bread on genes related to lipogenesis and fatty acid β-oxidation as shown in Figure 5A–D. Compared with the CON group, the mRNA expression levels of both SREBP-1c and FAS were significantly up-regulated ($p < 0.05$) in the liver tissues of rats in the OB group. Treatment with bee bread for 6 weeks significantly inhibited lipid synthesis as shown by the down-regulation ($p < 0.05$) of these mRNA factors and the effects were also comparable after intake of orlistat. Furthermore, our results also demonstrated significantly down-regulated ($p < 0.05$) PPARα and CPT1α mRNA expression levels in the OB group than those in the CON group, whereas, these effects were significantly abolished ($p < 0.05$) by both bee bread and orlistat treatments.

The present study also evaluated the relationship between the anti-lipid effect of bee bread and the levels of SIRT1 and AMPK in the liver as demonstrated in Figure 5E,F). Our results showed that rats in the OB group had notably suppressed ($p < 0.05$) SIRT1 and AMPK levels more than those in the CON group. Meanwhile, bee bread treatment markedly elevated ($p < 0.05$) these protein levels. Comparable outcomes ($p < 0.05$) were also demonstrated in the liver of rats in the orlistat group except for the SIRT1 level which was not significantly changed ($p > 0.05$) after 6 weeks of treatment.

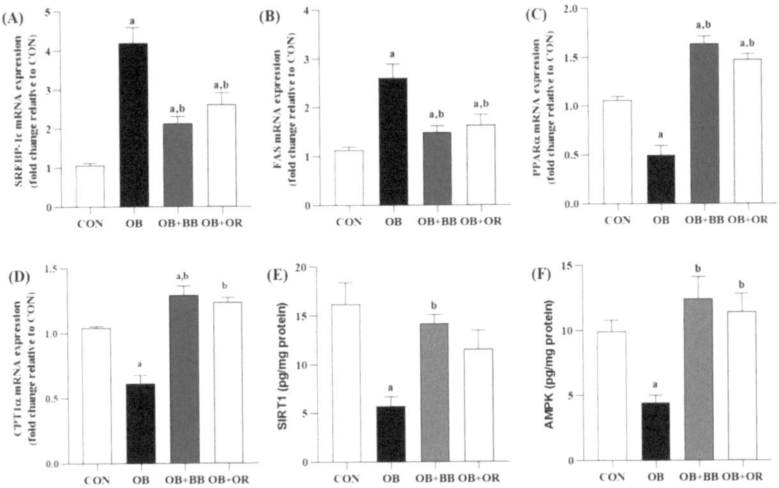

Figure 5. Effect of bee bread on hepatic lipid metabolism-related genes, and SIRT1 and AMPK levels. (**A–D**) The mRNA expression levels of SREBP-1c, FAS, CPT1α and SIRT1 were determined by RT-qPCR assays (n = 6/group). (**E,F**) The levels of SIRT1 and AMPK were detected using ELISA (n = 8/group). Values are presented as mean ± SEM. CON, control; OB, obese; OB + BB, obese + bee bread 0.5 g/kg body weight/day; OB + OR, obese + orlistat 10 mg/kg body weight/day; SREBP-1c, sterol regulatory element-binding protein-1c; FAS, fatty acid synthase; PPARα, peroxisome proliferator activated receptor α; CPT1α, carnitine palmitoyl transferase 1α; SIRT1, sirtuin 1; AMPK, AMP-activated protein kinase. One-way ANOVA, followed by Tukey post-hoc test. ^{a} $p < 0.05$ vs. CON group, ^{b} $p < 0.05$ vs. OB group.

4. Discussion

Obesity is a chronic metabolic disease that can affect the majority of the vital organs of the body, and studies have reported that obesity is a high-risk factor for the development of cardiovascular diseases [68], type 2 diabetes [69] and carcinoma [70]. Apart from increases in the size of adipocytes (hypertrophy) and number of adipocytes (hyperplasia), being overweight and obese also leads to the excessive accumulation of lipid in the liver, resulting in hepatic steatosis, the early stage of MAFLD [71]. We had previously reported excessive intake of HFD results in elevated liver fat contents, oxidative damage and abnormal inflammatory responses in obese rat models fed with HFD for 12 weeks. Subsequently, the supplementation of *H. itama* bee bread, concurrently given with the HFD for the same period exerted its hepatoprotective effect against the above parameters [40]. In addition, the latest study by Li et al. 2021 [41] also demonstrated the anti-lipogenic effect of bee bread against FAS and ACC levels in the HFD-induced fatty liver disease rat model. Hence, to date it is not known whether *H. itama* bee bread has a therapeutic effect in treating MAFLD after the induction of obesity in the animal model.

In the present study, we found that *H. itama* bee bread is rich in phenols (2-hydroxycinnamic acid, trans 3-hydroxycinnamic acid, trans ferulic acids) and flavonoids (caffeic acid, apigenin, kaempferol, quercetin, mangiferin). The highest amount of compound found in both aqueous and methanol extracts of this bee bread was trans 3-hydroxycinnamic acid, followed by 2-hydroxycinnamic acid and quercetin. This finding is consistent with the Indian and Romanian bee bread samples which demonstrated a relatively high amount of hydroxycinnamic acid derivatives, quercetin and kaempferol [72]. However, it is suggested that the exact amount of each phenolic compound present in the *H. itama* bee bread is further quantitated in future studies. Both hydroxycinnamic acid derivatives and quercetin exhibit numerous biological and pharmacological effects, such as anti-obesity, anti-lipid, antioxidant and anti-inflammatory activities, and the potential therapeutic benefits in experimental diabetes and hyperlipidemia [73–77]. Our results suggest that *H. itama* bee bread may serve as the treatment for obesity-related metabolic disorders including hyperglycemia, insulin resistance and hyperlipidemia in MAFLD. We also focused on and demonstrated the therapeutic effects of *H. itama* bee bread in inhibiting Keap1, activating the Nrf2 antioxidant pathway and then restoring the dysregulation of genes associated with hepatic lipid metabolism such as SREBP-1c, FAS, PPARα and CPT1α, as well as its modulators such as SIRT1 and AMPK, being consistent with its increase in glycogen accumulation and reduction in hepatic steatosis and progression towards NASH in the liver of obese male rats.

The present study demonstrated significant increases in the Lee obesity index and body weight gain in the obese group than those in the control group, which are consistent with our previously published reports [40,78]. Furthermore, excess energy or calorie intake following consumption of HFD led to an increased accumulation of lipids in the adipocytes, hence resulting in an increased Lee obesity index and body weight gain in this group, as reported by previous studies [33,37]. *H. itama* bee bread was able to reduce these parameters significantly although no changes in food and calorie intakes were observed in this group, compared to the obese group. These results demonstrated the anti-obesity effect of *H. itama* bee bread in which it increased weight loss without reducing the food intake; hence, these results were in line with previous published studies [33,37,40].

Indeed, many studies have reported the presence of a strong relationship between obesity and impaired insulin sensitivity including a reduction in the number of insulin receptors and receptor function defects, leading to hyperglycemia, hyperinsulinemia and insulin resistance [16,79–81]. The present study demonstrated that long-term intake of an HFD significantly elevated the fasting blood glucose and insulin levels in the obese group than those in the control group, meanwhile *H. itama* bee bread supplementation markedly reduced these parameters. In addition, the HOMA-IR index was also used in the present study, which is a common indicator to evaluate insulin sensitivity, insulin resistance and pancreatic β-cell function in diabetic patients [82,83]. Our present research showed that the HOMA-IR index was significantly increased in the obese group than in the control group, meanwhile, the HOMA-IR index was significantly reduced in the bee bread group compared with the obese group. These results showed that *H. itama* bee bread treatment alleviated obesity-induced hyperglycemia, hyperinsulinemia and insulin resistance, hence improving insulin sensitivity [40].

Moreover, hyperlipidemia is also commonly presented in obese patients and recognized as one of the hallmarks of MAFLD [84,85]. Our findings presented significant increased TG, TC and LDL-C levels and a significantly reduced HDL-C level in the obese group than those in the control group, hence indicating the presence of hyperlipidemia in this group, which is consistent with other previously reported studies in *Sprague Dawley* rats fed with HFD [86,87]. The bee bread-treated group showed significantly reduced serum levels of non-HDL lipids without affecting the level of HDL-C in this group, which is in agreement with a prior finding by Othman et al. 2021 [34]. The present study, however, contradicts our previous study using *H. itama* bee bread on MAFLD's rat protective model. The hypolipidemic effect of this bee bread might be attributed to the high amounts of phenols and flavonoids found in the *H. itama* bee bread sample as reported in a prior study [46]. Likewise, a former finding also reported that the potential of bee bread in reducing the non-HDL lipids might be through the action of saponin present in the bee bread which interacts with the dietary fat compositions and excretes out the lipids from the body via the feces [88].

Although the underlying mechanism for the development and progression of MAFLD is complex and multifactorial, it has been generally believed that MAFLD is a disease caused by a "second hit" with excessive lipid peroxidation and oxidative stress after hepatic fat accumulation (hepatic steatosis) that serves as the "first hit" [89]. Overfeeding of dietary fats causes the liver to become susceptible to toxins and oxidative stress, resulting in hepatocyte inflammation by the activation of oxidative stress, pro-inflammatory cytokines and hepatocyte mitochondrial dysfunction [90]. *H. itama* bee bread reduced TBARS and NO levels and increased the activities of SOD, CAT, GPx, GST and GR, as well as the levels of GSH and TAC in the liver of obese rats. These findings are correlated with previously reported studies using this bee bread which showed raised antioxidant enzyme activities in the liver [40], testis [39], aorta [35], heart [34] and renal [36] of the obese rats. The Nrf2 antioxidant pathway is a central cellular system in its defense against oxidative stress [91,92]. Suppression of Keap1 activity reduces the production of ROS, meanwhile Nrf2 knockout leads to an increased accumulation of ROS in mouse primary hepatocytes [93]. Elevated lipid peroxidation resulting from excess lipid accumulation in the liver promotes the dissociation of Nrf2 from Keap1 and translocation of Nrf2 from the cytoplasm into the nucleus to activate the antioxidant pathway by binding to the ARE in the promoter region of antioxidant and stimulates the expressions of antioxidant enzymes [94]. Recent studies have reported that HFD stimulates liver oxidative stress via the inhibition of Nrf2 nuclear translocation to suppress the expression of antioxidant enzymes [15,95]. Our present study demonstrated the up-regulated expression of Keap1 and the reduced translocation of Nrf2 from the cytoplasm into the nucleus in the liver of rats from the obese group. Meanwhile, *H. itama* bee bread was found to inhibit Keap1 and promote the translocation of Nrf2 from the cytoplasm into the nucleus. These results were in parallel with the increased activities of antioxidant enzymes such as SOD, CAT, GPx, GST and GR in the liver of the bee bread group. The increased activities of antioxidant enzymes could be due to increased expressions of antioxidant genes, the measurements of which should be undertaken in future studies. Keap1 suppression and Nrf2 antioxidant pathway activation with the high activities of antioxidant enzymes by *H. itama* bee bread might mitigate HFD-triggered ROS, subsequently effectively restoring hepatic redox status imbalance in the obese rats. The beneficial property of *H. itama* bee bread demonstrated in the present study might be attributed to its rich sources of phenolic compounds such as hydroxycinnamic acid derivatives, caffeic acid, gallic acid, ferulic acid, quercetin, apigenin, kaempferol and mangiferin, which are reported to possess antioxidant properties [44,75,96].

The imbalance between lipid acquisition (i.e., high fatty acid uptake and lipogenesis) and removal (i.e., low mitochondrial fatty acid oxidation and export of lipids) can result in an impaired lipid metabolism which later leads to several metabolic disorders including obesity, metabolic syndromes and MAFLD [97]. Of note, Nrf2 has been shown to inhibit ethanol-induced hepatic steatosis in mice [93,98]. Meanwhile, Keap1-knockout with Nrf2-enhanced mice suppressed ethanol-induced hepatic fat accumulation by reducing the expressions of SREBP-1 and SCD-1 genes in mice [93]. Similarly, Nrf2 deficiency down-regulates PPARα but increases SREBP-1c to enhance TG contents in the ethanol-exposed human hepatocytes cell line, whereas, increased Nrf2 expression reversed these malfunctions [98]. In our study, hyperinsulinemia with increased gene expression of SREBP-1c and its target gene FAS were found in the obese group compared to the control group. *H. itama* bee bread was shown to down-regulate these gene expression levels which might be accredited to reduced hyperinsulinemia in the bee bread-treated group. These findings are in line with a previous study using bee bread in HFD-induced fatty liver rats [41]. In addition, the decreased lipogenesis in this group following *H. itama* bee bread administration might explain the reduced contents of hepatic TG and TC in this group, which is in line with previous studies [16,99]. Our results demonstrated an impairment in hepatic fatty acid β-oxidation in the obese group, as indicated by the reductions in PPARα and CPT1α mRNA levels in this group. These results are in agreement with previous findings which reported that the disruption in fatty acid β-oxidation is associated with reductions in PPARα and CPT1α expressions or activities in humans and rodents with MAFLD [100–103]. Our study demonstrated that *H. itama* bee bread can enhance PPARα and CPT1α mRNA expression levels, hence increasing the fatty acid β-oxidation in the obese rats treated with bee bread. Furthermore, the reduced liver TG and TC contents in the bee bread group might also be attributed to the therapeutic ability of *H. itama* bee bread in increasing fatty acid β-oxidation and ATP production, as well as to reduce inflammation and oxidative stress [40]. Therefore, the down-regulation of *H. itama* bee bread on oxidative stress via the mediation of the Keap1/Nrf2 pathway might play a role in reducing liver lipid deposition in obese rats.

The liver of obese rats has reduced AMPK activation [104–106]. SIRT1 activates AMPK expression, promotes AMPK phosphorylation and suppresses SREBP-1c cleavage and nuclear translocation [20]. Polyphenols including mangiferin, resveratrol, quercetin and catechin, which were also found in the *H. itama* bee bread of the present study, have been reported to be one of the SIRT1 inducers and improve hepatic lipid metabolism via the promotion of the SIRT1/AMPK pathway in vitro and in vivo of the MAFLD model [99,107,108]. Moreover, treatment with bee pollen polysaccharide from *Rosa rugosa* alleviated hepatic steatosis and insulin resistance by promoting the phosphorylation of AMPK in HepG2 cells and HFD-induced animal models [109]. Previous findings have demonstrated that excessive lipid deposition in the liver can restrain the AMPK substrate ACC phosphorylation, promoting lipid synthesis [71]. A reduced level of AMPK may result in excessive hepatic lipid accumulation, accelerating steatosis and MAFLD, hence further demonstrating that AMPK is an energy sensor regulator to maintain lipid and glucose metabolism in the liver. It is also has been reported that the activation of AMPK promotes the Nrf2 antioxidant pathway [110,111]. In this study, we demonstrated that *H. itama* bee bread treatment significantly elevated SIRT1 and AMPK levels and the results were consistent with the reduced insulin resistance levels of lipogenic genes SREBP-1c and FAS, as well as levels of TG and TC in the liver, which might be accredited to its activation on Keap1/Nrf2 signaling pathway. However, it is suggested that the level of activated AMPK (p-AMPK) is measured in future studies to further support these findings.

Prolonged intake of HFD leads the liver to store more TGs, causing hepatic steatosis and stimulating the development of NASH, which can further progress to fibrosis and cirrhosis [112]. In the present study, the histopathological finding using H&E staining also revealed that the liver of rats in the obese group demonstrated pronounced hepatic steatosis, hepatocyte hypertrophy and hepatocellular inflammation, resulting in an elevated NAS score, hence indicating the presence of active NASH in this group. *H. itama* bee bread treatment mitigated hepatic steatosis and inflammation, significantly reducing the NAS score and preventing NASH in comparison with the rats in the obese group. These results further support our previous reported data on the beneficial protective effect of *H. itama* bee bread against NASH progression in the HFD-induced obese rats [40]. In addition, peripheral and hepatic insulin resistance, which are linked to the development of MAFLD, lead to insufficient suppression of hepatic gluconeogenesis, reduced glycogenesis and elevation in lipogenesis [113]. In the present study, we found that *H. itama* bee bread was able to significantly recover the reduced glycogen accumulation in the liver tissue of obese rats which might indicate that this bee bread could modulate hepatic glycogenesis, increase insulin sensitivity and maintain the metabolic function of the liver.

5. Conclusions

We found that treatment with *H. itama* bee bread inhibited Keap1 and activated the Nrf2 antioxidant pathway with a reduction in oxidative stress in obese rat livers. This therapeutic potential of *H. itama* bee bread subsequently improved lipid metabolism-related genes (SREBP-1c, FAS, PPARα and CPT1α) and its regulators SIRT1 and AMPK; hence, lipid accumulation and the progression of NASH were alleviated in the liver of obese rats. These improvements might be partly attributed to the presence of phenolic compounds in the *H. itama* bee bread which activate the Keap1/Nrf2 signaling pathway.

Supplementary Materials: The following are available online at https://www.mdpi.com/article/10.3390/antiox11112190/s1, Table S1: Lee obesity index and liver enzymes, Figure S1: Histology of liver.

Author Contributions: Conceptualization, M.M.; methodology, M.M., N.A.C.J. and W.S.W.G.; software, Z.Z.; validation, M.M. and N.A.C.J.; formal analysis, Z.Z.; investigation, Z.Z., Z.A.O. and J.B.S.; resources, M.M.; data curation, Z.Z., M.M. and N.A.C.J.; writing—original draft preparation, Z.Z.; writing—review and editing, Z.Z., M.M., N.A.C.J., W.S.W.G., K.M.F.M., N.N.Z. and K.A.K.; visualization, Z.Z.; supervision, M.M., W.S.W.G., N.A.C.J. and K.M.F.M. project administration, Z.Z.; funding acquisition, Z.Z., Z.A.O. and M.M. All authors have read and agreed to the published version of the manuscript.

Funding: This research was funded by the Universiti Sains Malaysia, Kubang Kerian, Kelantan, Malaysia (GIPS-PhD no 311/PPSP/4404813 and USM research projects no 1001.PPSP.8012378).

Institutional Review Board Statement: The study was conducted according to the guidelines of the Declaration of Helsinki, and approved by the USM Institutional Animal Care and Use Committee (USM IACUC) (USM/IACUC/2018/(113)(933)).

Informed Consent Statement: Not applicable.

Data Availability Statement: All of the data are contained within the article.

Acknowledgments: The authors would like to thank all the staff from the Physiology Laboratory and Central Research Laboratory (CRL) USM, for their technical assistant during the experimental study.

Conflicts of Interest: The authors declare no conflict of interest.

References

1. Chooi, Y.C.; Ding, C.; Magkos, F. The epidemiology of obesity. *Metabolism* **2019**, *92*, 6–10. [CrossRef] [PubMed]
2. Kyle, T.K.; Dhurandhar, E.J.; Allison, D.B. Regarding obesity as a disease: Evolving policies and their implications. *Endocrinol. Metab. Clin. N. Am.* **2016**, *45*, 511–520. [CrossRef]
3. Eslam, M.; Newsome, P.N.; Sarin, S.K.; Anstee, Q.M.; Targher, G.; Romer-Gomez, M.; Zelber-Sagi, S.; Wai-Sun Wong, V.; Dufor, J.F.; Schattenberg, J.M.; et al. A new definition for metabolic dysfunction-associated fatty liver disease: An international consensus statement. *J. Hepatol.* **2020**, *73*, 202–209. [CrossRef] [PubMed]
4. Eslam, M.; Sanyal, A.J.; George, J. MAFLD: A consensus-driven proposed nomenclature for metabolic associated fatty liver disease. *Gastroenterology* **2020**, *158*, 1999–2014.e1. [CrossRef] [PubMed]
5. García-Ruiz, C.; Fernández-Checa, J.C. Mitochondrial oxidative stress and antioxidants balance in fatty liver disease. *Hepatol. Commun.* **2018**, *2*, 1425–1439. [CrossRef]
6. Shin, S.K.; Cho, H.W.; Song, S.E.; Bae, J.H.; Im, S.S.; Hwang, I.; Ha, H.; Song, D.K. Ablation of catalase promotes non-alcoholic fatty liver via oxidative stress and mitochondrial dysfunction in diet-induced obese mice. *Pflugers Arch.-Eur. J. Physiol.* **2019**, *471*, 829–843. [CrossRef]
7. Lionetti, L.; Mollica, M.P.; Donizzetti, I.; Gifuni, G.; Sica, R.; Pignalosa, A.; Cavaliere, G.; Gaita, M.; De Filippo, C.; Zorzano, A.; et al. High-lard and high-fish-oil diets differ in their effects on function and dynamic behaviour of rat hepatic mitochondria. *PLoS ONE* **2014**, *9*, e92753. [CrossRef]
8. Masarone, M.; Rosato, V.; Dallio, M.; Gravina, A.G.; Aglitti, A.; Loguercio, C.; Federico, A.; Persico, M. Role of oxidative stress in pathophysiology of nonalcoholic fatty liver disease. *Oxidat. Med. Cell. Longev.* **2018**, *2018*, 9547613. [CrossRef]
9. Lotfi, A.; Saneei, P.; Hekmatdost, A.; Salehisahlabadi, A.; Shiranian, A.; Ghiasvand, R. The relationship between dietary antioxidant intake and physical activity rate with nonalcoholic fatty liver disease (NAFLD): A case-control study. *Clin. Nutr. ESPEN* **2019**, *34*, 45–49. [CrossRef]
10. Świderska, M.; Maciejczyk, M.; Zalewska, A.; Pogorzelska, J.; Flisiak, R.; Chabowski, A. Oxidative stress biomarkers in the serum and plasma of patients with non-alcoholic fatty liver disease (NAFLD). Can plasma AGE be a marker of NAFLD? Oxidative stress biomarkers in NAFLD patients. *Free Radic. Res.* **2019**, *53*, 841–850. [CrossRef]
11. Sohouli, M.H.; Fatahi, S.; Sayyari, A.; Olang, B.; Shidfar, F. Associations between dietary total antioxidant capacity and odds of non-alcoholic fatty liver disease (NAFLD) in adults: A case–control study. *J. Nutr. Sci.* **2020**, *9*, e48. [CrossRef] [PubMed]
12. Braud, L.; Battault, S.; Meyer, G.; Nascimento, A.; Gaillard, S.; de Sousa, G.; Rahmani, R.; Riva, C.; Armand, M.; Maixent, J.M.; et al. Antioxidant properties of tea blunt ROS-dependent lipogenesis: Beneficial effect on hepatic steatosis in a high fat-high sucrose diet NAFLD obese rat model. *J. Nutr. Biochem.* **2017**, *40*, 95–104. [CrossRef] [PubMed]
13. Nguyen, T.; Nioi, P.; Pickett, C.B. The Nrf2-antioxidant response element signaling pathway and its activation by oxidative stress. *J. Biol. Chem.* **2009**, *284*, 13291–13295. [CrossRef] [PubMed]
14. Sharma, R.S.; Harrison, D.J.; Kisielewski, D.; Cassidy, D.M.; McNeilly, A.D.; Gallagher, J.R.; Walsh, S.V.; Honda, T.; McCrimmon, R.J.; Dinkova-Kostova, A.T.; et al. Experimental nonalcoholic steatohepatitis and liver fibrosis are ameliorated by pharmacologic activation of Nrf2 (NF-E2 p45-related factor 2). *Cell. Mol. Gastroenterol. Hepatol.* **2018**, *5*, 367–398. [CrossRef] [PubMed]

15. Gao, G.; Xie, Z.; Li, E.; Yuan, Y.; Fu, Y.; Wang, P.; Zhang, X.; Qiao, Y.; Xu, J.; Hölscher, C.; et al. Dehydroabietic acid improves nonalcoholic fatty liver disease through activating the Keap1/Nrf2-ARE signaling pathway to reduce ferroptosis. *J. Nat. Med.* **2021**, *75*, 540–552. [CrossRef]
16. Pai, S.A.; Munshi, R.P.; Panchal, F.H.; Gaur, I.S.; Mestry, S.N.; Gursahani, M.S.; Juvekar, A.R. Plumbagin reduces obesity and nonalcoholic fatty liver disease induced by fructose in rats through regulation of lipid metabolism, inflammation and oxidative stress. *Biomed. Pharmacother.* **2019**, *111*, 686–694. [CrossRef]
17. Ferré, P.; Foufelle, F. Hepatic steatosis: A role for *de novo* lipogenesis and the transcription factor SREBP-1c. *Diabetes Obes. Metab.* **2010**, *12*, 83–92. [CrossRef]
18. Silva, A.K.S.; Peixoto, C.A. Role of peroxisome proliferator-activated receptors in non-alcoholic fatty liver disease inflammation. *Cell. Mol. Life Sci.* **2018**, *75*, 2951–2961. [CrossRef]
19. Day, E.A.; Ford, R.J.; Steinberg, G.R. AMPK as a therapeutic target for treating metabolic diseases. *Trends Endocrinol. Metab.* **2017**, *28*, 545–560. [CrossRef]
20. Karbasforooshan, H.; Karimi, G. The role of SIRT1 in diabetic retinopathy. *Biomed. Pharmacother.* **2018**, *97*, 190–194. [CrossRef]
21. Lopez, M. Hypothalamic AMPK and energy balance. *Eur. J. Clin. Investig.* **2018**, *48*, e12996. [CrossRef] [PubMed]
22. Liou, C.-J.; Wu, S.-J.; Shen, S.-C.; Chen, L.-C.; Chen, Y.-L.; Huang, W.-C. Phloretin ameliorates hepatic steatosis through regulation of lipogenesis and Sirt1/AMPK signaling in obese mice. *Cell Biosci.* **2020**, *10*, 114. [CrossRef] [PubMed]
23. Xia, S.F.; Shao, J.; Zhao, S.Y.; Qiu, Y.Y.; Teng, L.P.; Huang, W.; Wang, S.S.; Cheng, X.R.; Jiang, Y.Y. Niga-ichigoside F1 ameliorates high-fat diet-induced hepatic steatosis in male mice by Nrf2 activation. *Food Funct.* **2018**, *9*, 906–916. [CrossRef]
24. Mohd, N.; Mohd, F.; Sajap, A.S.; Rosliza, J.; Suri, R. Conservation and sustainable utilization of stingless bees for pollination services in agricultural ecosystems in Malaysia. In Proceedings of the International Seminar on Enhancement of Functional Biodiversity Relevant to Sustainable Food Production in ASPAC, Tsukuba, Japan, 9–11 November 2010; pp. 1–11. Available online: http://www.niaes.affrc.go.jp/sinfo/sympo/h22/1109/paper_04.pdf (accessed on 16 June 2022).
25. Mustafa, M.Z.; Yaacob, N.S.; Sulaiman, S.A. Reinventing the Honey Industry: Opportunities of the Stingless Bee. *Malays. J. Med. Sci.* **2018**, *25*, 1–5. [CrossRef]
26. Barajas, J.; Cortes-Rodriguez, M.; Rodríguez-Sandoval, E. Effect of temperature on the drying process of bee pollen from two zones of Colombia. *J. Food Process Eng.* **2012**, *35*, 134–148. [CrossRef]
27. Vásquez, A.; Olofsson, T.C. The lactic acid bacteria involved in the production of bee pollen and bee bread. *J. Apic. Res.* **2009**, *48*, 189–195. [CrossRef]
28. Andjelkovic, B.; Jevtic, G.; Markovic, J.; Mladenovic, M.; Pseva, V. Quality of honey bee bread collected in spring. In Proceedings of the International Symposium on Animal Science 2014, Belgrade, Serbia, 23–25 September 2014; pp. 450–454. Available online: http://arhiva.nara.ac.rs/handle/123456789/699 (accessed on 19 April 2021).
29. Khalifa, S.A.; Elashal, M.; Kieliszek, M.; Ghazala, N.E.; Farag, M.A.; Saeed, A.; Xiao, J.; Zou, X.; Khatib, A.; Göransson, U.; et al. Recent insights into chemical and pharmacological studies of bee bread. *Trends Food Sci. Technol.* **2020**, *97*, 300–316. [CrossRef]
30. Kieliszek, M.; Piwowarek, K.; Kot, A.M.; Błażejak, S.; Chlebowska-Śmigiel, A.; Wolska, I. Pollen and bee bread as new health-oriented products: A review. *Trends Food Sci. Technol.* **2018**, *71*, 170–180. [CrossRef]
31. Urcan, A.; Al-Marghitas, L.; Dezmirean, D.S.; Bobis, O.; Bonta v Murusen, C.L.; Margaoan, R. Chemical composition and biological activities of beebread—Review. *Bull. UASVM Anim. Sci. Biotechnol.* **2017**, *74*, 250–255. [CrossRef]
32. Baltrušaitytė, V.; Venskutonis, P.R.; Čeksterytė, V. Antibacterial activity of honey and beebread of different origin against *S. aureus* and *S. epidermidis*. *Food Technol. Biotechnol.* **2007**, *45*, 201–208. Available online: https://web.p.ebscohost.com/ehost/detail/detail?vid=0&sid=55dd179b-73d0-46bd-83dd-2a2087e2c989%40redis&bdata=#AN=25978276&db=fsr (accessed on 26 February 2018).
33. Othman, Z.A.; Ghazali, W.S.W.; Noordin, L.; Yusof, N.A.M.; Mohamed, M. Phenolic compounds and the anti-atherogenic effect of bee bread in high-fat diet-induced obese rats. *Antioxidants* **2020**, *9*, 33. [CrossRef] [PubMed]
34. Othman, Z.A.; Zakaria, Z.; Suleiman, J.B.; Nna, V.U.; Romli, A.C.; Ghazali, W.S.W.; Mohamed, M. Bee bread ameliorates vascular inflammation and impaired vasorelaxation in obesity-induced vascular damage rat model: The role of eNOS/NO/cGMP-signaling pathway. *Int. J. Mol. Sci.* **2021**, *22*, 4225. [CrossRef]
35. Othman, Z.A.; Zakaria, Z.; Suleiman, J.B.; Jalil, N.A.C.; Ghazali, W.S.W.; Mohamed, M. Bee bread attenuates the progression of atherosclerosis by activating Nrf2/Keap1 and modulating TNF-α/NF-κβ-associated mast cell migration and a mitochondrial-dependent apoptotic pathway in the obese rat model. *Food Funct.* **2022**, *13*, 8119–8130. [CrossRef]
36. Eleazu, C.; Suleiman, J.B.; Othman, Z.A.; Zakaria, Z.; Nna, V.U.; Hussain, N.H.N.; Mohamed, M. Bee bread attenuates high fat diet induced renal pathology in obese rats via modulation of oxidative stress, downregulation of NF-kB mediated inflammation and Bax signalling. *Arch. Physiol. Biochem.* **2020**, *128*, 1088–1104. [CrossRef] [PubMed]
37. Suleiman, J.B.; Abu Bakar, A.B.; Noor, M.M.; Nna, V.U.; Othman, Z.A.; Zakaria, Z.; Eleazu, C.O.; Mohamed, M. Bee bread mitigates downregulation of steroidogenic genes, decreased spermatogenesis, and epididymal oxidative stress in male rats fed with high-fat diet. *Am. J. Physiol. Endocrinol. Metab.* **2021**, *321*, e351–e366. [CrossRef] [PubMed]
38. Suleiman, J.B.; Nna, V.U.; Zakaria, Z.; Othman, Z.A.; Eleazu, C.O.; Bakar, A.B.A.; Ahmad, A.; Usman, U.Z.; Rahman, W.F.W.A.; Mohamed, M. Protective effects of bee bread on testicular oxidative stress, NF-κB-mediated inflammation, apoptosis and lactate transport decline in obese male rats. *Biomed. Pharmacother.* **2020**, *131*, 110781. [CrossRef] [PubMed]

39. Suleiman, J.B.; Mohamed, M.; Bakar, A.B.A.; Zakaria, Z.; Othman, Z.A.; NNa, V.U. Therapeutic effects of bee bread on obesity-induced testicular-derived oxidative stress, inflammation, and apoptosis in high-fat diet obese rat model. *Antioxidants* **2022**, *11*, 255. [CrossRef] [PubMed]
40. Zakaria, Z.; Othman, Z.A.; Suleiman, J.B.; Jalil, N.A.C.; Ghazali, W.S.W.; Nna, V.U.; Mohamed, M. Hepatoprotective effect of bee bread in metabolic dysfunction-associated fatty liver disease (MAFLD) rats: Impact on oxidative stress and inflammation. *Antioxidants* **2021**, *10*, 2031. [CrossRef]
41. Li, Z.; Huang, Q.; Liu, Y.; Peng, C.; Zeng, Z. Natural bee bread positively regulates lipid metabolism in rats. *Int. J. Agric. Sci. Food Technol.* **2021**, *7*, 266–271. [CrossRef]
42. Tanaka, M.; Sato, A.; Kishimoto, Y.; Mabashi-Asazuma, H.; Kondo, K.; Iida, K. Gallic acid inhibits lipid accumulation via AMPK pathway and suppresses apoptosis and macrophage-mediated inflammation in hepatocytes. *Nutrients* **2020**, *12*, 1479. [CrossRef]
43. Lu, J.; Meng, Z.; Cheng, B.; Liu, M.; Tao, S.; Guan, S. Apigenin reduces the excessive accumulation of lipids induced by palmitic acid via the AMPK signaling pathway in HepG2 cells. *Exp. Ther. Med.* **2019**, *18*, 2965–2971. [CrossRef] [PubMed]
44. Tian, C.; Liu, X.; Chang, Y.; Wang, R.; Lv, T.; Cui, C.; Liu, M. Investigation of the anti-inflammatory and antioxidant activities of luteolin, kaempferol, apigenin and quercetin. *Afr. J. Bot.* **2021**, *137*, 257–264. [CrossRef]
45. Bian, Y.; Lei, J.; Zhong, J.; Wang, B.; Wan, Y.; Li, J.; Liao, C.; He, Y.; Liu, Z.; Ito, K.; et al. Kaempferol reduces obesity, prevents intestinal inflammation, and modulates gut microbiota in high-fat diet mice. *J. Nutr. Biochem.* **2021**, *99*, 108840. [CrossRef]
46. Suleiman, J.B.; Mohamed, M.; Bakar, A.B.A.; Nna, V.U.; Zakaria, Z.; Othman, Z.A.; Aroyehun, A.B. Chemical Profile, Antioxidant Properties and Antimicrobial Activities of Malaysian *Heterotrigona itama* Bee Bread. *Molecules* **2021**, *26*, 4943. [CrossRef]
47. Othman, Z.A.; Noordin, L.; Omar, N.; Mohd Yusof, N.A.; Mohamaed, M. Protective Effects of orlistat on lipid profile, cardiac oxidative stress biomarkers and histology in high-fat diet-induced obese rats. *IIUM Med. J. Malays.* **2019**, *18*, 1–6. [CrossRef]
48. Bellinger, L.L.; Bernardis, L.L. Effect of dorsomedial hypothalamic nuclei knife cuts on ingestive behavior. *Am. J. Physiol. Integr. Comp. Physiol.* **1999**, *276*, R1772–R1779. [CrossRef]
49. Malafaia, A.B.; Nassif, P.A.N.; Ribas, C.A.P.M.; Ariede, B.L.; Sue, K.N.; Cruz, M.A. Obesity induction with high fat sucrose in rats. *ABCD Arquiros Bras. Cir. Dig.* **2013**, *26*, 17–21. [CrossRef]
50. Zaitone, S.A.; Essawy, S. Addition of a low dose of rimonabant to orlistat therapy decreases weight gain and reduces adiposity in dietary obese rats. *Clin. Exp. Pharmacol. Physiol.* **2012**, *39*, 551–559. [CrossRef]
51. Roza, N.A.; Possignolo, L.F.; Palanch, A.C.; Gontijo, J.A. Effect of long-term high-fat diet intake on peripheral insulin sensibility, blood pressure, and renal function in female rats. *Food Nutr. Res.* **2016**, *60*, 28536. [CrossRef]
52. Friedewald, W.T.; Levy, R.I.; Fredrickson, D.S. Estimation of the concentration of low-density lipoprotein cholesterol in plasma, without use of the preparative ultracentrifuge. *Clin. Chem.* **1972**, *18*, 499–502. [CrossRef]
53. Chatterjee, P.K.; Cuzzocrea, S.; Brown, P.A.; Zacharowski, K.; Stewart, K.N.; Mota-Filipe, H.; Thiemermann, C. Tempol, a membrane-permeable radical scavenger, reduces oxidant stress-mediated renal dysfunction and injury in the rat. *Kidney Int.* **2000**, *58*, 658–673. [CrossRef] [PubMed]
54. Al Batran, R.; Al-Bayaty, F.; Jamil Al-Obaidi, M.M.; Abdualkader, A.M.; Hadi, H.A.; Ali, H.M.; Abdulla, M.A. In vivo antioxidant and antiulcer activity of *Parkia speciosa* ethanolic leaf extract against ethanol-induced gastric ulcer in rats. *PLoS ONE* **2013**, *8*, e64751. [CrossRef] [PubMed]
55. Góth, L. A simple method for determination of serum catalase activity and revision of reference range. *Clin. Chim. Acta* **1991**, *196*, 143–151. [CrossRef]
56. Doğan, P.; Tanrikulu, G.; Soyuer, Ü.; Köse, K. Oxidative enzymes of polymorphonuclear leucocytes and plasma fibrinogen, ceruloplasmin, and copper levels in Behcet's disease. *Clin. Biochem.* **1994**, *27*, 413–418. [CrossRef]
57. Habig, W.H.; Pabst, M.J.; Jakoby, W.B. Glutathione S-transferases: The first enzymatic step in mercapturic acid formation. *J. Biol. Chem.* **1974**, *249*, 7130–7139. [CrossRef]
58. Luchese, C.; Pinton, S.; Nogueira, C.W. Brain and lungs of rats are differently affected by cigarette smoke exposure: Antioxidant effect of an organoselenium compound. *Pharmacol. Res.* **2009**, *59*, 194–201. [CrossRef] [PubMed]
59. Annuk, M.; Zilmer, M.; Lind, L.; Linde, T.; Fellström, B. Oxidative stress and endothelial function in chronic renal failure. *J. Am. Soc. Nephrol.* **2001**, *12*, 2747–2752. [CrossRef] [PubMed]
60. Koracevic, D.; Koracevic, G.; Djordjevic, V.; Andrejevic, S.; Cosic, V. Method for the measurement of antioxidant activity in human fluids. *J. Clin. Pathol.* **2001**, *54*, 356–361. [CrossRef]
61. Livak, K.J.; Schmittgen, T.D. Analysis of relative gene expression data using real-time quantitative PCR and the $2-\Delta\Delta CT$ method. *Methods* **2001**, *25*, 402–408. [CrossRef]
62. Skinner, R.C.; Warren, D.C.; Lateef, S.N.; Benedito, V.A.; Tou, J.C. Apple pomace consumption favorably alters hepatic lipid metabolism in young female Sprague-Dawley rats fed a western diet. *Nutrients* **2018**, *10*, 1882. [CrossRef]
63. Chen, J.; Deng, J.; Zhang, Y.; Yang, J.; He, Y.; Fu, W.; Xing, P.; Wan, H.T. Lipid-lowering effects of Danhong injection on hyperlipidemia rats. *J. Ethnopharmacol.* **2014**, *154*, 437–442. [CrossRef] [PubMed]
64. Miao, Y.F.; Li, J.; Zhang, Y.M.; Zhu, L.; Chen, H.; Yuan, L.; Hu, J.; Yi, X.L.; Wu, Q.T.; Wan, M.H.; et al. Sheng-jiang powder ameliorates obesity-induced pancreatic inflammatory injury via stimulating activation of the AMPK signalling pathway in rats: Basic Study. *World J. Gastroenterol.* **2018**, *24*, 4448–4461. [CrossRef] [PubMed]

65. Liang, W.; Menke, A.L.; Driessen, A.; Koek, G.H.; Lindeman, J.H.; Stoop, R.; Havekes, L.M.; Kleemann, R.; van den Hoek, A.M. Establishment of a general NAFLD scoring system for rodent models and comparison to human liver pathology. *PLoS ONE* **2014**, *9*, e115922. [CrossRef] [PubMed]
66. Kleiner, D.E.; Brunt, E.M.; Van Natta, M.; Behling, C.; Contos, M.J.; Cummings, O.W.; Ferrell, L.D.; Liu, Y.-C.; Torbenson, M.S.; Unalp-Arida, A.; et al. Design and validation of a histological scoring system for nonalcoholic fatty liver disease. *Hepatology* **2005**, *41*, 1313–1321. [CrossRef]
67. Hu, X.F.; Yao, J.; Gao, S.G.; Wang, X.S.; Peng, X.Q.; Yang, Y.T.; Feng, X.S. Nrf2 overexpression predicts prognosis and 5-FU resistance in gastric cancer. *Asian Pac. J. Cancer Prev.* **2013**, *14*, 5231–5235. [CrossRef]
68. Powell-Wiley, T.M.; Poirier, P.; Burke, L.E.; Despres, J.P.; Gordon-Larsen, P.; Lavie, C.J.; Lear, S.A.; Ndumele, C.E.; Neeland, I.J.; Sanders, P.; et al. Obesity and cardiovascular disease: A scientific statement from the American heart association. *Circulation* **2021**, *143*, e984–e1010. [CrossRef]
69. Pillon, N.J.; Loos, R.J.F.; Marshall, S.M.; Zierath, J.R. Metabolic consequences of obesity and type 2 diabetes: Balancing genes and environment for personalized care. *Cell* **2021**, *184*, 1530–1544. [CrossRef]
70. Friedenreich, C.M.; Ryder-Burbidge, C.; McNeil, J. Physical activity, obesity and sedentary behavior in cancer etiology: Epidemiologic evidence and biologic mechanisms. *Mol. Oncol.* **2020**, *15*, 790–800. [CrossRef]
71. Liou, C.-J.; Wei, C.-H.; Chen, Y.-L.; Cheng, C.-Y.; Wang, C.-L.; Huang, W.-C. Fisetin protects against hepatic steatosis through regulation of the Sirt1/AMPK and fatty acid β-oxidation signaling pathway in high-fat diet-induced obese mice. *Cell. Physiol. Biochem.* **2018**, *49*, 1870–1884. [CrossRef]
72. Urcan, A.C.; Criste, A.D.; Dezmirean, D.S.; Mărgăoan, R.; Caeiro, A.; Campos, M.G. Similarity of data from bee bread with the same taxa collected in India and Romania. *Molecules* **2018**, *23*, 2491. [CrossRef]
73. Alam, M.A.; Subhan, N.; Hossain, H.; Hossain, M.; Reza, H.M.; Rahman, M.M.; Ullah, M.O. Hydroxycinnamic acid derivatives: A potential class of natural compounds for the management of lipid metabolism and obesity. *Nutr. Metab.* **2016**, *13*, 27. [CrossRef] [PubMed]
74. Seo, M.-J.; Lee, Y.-J.; Hwang, J.-H.; Kim, K.-J.; Lee, B.-Y. The inhibitory effects of quercetin on obesity and obesity-induced inflammation by regulation of MAPK signaling. *J. Nutr. Biochem.* **2015**, *26*, 1308–1316. [CrossRef] [PubMed]
75. Teixeira, J.; Gaspar, A.; Garrido, E.M.; Garrido, J.; Borges, F. Hydroxycinnamic acid antioxidants: An electrochemical overview. *BioMed Res. Int.* **2013**, *2013*, 251754. [CrossRef] [PubMed]
76. Nagasaka, R.; Chotimarkorn, C.; Shafiqul, I.M.; Hori, M.; Ozaki, H.; Ushio, H. Anti-inflammatory effects of hydroxycinnamic acid derivatives. *Biochem. Biophys. Res. Commun.* **2007**, *358*, 615–619. [CrossRef] [PubMed]
77. Lesjak, M.; Beara, I.; Simin, N.; Pintac, D.; Majkic, T.; Bekvalac, K.; Orcic, D.; Mimica-Dukic, N. Antioxidant and anti-inflammatory activities of quercetin and its derivatives. *J. Funct. Foods* **2018**, *40*, 68–75. [CrossRef]
78. Zakaria, Z.; Othman, Z.A.; Bagi Suleiman, J.; Che Jalil, N.A.; Wan Ghazali, W.S.; Mohamed, M. Protective and therapeutic effects of orlistat on metabolic syndrome and oxidative stress in high-fat diet-induced metabolic dysfunction-associated fatty liver disease (MAFLD) in rats: Role on Nrf2 activation. *Vet. Sci.* **2021**, *8*, 274. [CrossRef]
79. Xu, H.; Zhou, Y.; Liu, Y.; Ping, J.; Shou, Q.; Chen, F.; Ruo, R. Metformin improves hepatic IRS2/PI3K/Akt signaling in insulin-resistant rats of NASH and cirrhosis. *J. Endocrinol.* **2016**, *229*, 133–144. [CrossRef]
80. Yang, Y.; Li, W.; Liu, Y.; Li, Y.; Gao, L.; Zhao, J.J. Alpha-lipoic acid attenuates insulin resistance and improves glucose metabolism in high fat diet-fed mice. *Acta. Pharmacol. Sin.* **2014**, *35*, 1285–1292. [CrossRef]
81. Qu, L.; Liu, Q.; Zhang, Q.; Liu, D.; Zhang, C.; Fan, D.; Deng, J.; Yang, H. Kiwifruit seed oil ameliorates inflammation and hepatic fat metabolism in high-fat diet-induced obese mice. *J. Funct. Foods* **2019**, *52*, 715–723. [CrossRef]
82. Nah, E.-H.; Cho, S.; Park, H.; Kim, S.; Cho, H.-I. Associations of complete blood count parameters with pancreatic beta-cell function and insulin resistance in prediabetes and type 2 diabetes mellitus. *J. Clin. Lab. Anal.* **2022**, *36*, e24454. [CrossRef]
83. Guo, H.; Ma, C.; Wu, X.; Pan, C. Functional status of pancreatic α and β cells in type 2 diabetes mellitus patients with different plasma triglyceride levels: A retrospective analysis. *Int. J. Endocrinol.* **2021**, *2021*, 1–6. [CrossRef]
84. Mansour-Ghanaei, F.; Joukar, F.; Mobaraki, S.N.; Mavaddati, S.; Hassanipourd, S.; Sepehrimaneshad, M. Prevalence of non-alcoholic fatty liver disease in patients with diabetes mellitus, hyperlipidemia, obesity and polycystic ovary syndrome: A cross-sectional study in north of Iran. *Diabetes Metab. Syndr. Clin. Res. Rev.* **2019**, *13*, 1591–1596. [CrossRef] [PubMed]
85. Chen, W.; Shao, S.; Cai, H.; Han, J.; Guo, T.; Fu, Y.; Yu, C.; Zhao, M.; Bo, T.; Yao, Z.; et al. Comparison of erythrocyte membrane lipid profiles between NAFLD patients with or without hyperlipidemia. *Int. J. Endocrinol.* **2020**, *2020*, 9501826. [CrossRef] [PubMed]
86. Begum, T.A.K.M.; Ramamurthy, V. The modulating effect of ferulic acid on high fat diet induced hyperlipidemia and obesity: A dose response study in male Sprague Dawleyrats. *Biomedicine* **2021**, *41*, 413–420. [CrossRef]
87. Feng, K.; Zhu, X.; Chen, T.; Peng, B.; Lu, M.; Zheng, H.; Huang, Q.; Ho, C.T.; Chen, Y.; Cao, Y. Prevention of obesity and hyperlipidemia by heptamethoxyflavone in high-fat diet-induced rats. *J. Agric. Food Chem.* **2019**, *67*, 2476–2489. [CrossRef] [PubMed]
88. Jeepipallia, S.P.K.; Du, B.; Sabitaliyevich, U.Y.; Xu, B. New insights into potential nutritional effects of dietary saponins in protecting against the development of obesity. *Food Chem.* **2020**, *318*, 126474. [CrossRef]
89. Day, C.P.; James, O.F. Steatohepatitis: A tale of two "hits"? *Gastroenterology* **1998**, *114*, 842–845. [CrossRef]

90. Lai, Y.S.; Chen, W.C.; Ho, C.T.; Lu, K.H.; Lin, S.H.; Tseng, H.C.; Lin, S.Y.; Sheen, L.Y. Garlic essential oil protects against obesity-triggered nonalcoholic fatty liver disease through modulation of lipid metabolism and oxidative stress. *J. Agric. Food Chem.* **2014**, *62*, 5897–5906. [CrossRef]
91. Cai, S.M.; Yang, R.Q.; Li, Y.; Ning, Z.W.; Zhang, L.L.; Zhou, G.S.; Luo, W.; Li, D.H.; Chen, Y.; Pan, M.X.; et al. Angiotensin-(1-7) improves liver fibrosis by regulating the NLRP3 inflammasome via redox balance modulation. *Antioxid. Redox Signal.* **2016**, *24*, 795–812. [CrossRef]
92. Wu, G.; Zhu, L.; Yuan, X.; Chen, H.; Xiong, R.; Zhang, S.; Cheng, H.; Shen, Y.; An, H.; Li, T.; et al. Britanin ameliorates cerebral ischemia–reperfusion injury by inducing the Nrf2 protective pathway. *Antioxid. Redox Signal.* **2017**, *27*, 754–768. [CrossRef]
93. Wu, K.C.; Liu, J.; Klaassen, C.D. Role of Nrf2 in preventing ethanol-induced oxidative stress and lipid accumulation. *Toxicol. Appl. Pharmacol.* **2012**, *262*, 321–329. [CrossRef]
94. Yang, M.; Jiang, Z.H.; Li, C.G.; Zhu, Y.J.; Li, Z.; Tang, Y.Z.; Ni, C.L. Apigenin prevents metabolic syndrome in high-fructose diet-fed mice by Keap1-Nrf2 pathway. *Biomed. Pharmacother.* **2018**, *105*, 1283–1290. [CrossRef] [PubMed]
95. Zhao, X.J.; Chen, L.; Zhao, Y.; Pan, Y.; Yang, Y.Z.; Sun, Y.; Jiao, R.Q.; Kong, L.D. *Polygonum cuspidatum extract* attenuates fructose-induced liver lipid accumulation through inhibiting Keap1 and activating Nrf2 antioxidant pathway. *Phytomedicine* **2019**, *63*, 152986. [CrossRef] [PubMed]
96. Xu, D.; Hu, M.J.; Wang, Y.Q.; Cui, Y.L. Antioxidant activities of quercetin and its complexes for medicinal application. *Molecules* **2019**, *24*, 1123. [CrossRef] [PubMed]
97. Zakaria, Z.; Othman, Z.A.; Nna, V.U.; Mohamed, M. The promising roles of medicinal plants and bioactive compounds on hepatic lipid metabolism in the treatment of non-alcoholic fatty liver disease in animal models: Molecular targets. *Arch. Physiol. Biochem.* **2021**, 1–17. [CrossRef]
98. Lu, C.; Zhang, F.; Xu, W.; Wu, X.; Lian, N.; Jin, H.; Chen, Q.; Chen, L.; Shao, J.; Wu, L.; et al. Curcumin attenuates ethanol-induced hepatic steatosis through modulating Nrf2/FXR signaling in hepatocytes. *IUBMB Life* **2015**, *67*, 645–658. [CrossRef]
99. Li, J.; Liu, M.; Yu, H.; Wang, W.; Han, L.; Chen, Q.; Ruan, J.; Wen, S.; Zhang, Y.; Wang, T. Mangiferin improves hepatic lipid metabolism mainly through its metabolite-norathyriol by modulating SIRT-1/AMPK/SREBP-1c signaling. *Front. Pharmacol.* **2018**, *9*, 201. [CrossRef] [PubMed]
100. Bonfleur, M.L.; Borck, P.C.; Ribeiro, R.A.; Caetano, L.C.; Soares, G.M.; Carneiro, E.M.; Balbo, S.L. Improvement in the expression of hepatic genes involved in fatty acid metabolism in obese rats supplemented with taurine. *Life Sci.* **2015**, *135*, 15–21. [CrossRef]
101. Kohjima, M.; Enjoji, M.; Higuchi, N.; Kato, M.; Kotoh, K.; Yoshimoto, T.; Fujino, T.; Yada, M.; Yada, R.; Harada, N.; et al. Re-evaluation of fatty acid metabolism-related gene expression in nonalcoholic fatty liver disease. *Int. J. Mol. Med.* **2007**, *20*, 351–358. [CrossRef] [PubMed]
102. Noh, H.; Lee, H.; Kim, E.; Mu, L.; Rhee, Y.K.; Cho, C.W.; Chung, J. Inhibitory effect of a *Cirsium setidens* extract on hepatic fat accumulation in mice fed a high-fat diet via the induction of fatty acid β-oxidation. *Biosci. Biotech. Biochem.* **2013**, *77*, 1424–1429. [CrossRef]
103. Zheng, H.; Li, S.; Ma, L.; Cheng, L.; Deng, C.; Chen, Z.; Xie, C.; Xiang, M.; Jiang, W.; Chen, L. A novel agonist of PPAR-γ based on barbituric acid alleviates the development of non-alcoholic fatty liver disease by regulating adipocytokine expression and preventing insulin resistance. *Eur. J. Pharmacol.* **2011**, *659*, 244–251. [CrossRef] [PubMed]
104. Lee, Y.; Lee, J.; Lee, M.S.; Chang, E.; Kim, Y. *Chrysanthemum morifolium* flower extract ameliorates obesity-induced inflammation and increases the muscle mitochondria content and AMPK/SIRT1 activities in obese rats. *Nutrients* **2021**, *13*, 3660. [CrossRef] [PubMed]
105. Sharma, A.; Anand, S.K.; Singh, N.; Dwarkanath, A.; Dwivedi, U.N.; Kakkar, P. Berbamine induced activation of the SIRT1/LKB1/AMPK signaling axis attenuates the development of hepatic steatosis in high-fat diet-induced NAFLD rats. *Food Funct.* **2021**, *12*, 892–909. [CrossRef] [PubMed]
106. Yao, H.; Tao, X.; Xu, L.; Qi, Y.; Yin, L.; Han, X.; Xu, Y.; Zheng, L.; Peng, J. Dioscin alleviates non-alcoholic fatty liver disease through adjusting lipid metabolism via SIRT1/AMPK signaling pathway. *Pharmacol. Res.* **2018**, *131*, 51–60. [CrossRef] [PubMed]
107. Rafiei, H.; Omidian, K.; Bandy, B. Dietary polyphenols protect against oleic acid-induced steatosis in an in vitro model of NAFLD by modulating lipid metabolism and improving mitochondrial function. *Nutrients* **2019**, *11*, 541. [CrossRef]
108. Tian, Y.; Ma, J.; Wang, W.; Zhang, L.; Xu, J.; Wang, K.; Li, D. Resveratrol supplement inhibited the NF-κB inflammation pathway through activating AMPKα-SIRT1 pathway in mice with fatty liver. *Mol. Cell. Biochem.* **2016**, *422*, 75–84. [CrossRef]
109. Li, X.; Gong, H.; Yang, S.; Yang, L.; Fan, Y.; Zhou, Y. Pectic bee pollen polysaccharide from *Rosa rugosa* alleviates diet-induced hepatic steatosis and insulin resistance via induction of AMPK/mTOR-mediated autophagy. *Molecules* **2017**, *22*, 699. [CrossRef]
110. Liu, X.M.; Peyton, K.J.; Shebib, A.R.; Wang, H.; Korthuis, R.J.; Durante, W. Activation of AMPK stimulates heme oxygenase-1 gene expression and human endothelial cell survival. *Am. J. Physiol. Heart Circ. Physiol.* **2011**, *300*, H84–H93. [CrossRef]
111. Mo, C.; Wang, L.; Zhang, J.; Numazawa, S.; Tang, H.; Tang, X.; Han, X.; Li, J.; Yang, M.; Wang, Z.; et al. The crosstalk between Nrf2 and AMPK signal pathways is important for the anti-inflammatory effect of berberine in LPS-stimulated macrophages and endotoxin-shocked mice. *Antioxid. Redox Signal.* **2014**, *20*, 574–588. [CrossRef]
112. Buzzetti, E.; Pinzani, M.; Tsochatzis, E.A. The multiple-hit pathogenesis of non-alcoholic fatty liver disease (NAFLD). *Metabolism* **2016**, *65*, 1038–1048. [CrossRef]
113. Tilg, H.; Moschen, A.R.; Roden, M. NAFLD and diabetes mellitus. *Nat. Rev. Gastroenterol. Hepatol.* **2017**, *14*, 32–42. [CrossRef] [PubMed]

antioxidants

Review

Bee Pollen as Functional Food: Insights into Its Composition and Therapeutic Properties

Asmae El Ghouizi [1], Meryem Bakour [1,2], Hassan Laaroussi [1], Driss Ousaaid [1], Naoual El Menyiy [3], Christophe Hano [4,*] and Badiaa Lyoussi [1,*]

[1] Laboratory of Natural Substances, Pharmacology, Environment, Modeling, Health, and Quality of Life, Faculty of Sciences Dhar El Mahraz, University Sidi Mohamed Ben Abdellah, Fez 30000, Morocco
[2] The Higher Institute of Nursing Professions and Health Techniques, Fez 30000, Morocco
[3] Laboratory of Pharmacology, National Agency of Medicinal and Aromatic Plants, Taounate 34025, Morocco
[4] Department of Chemical Biology, Eure et Loir Campus, University of Orleans, 28000 Chartres, France
* Correspondence: hano@univ-orleans.fr (C.H.); badiaa.lyoussi@usmba.ac.ma (B.L.)

Abstract: Bee pollen is a hive product made up of flower pollen grains, nectar, and bee salivary secretions that beekeepers can collect without damaging the hive. Bee pollen, also called bee-collected pollen, contains a wide range of nutritious elements, including proteins, carbs, lipids, and dietary fibers, as well as bioactive micronutrients including vitamins, minerals, phenolic, and volatile compounds. Because of this composition of high quality, this product has been gaining prominence as a functional food, and studies have been conducted to show and establish its therapeutic potential for medical and food applications. In this context, this work aimed to provide a meticulous summary of the most relevant data about bee pollen, its composition—especially the phenolic compounds—and its biological and/or therapeutic properties as well as the involved molecular pathways.

Keywords: bee pollen; composition; medicinal properties; functional food

Citation: El Ghouizi, A.; Bakour, M.; Laaroussi, H.; Ousaaid, D.; El Menyiy, N.; Hano, C.; Lyoussi, B. Bee Pollen as Functional Food: Insights into Its Composition and Therapeutic Properties. *Antioxidants* **2023**, *12*, 557. https://doi.org/10.3390/antiox12030557

Academic Editors: Josipa Vlainić and Ivana Tlak Gajger

Received: 21 January 2023
Revised: 15 February 2023
Accepted: 17 February 2023
Published: 23 February 2023

1. Introduction

The pollen grain is the male flower's reproductive organ; in other words, it carries the organ that carries the male gametes to their progenitor cells. It is produced and spread by higher plants as part of their reproductive process [1]. Bee pollen is the final result of the agglutination of pollen grains harvested by worker bees, held together by nectar and/or honey, and gland secretions, and collected at the hive entrance [2]. It is among the most important bee products gaining popularity as a functional food due to its high concentration of bioactive compounds known for their benefits for both mental and physical health, such as proteins, dietary fibers, lipids, carbohydrates, and minerals [3]. Furthermore, bee pollen is known as "the perfectly complete food" due to its strong antioxidant potential and the presence of antioxidant compounds such as polyphenols, flavonoids, carotenoids, and vitamins (A, C, and E) giving this product a high antioxidant potential and making it the latest trend in dietary supplements [4]. Because of this high load of natural bioactive molecules, various scientific studies have reported that bee pollen possesses a wide spectrum of biological properties such as antioxidant [5], hypoglycemic [6], anti-inflammatory [7], antibacterial [8], and anticancer [9]. As a result of all of this, the German Federal Ministry of Health has formally acknowledged bee pollen as a drug. [10]. This paper aims to provide a comprehensive overview of bee pollen in terms of bee pollen harvesting, chemical and nutritional content, and, finally, its biological and therapeutic characteristics. Furthermore, we intended to unveil for the first time the proposed mechanism of action and the involved biomolecular pathways of the various bioactive compounds of bee pollen that are responsible for the improvement of oxidative stress and associated health conditions.

2. Methodology

The following online databases were employed to collect the literature data for this paper: Science Direct, Google Scholar, Web of Science, Pub-Med, and Scopus, using the keywords: "bee pollen"; "bee-collected pollen"; "chemical composition of bee pollen"; "therapeutic effect of bee pollen"; "functional effect of bee pollen", "protective effect of bee pollen", and "nutritional value of bee pollen". After collecting and reviewing all selected articles, their general ideas were summarized and used in this review.

3. Bee Pollen: From Flowers to the Hive

It is commonly known that nectar and pollen grains are the main sources of nutrients required for the survival and health of bee colonies. Nectar is mainly produced by the nectariferous glands of plants and serves as the raw material for honey production [11], whereas pollen grains represent the plant's male gametophyte. They take the form of fine dust with tiny particles that vary in color based on the floral origin [12]. Flower nectar provides bees with carbon and nitrogen, while pollen grains provide other dietary components such as lipids, proteins, vitamins, and minerals [13,14].

The flower–bee interaction is a mutualistic relationship in which the flowers reproduce sexually and the bees feed on nectar and pollen [14]. In this respect, plants adopt many techniques to be pollinated and reproduce, including the coloration of their petals and the emission of scents known as pheromones [15]. Worker honeybees perform hundreds of flights to blooms to collect the necessary amount of pollen and nectar [16]. The honeybees' behavior of gathering nectar and pollen is known as "foraging" and it is highly vital and necessary for the survival of bee colonies. Foraging is a behavior that develops in worker bees between the ages of one and two weeks [17]. Young foraging bees conduct many scouting flights to become acquainted, and at the age of 21 days, they will leave their hives to seek and gather nectar, honeydew, pollen, water, and many essential elements, as well as the resin used to maintain the hive's asepsis [18,19]. Foraging bees use their proboscis to collect nectar and water by pumping and capillarity; the liquids are stored in the foragers' crops until they are discharged to the other workers in the hive [20], who then use their hind legs coated with short stiff hairs called "scopae" to squeeze the collected pollen grains into pollen balls using their saliva and honey, which they finally place into their pollen baskets [16]. The majority of bees have developed specialized mechanisms for transporting pollen to their nests and have adjusted their grooming behaviors to transfer pollen from their bodies to the hive [21].

When the foraging bees arrive at the hive, they cover the pollen balls with saliva, then compact the alveoli with a layer of pollen balls and honey, and finally cap them with a layer of wax [22].

Lactic fermentation occurs at this stage due to the participation of lactic bacteria strains that proliferate inside the hives [14]. These bacteria are *Pseudomonas*, which consumes oxygen and creates an anaerobic environment; *Lactobacillus*, which converts carbohydrates into lactic acid; and finally, *Saccharomyces*, which ensures the metabolism of the rest of the sugars that exist in the medium. These reaction chains reduce the environment's pH, prevent pollen germination, and improve bee pollen absorption capacity and nutritional value [23]. When bee-collected pollen is completely fermented, it becomes "bee bread," which provides additional proteins for bees, particularly during the period of royal jelly production, as well as nutrition for larvae, and future workers who are fed a diet of pollen, honey, and a small quantity of royal jelly [18].

4. Chemical Composition

Bee pollen is one of the magical superfoods due to its extremely wide range of nutritional compounds and microelements. However, this composition may be affected by botanical origin, harvesting season, and storage methods (freeze-drying duration). Considering this large variability, its nutritional and chemical composition has been extensively studied, summarized, and standardized. In this section, we assembled the main macro- and micronutrients of bee-collected pollen.

4.1. Main Compound

4.1.1. Water

Several studies have succeeded in quantifying the water content in bee pollen samples, the results obtained being largely variable and dependent on storage conditions (fresh or dried bee pollen), botanical, and geographical origins [24].

In fresh bee-collected pollen, water content varies between 20 and 30% [2]. However, this high humidity is considered a favorable environment for bacterial and fungal growth [25–27]. As a result, the bee pollen freezing process must begin immediately after harvesting [28]. Other researchers prefer the nitrogen processing of fresh bee pollen to preserve the optimal microbiological and nutritional properties [29]. Meanwhile, the maximum water content allowed in dried bee pollen depends on the country and must not exceed 4% according to the Brazilian legislation, 6% in Poland and Switzerland, 8% in Argentina, and 10% in Bulgaria [2,24,30–33]. Thus, the moisture content can be used as a bee pollen quality criterion.

4.1.2. Protein Content

The production of protein for human consumption as well as defining the need for protein, evaluating its quality to meet human needs, and managing the consequences of variations in dietary protein intake are considered major public health issues. Consequently, it is necessary to ensure a sufficient daily supply of protein of good biological quality, since the human body is unable to store it. Therefore, bee pollen, which has high protein content, could guarantee an ideal diet in terms of protein requirement. In addition, pollen is the principal source of protein for bees, providing the necessary elements for their longevity, organ development, larva growth, and body size [34,35]. Proteins also provide essential substances for royal jelly production in the hive [36].The protein content is highly variable between plant species and harvesting geographic areas [37] and varies between 10 and 40% of pollen dry weight [2].

4.1.3. Amino Acids

Amino acids are crucial not only for protein synthesis but also for the biosynthesis of hormones as well as other molecules with a biological role. Nutritionally, two kinds of amino acids are distinguished, namely, the essential amino acids that the human organism cannot synthesize, and non-essential amino acids that our body has all the machinery to synthesize. In the same context, bee-collected pollen is often considered the "most natural perfect food" because it is a great source of all essential amino acids needed in honey bee and human nutrition [38,39]. This content varies strongly from species to species and depends on botanical and geographical origin, climatic conditions, and nutrient availability in the plant [37,40,41]. Therefore, the amino acid amount can be used as an indicator of freshness, storage, and drying process adequacy [3,42].

The total amino acid content in bee pollen has been quantified by many researchers, and it generally ranges between 108.1 and 287.7 mg/g of bee pollen [43]. Concerning the amino acid profile, De-Melo and Almeida-Muradian have reported twenty-five amino acids, eight of which are essential (valine, leucine, isoleucine, lysine, phenylalanine, threonine, histidine, and methionine). Tryptophan is usually undetectable because of the hydrolysis method employed in the determination of the amino acids. However, tryptophan was detected in Chinese, Slovenian, Spanish, and Italian bee pollen using specific high-performance liquid chromatography methods [44–46]. The remaining amino acids are non-essential, such as aspartic acid, alanine, glycine, glutamine, arginine, asparagine, glutamic acid, serine, tyrosine, cystine, cysteine, γ-aminobutyric acid (GABA), ornithine, proline, and homoserine [3]. It has been reported that proline is the most abundant amino acid in dried bee pollen from many countries, while glutamic acid is the main amino acid in freshly collected bee pollen [3,44,47].

4.1.4. Carbohydrates

Bee pollen is composed of pollen mixed with nectar, and the bee's salivary secretions. Carbohydrates are the major fraction of bee pollen (13–55%), they are mainly polysaccharides and cell wall material [2]. Carbohydrates can be affected by botanical and geographical origin, harvesting methods, and conditioning processes such as high temperature when drying fresh bee pollen [48].

- Sugar

Sugar content is the most important quality parameter in bee pollen characterization studies and should not be less than 40 % [2]. Nevertheless, sugars are generally neglected or included in the total carbohydrates which also regroups the dietary fiber and starch [49]. The sugar composition of bee pollen has been assessed by many studies either as reducing sugar [50–54] or as individual sugars [41,44,49,55,56] and all showed a predominance of glucose and fructose as monosaccharides which represented the major amount of sugar fraction. Sucrose, maltose, trehalose, turanose, and melezitose have been identified in previous studies [49,57]. The sugar content and profile can be considerably influenced by nectar added by bees during the packaging and storage of bee-collected pollen [50,58]. Floral source, drying process, and extraction methods can also greatly affect the sugar content [48,56]

Mannitol is a polyol previously identified and isolated in high concentrations from bee pollen collected by stingless bee *Melipona subnitida* from Jandaíra, Brazil, *Tetragonula biroi Friese* from the Philippines, and *Trigona* from Malaysia [41,59,60]. According to these authors, the significant amount of mannitol did not depend on the floral origin, and it is supposed that the previous stingless bee species are capable of converting the glucose and fructose mainly found in flowers into mannitol via their salivary enzymes.

On this basis, sugars can be considered as an additional parameter for establishing quality standards for bee pollen.

- Dietary fibers

Dietary fibers describe the soluble and insoluble fraction of fibers from plant-based foods, which include hemicellulose, cellulose, lignin, oligosaccharides, pectins, gums, and waxes; these compounds are resistant to digestive enzymes, thus they are neither hydrolyzed nor absorbed in the intestinal tract [61,62]. Many recent studies have supported the crucial physiological role of dietary fibers in the human body; indeed, they are involved in type 2 diabetes management by the selective promotion of certain gut microbiota [63,64]. A high-fiber diet was found to be effective in many conditions such as obesity-related disorders, cardiovascular diseases, constipation, inflammatory bowel diseases, and colon cancers [65–69]. Regarding this, bee pollen can be a good source of dietary fiber, especially crude fibers. Despite the importance of bee pollen in the human diet, few characterization studies have focused on the determination of the dietary fiber content of bee pollen. According to Compos et al., total dietary fiber should range between 0.3 and 20 g/100 g of bee pollen dry weight [2]. Dietary fiber content has been reported by a few studies. For instance, a recent study carried out on Slovenian bee pollen showed a range of 10–21.4 g/100 g dry weight bee pollen with 73–82% of crude fiber [49]. Brazilian researchers reported an average of 3.6 ± 1.4 g/100 g of dry weight bee pollen [54]. Colombian bee pollen has also been characterized, and the results showed an average of 14.5 ± 3.5 g/100 g dry weight [70], while El-Kazafy recorded that different Egyptian bee pollen showed a range of 0.15 ± 0.01 and 1.70 ± 0.02 g/100 g dry weight bee pollen [47]. Dietary fiber content may vary according to the botanical origin and methods used during hydrolysis.

4.1.5. Lipid and Fat Content

Physiologically, the human body uses a variety of biosynthetic pathways to synthesize lipids; however, some important lipids cannot be obtained through biosynthesis and must be obtained from food. Essential fatty acids (especially omega-3 fatty acids) are involved in many biological functions and play an important role in the prevention of inflammatory and cardiovascular diseases and hormone-dependent tumors. [71,72]. Indeed, bee pollen can be a great source of these compounds since they are crucial for royal jelly production [73]. The lipidic fraction is most attractive for bees, and thus plants with high lipid concentration pollen are more frequently visited [37]. According to Campos et al., lipid content ranges between 1–13 g/100 g [2], while De-Melo and Almeida-Muradian, reported that the total lipid fraction can reach 22 g/100 g [3]. In the same context, Thakur et al. and De-Melo et al. have reported a huge variation of lipid content in monofloral bee pollen from different countries; *Brassica napus* bee pollen from Brazil, China, India, and Greece showed a total lipid content of 7.4%, 6.6 %, 12.38%, and 7.76% respectively; Cistus bee pollen from Italy, Spain, and Greece showed a total lipid content of −1.9%, 7.2%, and 3.80% respectively. *Cocos nucifera* bee pollen from India and Brazil showed 10.43% and 4.6–5.1% total lipid content [3,73].

The lipid profile of bee pollen has been barely investigated while most research studies have focused on the protein, carbohydrate, and antioxidant content. According to Ares et al., carotenoids, steroids, and fatty acids are the main constituents of bee pollen's total lipid fraction [52]. A study conducted by Li et al. on three monofloral bee pollen samples from China suggested the presence of nine lipid classes, including triglycerides and fatty acids [74].

The fatty acid profile of bee pollen varies between saturated fatty acids, which include mainly the myristic, stearic, and palmitic acids, and unsaturated fatty acids, which include the oleic, α-linolenic (omega-3), and linoleic (omega-6) acids. This fraction is the most dominant in bee pollen [4,44,73,75–77]. Other lipid classes such as phospholipids, triterpenes (oleanolic and ursolic acids), and plant sterols (β-sitosterol) have been isolated from bee pollen in smaller amounts [4,36,76]. All these studies have reported that the lipid content in the studied bee pollen depends on botanical origin, harvesting season, drying, storage, and beekeeping methods.

4.2. Micronutrients

Micronutrients include minerals and vitamins that are not involved in the energetic balance but are essential for all chemical reactions and for the maintenance of life. Micronutrients are required in small amounts by the body for its growth and development from birth to old age. In a recent report (2020), most food and health organizations estimated that more than two billion people globally suffer from micronutrient deficiency occurring due to an insufficient intake or impaired absorption of vitamins and minerals [78,79]. Generally, micronutrient deficiency is considered a global health concern for all ages. During pregnancy, this deficiency has a devastating effect on both the mother and her fetus, being associated with anemia, hypertension, gestational diabetes, thyroid disorders, obstetric complications, and failure in the growth and development of the fetus, among other conditions [80–84]. During childhood, micronutrient insufficiency may affect the mental and physical development of the children and increase their vulnerability to and exacerbation of diseases such as impaired host defense and infections, developmental disabilities, autism, ocular disorders, and general loss of energy and potential [85–87]. The elderly population is also vulnerable to micronutrient deficiencies, which can lead to many age-related diseases such as mild cognitive decline, high risk of type 2 diabetes, cardiovascular diseases, acute respiratory infections especially coronavirus infection, and immune function impairments [88–91]. Based on what has been stated above, it is clear that maintaining a micronutrient-rich diet may provide enough protection against all of these pathologies. This protection can be guaranteed by consuming a large variety of well-balanced and rich natural products such as bee pollen.

4.2.1. Minerals

There are around twenty minerals that are essential in the human diet and are classified as macro-elements and oligo-elements, also known as trace elements. Mineral deficiency in the human body causes several metabolic problems and severe developmental defects in pregnancy, and significantly affects the individual's wellness and economic output [92]. Bee pollen is a good source of essential minerals for the development of bees as well as humans, which represent 2–6% of its content, with about 25 elements [2]. This makes bee pollen an interesting value-added product.

Potassium (K) is the principal mineral element found in high concentrations in bee pollen (400–2000 mg/100 g of bee pollen), and 15 g of bee pollen covers up to 25% of the recommended daily intake (RDI) of this element (2000 mg/day). Phosphorus (P) is the second element mainly present in bee pollen (0.80–6 mg/100 g of bee pollen), covering 16% of the RDI (1000 mg/day) of 15 g of bee pollen. The third important element is magnesium (Mg; 20–300 mg/100 g of bee pollen) which covers up to 23% of the RDI of this element (350 mg/day) of 15 g of bee pollen. Calcium (Ca) is also widely present in pollen (20–300 mg/100 g of pollen) and covers 7% of the RDI (1100 mg/day) of calcium. These elements are known for their crucial role in bone tissue formation by maintaining the proper osmotic pressure of blood as well as cellular fluids. Iron (Fe), zinc (Zn), copper (Cu), and manganese (Mn) are also microelements present in large quantities in bee pollen, covering up to 37%, 79%, 36%, and 85% of the respective RDIs. These trace elements play an important role in blood formation and also in the growth, development, and reproduction process [2,28,93,94].

There are other trace elements such as cobalt (Co), selenium (Se), molybdenum (Mo), and boron (B) which have been identified in bee pollen from different countries. Adequate intake of these trace elements is necessary to support bone and brain health, they play a key role in the maintenance of vitamin structure, reproduction, thyroid hormone metabolism, DNA synthesis, and protection against oxidative damage and infections [3,95–97].

Sodium (Na) is also a macro-element present in bee pollen; however, its content remains below 2 g/kg with a high K/Na ratio, and this ratio makes bee pollen beneficial and safe for daily diets with a good electrolyte balance [3]. The mineral content of bee pollen is recommended as a distinct marker of its floral and geographical origin as well as its quality [5,73].

4.2.2. Vitamins

Vitamins are a class of nutrients or organic compounds essential for the body not synthesized by humans, except for vitamins D, K, and biotin (B7), where vitamin D is synthesized in the body by irradiating skin sterols with UV rays, while biotin and vitamin K are present in certain foods but can also be synthesized by the human intestinal flora. Therefore, the essential vitamins must be daily ingested from food to prevent metabolic disorders related to vitamin deficiencies due to their major role in the synthesis of vital cofactors, enzymes, and metabolic reactions based on coenzymes [98–100]. Bee pollen is considered a "vitamin bomb" due to the presence of almost all vitamins with an average of 0.02–0.7% of its total content, with a higher amount of water-soluble than fat-soluble vitamins [4,101]. Table 1 summarizes the different vitamins identified in different samples of bee pollen with different floral species and geographical origins.

Table 1. Summary of the different vitamins isolated from bee pollen.

Identified Vitamins	Apiaries	Floral Origin	Isolation Methods	References
A, B1, B2, B5, B6, B7, B12, C, E, K2	Turkey	monofloral bee pollen of *Rhododendron ponticum*	HPLC-FLD HPLC-UV	[102]
β-Carotene Vit. C	Portugal	Polyfloral bee pollen of: *Rubus* spp.; *Castanea; sativa; Cytisus* spp.; *Quercus* spp.; *Echium* spp.; *Prunus* spp.; *Leontondon* spp.; *Eucalyptus* spp.; *Erica* spp.; *Cistus* spp.; *Trifolium* spp.	(NH4)2SO4 for β-Carotene AOAC for vitamin-C	[103]
B2, B3, B6, B9	Italy	Polyfloral bee pollen of: *Prunu; Erica; Brassicaceae; Rubus; Viburnum Viburnum; Trifolium pratense; Asteraceae T.; Eucalyptus; Rosa* spp.	Fluorescence spectroscopy (Bulk analysis)	[104]
B1, B2, B6	Brazil	Polyfloral bee pollen of: *Arecaceae; Cecropia; Cestrum; Cyperaceae; Eucalyptus; Ilex; Myrcia; Piper; Vernonia; Trema*	HPLC	[105]
B3 (Niacin); B6 (Pyridoxine) B9 (Folic acid) B12 (Cobalamin)	Saudi Arabia	Monofloral bee pollen of: *alfalfa*; date palm; *rape*; summer squash; sunflower	HPLC	[75]
C; E; Provit. A (β-carotene)	Brazil	ND	Vit C: AOAC Vit E: HPLC β-carotene: OCC	[106]

ND: Not determined.

- water-soluble vitamins

B vitamin group (thiamin (B1), riboflavin (B2), PP vitamin or niacin (B3), pantothenic acid (B5), pyridoxine (B6), biotin (B7), folic acid (B9), and cobalamin (B12)) are the most commonly identified class of water-soluble vitamins in bee pollen (Table 2) [29,75,104,105,107]. Vitamin C or L-ascorbic acid is marginally identified because of its deterioration by thermal pretreatments [28]. Water-soluble vitamins, such as B-complex, are not normally stored in the body in significant amounts, necessitating a daily intake of these vitamins. This group of vitamins plays a key role in host immunity, dermatology, and cellular energy production (B1, B2); they facilitate the production of amino acids and improve their metabolism (B6); and help the body to convert carbohydrates into glucose (B3 or PP). Water-soluble vitamin deficiencies triggered by malnutrition can be the origin of certain metabolic and nervous pathologies [108]. Regarding its high content of water-soluble vitamins, bee pollen could be one of their potential sources.

- Fat-soluble vitamins

Bee pollen contains fat-soluble vitamins such as vitamin D, K, E, and A (β-Carotene) with low and variable amounts depending on the botanical origin and the season of collection [73,109,110].

The family of vitamins E is commonly called tocochromanols (tocopherols and tocotrienols), and pollen contains in particular the group of tocopherols (α tocopherol, β-tocopherol, γ-tocopherol, and δ tocopherol) with a dominance of α and γ-tocopherol [56,111]. Vitamin K2 (Menaquinone-4), and two types of vitamin A (β-carotene and retinol) were also detected by Bayram et al. [102]. Although there are no identification studies of vitamin D listed in the international literature, Campos, Komosinska-Vassev, and Khalifa et al. cited the presence of vitamin D in their review articles [36,110,112]. Fat-soluble vitamins are involved in a multitude of physiological processes such as vision, bone health, immune function, and coagulation [113].

4.2.3. Carotenoids

Carotenoids are a highly diversified group of yellow- to red-colored polyenes responsible for the colors in many plant-derived products and play an important role in human health [94]. In bee pollen, β-carotene is the most frequently identified of this class. It is an antioxidant provitamin with good effects on human health (anti-tumor, anti-leukemic, and beneficial against cardiovascular diseases) [114]. Besides β-carotene, other carotenoids such as lutein and cryptoxanthin, zeaxanthin, β-cryptoxanthin, and α-carotene have been identified [60,115]. The carotenoid content varies according to the botanical origin, harvest period (climate conditions), drying, and storage condition [2,52,107].

4.3. Pollen Probiotics

Throughout history, bee pollen has been considered a complete food with many therapeutic virtues, and for this reason, it has been the subject of numerous and diverse biochemical and microbiological studies. Many studies have been concerned with the anti-microbial activities of bee pollen, but little is known about its microbiome.

P. Percie du Sert has reported that bees raise lactic ferments in the nectar stored inside the hive, and this bacteria-rich nectar will be used during their flight to stick pollen grains on their forelegs, which explains the presence of lactic acid bacteria in freshly harvested pollen [29]. Several studies have demonstrated that honey bees and bumblebees seem to have a simple intestinal bacterial fauna which includes acidophilic bacteria, mainly from the Lactobacillus family such as *Lactobacillus kunkeii, Lactobacillus plantarum, Lactobacillus fermentum; lactobacillus kunkeii; Lactobacillus plantarum; Lactobacillus fermentum; Lactococcus slactis; Pediococcus acidi lactici; Pediococcus pentosaceus; Lactobacillus ingluviei*; and *Weissella cibaria* [116–119]. It is now quite clear that the fresh bee pollen bacterial community comes from the specific bacterial fauna of the bee intestinal gut.

Previous in vitro studies have reported the beneficial effect of the isolated probiotic strains of fresh pollen against pathogenic Gram-positive and Gram-negative bacteria through the high production of bacteriocins, such as organic acids, which can be of major importance in fighting human infectious diseases [116,120,121]. Lactobacillus strains isolated from fresh bee pollen can survive under human digestive tract conditions such as low pH, and bile salts. In the same context, the high hydrophobicity and autoaggregation of fresh bee pollen lactobacillus strains are necessary characteristics for the bacterial adhesion to the host system, and its protection through biofilm formation over the host intestinal tissue, making these bacteria promising candidates for use as novel probiotics in the food and pharmaceutical industries [122]. In a recent Turkish study, another bacterial strain known as "fructophilic lactic acid bacteria (FLAB)" was isolated from fresh bee pollen and bee bread samples. FLAB are lactic acid bacteria (LAB) that prefer fructose over glucose as a carbon source and have been isolated from ecological fructose-rich niches including flowers and fruits, as well as the gastrointestinal tracts (GIT) of several fructose-based diet insects such as honeybees [123]. Several studies have shown the beneficial uses of probiotics in humans against type 2 diabetes, obesity, inflammation, tumors, allergy, metabolic disorders, and infectious diseases [124–127]. Due to this large bacterial population, bee pollen may have a promising role in the food and pharmaceutical industries.

4.4. Phenolic Profile and In Vitro Antioxidant Potential of Bee Pollen

4.4.1. Volatile Compounds of Bee Pollen

The volatile content of bee pollen is rarely studied; indeed, two recent studies on samples from Lithuania, China, and Spain showed the presence of 42 different volatile compounds, mainly nonanal, dodecane, tridecane, hexane, 6-methyl-5-hepten2-one, methyl butanoic acid, limonene, and styrene [113,128]. These compounds are mainly found in flowers and participate in the attracting behavior of pollinators [129]. In other studies (from Greece and Poland), the results showed the presence of aldehydes, ketones, terpenoids, and minor amounts of furfural [115,130]. Bee pollen's aromatic profile is related to the botanical and geographical origins as well as climatic conditions and bee species [28,113].

4.4.2. Phenolic Profile of Bee Pollen

Bioactive compounds of bee pollen constitute an important quality criterion. Bee pollen as a natural product has gained great scientific interest due to its beneficial properties [131]. Interestingly, bioactive compounds may counteract the installation and/or development of different pathologies [132]. Therefore, the determination of the phenolic profile of bee pollen is considered the first step toward the standardization and prediction of the usefulness of this beehive product. The analysis of their composition revealed that the polyphenolic content presented an average of 3% to 5% of their composition, depending on the botanical origin of the bee pollen [133].

Phenolic acids represent an average of 0.19% of bee pollen, and their properties are mainly related to their structure [132]. Phenolic acids could be divided into benzoic acids, phenylacetic acid, and cinnamic acids, which are of the greatest interest and exert a good antioxidant activity as compared to the other phenolic acids groups. The main molecules of phenolic acids are chlorogenic acid, gallic acid, cinnamic acid, ferulic acid [103], hydroxycinnamic acid, and coumaric acid [134]. Flavonoids, on the other hand, are the most important group of polyphenols found in bee pollen with an average of 0.25% and 1.4% of its total composition; they are an excellent indicator of bee pollen quality [134]. Flavonoids are mostly found in bee pollen as glycosides (flavonoids associated with sugar units), with flavonol glycosides being the most abundant. However, the presence of glycoside bonds decreases the antioxidant activity of flavonols because of the steric effects, the reason why the content of bee pollen in free flavonoids is a good quality criterion [135,136]. The main flavonols identified in bee pollen are quercetin, kaempferol, and rutin. The main flavones are represented by apigenin, chrysin, and luteolin, and flavanones are represented by naringenin and pinocembrin, while genistein is the major isoflavone identified in bee pollen [132]. Additionally, resveratrol, the most important stilbene has been isolated by Ares et al. [137]. Table 2 summarizes the different isolated phenolic compounds with the different extraction techniques.

Table 2. Summary of the different methods used to determine the different active molecules of pollen.

Country	Floral Origin	Techniques	Phenolic Compounds Name	References
Turkey	ND	HPLC-PDA detector	gallic acid, 3,4-hydroxybenzoic acid, (+)-catechin1,2-dihydroxy-benzene, syringic acid, caffeic acid, rutin trihydrate, p-coumaric acid, trans-ferulic acid, apigenin 7 glucoside, resveratrol, quercetin, trans-cinnamic acid, naringenin, kaempferol, isorhamnetin	[131]
Turkey	*Castanea* spp.	HPLC-DAD	rosmarinic acid, vitexin, hyperoside, pinocembrin, trans-chalcone, apigenin, protocatechuic acid, galangin	[133]
Portugal	Monofloral: *Rubus* spp., *Cystisus* spp., *Quercus* spp., *Prunus* spp., *Leontondon* spp., *Cistus* spp., and *Trifolium* spp. Heterofloral: Castanea sativa and Echium spp. and ii) *Erica* spp., and *Eucalyptus* spp.	UHPLC-DAD-ESI-MS	coumaroyl quinic acid, myricetin-O-rutinoside, luteolin-O-dihexoside, quercetin-O-dihexoside, myricetin-O-hexoside, myricetin-O-(malonyl)rutinoside, isorhamnetin-O-dihexoside, quercetin-O-hexosyl-pentoside, quercetin-O-rutinoside isomer 1, quercetin-O-rutinoside isomer 2, luteolin-di-O-hexosyl-rhamoside, quercetin-O-(malonyl)rutinoside, isorhamnetin-O-rutinoside, hydroxybenzoyl myricetin, quercetin-O-(malonyl)hexoside, quercetin derivative, quercetin-O-rhamnoside, isorhamnetin-O-(malonyl)hexoside isomer 1, luteolin-O-(malonyl)hexoside, myricetin, isorhamnetin-O-(malonyl)hexoside isomer 2, myricetin-O-dihydroferuloyl protocatechuic acid, myricetin-O-acetyl hydroxybenzoyl protocatechuic acid-isomer 1, myricetin-O-acetyl hydroxybenzoyl protocatechuic acid isomer 2, quercetin-O-acetyl hydroxybenzoyl protocatechuic acid isomer 1, myricetin-O-acetyl hydroxybenzoyl hydroxybenzoic acid isomer 2, quercetin-O-acetyl hydroxybenzoyl hydroxybenzoic acid isomer 1, quercetin-O-acetyl hydroxybenzoyl hydroxybenzoic acid isomer 2, O-dihydroxy benzoyl acetyl malonyl coumaric acid flavonoid derivative	[103]

Table 2. *Cont.*

Country	Floral Origin	Techniques	Phenolic Compounds Name	References
China	*Rosa rugosa*	UPLC-ESI-QTOF-MS/MS	isorhamnetin 3-O-diglucoside, sorhamnetin-3-O-coumaroyl glucoside, isorhamnetin-3-O-6-O-acetyl-β-D-glucopyranosy, kaempferol-3-O-neohesperidoside, N′,N″,N‴-Tricaffeoyl spermidine, N′,N″,N‴-Dicaffeoyl p-coumaroyl spermidine, N′,N″,N‴-Di-p-coumaroyl caffeoyl spermidine, N′,N″,N‴-Tri-p-coumaroyl spermidine	[134]
Chile	*Brassica rapa* and *Eschscholzia californica*	HAPLC-DAD	syringic acid, coumaric acid, sinapic acid, ferulic acid, cinnamic acid, abscisic acid, catechin, myricetin, quercetin, apigenin, kaempferol, naringenin, rhamnetin	[135]
Brazil	*Eucalyptus marginata; Corymbia calophylla*	HPLC	gallic acid, 4-hydroxyphenylacetic acid, rutin, resveratrol, myricetin, quercetin-3-O-glucopyranoside, kaempferol-3-O-glucoside, kaempferol-3-O-rutinoside, naringenin, quercetin, phloretine, kaempferol	[136]
Morocco	*Coriandrum sativum*	HPLC/DAD/ESI-MSn	myricetin-3-O-rutinoside, quercetin-diglucoside, quercetin-3-O-rutinoside, kaempferol-3-O-rutinoside, isorhamnetin-3-O-rutinoside, isorhamnetin-O-pentosylhexoside, kaempferol-diglucuronide, isorhamnetin-3-O-glucoside, quercetin-3-O-rhamnoside, ellagic acid, N1-p-coumaroyl-N5, N10-dicaffeoylspermidinea, N1, N10-di-p-coumaroyl-N5-caffeoylspermidine, luteolin, quercetin-3-methyl-ether, N1, N5-di-p-coumaroyl-N10-caffeoylspermidine, N1, N5, N10-tri-pcoumaroylspermidine, N1, N5, N10-tri-pcoumaroylspermidine, N1, N5, N10-tri-pcoumaroylspermidine, N1, N5, N10-tri-pcoumaroylspermidine	[138]
Romania	*Hedera, Helianthus, Cistus, Cornus, Brassica, Gledistia, Hedysarum, Trifolium, Castanea, lamium, Magnolia, Fraxinus, Papaver, Crataegus, Prunus, Rubus, and Cordiandrum*	HPLC-DAD	gallic acid, protocatechuic acid, p-hydroxybenzoic acid, vanillic acid, caffeic acid, chlorogenic acid, p-coumaric acid, rosmarinic acid, myricetin, luteolin, quercetin, kaempferol	[139]
Italy	*Cistus ladanifer, Echium, Achillea, Quercus ilex, Rubus,* Pinaceae, *Filipendula, Trifolium incarnatum, Trifolium pratense, Trifolium repens, Prunus, Pyrus, Malus,* and *Oxalis*	UHPLC-ESI-QTOF-MS	cyanidin 3-O-xyloside/arabinoside, delphinidin 3-O-(60 ′-p-coumaroyl-glucoside), petunidin 3-O-arabinoside, pelargonidin 3-O-glucoside, delphinidin 3-O-glucoside, delphinidin 3-O-glucosyl-glucoside, delphinidin 3-O-rutinoside, cyanidin 3-O-sophoroside, naringin 60-malonate, naringin 40-O-glucoside, naringenin 7-O-glucoside, apigenin 7-O-(60′-malonyl-apiosyl-glucoside), tetramethylscutellarein, luteolin 7-O-glucuronide, apigenin 6-C-glucoside, kaempferol 3-O-glucuronide, quercetin 3-O-rutinoside, kamepferol 3,7-O-diglucoside, quercetin 3-O-galactoside 7-O-rhamnoside, quercetin 3-O-rhamnosyl-galactoside, kaempferol 3-O-sophoroside, 3,7-Dimethylquercetin, dihydroquercetin, formononetin, genistin, gallic acid ethyl ester, syringic acid, caffeic acid 4-O-glucoside, caffeoyl glucose, feruloyl glucose, caffeic acid, sesamol, hydroxytyrosol 4-O-glucoside, curcumin, and carnosic acid	[140]
Colombia	*Cistus ladanifer; Echium Achillea; Taraxacum; Carduus; Cirsium; Vicia; Quercus ilex; Rubus; Pinaceae; Filipendula; Trifolium incarnatum, Trifolium pratense; Trifolium repens; Prunus, Pyrus; Malus* and *Oxalis*	UHPLC-DAD	Caffeic acid, ferulic acid, S-N1,5,10-tri-ferulic acid isorhamnetin, kaempferol, luteolin, myricetin, p-coumaric acid, SP-N1,5,10,14-tetra-p-coumaric acid, pinobanskin, quercetin, spermidine, spermine, 4-methyl gallic acid, apigenin, amentoflavone, N1-caffeoyl-N5,10-di-p-coumaroyl-spermidine, and N1,10-di-pcoumaroyl-N5-caffeoyl-spermidine.	[44]

ND: Not determined.

4.4.3. In Vitro Antioxidant Activity of Bee Pollen

Oxidative stress is involved in the development of several pathologies such as diabetes, Alzheimer's, cancer, atherosclerotic, and other disorders [141]. The use of natural products such as beehive products including honey, pollen, royal jelly, and propolis as a source of antioxidant molecules has been supported and suggested to protect human cells from the effects of oxidative stress by numerous scientific studies [142]. Moreover, the antioxidant activities of bee pollen have been evaluated in several works using well-known techniques such as DPPH, ABTS, β- carotene, FRAP, CUPRAC, NO, and TAC assays [132,138,140,143–170]. Table 3 summarizes all studies that have evaluated the antioxidant activity of bee pollen. It presents the pollen origin, the type of extract, the methods used, and the main results. Indeed, El Ghouizi et al. [138] evaluated in vitro the antioxidant activity of the aqueous extract of Moroccan fresh bee pollen and revealed an important scavenging capacity against DPPH and FRAP with IC_{50} values of 0.39 ± 0.13 mg/mL and 0.54 ± 0.53 mg/mL, respectively.

Table 3. Summary of the different studies of the antioxidant activity of bee pollen.

Country	Botanical Origin	Extracts	Used Methods	Key Results	References
Morocco	*Coriandrum sativum*	Aqueous extract	DPPH	IC_{50} = 0.39 ± 0.13 mg/mL	[138]
			Ferric reducing power	IC_{50} = 0.54 ± 0.53 mg/mL	
			Total antioxidant capacity	56.92 ± 0.21 mg AAE/g	
Algeria	Monofloral samples: *wild carrot, rosemary*, and *eucalyptus*	Methanolic extract	Molybdate ion reduction Assay	101.58 ugGAE/g	[165]
Brazil	Monofloral: *Brassica genus; Brassica rapa; Astrocaryum; Aculeatissimum; Cocos nucifera; Myrcia; Alternanthera; M. scabrella; Eucalyptus; Coffea; M. scabrella; M. verrucosa; Eupatorium; Syagrus; A. aculeatissimum; Eupatorium; Myrcia; Cecropia; Myrcia; Alternanthera; M. caesalpiniifolia; Montanoa; Asteraceae; C. nucifera; Machaerium; M. caesalpiniifolia; Myrcia; Anadenanthera; Cecropia; Schinus; Ilex; Ricinus*	Ethanolic extract	DPPH	140 ± 5 mmol TE/g	[150]
			ORAC	563 ± 15 mmol TE/g	
Brazil	*Mimosa misera, Mimosa caesalpinifolia, Eythrina velutina, Ziziphus lotus, Prosopis juliflora, Mimosa tenuiflora, Piptadenia macrocarpa, Cautarea hexandra, Hyptis suavelens, Cautarea hexandra*, and *Maytenus rigida*	Ethanolic extract	β-carotene bleaching method	Antioxidant activity = 83.3%	[168]
Brazil	Heterofloral: *Arecaceae, Asteraceae baccharis*, and *Asteraceae eupatorium*	Hydroethanolic extract	FRAP	131.47 ± 75.08 mg GA eq/g	[145]
			DPPH	% inhibition = 72.46 ± 5.25%	
Brazil	*Arecaceae; Asteraceae baccharis; Asteraceae eupatorium; Brassicaceae*	Lyophilized extract	ABTS	120.10 ± 0.21 mmol TEAC/g	[145]
			DPPH	Antioxidant activity = 54.42 ± 0.23%	
			FRAP	60.64 ± 0.63 mmol of Fe2þ/g	
			β-carotene/linoleic acid Assay	Antioxidant activity = 91.93 ± 0.22%	
Brazil	ND	Hydroethanolic extract	DPPH	EC_{50} = 0.86 mg/mL	[151]
			FRAP	123.4 mgGAEq.100 g^{-1}	
			β-carotene/linoleic acid Assay	Antioxidant activity = 83.3%	

Table 3. *Cont.*

Country	Botanical Origin	Extracts	Used Methods	Key Results	References
Chile	*Tilia Tuan* Szyszyl	Aqueous extract	DPPH	IC_{50} = 2.36 mg/mL	[161]
			Superoxide-scavenging activity	IC_{50} = 2.29 mg/mL	
		Methanolic extract	DPPH	IC_{50} = 1.72 mg/mL	
			Superoxide-scavenging activity	IC_{50} = 3.48 mg/mL	
China	*Agastache rugosatache* rugosa *Brassica napus* L. *Camellia japonica* L. *Crataegus pinnatifi* *Dendranthema indicum* L. *Fagopyrum esculentum moench* *Helianthus annuus* L. *Nelumbo nucifera Gaertn.* *Phellodendron amurensis* *Prunus armeniaca* *Prunus persica* L. *Rosa rugosa Thunb.* *Schisandra chinensis* *Taraxacum mongolicum*	Hydroethanolic extract	ABTS	1.06 ± 0.02 mmol TE g^{-1}	[169]
			DPPH	IC_{50} = 1.28 ± 0.03 mg/mL	
			Reducing power	Antioxidant activity = 70.55 ± 0.00%	
China	*Lotus uligionosus*, *Escallonia rubra*	Aqueous extract	DPPH	119.9 eq/g	[157]
			Reducing power	69.5 eq/g	
Egypt	*Trifolium alexandrinum* L.	Ethanolic extract	DPPH	Antioxidant activity = 90%	[144]
		Petroleum ether	DPPH	Antioxidant activity = 75%	
		Dichloromethane	DPPH	Antioxidant activity = 63%	
		Ethyl acetate	DPPH	Antioxidant activity = 79%	
Egypt	*Zea mays*	Methanolic extract	DPPH	Antioxidant activity = 59%	[162]
			ABTS	Antioxidant activity = 76.51%	
Greece	Monofloral sample: *Brassica* sp. Heterofloral sample: *Cistus* sp. (Cistaceae), *Verbascum* sp. (Scrophulariaceae), *Trifolium* sp. (Leguminosae), *Prunus* sp. (Rosaceae), *Rubus* sp. (Rosaceae), *Asphodelus* sp. (Liliaceae), and *Persea americana* (Lauraceae)	Aqueous extract	DPPH	IC_{50} = 233.3 ± 6.1 µg/mL	[147]
			ABTS	IC_{50} = 56.2 ± 0.8 µg/mL	
Italy	Genus: *Hedera, Helianthus, Cistus, Cornus, Brassica, Gledistia, Hedysarum, Trifolium, Castanea, lamium, Magnolia, Fraxinus, Papaver, Crataegus, Prunus, Rubus, and Cordiandrum*	Aqueous/ methanol extract	ORAC	839.5 ± 49.5 µmol TE g^{-1} DW	[140]
			ABTS	224.6 ± 18.6 µmol TE g^{-1} DW	
			DPPH	134.7 ± 4.3 µmol TE g^{-1} DW	

Table 3. *Cont.*

Country	Botanical Origin	Extracts	Used Methods	Key Results	References
Korea	Monofloral samples: *Quercus palustris, Actinidia arguta, Robinia pseudoacacia, and Amygdalus persica.*	Ethanolic extract	DPPH	EC_{50} = 292.0 ± 13.05 µg/mL	[170]
Portugal	*Cistus ladanifer, Echium* spp., *Apiaceae,* and *Cistaceae*	Hydroethanolic extract	DPPH	EC_{50} = 2.62 ± 0.09 mg/mL	[146]
			Reducing power Assay	6.51 ± 0.30 mg GAE/mL	
Portugal	*Cistacae Boraginacae, Rosaceae, Fagaceae, Asteraceae, Fabaceae, Ericaceae, Mimosaceae,* and *Myrtaceae.*	Methanolic extract	DPPH	EC_{50} = 3.0 ± 0.7 mg/mL	[155]
			β-carotene bleaching Assays	EC_{50} = 4.6 mg/mL	
Spain	*Cistaceae, Fabaceae, Cistaceae, Ericaceae, Fabaceae, Cistaceae, Ericaceae, and Boraginaceae*	Methanolic extract	DPPH	EC_{50} = 2.98 ± 0.47 mg/mg extract	[164]
			TBARS	EC_{50} = 0.35 ± 0.02 mg/mg extract	
Turkey	ND	Methanolic extract	FRAP	11.77 ± 0.63–105.06 ± 0.59 mmol Trolox/g pollen	[167]
			DPPH	SC_{50} = 0.65–8.20 mg/mL	
			CUPRAC Assay	33.1 ± 0.4–91.8 ± 1.8 mmol Trolox/g pollen	
Turkey	*Centaurea* sp, *Lotus* sp., *Coronilla* sp., *Centaurea* sp., *Scabiosa* sp., *Euphorbia* sp., *Echium* sp., *Coronilla* sp., *Teucrium* sp., *Crepis* sp., and *Castanea sativa*	Ethanolic extract	ABTS	0.373 ± 0.015–5.980 ± 0.100 mg TEAC/g	[160]
			DDPH Assays	1.293 ± 0.031–3.85 ± 0.030 mg TEAC/g	
Turkey	Commercial bee pollen	Extractable fraction	CUPRAC Assay	6.25–64.88 µmol TE/g	[152]
			ABTS	6.20–38.20 µmol TE/g	
			DPPH	0.44–22.45 µmol TE/g	
		Hydrolysable fraction	CUPRAC Assay	69.16–192.96 µmol TE/g	
			ABTS	37.63–80.49 µmol TE/g	
			DPPH	33.21–62.37 µmol TE/g	
		Bio-accessible fraction	CUPRAC Assay	83.24–257.27 µmol TE/g	
			ABTS	48.96–111.40 µmol TE/g	
			DPPH	35.69–83.84 µmol TE/g	
Turkey	ND	Methanolic extract	CUPRAC Assay	0.02 ± 0.02–0.24 ± 0.04 mmol Trolox/g	[166]
			FRAP	8.69 ± 1.64–84.89 ± 10.09µmol FeSO4.7H2O/g	
			DPPH	SC_{50} =0.47 ± 0.51–0.84 ± 0.17 mg/mL	

Table 3. *Cont.*

Country	Botanical Origin	Extracts	Used Methods	Key Results	References
Poland	*Aesculus hippocastanum, Chamerion angustifolium, Lamium purpureum, Lupinus polyphyllus, Malus domestica, Phacelia tanacetifolia, Pyrus communis, Robinia pseudoacacia, Sinapis alba, Taraxacum officinale, Trifolium* sp., and *Zea mays.*	Pepsin-digested extract	DPPH	EC_{50} = 20.912 ± 0.821 µL/mL	[159]
			ABTS	1.752 ± 0.024 mmol Trolox/g	
Malaysia	ND	Ethanolic extract	DPPH	Antioxidant activity = 39%	[163]
Serbia	*Helianthus annuus* L.	Methanolic extract	ABTS	Antioxidant activity = 95.5%	[158]
			FRAP	A700 nm = 0.738	
		Ethanolic extract	ABTS	Antioxidant activity = 75%	
			FRAP	A700 nm = 0.485	
Slovakia	*Helianthus annuus* L.	Ethanolic extract	DPPH	Antioxidant activity = 47.97 ± 0.29–50.46 ± 0.43%	[153]
Slovakia	Monofloral samples: *Brassica napus* L. *var. napus, Helianthus annuus* L., *Papaver somniferum* L., *Phacelia tanacetifolia* L., *Robinia pseudoacacia* L., and *Trifolium repens* L.	Methanolic extract	ABTS	0.83 ± 0.10–2.08 ± 0.25 mm/l	[148]
			DPPH	Antioxidant activity = 25.96 ± 1.61–93.69 ± 5.80%	
		Aqueous extract	DPPH	Antioxidant activity = 19.66 ± 1.06–50.29 ± 3.05%	
Slovakia	Monofloral samples: *Brassica napus* subsp. *napus* L, *Papaver somniferum* L., *and Helianthus annuus* L	Ethanolic extract	DPPH	Antioxidant activity = 70.05± 17.17%	[154]
			Reduction power	3575.56 ± 749.04 µg. mL^{-1}	
Thailand	Commercial bee pollen	Ethanolic extract	DPPH	40.69 ± 3.01 mg GAE/g extract	[156]
		Aqueous extract	DPPH	21.27 ± 2.63 mg GAE/g extract	
Bosnia and Herzegovina	*Poaceae* spp., Trifolium spp., *Zea mays*, and *Plantago* spp.	Methanolic extract	DPPH	IC_{50} = 1.43 ± 0.00 mg/g	[143]
			FRAP	4.111 ± 0.136 mmol Fe^{+2}/g	
			ABTS	Antioxidant activity = 86.13 ± 2.28%	

ND: Not determined.

In Brazil, several authors have investigated bee pollen samples from several botanical and geographical origins for their antioxidant proprieties using four antioxidant assays (DPPH, ORAC, β-carotene, and FRAP) and the results revealed that the bee pollen extracts exhibited important antioxidant activity in all tests with a significant difference among them, and a potential correlation between this activity and polyphenolic composition, which in turn varied depending on the geographical and botanical origins of the plant visited by the bees [145,150,151,168,171].

In Egypt, authors have investigated the antioxidant activity of ethanol, methanol, petroleum ether, dichloromethane, and ethyl acetate extracts of bee pollen samples from monofloral sources (*Zea mays* and *Trifolium alexandrinum* L) by two in vitro methods (ABTS and DPPH). The results showed an interesting anti-DPPH activity of ethanol extract with a percentage activity of 90%, while the methanol extract revealed a strong activity using the ABTS test with a percentage activity of 76.51%. Based on the findings, the high antioxidant activity of ethanol extract could be related to its major compounds including catechin, quercetin, and caffeic and gallic acid [144,162]. These results are confirmed by studies on Chinese bee pollen. In those works, authors investigated the antioxidant capacity of aqueous, ethanol, and methanol extracts using DPPH, ABTS, superoxide-scavenging activity, and reducing power and showed that ethanol extract of bee pollen has a good anti-DPPH effect with IC_{50} = 1.28 ± 0.03 mg/mL, and an important anti-ABTS (1.06 ± 0.02 mmol TE g^{-1}) and reduction power activity (70.55%). The methanol extract showed also an antioxidant effect with an anti-DPPH IC_{50} value of 1.72 mg/mL, and IC_{50} = 3.48 mg/mL for superoxide-scavenging activity [157,161,169]. Methanol extracts of twenty-two bee pollen samples from different floral origins in Portugal also showed an interesting antioxidant effect, which manifests in scavenger activity of the free DPPH• and β-carotene in bleaching assays with mean values of 3.0 ± 0.7 mg/mL and 4.6 mg/mL, respectively [155].

On the other hand, several authors have reported the antioxidant activity of Turkish bee pollen extracts based on different geographical origins [160,167], bee races (*Apis mellifera caucasica*, *Apis mellifera anatoliaca*, *Apis mellifera syriaca*, and *Apis mellifera carnica*) [166], and extraction methods as well as the storage conditions [152]. The authors showed important antiradical activity of bee pollen which is highly affected by chemical composition, plant origins, geographical origin, and storage conditions.

Other research works reported a variable antioxidant effect of different bee pollen extracts including those from Algeria, Greece, Italy, Korea, Spain, Slovakia, Malaysia, Serbia, Thailand, and Poland, and this variability can be significantly related to a variety of botanical and geographical origins [132,140,143,147,148,153,154,156,158,159,163–165,170].

5. Therapeutic Properties of Bee Pollen against Oxidative Stress-Related Diseases

The use of bee pollen in traditional medicine dates to ancient times and is attested by books of Arab and Jewish doctors such as Ibn al-Beithar and Maimonides in the early 1100s, where they described bee pollen as an aphrodisiac, sedative, and effective for the stomach, the heart, and intestines [172,173].

5.1. Antioxidative Properties

Oxidative stress is caused by an imbalance between free radical production and antioxidant defense systems, resulting in an accumulation of reactive oxygen species (ROS). In turn, ROS interact with various cytoplasmic components such as proteins, membrane lipids, and DNA [174]. As a result, ROS induce serious cell damage and participate in the development of many chronic illnesses such as diabetes and associated complications, arthritis, Parkinson, and Alzheimer's [175]. Bee pollen is one of the natural antioxidant-rich products mainly used against oxidative stress and related pathologies. In this context, Kawther and coworkers have proved that the administration of bee pollen extract (250 mg/kg b.w) attenuated oxidative stress induced by protein [176]. It has been demonstrated that the anti-oxidative activity of bee pollen is attributed to its content of secondary metabolites including, vitamin E, vitamin C, carotenoids, and phenolic compounds [149]. Thanks to its phenolic hydroxyl group, the flavonoids present in bee pollen can scavenge ROS and free radicals, and inactivate electrophiles [164]. As reported in Table 4, several studies have reported the antioxidative properties of various phenolic compounds and their involved mechanisms. Pari et al. reported that the sub-chronic administration of Caffeic acid (6 mg/kg b.w) improved the oxidative stress caused by alcohol-induced toxicity in rats by increasing non-enzymic antioxidant defense systems, and by preventing lipid peroxidation [177]. Likewise, cinnamic acid occurs in the antioxidative process by modulating lipid metabolism and boosting GSH, SOD, and CAT enzyme activities as well as scavenging and decreasing ROS production [178]. In addition, the oral administration of rutin at a dose of 50 and 100 mg/kg/b.w for 20 days enhances the production of antioxidant enzymes, decreases serum toxicity markers, and downregulates COX, 2p38-, MAPK, i-NOS, and the NF-κB signaling pathway [179]. Furthermore, a recent study showed that quercetin weakened oxidative stress and decreased the expression of TNFα, IL-1β, and IL-6 [180]. Similarly, treatment with luteolin minimized the oxidative stress through multiple mechanisms: (a) up-regulation of the Nrf-2 pathway, (b) enhancement of HO-1 expression, (c) increase in GSH, SOD, and GPX activities, and (d) decrease in MDA levels [181]. Pinocembrin decreased oxidative stress, apoptotic, and inflammatory markers [182] Figure 1.

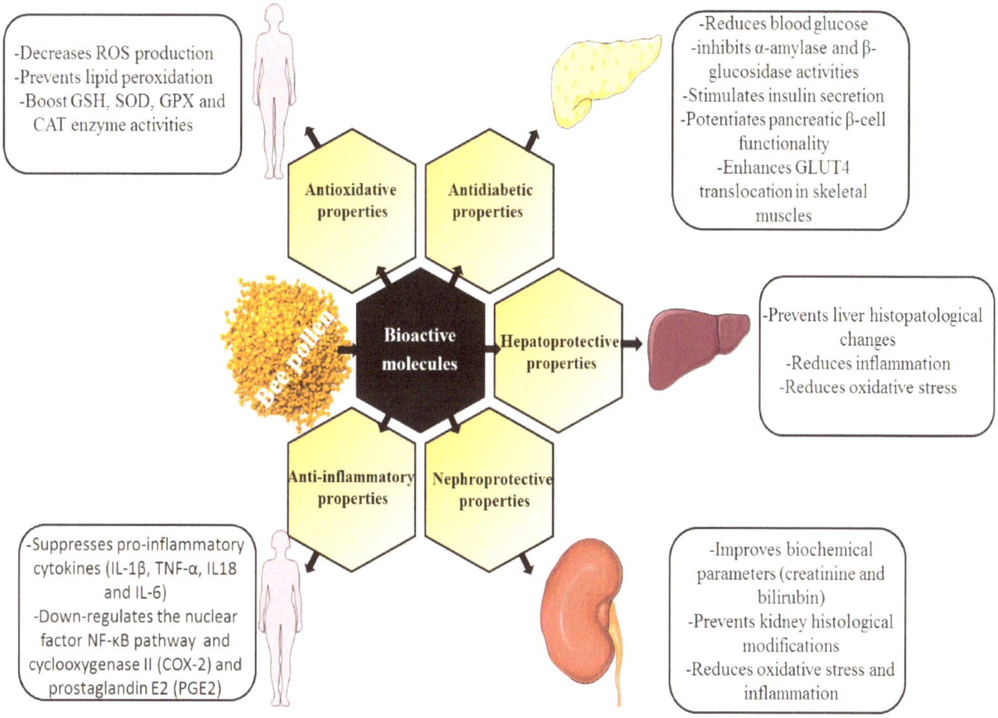

Figure 1. Schematic presentation of the therapeutic properties of bee pollen.

5.2. Antidiabetic and Anti-Hyperglycemic Properties

Diabetes mellitus is an endocrine disorder characterized either by insufficient insulin secretion and/or its defective utilization [183]. The antidiabetic/anti-hyperglycemic activity of bee pollen has been previously studied. According to Nema et al., bee pollen administration at 100 mg/kg body weight/day for 4 weeks lowered blood glucose and prevented pituitary–testicular axis dysfunction [6]. Furthermore, bee pollen exhibited a potent anti-hyperglycemic activity in patients with insulin-independent diabetes mellitus (T2DM) [184]. Moreover, thanks to its bioactive constituents, bee pollen exerts its anti-hyperglycemic effect by modulating glucose uptake and inhibiting α-amylase and β-glucosidase activities, leading to the management of diabetes and its serious complications [185]. The anti-diabetic properties of many individual phenolic compounds present in bee pollen have already been investigated. Adisakwattana and coworkers enunciated that cinnamic acid administered orally at a dose of 50 mg/kg/day for 5 weeks stimulates insulin and adiponectin secretions, increases hepatic glycolysis, improves glucose uptake, potentiates pancreatic β-cell functionality, and decreases protein glycation [186]. Rutin regulates glycemia and ensures its anti-diabetic effect through the inhibition of the polyol signaling pathway as well as via the modulation of lipid metabolism and the prevention of lipid peroxidation [187]. Apigenin facilitates and enhances GLUT4 translocation in skeletal muscles either by up-regulating the AMP-activated protein kinase pathway or by activating the insulin signaling pathway, which leads to glucose uptake and thus hypoglycemia. A paper published by Alkhalidy et al. explored the anti-diabetic effect of kaempferol against streptozotocin-induced diabetes in rats and found that chronic administration of kaempferol (50 mg/kg-b.w) reduced hepatic glucose production, increased hexokinase activity, decreased hepatic pyruvate carboxylase activity, and inhibited the gluconeogenesis pathway [188], as shown in Figure 1.

5.3. Hepatoprotective Properties

Bee pollen extract has been found to possess a potent hepato-protective effect. Cheng and coworkers reported that bee pollen extract administration increased tissue catalase SOD and GSH-Px activity, and prevents liver histological changes induced by carbon tetrachloride treatment in mice. This suggests the potential role of bee pollen in preventing hepatocellular changes associated with exposure to xenobiotics. The hepatoprotective capacity of bee pollen is largely attributed to its rich content of natural antioxidants such as phenols and flavonoids [189]. The hepatoprotective ability of phenolic compounds has been explored in numerous studies (Table 3). Malayeri et al. showed that the co-administration of a single dose of naringenin (50 mg/kg/b.w) boosted enzymatic and non-enzymatic antioxidant activities, and reduced NO, TNF-α, and IL-6 levels [190]. Yang and coworkers indicated that ferulic acid exhibits its protective role against CCL4-caused acute oxidative liver damage in rats via the up-regulation of p-JNK, p-p38 MAPK, and Bcl-2 signaling pathways and thus, decreased the expression of pro-inflammatory mediators of hepatic-toxicity TNF-α and IL-1β [191]. Owumi et al. recently emphasized that protocatechuic acid protects against methotrexate-induced liver dysfunction via enhancing enzyme antioxidant defense mechanisms and decreasing oxidative stress and free radical production, which was confirmed by biochemical analysis and histopathological investigations [192]. Vanderson et al. demonstrated that caffeic acid treatment improves oxidative stress and kidney dysfunction mediated by ethanol in a rat model. This was attributed to the down-regulation of CYP2E1 and the protection of DNA against oxidative damage [193]. Ebrahimi and coworkers proved that ellagic acid treatment reduced oxidative damage and liver ultrastructure changes in methotrexate-induced mitochondrial dysfunction and liver toxicity in rats [194] (see Figure 1).

5.4. Nephroprotective Properties

Owing to its rich content of various bioactive molecules, bee pollen has been found to have a potent nephroprotective activity. In a rat model, it was reported that bee pollen extract improved biochemical parameters (creatinine and bilirubin), increased the antioxidant defense system (SOD, CAT, and GSH), lowered oxidative stress biomarkers (MDA and iNOS), and prevented kidney histological effects induced by cisplatin. Intraperitoneal administration of apigenin reduced COXI, COXII, and MDA levels, and increased kidney GSH levels [195]. Recently, Owumi et al. reported that protocatechuic acid exhibits its reno-protective activity by increasing the activity of antioxidant enzymes (SOD, CAT, GSH, and GPX) and decreasing the renal (RNOS and LPO) levels, as well as reducing the inflammation biomarkers NO, TNF-α, and IL-1β levels in renal tissue [192]. Another study found that naringenin (100 mg/kg/b.w) reduced oxidative stress and prevented lipid peroxidation in rats after cyclosporine treatment [196]. According to Chowdhury and colleagues, ferulic acid prevented hyperglycemia-induced kidney damage and oxidative stress in rats; this was related to the modification of AGEs, MAPKs (p38 and JNK), and the NF-B signaling pathways by this acid [197]. Mohammed and coworkers evidenced that ellagic acid stimulated the expression of SIRT1, decreased P53 protein levels, reduced ROS production, and enhanced enzymatic and non-enzymatic antioxidant systems [198]. This was demonstrated through improved biochemical values (creatinine, urea, and uric acid), kidney histopathological tissue, and renal biomarker stress (MDA, GSH, CAT). Another study showed that pinocembrin treatment mitigates gentamicin-induced inflammation and renal toxicity via the modulation of Nrf2/HO-1 and NQO1 pathways [199]. This could suggest its potent ability to reduce oxidative stress and inflammation-related nephrocellular dysfunctions. Shanmugam et al. reported that oral administration of Kaempferol (100 mg/kg/day/b.w) exerts its nephroprotective action by inhibiting RhoA/Rho kinase-mediated inflammatory pathway [200] Figure 1.

5.5. Anti-Inflammatory Properties

Numerous scientific studies have indicated that bee pollen has a potent anti-inflammatory impact. Indeed, it has been shown that flavonoids and phenolic acids play a major role in the anti-inflammatory activity of bee pollen extracts. As indicated in Table 3, several individual phenolic components have shown anti-inflammatory effects through different signaling pathways. Indeed, phenolic acids including caffeic acid, ferulic acid, and cinnamic acid are documented as potent inhibitors of tumor necrosis factor (TNF) and thus down-regulation of the nuclear factor NF-κB pathway (proinflammatory signaling pathway) [201–203]. In addition, ellagic acid has been found to inhibit nitric oxide (NO), TNF-α, and IL-6, and induce the down-regulation of cyclooxygenase II (COX-2) and prostaglandin E2 (PGE2) [204]. The anti-inflammatory activity of bee pollen is also linked to Galangin, chrysin, quercetin, resveratrol, kaempferol, and other flavonoid molecules. Choi and coworkers have shown that galangin inhibits the expression of iNOS, COX-2, and the release of pro-inflammatory cytokines such as IL-1β and TNF-α [205]. In the same context, the findings of Li, Zhipeng, et al. showed that chrysin improves inflammatory reaction through the inhibition of NO, prostaglandin E2, and the NF-κB signaling pathway [206]. Quercetin and resveratrol exhibit their anti-inflammatory actvity via the down-regulation of the NF-κB pathway and the inhibition of COX-1 and COX-2 activities [207,208]. Park and coworkers reported that Kaempferol ensures its protective effect in aged kidney tissues via the suppression of pro-inflammatory cytokines (IL-1β, TNF-α, IL-18, and IL-6) [209]. These data revealed that kaempferol could have the ability to attenuate age-related chronic inflammatory reactions Figure 1.

Table 4. Summary of the different pharmacological properties of different phenolic compounds found in bee pollen.

Molecules	Dosage, Route, and Exposure Duration	Pharmacological Properties	Involved Mechanisms	References
Caffeic acid	6 mg/kg/day, orally for 45 days.	Anti-oxidative properties	↑Non-enzymic antioxidants, ↓lipid peroxidation, and ↓TBARS level.	[177]
Cinnamic acid	20 mg/kg/day, i.p for 40 days.		↓lipid peroxidation, ↓ROS production ↑GSH, ↑SOD, and ↑CAT levels.	[178]
Ferulic acid	25 mg/kg/day, orally for 10 days.		↓lipid peroxidation, ↓ROS levels, and ↓N-acetyl-β-glucosminidase activity.	[210]
Ellagic acid	10 and 30 mg/kg/day for 30 days.		Enhances the concentration of enzymatic antioxidant levels (SOD, CAT, and GPx), and ↓ MDA, ↓TNF-α, and ↓IL-1β.	[211]
Quercetin	50 mg/kg/day, i.p for 21 days.		Increases GSH level, SOD, GR, G, P, and CAT activity, and decreases the expression of TNFα, IL-1β, and IL-6.	[180]
Kaempferol	100 mg/ kg/day, i.p for 6 weeks.		Inhibits the activity of ASK1/MAPK signaling pathways (JNK1/2 and p38).	[212]
Galangin	8 mg/kg/day, i.p for 45 days.		↓lipid peroxidation, ↑enzymatic and non-enzymatic antioxidants.	[213]
Chrysin	30 mg/kg/day, orally, for 14 days.		↑GSH, ↓TBARS, ↓XO, and ↓NADPH levels	[214]
Protocatechuic acid	100 mg/kg/day, i.p for 7 days.		Prevents lipid peroxidation and the formation of NO, and enhances antioxidant enzymes.	[215]
Apigenin	0.625, 1.25, and 2.5 mg/mL.		↓oxidative stress, GSH level ↑SOD activity, ↓IL-6, and ↓NF-κB levels.	[216]
Luteolin	100 and 200 mg/kg/ day, orally for 28 days.		↓MDA, ↑GSH, ↑SOD ↑GPX ↑Nrf2, and ↑HO-1 Expressions.	[181]
Rutin	50 and 100 mg/kg/day, orally for 20 days.		↑Production of antioxidant enzymes, ↓serum toxicity markers, and downregulation of (COX, 2p38-, MAPK, i-NOS, and NF-κB).	[179]
Naringenin	50 mg/kg/day, orally for 8 weeks.		Minimizes oxidative stress and enhances CAT, SOD GSH, and GPx levels.	[217]
Pinocembrin	10 mg/kg/day, orally for 7 days.		Decreases oxidative stress, and apoptotic and inflammatory markers.	[182]

Table 4. *Cont.*

Molecules	Dosage, Route, and Exposure Duration	Pharmacological Properties	Involved Mechanisms	References
Caffeic acid	100 mg/kg/day, orally for 4 weeks.	Antidiabetic and anti-hyperglycemic properties	IL-6, ↓ IL-1β, ↓ TNF-α, ↓ MCP-1, ↓HbA1c, ↓ UGA, ↓ sorbitol, ↓ fructose, and ↑AMPKα2.	[218]
Ferulic acid	10 mg/kg/day, orally for 15 days.		Down-regulation of NF- κB pathway.	[219]
Cinnamic acid	50 mg/kg/day, orally for 5 weeks.		↑insulin secretion, ↑hepatic glycolysis, ↑adiponectin secretion ↑glucose uptake, ↑pancreatic β-cell functionality, and ↓protein glycation.	[186]
Ellagic acid	250 mg/kg/day, orally for 28 days.		↑ insulin secretion, ↑β-cell number, ↑plasma total antioxidants, and ↑glucose intolerance.	[220]
Quercetin	10 and 30 mg/kg/day, i.p for 14 days.		↑GLUTs, ↑IR-P, ↑GLUT4, ↑Glucose uptake, ↑pancreatic cell-β generation, ↑glucokinase activity, ↓α-glucosidase activity.	[221]
Kaempferol	50 mg/kg/day, orally for 12 weeks. 200 mg/kg/day, orally for 14 days.		↓hepatic glucose production, ↑hexokinase activity, ↓hepatic pyruvate carboxylase activity, and gluconeogenesis. ↑GLP-1 and insulin release, ↑ cAMP, and Ca2+ intracellular levels.	[188,222]
Galangin	4, 8, and 16 mg/kg/day, orally for 45 days.		Inhibition of DPP-4, ↓oxidative stress, and ↑antioxidant status.	[213]
Chrysin	40 mg/kg/day, orally for 16 weeks.		Inhibition of the TNF-α pathway, ↓secretion of pro-inflammatory cytokines, and ↓glucose and lipid peroxidation levels.	[223]
Protocatechuic acid	50 and 100 mg/kg/day, orally for 7 days.		↑insulin sensitivity, ↓insulin resistance, ↓gluconeogenesis, and ↑glucose uptake.	[224]
Apigenin	1.5 mg/kg/day, i.p for 28 days.		Enhances GLUT4 translocation.	[225]
Luteolin	10 mg/kg/day, orally for 24 weeks.		Reduces oxidative stress and inhibits the STAT3 pathway.	[226]
Rutin	90 mg/kg/day, orally for 10 weeks.		Inhibition of polyol pathway, oxidative stress, and lipid peroxidation.	[187]
Naringenin	50 and 100 mg/kg/day, orally for 6 weeks.		Improvement of glucose and lipid metabolism, and ↓insulin resistance.	[227]
Pinocembrin	50 mg/ kg/day, orally for 10 days.		↓ NF-κB and TNF-α levels.	[228]
Resveratrol	12 mg/kg/day, orally for 1 5 days.		Down-regulation of NF- κB pathway.	[219]
Caffeic acid	100 mg/kg/day, orally for 4 days.	Hepato-protective properties	Downregulation of CYP2E1 and the protection of DNA against oxidative damage.	[193]
Cinnamic acid	20 mg/kg/day, orally for 10 days.		↓NF-kB and ↓iNOS activities.	[229]
Ellagic acid	5 and 10 mg/kg/day, orally for 10 days		Up-Regulation of Nrf2 and HO-1 expression and inhibition of NF-κB signaling pathway.	[194]
Quercetin	20, 40, and 80 mg/kg/day, orally for 7 days.		Modulation of the expression of nuclear orphan receptors (CAR, PXR) and cytochrome P450 enzymes (CYP1A2, CYP2E1, CYP2D22, CYP3A11).	[230]
Kaempferol	20 mg/kg, twice a day, orally for 28 days.		↓CYP2E1 activity and ↓ROS production.	[231]
Galangin	15, 3,0, and 60 mg/kg/day, orally for 15 days.		Activation of Nrf2 and HO-1 signaling pathway.	[232]
Chrysin	25 or 50 mg/kg, orally for 6 days.		Decreases the expression of COX-2, iNOS.	[233]
Ferulic acid	25, 50, and 100 mg/kg/day, orally for 7 days.		↓ the expression TNF-α and IL-1β, upregulation of p-JNK, p-p38 MAPK, and Bcl-2.	[191]
Protocatechuic acid	25 and 50 mg/kg/day, i.p for 7 days.		↓ oxidant species ↑antioxidant enzymes	[192]
Apigenin	10 mg/kg/day, orally for 3 weeks.		Enhances antioxidant defense mechanisms and decreases lipid peroxidation.	[234]
Luteolin	100 mg/kg/day, i.p for 7 days.		Modulation of Nrf2/HO-1 pathway and ↓oxidative stress.	[235]
Rutin	20 mg/kg/day, orally for 15 days.		↑Antioxidant profile and regulation of Na+/K+ ATPase activity.	[236]
Naringenin	50 mg/kg/day, orally for 10 days.		↑the enzymatic and non-enzymatic antioxidant levels, ↓NO, TNF-α, and IL-6 levels.	[190]
Pinocembrin	50 and 75 mg/kg/day, i.p for 10 days.		Modulation of Nrf2/HO-1 and NQO1 pathways.	[199]
Resveratrol	50 and 100 mg/kg/day, orally for 28 days		Modulation of SIRT1 and p53 pathways.	[237]

Table 4. *Cont.*

Molecules	Dosage, Route, and Exposure Duration	Pharmacological Properties	Involved Mechanisms	References
Caffeic acid	100 mg/kg/day, orally for 14 days.	Nephroprotective properties	Enhances the antioxidant defense system and reduces lipid peroxidation.	[238]
Ferulic acid	50 mg/kg/day, orally for 8 weeks.		Modulation of AGEs, MAPKs (p38 and JNK), and NF-κB pathways, and ↓oxidative stress.	[197]
Cinnamic acid	50 mg/kg /day, orally for 7 days.		antioxidant expression GSH levels, SOD, CAT, and GPx activities.	[239]
Ellagic acid	10 mg/kg/day, orally for 30 days.		Stimulates the expression of SIRT1, ↓P53 protein level, ↓ROS production, and ↑enzymatic and non-enzymatic antioxidant system.	[198]
Quercetin	10 mg/kg/day, i.p for 10 weeks.		↑antioxidant expression and ↓lipid peroxidation.	[240]
Kaempferol	100 mg/kg/day, orally for 28 days.		Inhibits RhoA/Rho Kinase mediated inflammatory pathway.	[241]
Chrysin	30 and 100 mg/kg, ip for 26 days.		↑iNOS and PKC Levels, and ↓AGEs and RAGE.	[242]
Protocatechuic acid	25 and 50 mg/kg/day, i.p for 7 days.		↓ oxidant species ↑antioxidant enzymes.	[192]
Apigenin	3 mg/kg/day, i.p for 7 days.		Reduces COXI and COXII, MDA levels and increases GSH level.	[195]
Luteolin	10 and 20 mg/kg/day, orally for 4 weeks.		Inhibition of RIP140/NF-κB pathway.	[243]
Rutin	100 mg/kg/day, orally for 14 days.		Suppresses NF-κB activation and TGF-β1/Smad3 signaling.	[244]
Naringenin	100 mg/kg/day, orally for 45 days.		↓ oxidative stress and lipid peroxidation levels.	[196]
Pinocembrin	50 and 75 mg/kg/day, i.p for 10 days.		Modulation of Nrf2/HO-1 and NQO1 pathways.	[199]
Resveratrol	20 mg/kg/day, orally for 40 weeks.		Modulation of the NF-κB signaling pathway.	[245]
Caffeic acid	50 mg/kg/day, orally for 21 days.	Anti-inflammatory properties	inhibition of NO, prostaglandin E2, and NF-κB signaling pathways.	[201]
Ferulic acid	100 mg/kg/day, orally for 6 weeks.		Inhibition of NADPH oxidase and NF-κB pathway.	[202]
Cinnamic acid	60 mg/kg/day, orally for 21 days.		Down-regulation of the NLRP3, NF- κB, and ASK1/MAPK signaling pathways.	[203]
Ellagic acid	1, 3, 10, and 30 mg/kg, i.p for 5 h.		Suppression of NF-κB pathway and NO, TNF-α, IL-6, COX-2 activity, and PGE2.	[246]
Quercetin	1 mg/kg/day, orally for 15 days.		Down-regulation of the NF-κB pathway.	[208]
Kaempferol	2 and 4 mg/kg/day for 10 days.		Decreases the synthesis of IL-1β, TNF-α, IL-18, and IL-6.	[209]
Galangin	50 mg/kg per day, orally for 4 days.		Inhibits the expression of iNOS, COX-2, and pro-inflammatory cytokines.	[205]
Chrysin	40 mg/kg/day, orally for 16 weeks.		inhibition of NO, prostaglandin E2, and NF-κB signaling pathways.	[206]
Protocatechuic acid	20 mg/kg/day, orally for 8 weeks.		↓IL-1β, ↓IL-6, and ↓TNF-α synthesis pathways.	[247]
Apigenin	20 and 40 mg/kg/day, orally for 28 days.		↓TNF-α and IL-6 production.	[248]
Luteolin	100 mg/kg, i.p for 6 h.		↑HO-1 expression, ↑IL-10, ↓TNF-α, and ↓IL-6 levels.	[249]
Rutin	30 mg/kg/day, orally for 14 days.		Inhibition of p38-MAPK pathway.	[250]
Naringenin	5, 10, and 20 mg/kg/day, for 16 days.		Up-regulation of Nrf-2/HO-1pathway and ↓NF-kB mRNA expression.	[251]
Pinocembrin	50 mg/kg/day, i.p for 24 days.		Down-regulation of NF-kB pathway.	[252]
Resveratrol	10 or 50 mg/kg/day, orally for 28 days.		Inhibition of COX-1 and COX-2 activities	[207]

↑ represent increases; ↓ represent decreases.

5.6. Other Beneficial Effects of Bee Pollen

Bee pollen has a broad spectrum of pharmacological effects and provides a promising area for researchers interested in the therapeutic effects of natural products, particularly hive products. The rich composition of probiotics, proteins, macro-, and micronutrients in bee pollen has been related to its positive effect on morphological development (thickness of epithelium) and functioning (absorption) of the small intestine, leading to the proper functioning of the gastrointestinal tract [217,253].

Bee pollen is also known to modulate the secretory activity (release of IGF-I growth factor and progesterone and estradiol steroid hormones) and apoptotic activity of the ovary in rats. In postmenopausal women with breast cancer, bee pollen can improve menopausal-related symptoms when used in association with antihormonal treatment; it is also beneficial for women who suffer from post-menopausal disorders [254].

Studies have demonstrated the positive effects of the phenolic and probiotic content of bee pollen on preventing metabolic syndrome by reducing body and liver weight gain, decreasing fasting blood glucose, and lipid accumulation in serum and liver, which can be explained through the regulation of intestinal microbiota [255].

In many cultures, bee pollen has long been used by women to maintain their beauty and whiten their skin. Since more than 70% of bee pollen composition is active, (proteins, carbohydrates, lipids/fatty acids, phenolic compounds, and vitamins), the cosmeceutical properties of bee pollen in the laboratory have been studied and researchers have demonstrated that pollen can boost protective mechanisms against skin aging (polyphenols, vitamin E, C), skin dryness (sugars and fatty acids), ultraviolet radiation (carotenoids), oxidative damage (polyphenols), and inflammation and melanogenesis, which are involved in human skin damage [256,257]. These scientific pieces of evidence are turning cosmetologists' attention toward introducing bee pollen into their beauty products and formulations, and guaranteeing better quality and functionality.

6. Conclusions

As has been shown so far, pollen grains are microscopic vegetal cells produced and dispersed during the process of plant reproduction. In the hive, pollen grains transform into bee bread through the fermentation process and become accessible for human consumption because of their complete composition of macronutrients such as proteins, carbohydrates, lipids, and micronutrients such as minerals, vitamins, and phenolic compounds. This diversified composition affords a wide range of pharmacological and biological properties to the bee pollen. Nonetheless, due to the significant variability of its composition, which is affected by several factors, bee pollen application in phytomedicine remains quite limited. On this basis, scientists and professionals should pay closer attention to a few key aspects: (i) standardization should be expanded to include the phenolic composition and nutritional value of different types of bee-collected pollen, especially monofloral pollen; (ii) more quality-control research is needed to encourage beekeepers to produce clean, safe, and economically valuable bee pollen; (iii) as bee pollen is partially digested by human digestive enzymes, more pharmacological and biochemical studies are necessary to enhance the bioavailability of bee pollen bioactive compounds and capitalize on bee pollen's biological importance; (iv) considering the techno-functional value of bee pollen as a superfood, it can be potentially used as a good ingredient in the food and pharmaceutical industries for the production of novel bee pollen-enriched food products or dietary supplements; (v) eventually, considering bee pollen's techno-functional value and biological properties, more clinical trials should be conducted to investigate the beneficial effect of this superfood on human health and to encourage the food and pharmaceutical industries to develop and manufacture novel bee pollen-enriched food products and dietary supplements.

Author Contributions: Conceptualization, A.E.G. and M.B.; writing—original draft preparation, A.E.G., M.B., D.O., H.L. and N.E.M.; writing—review and editing, C.H. and B.L.; supervision, C.H. and B.L. All authors have read and agreed to the published version of the manuscript.

Funding: This study was partially funded by the SNAMOPEQ laboratory, FSDM, USMBA, Fez, Morocco.

Institutional Review Board Statement: Not applicable.

Informed Consent Statement: Not applicable.

Data Availability Statement: Data are available upon request.

Conflicts of Interest: The authors declare no conflict of interest.

References

1. Knox, R.B. The Pollen Grain. In *Embryology of Angiosperms*; Johri, B.M., Ed.; Springer: Berlin/Heidelberg, Germany, 1984; pp. 197–271. ISBN 978-3-642-69302-1.
2. Campos, M.G.R.; Bogdanov, S.; de Almeida-Muradian, L.B.; Szczesna, T.; Mancebo, Y.; Frigerio, C.; Ferreira, F. Pollen Composition and Standardisation of Analytical Methods. *J. Apic. Res.* **2008**, *47*, 154–161. [CrossRef]
3. De-Melo, A.A.M.; de Almeida-Muradian, L.B. Chemical Composition of Bee Pollen. In *Bee Products-Chemical and Biological Properties*; Alvarez-Suarez, J.M., Ed.; Springer International Publishing: Cham, Switzerland, 2017; pp. 221–259. ISBN 978-3-319-59688-4.
4. Kieliszek, M.; Piwowarek, K.; Kot, A.M.; Błażejak, S.; Chlebowska-Śmigiel, A.; Wolska, I. Pollen and Bee Bread as New Health-Oriented Products: A Review. *Trends Food Sci. Technol.* **2018**, *71*, 170–180. [CrossRef]
5. Asmae, E.G.; Nawal, E.M.; Bakour, M.; Lyoussi, B. Moroccan Monofloral Bee Pollen: Botanical Origin, Physicochemical Characterization, and Antioxidant Activities. *J. Food Qual.* **2021**, *2021*, 1–10. [CrossRef]
6. Mohamed, N.A.; Ahmed, O.M.; Hozayen, W.G.; Ahmed, M.A. Ameliorative Effects of Bee Pollen and Date Palm Pollen on the Glycemic State and Male Sexual Dysfunctions in Streptozotocin-Induced Diabetic Wistar Rats. *Biomed. Pharm.* **2018**, *97*, 9–18. [CrossRef]
7. Maruyama, H.; Sakamoto, T.; Araki, Y.; Hara, H. Anti-Inflammatory Effect of Bee Pollen Ethanol Extract from Cistus Sp. of Spanish on Carrageenan-Induced Rat Hind Paw Edema. *BMC Complement. Altern. Med.* **2010**, *10*, 30. [CrossRef]
8. Kaškonienė, V.; Adaškevičiūtė, V.; Kaškonas, P.; Mickienė, R.; Maruška, A. Antimicrobial and Antioxidant Activities of Natural and Fermented Bee Pollen. *Food Biosci.* **2020**, *34*, 100532. [CrossRef]
9. Wang, B.; Diao, Q.; Zhang, Z.; Liu, Y.; Gao, Q.; Zhou, Y.; Li, S. Antitumor Activity of Bee Pollen Polysaccharides from Rosa Rugosa. *Mol. Med. Rep.* **2013**, *7*, 1555–1558. [CrossRef]
10. Prieto, P.; Pineda, M.; Aguilar, M. Spectrophotometric Quantitation of Antioxidant Capacity through the Formation of a Phosphomolybdenum Complex: Specific Application to the Determination of Vitamin E. *Anal. Biochem.* **1999**, *269*, 337–341. [CrossRef]
11. Roy, R.; Schmitt, A.J.; Thomas, J.B.; Carter, C.J. Nectar Biology: From Molecules to Ecosystems. *Plant Sci.* **2017**, *262*, 148–164. [CrossRef]
12. Wodehouse, R.P. Pollen Grains. In *Their Structure, Identification and Significance in Science and Medicine*, 1st ed.; McGraw-Hill Book Company: New York, NY, USA, 1935.
13. Dolezal, A.G.; Toth, A.L. Feedbacks between Nutrition and Disease in Honey Bee Health. *Curr. Opin. Insect Sci.* **2018**, *26*, 114–119. [CrossRef]
14. Nepi, M.; Grasso, D.A.; Mancuso, S. Nectar in Plant–Insect Mutualistic Relationships: From Food Reward to Partner Manipulation. *Front. Plant Sci.* **2018**, *9*, 1063. [CrossRef]
15. Glover, B.J. Pollinator Attraction: The Importance of Looking Good and Smelling Nice. *Curr. Biol.* **2011**, *21*, R307–R309. [CrossRef]
16. Garbuzov, M.; Ratnieks, F.L.W. Quantifying Variation among Garden Plants in Attractiveness to Bees and Other Flower-Visiting Insects. *Funct. Ecol.* **2014**, *28*, 364–374. [CrossRef]
17. Page, R.E.; Scheiner, R.; Erber, J.; Amdam, G.V. The Development and Evolution of Division of Labor and Foraging Specialization in a Social Insect (*Apis Mellifera* L.). *Curr. Top. Dev. Biol.* **2006**, *74*, 253–286. [CrossRef]
18. Abou-Shaara, H.F. The Foraging Behaviour of Honey Bees, Apis Mellifera: A Review. *Vet. Med.* **2014**, *59*, 1–10. [CrossRef]
19. Vance, J.T.; Williams, J.B.; Elekonich, M.M.; Roberts, S.P. The Effects of Age and Behavioral Development on Honey Bee (Apis Mellifera) Flight Performance. *J. Exp. Biol.* **2009**, *212*, 2604–2611. [CrossRef]
20. Seeley, T.D. *The Wisdom of the Hive: The Social Physiology of Honey Bee Colonies*; Harvard University Press: Cambridge, MA, USA; London, UK, 1995.
21. Thorp, R.W. The Collection of Pollen by Bees. In *Pollen and Pollination*; Dafni, A., Hesse, M., Pacini, E., Eds.; Springer: Vienna, Austria, 2000; pp. 211–223. ISBN 978-3-7091-6306-1.
22. Nicolson, S.W.; Da Silva Das Neves, S.; Human, H.; Pirk, C.W.W. Digestibility and Nutritional Value of Fresh and Stored Pollen for Honey Bees (Apis Mellifera Scutellata). *J. Insect Physiol.* **2018**, *107*, 302–308. [CrossRef]
23. Dozuotų, B.D.T.I.J.; Kūrimas, F. Investigation of Bee Bread and Development of Its Dosage Forms. *Proteins* **2014**, *24*, 20.30–21.70.
24. Radev, Z. Water Content of Honey Bee Collected Pollen from 50 Plants from Bulgaria. *Mod. Concepts Dev. Agron.* **2018**, *3*, 314–316. [CrossRef]
25. Mauriello, G.; De Prisco, A.; Di Prisco, G.; La Storia, A.; Caprio, E. Microbial Characterization of Bee Pollen from the Vesuvius Area Collected by Using Three Different Traps. *PLoS ONE* **2017**, *12*, e0183208. [CrossRef]
26. Kaèániová, M.; Fikselová, M.; Hašèík, P.; Kòazovická, V.; Nô, J.; Fatrcová-Šrámková, K. Changes in microflora of bee pollen treated with uv light and freezing during storage. *Ecol. Chem. Eng. A* **2010**, *17*, 89–94.
27. Mekki, I. *Microbial Contamination of Bee Pollen and Impact of Preservation Methods*; Univérsité libre de Tunis: Tunis, Tunisia, 2019.
28. Campos, M.G.R.; Frigerio, C.; Lopes, J.; Bogdanov, S. What Is the Future of Bee-Pollen? *J. ApiProd. ApiMed. Sci.* **2010**, *2*, 131–144. [CrossRef]
29. Percie du Sert, P. Les pollens apicoles. *Phytothérapie* **2009**, *7*, 75–82. [CrossRef]
30. Bogdanov, S. Quality and Standards of Pollen and Beeswax. *APIACTA* **2004**, *38*, 334–341.

31. Martins, M.C.T.; Morgano, M.A.; Vicente, E.; Baggio, S.R.; Rodriguez-Amaya, D.B. Physicochemical composition of bee pollen from eleven brazilian states. *J. Apic. Sci.* **2011**, *55*, 11.
32. De Melo, I.L.P.; de Almeida-Muradian, L.B. Comparison of Methodologies for Moisture Determination on Dried Bee Pollen Samples. *Food Sci. Technol.* **2011**, *31*, 194–197. [CrossRef]
33. Coronel, B.B.; Grasso, D.; Pereira, S.C.; Fernández, G. Caracterización bromatológica del polen apícola argentino. *Ciencia Docencia y Tecnología* **2004**, *38*, 145–181.
34. Vanderplanck, M.; Leroy, B.; Wathelet, B.; Wattiez, R.; Michez, D. Standardized Protocol to Evaluate Pollen Polypeptides as Bee Food Source. *Apidologie* **2014**, *45*, 192–204. [CrossRef]
35. Frias, B.E.D.; Barbosa, C.D.; Lourenço, A.P. Pollen Nutrition in Honey Bees (Apis Mellifera): Impact on Adult Health. *Apidologie* **2016**, *47*, 15–25. [CrossRef]
36. Komosinska-Vassev, K.; Olczyk, P.; Kaźmierczak, J.; Mencner, L.; Olczyk, K. Bee Pollen: Chemical Composition and Therapeutic Application. *Evid. -Based Complement. Altern. Med.* **2015**, *2015*, 1–6. [CrossRef]
37. Negrão, A.F.; Orsi, R.O. Harvesting Season and Botanical Origin Interferes in Production and Nutritional Composition of Bee Pollen. *An. Acad. Bras. Ciênc.* **2018**, *90*, 325–332. [CrossRef]
38. Hendriksma, H.P.; Pachow, C.D.; Nieh, J.C. Effects of Essential Amino Acid Supplementation to Promote Honey Bee Gland and Muscle Development in Cages and Colonies. *J. Insect Physiol.* **2019**, *117*, 103906. [CrossRef]
39. Pizzorno, J.E.; Murray, M.T. *Textbook of Natural Medicine–E-Book*; Elsevier Health Sciences: Amsterdam, The Netherlands, 2020; ISBN 978-0-323-52380-6.
40. Kaur, R.; Nagpal, A.K.; Katnoria, J.K. Amino Acids Composition of Pollen Grains of Some Medicinally Important Plant Species. *J. Chem. Pharm. Res.* **2015**, *7*, 122–127.
41. Da Silva, G.R.; da Natividade, T.B.; Camara, C.A.; da Silva, E.M.S.; dos Santos, F.d.A.R.; Silva, T.M.S. Identification of Sugar, Amino Acids and Minerals from the Pollen of Jandaíra Stingless Bees (*Melipona Subnitida*). *Food Nutr. Sci.* **2014**, *2014*. [CrossRef]
42. Canale, A.; Benelli, G.; Castagna, A.; Sgherri, C.; Poli, P.; Serra, A.; Mele, M.; Ranieri, A.; Signorini, F.; Bientinesi, M.; et al. Microwave-Assisted Drying for the Conservation of Honeybee Pollen. *Materials* **2016**, *9*, 363. [CrossRef]
43. Alvarez-Suarez, J.M. *Bee Products-Chemical and Biological Properties*; Springer: New York, NY, USA, 2017; ISBN 978-3-319-59689-1.
44. Gardana, C.; Del Bo', C.; Quicazán, M.C.; Corrrea, A.R.; Simonetti, P. Nutrients, Phytochemicals and Botanical Origin of Commercial Bee Pollen from Different Geographical Areas. *J. Food Compos. Anal.* **2018**, *73*, 29–38. [CrossRef]
45. Lilek, N.; Gonzales, A.P.; Boži, J.; Borovšak, A.K.; Bertoncelj, J. Chemical Composition and Content of Free Tryptophan in Slovenian Bee Pollen. *J. Food Nutr. Res.* **2015**, *54*, 11.
46. Yang, K.; Wu, D.; Ye, X.; Liu, D.; Chen, J.; Sun, P. Characterization of Chemical Composition of Bee Pollen in China. *J. Agric. Food Chem.* **2013**, *61*, 708–718. [CrossRef]
47. Taha, E.-K.A.; Al-Kahtani, S.; Taha, R. Protein Content and Amino Acids Composition of Bee-Pollens from Major Floral Sources in Al-Ahsa, Eastern Saudi Arabia. *Saudi J. Biol. Sci.* **2017**. [CrossRef]
48. Isik, A.; Ozdemir, M.; Doymaz, I.; Isik, A.; Ozdemir, M.; Doymaz, I. Effect of Hot Air Drying on Quality Characteristics and Physicochemical Properties of Bee Pollen. *Food Sci. Technol.* **2019**, *39*, 224–231. [CrossRef]
49. Bertoncelj, J.; Polak, T.; Pucihar, T.; Lilek, N.; Borovšak, A.K.; Korošec, M. Carbohydrate Composition of Slovenian Bee Pollens. *Int. J. Food Sci. Technol.* **2018**, *53*, 1880–1888. [CrossRef]
50. Anjos, O.; Paula, V.; Delgado, T.; Estevinho, L. Influence of the Storage Conditions on the Quality of Bee Pollen. *Zemdirb. Agric.* **2019**, *106*, 87–94. [CrossRef]
51. Wu, W.; Wang, K.; Qiao, J.; Dong, J.; Li, Z.; Zhang, H. Improving Nutrient Release of Wall-Disrupted Bee Pollen with a Combination of Ultrasonication and High Shear Technique. *J. Sci. Food Agric.* **2019**, *99*, 564–575. [CrossRef]
52. Ares, A.M.; Valverde, S.; Bernal, J.L.; Nozal, M.J.; Bernal, J. Extraction and Determination of Bioactive Compounds from Bee Pollen. *J. Pharm. Biomed. Anal.* **2018**, *147*, 110–124. [CrossRef]
53. de Arruda, V.A.S.; Vieria dos Santos, A.; Figueiredo Sampaio, D.; da Silva Araújo, E.; de Castro Peixoto, A.L.; Estevinho, M.L.F.; Bicudo de Almeida-Muradian, L. Microbiological Quality and Physicochemical Characterization of Brazilian Bee Pollen. *J. Apic. Res.* **2017**, *56*, 231–238. [CrossRef]
54. Bárbara, M.S.; Machado, C.S.; Sodré, G.D.S.; Dias, L.G.; Estevinho, L.M.; De Carvalho, C.A.L. Microbiological Assessment, Nutritional Characterization and Phenolic Compounds of Bee Pollen from Mellipona Mandacaia Smith, 1983. *Molecules* **2015**, *20*, 12525–12544. [CrossRef]
55. Kalaycıoğlu, Z.; Kaygusuz, H.; Döker, S.; Kolaylı, S.; Erim, F.B. Characterization of Turkish Honeybee Pollens by Principal Component Analysis Based on Their Individual Organic Acids, Sugars, Minerals, and Antioxidant Activities. *LWT* **2017**, *84*, 402–408. [CrossRef]
56. Sattler, J.A.G.; de Melo, I.L.P.; Granato, D.; Araújo, E.; da Silva de Freitas, A.; Barth, O.M.; Sattler, A.; de Almeida-Muradian, L.B. Impact of Origin on Bioactive Compounds and Nutritional Composition of Bee Pollen from Southern Brazil: A Screening Study. *Food Res. Int.* **2015**, *77*, 82–91. [CrossRef]
57. Szczęsna, T. Study on the Sugar Composition of Honeybee-Collected Pollen. Available online: http://agro.icm.edu.pl/agro/element/bwmeta1.element.dl-catalog-2bfe8051-7503-4faa-b7c0-552c257dd9a3 (accessed on 19 October 2020).
58. Liolios, V.; Tananaki, C.; Dimou, M.; Kanelis, D.; Rodopoulou, M.-A.; Thrasyvoulou, A. Exploring the Sugar Profile of Unifloral Bee Pollen Using High Performance Liquid Chromatography. *J. Food Nutr. Res.* **2018**, *57*, 11.

59. Belina-Aldemita, M.D.; Opper, C.; Schreiner, M.; D'Amico, S. Nutritional Composition of Pot-Pollen Produced by Stingless Bees (Tetragonula Biroi Friese) from the Philippines. *J. Food Compos. Anal.* **2019**, *82*, 103215. [CrossRef]
60. Omar, W.A.W.; Yahaya, N.; Ghaffar, Z.A.; Fadzilah, N.H. GC-MS Analysis of Chemical Constituents in Ethanolic Bee Pollen Extracts from Three Species of Malaysian Stingless Bee. *J. Apic. Sci.* **2018**, *62*, 275–284. [CrossRef]
61. Li, Y.O.; Komarek, A.R. Dietary Fibre Basics: Health, Nutrition, Analysis, and Applications. *Food Qual. Saf.* **2017**, *1*, 47–59. [CrossRef]
62. Trowell, H.; Burkitt, D.; Heaton, K. *Dietary Fibre, Fibre-Depleted Foods and Disease*; Academic Press: Orlando, FL, USA, 1985.
63. Makki, K.; Deehan, E.C.; Walter, J.; Bäckhed, F. The Impact of Dietary Fiber on Gut Microbiota in Host Health and Disease. *Cell Host Microbe* **2018**, *23*, 705–715. [CrossRef]
64. Zhao, L.; Zhang, F.; Ding, X.; Wu, G.; Lam, Y.Y.; Wang, X.; Fu, H.; Xue, X.; Lu, C.; Ma, J.; et al. Gut Bacteria Selectively Promoted by Dietary Fibers Alleviate Type 2 Diabetes. *Science* **2018**, *359*, 1151–1156. [CrossRef]
65. Bishehsari, F.; Engen, P.A.; Preite, N.Z.; Tuncil, Y.E.; Naqib, A.; Shaikh, M.; Rossi, M.; Wilber, S.; Green, S.J.; Hamaker, B.R.; et al. Dietary Fiber Treatment Corrects the Composition of Gut Microbiota, Promotes SCFA Production, and Suppresses Colon Carcinogenesis. *Genes* **2018**, *9*, 102. [CrossRef]
66. Dayib, M.; Larson, J.; Slavin, J. Dietary Fibers Reduce Obesity-Related Disorders: Mechanisms of Action. *Curr. Opin. Clin. Nutr. Metab. Care* **2020**, *23*, 445–450. [CrossRef]
67. McRae, M.P. Dietary Fiber Is Beneficial for the Prevention of Cardiovascular Disease: An Umbrella Review of Meta-Analyses. *J. Chiropr. Med.* **2017**, *16*, 289–299. [CrossRef]
68. Pituch-Zdanowska, A.; Banaszkiewicz, A.; Albrecht, P. The Role of Dietary Fibre in Inflammatory Bowel Disease. *Prz. Gastroenterol.* **2015**, *10*, 135–141. [CrossRef]
69. Merenkova, S.P.; Zinina, O.V.; Stuart, M.; Okuskhanova, E.K.; Androsova, N.V. Effects of dietary fiber on human health: A review. *Human. Sport. Med.* **2020**, *20*, 106–113. [CrossRef]
70. Fuenmayor, B.C.; Zuluaga, D.C.; Díaz, M.C.; de Quicazán, C.M.; Cosio, M.; Mannino, S. Evaluation of the Physicochemical and Functional Properties of Colombian Bee Pollen. *Revista MVZ Córdoba* **2014**, *19*, 4003–4014. [CrossRef]
71. Calder, P.C. Omega-3 Fatty Acids and Inflammatory Processes: From Molecules to Man. *Biochem. Soc. Trans.* **2017**, *45*, 1105–1115. [CrossRef]
72. La Guardia, M.; Giammanco, S.; Di Majo, D.; Tabacchi, C.; Tripoli, E.; Giammanco, M. Omega 3 Fatty Acids: Biological Activity and Effects on Human Health. *Panminerva. Med.* **2005**, *47*, 245–257.
73. Thakur, M.; Nanda, V. Composition and Functionality of Bee Pollen: A Review. *Trends Food Sci. Technol.* **2020**, *98*, 82–106. [CrossRef]
74. Li, Q.; Liang, X.; Zhao, L.; Zhang, Z.; Xue, X.; Wang, K.; Wu, L. UPLC-Q-Exactive Orbitrap/MS-Based Lipidomics Approach To Characterize Lipid Extracts from Bee Pollen and Their in Vitro Anti-Inflammatory Properties. *J. Agric. Food Chem.* **2017**, *65*, 6848–6860. [CrossRef]
75. AL-Kahtani, S.N. Fatty Acids and B Vitamins Contents in Honey Bee Collected Pollen in Relation to Botanical Origin. *Sci. J. King Faisal Univ.* **2017**, *18*, 41–48.
76. Conte, G.; Benelli, G.; Serra, A.; Signorini, F.; Bientinesi, M.; Nicolella, C.; Mele, M.; Canale, A. Lipid Characterization of Chestnut and Willow Honeybee-Collected Pollen: Impact of Freeze-Drying and Microwave-Assisted Drying. *J. Food Compos. Anal.* **2017**, *55*, 12–19. [CrossRef]
77. Dong, J.; Yang, Y.; Wang, X.; Zhang, H. Fatty Acid Profiles of 20 Species of Monofloral Bee Pollen from China. *J. Apic. Res.* **2015**, *54*, 503–511. [CrossRef]
78. Bailey, R.L.; West, K.P., Jr.; Black, R.E. The Epidemiology of Global Micronutrient Deficiencies. *ANM* **2015**, *66*, 22–33. [CrossRef]
79. FAO; IFAD; UNICEF; WFP; WHO. The State of Food Security and Nutrition in the World 2020. In *Transforming Food Systems for Affordable Healthy Diets*; FAO: Rome, Italy, 2020. [CrossRef]
80. Gernand, A.D.; Schulze, K.J.; Stewart, C.P.; West, K.P.; Christian, P. Micronutrient Deficiencies in Pregnancy Worldwide: Health Effects and Prevention. *Nat. Rev. Endocrinol.* **2016**, *12*, 274–289. [CrossRef]
81. Kangalgil, M.; Sahinler, A.; Kırkbir, I.B.; Ozcelik, A.O. Associations of Maternal Characteristics and Dietary Factors with Anemia and Iron-Deficiency in Pregnancy. *J. Gynecol. Obstet. Hum. Reprod.* **2021**, *50*, 102137. [CrossRef]
82. Khayat, S.; Fanaei, H.; Ghanbarzehi, A. Minerals in Pregnancy and Lactation: A Review Article. *J. Clin. Diagn. Res.* **2017**, *11*, QE01–QE05. [CrossRef]
83. Looman, M.; Schoenaker, D.A.J.M.; Soedamah-Muthu, S.S.; Mishra, G.D.; Geelen, A.; Feskens, E.J.M. Pre-Pregnancy Dietary Micronutrient Adequacy Is Associated with Lower Risk of Developing Gestational Diabetes in Australian Women. *Nutr. Res.* **2019**, *62*, 32–40. [CrossRef]
84. Moreno-Reyes, R. Micronutrient Deficiencies and the Thyroid in Pregnancy. *Placenta* **2017**, *51*, 102. [CrossRef]
85. Ibrahim, M.K.; Zambruni, M.; Melby, C.L.; Melby, P.C. Impact of Childhood Malnutrition on Host Defense and Infection. *Clin. Microbiol. Rev.* **2017**, *30*, 919–971. [CrossRef]
86. Ritchie, H.; Roser, M. Micronutrient Deficiency. *Our World in Data*, August 2017.
87. Tan, M.; Yang, T.; Zhu, J.; Li, Q.; Lai, X.; Li, Y.; Tang, T.; Chen, J.; Li, T. Maternal Folic Acid and Micronutrient Supplementation Is Associated with Vitamin Levels and Symptoms in Children with Autism Spectrum Disorders. *Reprod. Toxicol.* **2020**, *91*, 109–115. [CrossRef]

88. Gorji, A.; Khaleghi Ghadiri, M. Potential Roles of Micronutrient Deficiency and Immune System Dysfunction in the Coronavirus Disease 2019 (COVID-19) Pandemic. *Nutrition* **2021**, *82*, 111047. [CrossRef]
89. Hoffman, R. Micronutrient Deficiencies in the Elderly–Could Ready Meals Be Part of the Solution? *J. Nutr. Sci.* **2017**, *6*, e2. [CrossRef]
90. Song, E.K.; Moser, D.K.; Kang, S.-M.; Lennie, T.A. Association of Depressive Symptoms and Micronutrient Deficiency With Cardiac Event–Free Survival in Patients With Heart Failure. *J. Card. Fail.* **2015**, *21*, 945–951. [CrossRef]
91. Wang, M.X.; Koh, J.; Pang, J. Association between Micronutrient Deficiency and Acute Respiratory Infections in Healthy Adults: A Systematic Review of Observational Studies. *Nutr. J.* **2019**, *18*, 80. [CrossRef]
92. Shankar, A.H. 140-Mineral Deficiencies. In *Hunter's Tropical Medicine and Emerging Infectious Disease*, 9th ed.; Magill, A.J., Hill, D.R., Solomon, T., Ryan, E.T., Eds.; W.B. Saunders: London, UK, 2013; pp. 1003–1010. ISBN 978-1-4160-4390-4.
93. Bogdanov, S. Pollen: Nutrition, Functional Properties, Health. *Bee Prod. Sci.* **2012**, 20–350.
94. Li, Q.-Q.; Wang, K.; Marcucci, M.C.; Sawaya, A.C.H.F.; Hu, L.; Xue, X.-F.; Wu, L.-M.; Hu, F.-L. Nutrient-Rich Bee Pollen: A Treasure Trove of Active Natural Metabolites. *J. Funct. Foods* **2018**, *49*, 472–484. [CrossRef]
95. Adaškevičiūtė, V.; Kaškonienė, V.; Kaškonas, P.; Barčauskaitė, K.; Maruška, A. Comparison of Physicochemical Properties of Bee Pollen with Other Bee Products. *Biomolecules* **2019**, *9*, 819. [CrossRef]
96. Altunatmaz, S.S.; Tarhan, D.; Aksu, F.; Barutçu, U.B.; Or, M.E. Mineral Element and Heavy Metal (Cadmium, Lead and Arsenic) Levels of Bee Pollen in Turkey. *Food Sci. Technol.* **2017**, *37*, 136–141. [CrossRef]
97. Sattler, J.A.G.; De-Melo, A.A.M.; Nascimento, K.S.D.; Melo, I.L.P.D.; Mancini-Filho, J.; Sattler, A.; Almeida-Muradian, L.B.D. Essential Minerals and Inorganic Contaminants (Barium, Cadmium, Lithium, Lead and Vanadium) in Dried Bee Pollen Produced in Rio Grande Do Sul State, Brazil. *Food Sci. Technol.* **2016**, *36*, 505–509. [CrossRef]
98. Mellidou, I.; Georgiadou, E.C.; Kaloudas, D.; Kalaitzis, P.; Fotopoulos, V.; Kanellis, A.K. Chapter 17-Vitamins. In *Postharvest Physiology and Biochemistry of Fruits and Vegetables*; Yahia, E.M., Ed.; Woodhead Publishing: Cambridge, UK, 2019; pp. 359–383. ISBN 978-0-12-813278-4.
99. Ravisankar, P.; Reddy, A.A.; Nagalakshmi, B.; Sai, O.; Kumar, B.V.; Anvith, P.S. The Comprehensive Review on Fat Soluble Vitamins. *IOSR J. Pharm.* **2015**, *5*, 12–28.
100. Yoshii, K.; Hosomi, K.; Sawane, K.; Kunisawa, J. Metabolism of Dietary and Microbial Vitamin B Family in the Regulation of Host Immunity. *Front. Nutr.* **2019**, *6*, 48. [CrossRef]
101. Farag, S.; El-Rayes, T. Effect of Bee-Pollen Supplementation on Performance, Carcass Traits and Blood Parameters of Broiler Chickens. *Asian J. Anim. Vet. Adv.* **2016**, *11*, 168–177. [CrossRef]
102. Ecem Bayram, N. Vitamin, Mineral, Polyphenol, Amino Acid Profile of Bee Pollen from Rhododendron Ponticum (Source of "Mad Honey"): Nutritional and Palynological Approach. *Food Meas.* **2021**, *15*, 2659–2666. [CrossRef]
103. Estevinho, L.M.; Dias, T.; Anjos, O. Influence of the Storage Conditions (Frozen vs. Dried) in Health-Related Lipid Indexes and Antioxidants of Bee Pollen. *Eur. J. Lipid Sci. Technol.* **2019**, *121*, 1800393. [CrossRef]
104. Barbieri, D.; Gabriele, M.; Summa, M.; Colosimo, R.; Leonardi, D.; Domenici, V.; Pucci, L. Antioxidant, Nutraceutical Properties, and Fluorescence Spectral Profiles of Bee Pollen Samples from Different Botanical Origins. *Antioxidants* **2020**, *9*, 1001. [CrossRef]
105. de Arruda, V.A.S.; Pereira, A.A.S.; de Freitas, A.S.; Barth, O.M.; de Almeida-Muradian, L.B. Dried Bee Pollen: B Complex Vitamins, Physicochemical and Botanical Composition. *J. Food Compos. Anal.* **2013**, *29*, 100–105. [CrossRef]
106. de Melo, I.L.P.; de Almeida-Muradian, L.B. Stability of Antioxidants Vitamins in Bee Pollen Samples. *Quím. Nova* **2010**, *33*, 514–518. [CrossRef]
107. De-Melo, A.A.M.; Estevinho, M.L.M.F.; Sattler, J.A.G.; Souza, B.R.; da Freitas, A.S.; Barth, O.M.; Almeida-Muradian, L.B. Effect of Processing Conditions on Characteristics of Dehydrated Bee-Pollen and Correlation between Quality Parameters. *LWT-Food Sci. Technol.* **2016**, *65*, 808–815. [CrossRef]
108. Chawla, J.; Kvarnberg, D. Chapter 59-Hydrosoluble Vitamins. In *Handbook of Clinical Neurology*; Biller, J., Ferro, J.M., Eds.; Neurologic Aspects of Systemic Disease Part II; Elsevier: Amsterdam, The Netherlands, 2014; Volume 120, pp. 891–914.
109. Aylanc, V.; Falcão, S.I.; Ertosun, S.; Vilas-Boas, M. From the Hive to the Table: Nutrition Value, Digestibility and Bioavailability of the Dietary Phytochemicals Present in the Bee Pollen and Bee Bread. *Trends Food Sci. Technol.* **2021**, *109*, 464–481. [CrossRef]
110. Khalifa, S.A.M.; Elashal, M.H.; Yosri, N.; Du, M.; Musharraf, S.G.; Nahar, L.; Sarker, S.D.; Guo, Z.; Cao, W.; Zou, X.; et al. Bee Pollen: Current Status and Therapeutic Potential. *Nutrients* **2021**, *13*, 1876. [CrossRef]
111. Oliveira, K.C.L.S.; Moriya, M.; Azedo, R.A.B.; de Almeida-Muradian, L.B.; Teixeira, E.W.; Alves, M.L.T.M.F.; Moreti, A.C. de C.C. Relationship between Botanical Origin and Antioxidants Vitamins of Bee-Collected Pollen. *Quím. Nova* **2009**, *32*, 1099–1102. [CrossRef]
112. Campos, M.G.; Cunha, A.; Markham, K.R. Bee-Pollen: Composition, Properties, and Applications. In *Bee Products*; Mizrahi, A., Lensky, Y., Eds.; Springer: Boston, MA, USA, 1997; pp. 93–100. ISBN 978-1-4757-9373-4.
113. Kaškonienė, V.; Ruočkuvienė, G.; Kaškonas, P.; Akuneca, I.; Maruška, A. Chemometric Analysis of Bee Pollen Based on Volatile and Phenolic Compound Compositions and Antioxidant Properties. *Food Anal. Methods* **2015**, *8*, 1150–1163. [CrossRef]
114. Milani, A.; Basirnejad, M.; Shahbazi, S.; Bolhassani, A. Carotenoids: Biochemistry, Pharmacology and Treatment. *Br. J. Pharmacol.* **2017**, *174*, 1290–1324. [CrossRef]
115. Karabagias, I.K.; Karabagias, V.K.; Karabournioti, S.; Badeka, A.V. Aroma Identification of Greek Bee Pollen Using HS-SPME/GC–MS. *Eur. Food Res. Technol.* **2021**, *247*, 1781–1789. [CrossRef]

116. Belhadj, H.; Harzallah, D.; Bouamra, D.; Khennouf, S.; Dahamna, S.; Ghadbane, M. Phenotypic and Genotypic Characterization of Some Lactic Acid Bacteria Isolated from Bee Pollen: A Preliminary Study. *Biosci. Microbiota Food Health* **2014**, *33*, 11–23. [CrossRef]
117. Friedle, C.; D'Alvise, P.; Schweikert, K.; Wallner, K.; Hasselmann, M. Changes of Microorganism Composition in Fresh and Stored Bee Pollen from Southern Germany. *Environ. Sci. Pollut. Res.* **2021**, *28*, 47251–47261. [CrossRef]
118. Gilliam, M. microbiology of pollen and bee BREAD: The yeasts. *Apidologie* **1979**, *10*, 43–53. [CrossRef]
119. Tajabadi, N.; Mardan, M.; Saari, N.; Mustafa, S.; Bahreini, R.; Manap, M.Y.A. Identification of Lactobacillus Plantarum, Lactobacillus Pentosus and Lactobacillus Fermentum from Honey Stomach of Honeybee. *Braz. J. Microbiol.* **2013**, *44*, 717–722. [CrossRef]
120. Belhadj, H.; Harzallah, D.; Khennouf, S.; Dahamna, S.; Bouharati, S.; Baghiani, A. Isolation, identification and antimicrobial activity of lactic acid bacteria from algerian honeybee collected pollen. *Acta Hortic.* **2010**, 51–58. [CrossRef]
121. Vásquez, A.; Olofsson, T.C. The Lactic Acid Bacteria Involved in the Production of Bee Pollen and Bee Bread. *J. Apic. Res.* **2009**, *48*, 189–195. [CrossRef]
122. Belhamra, Z.; Harzallah, D.; Naili, O.; Belhadj, H. Probiotic Potential of Lactobacillus Strains Isolated from Fresh Bee Pollen. *Pharm. Lett.* **2016**, *8*, 357–365.
123. Filannino, P.; Di Cagno, R.; Addante, R.; Pontonio, E.; Gobbetti, M. Metabolism of Fructophilic Lactic Acid Bacteria Isolated from the Apis Mellifera L. Bee Gut: Phenolic Acids as External Electron Acceptors. *Appl. Environ. Microbiol.* **2016**, *82*, 6899–6911. [CrossRef]
124. George Kerry, R.; Patra, J.K.; Gouda, S.; Park, Y.; Shin, H.-S.; Das, G. Benefaction of Probiotics for Human Health: A Review. *J. Food Drug Anal.* **2018**, *26*, 927–939. [CrossRef]
125. Goldin, B.R.; Gorbach, S.L. Clinical Indications for Probiotics: An Overview. *Clin. Infect. Dis.* **2008**, *46*, S96–S100. [CrossRef]
126. Petrof, E.O. Probiotics and Gastrointestinal Disease: Clinical Evidence and Basic Science. *Anti-Inflamm. Anti-Allergy Agents Med. Chem. (Former. Curr. Med. Chem. -Anti-Inflamm. Anti-Allergy Agents)* **2009**, *8*, 260–269. [CrossRef]
127. Sun, Z.; Sun, X.; Li, J.; Li, Z.; Hu, Q.; Li, L.; Hao, X.; Song, M.; Li, C. Using Probiotics for Type 2 Diabetes Mellitus Intervention: Advances, Questions, and Potential. *Crit. Rev. Food Sci. Nutr.* **2020**, *60*, 670–683. [CrossRef]
128. Kaškonienė, V.; Kaškonas, P.; Maruška, A. Volatile Compounds Composition and Antioxidant Activity of Bee Pollen Collected in Lithuania. *Chem. Pap.* **2015**, *69*, 291–299. [CrossRef]
129. Klatt, B.K.; Burmeister, C.; Westphal, C.; Tscharntke, T.; von Fragstein, M. Flower Volatiles, Crop Varieties and Bee Responses. *PLoS ONE* **2013**, *8*, e72724. [CrossRef]
130. Starowicz, M.; Hanus, P.; Lamparski, G.; Sawicki, T. Characterizing the Volatile and Sensory Profiles, and Sugar Content of Beeswax, Beebread, Bee Pollen, and Honey. *Molecules* **2021**, *26*, 3410. [CrossRef]
131. Özcan, M.M.; Aljuhaimi, F.; Babiker, E.E.; Uslu, N.; Ceylan, D.A.; Ghafoor, K.; Özcan, M.M.; Dursun, N.; Ahmed, I.M.; Jamiu, F.G.; et al. Determination of Antioxidant Activity, Phenolic Compound, Mineral Contents and Fatty Acid Compositions of Bee Pollen Grains Collected from Different Locations. *J. Apic. Sci.* **2019**, *63*, 69–79. [CrossRef]
132. Rzepecka-Stojko, A.; Stojko, J.; Kurek-Górecka, A.; Górecki, M.; Kabała-Dzik, A.; Kubina, R.; Moździerz, A.; Buszman, E. Polyphenols from Bee Pollen: Structure, Absorption, Metabolism and Biological Activity. *Molecules* **2015**, *20*, 21732–21749. [CrossRef]
133. Şahin, S.; Karkar, B. The Antioxidant Properties of the Chestnut Bee Pollen Extract and Its Preventive Action against Oxidatively Induced Damage in DNA Bases. *J. Food Biochem.* **2019**, *43*, e12888. [CrossRef]
134. Yang, Y.; Zhang, J.; Zhou, Q.; Wang, L.; Huang, W.; Wang, R. Effect of Ultrasonic and Ball-milling Treatment on Cell Wall, Nutrients, and Antioxidant Capacity of Rose (Rosa Rugosa) Bee Pollen, and Identification of Bioactive Components. *J. Sci. Food Agric.* **2019**, *99*, 5350–5357. [CrossRef]
135. Bridi, R.; Atala, E.; Pizarro, P.N.; Montenegro, G. Honeybee Pollen Load: Phenolic Composition and Antimicrobial Activity and Antioxidant Capacity. *J. Nat. Prod.* **2019**, *82*, 559–565. [CrossRef]
136. Santa Bárbara, M.F.; Moreira, M.M.; Machado, C.S.; Chambó, E.D.; Pascoal, A.; de Carvalho, C.A.L.; da Silva Sodré, G.; Delerue-Matos, C.; Estevinho, L.M. Storage Methods, Phenolic Composition, and Bioactive Properties of Apis Mellifera and Trigona Spinipes Pollen. *J. Apic. Res.* **2021**, *60*, 99–107. [CrossRef]
137. Ares, A.M.; Soto, M.E.; Nozal, M.J.; Bernal, J.L.; Higes, M.; Bernal, J. Determination of Resveratrol and Piceid Isomers in Bee Pollen by Liquid Chromatography Coupled to Electrospray Ionization-Mass Spectrometry. *Food Anal. Methods* **2015**, *8*, 1565–1575. [CrossRef]
138. El Ghouizi, A.; El Menyiy, N.; Falcão, S.I.; Vilas-Boas, M.; Lyoussi, B. Chemical Composition, Antioxidant Activity, and Diuretic Effect of Moroccan Fresh Bee Pollen in Rats. *Vet World* **2020**, *13*, 1251–1261. [CrossRef]
139. Oroian, M.; Ursachi, F.; Dranca, F. Ultrasound-Assisted Extraction of Polyphenols from Crude Pollen. *Antioxidants* **2020**, *9*, 322. [CrossRef]
140. Rocchetti, G.; Castiglioni, S.; Maldarizzi, G.; Carloni, P.; Lucini, L. UHPLC-ESI-QTOF-MS Phenolic Profiling and Antioxidant Capacity of Bee Pollen from Different Botanical Origin. *Int. J. Food Sci. Technol.* **2019**, *54*, 335–346. [CrossRef]
141. Denisow, B.; Denisow-Pietrzyk, M. Biological and Therapeutic Properties of Bee Pollen: A Review: Bee Pollen and Medicine. *J. Sci. Food Agric.* **2016**, *96*, 4303–4309. [CrossRef]
142. El-Guendouz, S.; Lyoussi, B.; Miguel, M.G. Insight into the Chemical Composition and Biological Properties of Mediterranean Royal Jelly. *J. Apic. Res.* **2020**, *59*, 890–909. [CrossRef]

143. Aličić, D.; Flanjak, I.; Ačkar, Đ.; Jašić, M.; Babić, J.; Šubarić, D. Physicochemical Properties and Antioxidant Capacity of Bee Pollen Collected in Tuzla Canton (B&H). *J. Cent. Eur. Agric.* **2020**, *21*, 42–50. [CrossRef]
144. AbdElsalam, E.; Foda, H.S.; Abdel-Aziz, M.S.; Abd, F.K. Antioxidant and Antimicrobial Activities of Egyptian Bee Pollen. *Middle East J. Appl. Sci.* **2018**, *8*, 1248–1255.
145. de Florio Almeida, J.; dos Reis, A.S.; Heldt, L.F.S.; Pereira, D.; Bianchin, M.; de Moura, C.; Oviedo, M.V.P.; Haminiuk, C.W.I.; Ribeiro, I.S.; da Luz, C.F.P.; et al. Lyophilized Bee Pollen Extract: A Natural Antioxidant Source to Prevent Lipid Oxidation in Refrigerated Sausages. *LWT-Food Sci. Technol.* **2017**, *76*, 299–305. [CrossRef]
146. Anjos, O.; Fernandes, R.; Cardoso, S.M.; Delgado, T.; Farinha, N.; Paula, V.; Estevinho, L.M.; Carpes, S.T. Bee Pollen as a Natural Antioxidant Source to Prevent Lipid Oxidation in Black Pudding. *LWT* **2019**, *111*, 869–875. [CrossRef]
147. Atsalakis, E.; Chinou, I.; Makropoulou, M.; Karabournioti, S.; Graikou, K. Evaluation of Phenolic Compounds in Cistus Creticus Bee Pollen from Greece. Antioxidant and Antimicrobial Properties. *Nat. Prod. Commun.* **2017**, *12*, 1934578X1701201141. [CrossRef]
148. Bleha, R.; Shevtsova, T.; Kruzik, V.; Brindza, J.; Sinica, A. Morphology, Physicochemical Properties and Antioxidant Capacity of Bee Pollens. *Czech J. Food Sci.* **2019**, *37*, 1–8. [CrossRef]
149. Carpes, S.T.; Begnini, R.; de Alencar, S.M.; Masson, M.L. Study of Preparations of Bee Pollen Extracts, Antioxidant and Antibacterial Activity. *Ciênc. Agrotec.* **2007**, *31*, 1818–1825. [CrossRef]
150. De-Melo, A.A.M.; Estevinho, L.M.; Moreira, M.M.; Delerue-Matos, C.; de Freitas, A.D.S.; Barth, O.M.; Almeida-Muradian, L.B. de A Multivariate Approach Based on Physicochemical Parameters and Biological Potential for the Botanical and Geographical Discrimination of Brazilian Bee Pollen. *Food Biosci.* **2018**, *25*, 91–110. [CrossRef]
151. Duarte, A.W.F.; dos Vasconcelos, M.R.S.; Oda-Souza, M.; de Oliveira, F.F.; López, A.M.Q. Honey and Bee Pollen Produced by Meliponini (Apidae) in Alagoas, Brazil: Multivariate Analysis of Physicochemical and Antioxidant Profiles. *Food Sci. Technol.* **2018**, *38*, 493–503. [CrossRef]
152. Dulger Altiner, D.; Sandſkcſ Altunatmaz, S.; Sabuncu, M.; Aksu, F.; Sahan, Y. In-Vitro Bioaccessibility of Antioxidant Properties of Bee Pollen in Turkey. *Food Sci. Technol.* **2020**, *41*, 133–141. [CrossRef]
153. Fatrcová-Šramková, K.; Nôžková, J.; Máriássyová, M.; Kačániová, M. Biologically Active Antimicrobial and Antioxidant Substances in the Helianthus Annuus L. Bee Pollen. *J. Environ. Sci. Health Part B* **2016**, *51*, 176–181. [CrossRef]
154. Fatrcová-Šramková, K.; Nôžková, J.; Kačániová, M.; Máriássyová, M.; Rovná, K.; Stričík, M. Antioxidant and Antimicrobial Properties of Monofloral Bee Pollen. *J. Environ. Sci. Health Part B* **2013**, *48*, 133–138. [CrossRef]
155. Feás, X.; Vázquez-Tato, M.P.; Estevinho, L.; Seijas, J.A.; Iglesias, A. Organic Bee Pollen: Botanical Origin, Nutritional Value, Bioactive Compounds, Antioxidant Activity and Microbiological Quality. *Molecules* **2012**, *17*, 8359–8377. [CrossRef]
156. Futui, W.; Thongwai, N. Antimicrobial and Antioxidant Activities, Total Phenolic and Flavonoid Contents of Bee Pollen Crude Extracts. *Int. J. Biosci. Biochem. Bioinform* **2020**, *10*, 42–48. [CrossRef]
157. Jin, T.-Y.; Saravanakumar, K.; Wang, M.-H. In Vitro and in Vivo Antioxidant Properties of Water and Methanol Extracts of Linden Bee Pollen. *Biocatal. Agric. Biotechnol.* **2018**, *13*, 186–189. [CrossRef]
158. Kostić, A.Ž.; Milinčić, D.D.; Gašić, U.M.; Nedić, N.; Stanojević, S.P.; Tešić, Ž.L.; Pešić, M.B. Polyphenolic Profile and Antioxidant Properties of Bee-Collected Pollen from Sunflower (*Helianthus Annuus* L.) Plant. *LWT* **2019**, *112*, 108244. [CrossRef]
159. Leja, M.; Mareczek, A.; Wyżgolik, G.; Klepacz-Baniak, J.; Czekońska, K. Antioxidative Properties of Bee Pollen in Selected Plant Species. *Food Chem.* **2007**, *100*, 237–240. [CrossRef]
160. Mayda, N.; Özkök, A.; Ecem Bayram, N.; Gerçek, Y.C.; Sorkun, K. Bee Bread and Bee Pollen of Different Plant Sources: Determination of Phenolic Content, Antioxidant Activity, Fatty Acid and Element Profiles. *Food Meas.* **2020**, *14*, 1795–1809. [CrossRef]
161. Mejías, E.; Montenegro, G. The Antioxidant Activity of C Hilean Honey and Bee Pollen Produced in the L Laima V Olcano's Zones. *J. Food Qual.* **2012**, *35*, 315–322. [CrossRef]
162. Mohdaly, A.A.A.; Mahmoud, A.A.; Roby, M.H.H.; Smetanska, I.; Ramadan, M.F. Phenolic Extract from Propolis and Bee Pollen: Composition, Antioxidant and Antibacterial Activities. *J. Food Biochem.* **2015**, *39*, 538–547. [CrossRef]
163. Nurdianah, H.F.; Firdaus, A.A.; Azam, O.E.; Adnan, W.W. Antioxidant Activity of Bee Pollen Ethanolic Extracts from Malaysian Stingless Bee Measured Using DPPH-HPLC Assay. *Int. Food Res. J.* **2016**, *23*, 403.
164. Pascoal, A.; Rodrigues, S.; Teixeira, A.; Feás, X.; Estevinho, L.M. Biological Activities of Commercial Bee Pollens: Antimicrobial, Antimutagenic, Antioxidant and Anti-Inflammatory. *Food Chem. Toxicol.* **2014**, *63*, 233–239. [CrossRef]
165. Rebiai, A.; Lanez, T. Chemical Composition and Antioxidant Activity of *Apis Mellifera* Bee Pollen from Northwest Algeria. *J. Fundam. Appl. Sci.* **2015**, *4*, 155. [CrossRef]
166. Saral, Ö.; Kilicarslan, M.; Şahin, H.; Yildiz, O.; Dincer, B. Evaluation of Antioxidant Activity of Bee Products of Different Bee Races in Turkey. *Turk. J. Vet. Anim. Sci.* **2019**, *43*, 441–447. [CrossRef]
167. Ulusoy, E.; Kolayli, S. Phenolic Composition and Antioxidant Properties of Anzer Bee Pollen. *J. Food Biochem.* **2014**, *38*, 73–82. [CrossRef]
168. dos Vasconcelos, M.R.S.; Duarte, A.W.F.; Gomes, E.P.; da Silva, S.C.; López, A.M.Q. Physicochemical Composition and Antioxidant Potential of Bee Pollen from Different Botanical Sources in Alagoas, Brazil. *Ciência e Agrotecnologia* **2017**, *41*, 447–458. [CrossRef]
169. Zhang, H.; Wang, X.; Wang, K.; Li, C. Antioxidant and Tyrosinase Inhibitory Properties of Aqueous Ethanol Extracts from Monofloral Bee Pollen. *J. Apic. Sci.* **2015**, *59*, 119–128. [CrossRef]

170. Zou, Y.; Hu, J.; Huang, W.; Zhu, L.; Shao, M.; Dordoe, C.; Ahn, Y.-J.; Wang, D.; Zhao, Y.; Xiong, Y. The Botanical Origin and Antioxidant, Anti-BACE1 and Antiproliferative Properties of Bee Pollen from Different Regions of South Korea. *BMC Complement. Med. Ther.* **2020**, *20*, 236. [CrossRef]
171. Carpes, S.T.; Prado, A.; Moreno, I.A.M.; Mourão, G.B.; de Alencar, S.M.; Masson, M.L. Avaliação do potencial antioxidante do pólen apícola produzido na região sul do Brasil. *Química Nova* **2008**, *31*, 1660–1664. [CrossRef]
172. Crane, E. The Past and Present Importance of Bee Products to Man. In *Bee Products: Properties, Applications, and Apitherapy*; Mizrahi, A., Lensky, Y., Eds.; Springer: Boston, MA, USA, 1997; pp. 1–13. ISBN 978-1-4757-9371-0.
173. Mizrahi, A.; Lensky, Y. (Eds.) *Bee Products: Properties, Applications, and Apitherapy*; Springer: New York, NY, USA, 1997; ISBN 978-0-306-45502-5.
174. Arivarasu, N.A.; Priyamvada, S.; Mahmood, R. Oral Administration of Caffeic Acid Ameliorates the Effect of Cisplatin on Brush Border Membrane Enzymes and Antioxidant System in Rat Intestine. *Exp. Toxicol. Pathol.* **2013**, *65*, 21–25. [CrossRef]
175. Derochette, S.; Franck, T.; Mouithys-Mickalad, A.; Deby-Dupont, G.; Neven, P.; Serteyn, D. Intra- and Extracellular Antioxidant Capacities of the New Water-Soluble Form of Curcumin (NDS27) on Stimulated Neutrophils and HL-60 Cells. *Chem. Biol. Interact* **2013**, *201*, 49–57. [CrossRef]
176. Aabed, K.; Shafi Bhat, R.; Moubayed, N.; Al-Mutiri, M.; Al-Marshoud, M.; Al-Qahtani, A.; Ansary, A. Ameliorative Effect of Probiotics (Lactobacillus Paracaseii and Protexin®) and Prebiotics (Propolis and Bee Pollen) on Clindamycin and Propionic Acid-Induced Oxidative Stress and Altered Gut Microbiota in a Rodent Model of Autism. *Cell Mol. Biol. (Noisy-le-grand)* **2019**, *65*, 1–7. [CrossRef]
177. Pari, L.; Karthikesan, K. Protective Role of Caffeic Acid against Alcohol-induced Biochemical Changes in Rats. *Fundam. Clin. Pharmacol.* **2007**, *21*, 355–361. [CrossRef]
178. Hemmati, A.A.; Alboghobeish, S.; Ahangarpour, A. Effects of Cinnamic Acid on Memory Deficits and Brain Oxidative Stress in Streptozotocin-Induced Diabetic Mice. *Korean J. Physiol. Pharmacol.* **2018**, *22*, 257–267. [CrossRef]
179. Nafees, S.; Rashid, S.; Ali, N.; Hasan, S.K.; Sultana, S. Rutin Ameliorates Cyclophosphamide Induced Oxidative Stress and Inflammation in Wistar Rats: Role of NFκB/MAPK Pathway. *Chem. Biol. Interact.* **2015**, *231*, 98–107. [CrossRef]
180. Ince, E. The Protective Effect of Quercetin in the Alcohol-Induced Liver and Lymphoid Tissue Injuries in Newborns. *Mol. Biol. Rep.* **2020**, *47*, 451–459. [CrossRef]
181. Alekhya Sita, G.J.; Gowthami, M.; Srikanth, G.; Krishna, M.M.; Rama Sireesha, K.; Sajjarao, M.; Nagarjuna, K.; Nagarjuna, M.; Chinnaboina, G.K.; Mishra, A. Protective Role of Luteolin against Bisphenol A-induced Renal Toxicity through Suppressing Oxidative Stress, Inflammation, and Upregulating Nrf2/ARE/HO-1 Pathway. *IUBMB Life* **2019**, *71*, 1041–1047. [CrossRef]
182. Saad, M.A.; Salam, R.M.A.; Kenawy, S.A.; Attia, A.S. Pinocembrin Attenuates Hippocampal Inflammation, Oxidative Perturbations and Apoptosis in a Rat Model of Global Cerebral Ischemia Reperfusion. *Pharmacol. Rep.* **2015**, *67*, 115–122. [CrossRef]
183. Afsharpour, F.; Javadi, M.; Hashemipour, S.; Koushan, Y. Propolis Supplementation Improves Glycemic and Antioxidant Status in Patients with Type 2 Diabetes: A Randomized, Double-Blind, Placebo-Controlled Study. *Complement. Ther. Med.* **2019**, *43*, 283–288. [CrossRef]
184. Senyuk, B.; Boreyko, L.; Yurnyuk, S. Correction of clinical and biochemical parameters using bee pollen in patients with diabetes. *Unity Sci. Int. Sci. Period. J.* **2016**, 137–139.
185. Daudu, O.M. Bee Pollen Extracts as Potential Antioxidants and Inhibitors of α-Amylase and α-Glucosidase Enzymes In Vitro Assessment. *J. Apic. Sci.* **2019**, *63*, 315–325. [CrossRef]
186. Adisakwattana, S. Cinnamic Acid and Its Derivatives: Mechanisms for Prevention and Management of Diabetes and Its Complications. *Nutrients* **2017**, *9*, 163. [CrossRef]
187. Lu, Q.; Hao, M.; Wu, W.; Zhang, N.; Isaac, A.T.; Yin, J.; Zhu, X.; Du, L.; Yin, X. Antidiabetic Cataract Effects of GbE, Rutin and Quercetin Are Mediated by the Inhibition of Oxidative Stress and Polyol Pathway. *Acta Biochim. Pol.* **2018**, *65*, 35–41. [CrossRef]
188. Alkhalidy, H.; Moore, W.; Wang, Y.; Luo, J.; McMillan, R.P.; Zhen, W.; Zhou, K.; Liu, D. The Flavonoid Kaempferol Ameliorates Streptozotocin-Induced Diabetes by Suppressing Hepatic Glucose Production. *Molecules* **2018**, *23*, 2338. [CrossRef]
189. Cheng, N.; Ren, N.; Gao, H.; Lei, X.; Zheng, J.; Cao, W. Antioxidant and Hepatoprotective Effects of Schisandra Chinensis Pollen Extract on CCl4-Induced Acute Liver Damage in Mice. *Food Chem. Toxicol.* **2013**, *55*, 234–240. [CrossRef]
190. Malayeri, A.; Badparva, R.; Mombeini, M.A.; Khorsandi, L.; Goudarzi, M. Naringenin: A Potential Natural Remedy against Methotrexate-Induced Hepatotoxicity in Rats. *Drug Chem. Toxicol.* **2022**, *45*, 491–498. [CrossRef]
191. Yang, C.; Li, L.; Ma, Z.; Zhong, Y.; Pang, W.; Xiong, M.; Fang, S.; Li, Y. Hepatoprotective Effect of Methyl Ferulic Acid against Carbon Tetrachloride-Induced Acute Liver Injury in Rats. *Exp. Ther. Med.* **2018**, *15*, 2228–2238. [CrossRef]
192. Owumi, S.; Ajijola, I.; Agbeti, O. Hepatorenal Protective Effects of Protocatechuic Acid in Rats Administered with Anticancer Drug Methotrexate. *Hum. Exp. Toxicol.* **2019**, *38*, 1254–1265. [CrossRef]
193. Bispo, V.S.; Dantas, L.S.; CHAVES, A.B.; Pinto, I.F.; SILVA, R.P.; Otsuka, F.A.; Santos, R.B.; Santos, A.C.; Trindade, D.J.; Matos, H.R. Reduction of the DNA Damages, Hepatoprotective Effect and Antioxidant Potential of the Coconut Water, Ascorbic and Caffeic Acids in Oxidative Stress Mediated by Ethanol. *Anais da Academia Brasileira de Ciências* **2017**, *89*, 1095–1109. [CrossRef]
194. Ebrahimi, R.; Sepand, M.R.; Seyednejad, S.A.; Omidi, A.; Akbariani, M.; Gholami, M.; Sabzevari, O. Ellagic Acid Reduces Methotrexate-Induced Apoptosis and Mitochondrial Dysfunction via up-Regulating Nrf2 Expression and Inhibiting the IkBα/NFkB in Rats. *DARU J. Pharm. Sci.* **2019**, *27*, 721–733. [CrossRef]

195. Hassan, S.M.; Khalaf, M.M.; Sadek, S.A.; Abo-Youssef, A.M. Protective Effects of Apigenin and Myricetin against Cisplatin-Induced Nephrotoxicity in Mice. *Pharm. Biol.* **2017**, *55*, 766–774. [CrossRef]
196. Said, S.E.; Eldin, H.O.H.; Meghawry, E.K.A.E. Renoprotective effects of naringenin and olive oil against cyclosporine- induced nephrotoxicity in rats. *Iran. J. Toxicol.* **2016**, *10*, 27–37.
197. Chowdhury, S.; Ghosh, S.; Das, A.K.; Sil, P.C. Ferulic Acid Protects Hyperglycemia-Induced Kidney Damage by Regulating Oxidative Insult, Inflammation and Autophagy. *Front. Pharmacol.* **2019**, *10*, 27. [CrossRef]
198. Mohammed, E.T.; Hashem, K.S.; Abdelazem, A.Z.; Foda, F.A.M.A. Prospective Protective Effect of Ellagic Acid as a SIRT1 Activator in Iron Oxide Nanoparticle-Induced Renal Damage in Rats. *Biol. Trace. Elem. Res.* **2020**, *198*, 177–188. [CrossRef]
199. Promsan, S.; Jaikumkao, K.; Pongchaidecha, A.; Chattipakorn, N.; Chatsudthipong, V.; Arjinajarn, P.; Pompimon, W.; Lungkaphin, A. Pinocembrin Attenuates Gentamicin-Induced Nephrotoxicity in Rats. *Can. J. Physiol. Pharmacol.* **2016**, *94*, 808–818. [CrossRef]
200. Shanmugam, K.R.; Shanmugam, B.; Subbaiah, G.V.; Ravi, S.; Reddy, K.S. Medicinal Plants and Bioactive Compounds for Diabetes Management: Important Advances in Drug Discovery. *Curr. Pharm. Des.* **2021**, *27*, 763–774. [CrossRef]
201. Shin, K.-M.; Kim, I.-T.; Park, Y.-M.; Ha, J.; Choi, J.-W.; Park, H.-J.; Lee, Y.S.; Lee, K.-T. Anti-Inflammatory Effect of Caffeic Acid Methyl Ester and Its Mode of Action through the Inhibition of Prostaglandin E2, Nitric Oxide and Tumor Necrosis Factor-α Production. *Biochem. Pharmacol.* **2004**, *68*, 2327–2336. [CrossRef]
202. Bumrungpert, A.; Lilitchan, S.; Tuntipopipat, S.; Tirawanchai, N.; Komindr, S. Ferulic Acid Supplementation Improves Lipid Profiles, Oxidative Stress, and Inflammatory Status in Hyperlipidemic Subjects: A Randomized, Double-Blind, Placebo-Controlled Clinical Trial. *Nutrients* **2018**, *10*, 713. [CrossRef]
203. Abozaid, O.A.R.; Moawed, F.S.M.; Ahmed, E.S.A.; Ibrahim, Z.A. Cinnamic Acid Nanoparticles Modulate Redox Signal and Inflammatory Response in Gamma Irradiated Rats Suffering from Acute Pancreatitis. *Biochim. et Biophys. Acta (BBA)-Mol. Basis Dis.* **2020**, *1866*, 165904. [CrossRef]
204. Khanal, L.; Yadav, P.; Baral, P.; Shah, R.; Rauniar, G.P. Effect of Local Bee Honey on Dihydrofolate Reductase Enzyme Inhibitor-Induced Mucositis: A Histological Study on Albino Wistar Rats. *Indian J. Dent. Res.* **2019**, *30*, 708. [CrossRef]
205. Choi, M.-J.; Lee, E.-J.; Park, J.-S.; Kim, S.-N.; Park, E.-M.; Kim, H.-S. Anti-Inflammatory Mechanism of Galangin in Lipopolysaccharide-Stimulated Microglia: Critical Role of PPAR-γ Signaling Pathway. *Biochem. Pharmacol.* **2017**, *144*, 120–131. [CrossRef]
206. Li, Z.; Chu, S.; He, W.; Zhang, Z.; Liu, J.; Cui, L.; Yan, X.; Li, D.; Chen, N. A20 as a Novel Target for the Anti-Neuroinflammatory Effect of Chrysin via Inhibition of NF-KB Signaling Pathway. *Brain Behav. Immun.* **2019**, *79*, 228–235. [CrossRef]
207. Chen, X.; Lu, J.; An, M.; Ma, Z.; Zong, H.; Yang, J. Anti-inflammatory Effect of Resveratrol on Adjuvant Arthritis Rats with Abnormal Immunological Function via the Reduction of Cyclooxygenase-2 and Prostaglandin E2. *Mol. Med. Rep.* **2014**, *9*, 2592–2598. [CrossRef]
208. Comalada, M.; Camuesco, D.; Sierra, S.; Ballester, I.; Xaus, J.; Gálvez, J.; Zarzuelo, A. In Vivo Quercitrin Anti-Inflammatory Effect Involves Release of Quercetin, Which Inhibits Inflammation through down-Regulation of the NF-KB Pathway. *Eur. J. Immunol.* **2005**, *35*, 584–592. [CrossRef]
209. Park, M.J.; Lee, E.K.; Heo, H.-S.; Kim, M.-S.; Sung, B.; Kim, M.K.; Lee, J.; Kim, N.D.; Anton, S.; Choi, J.S.; et al. The Anti-Inflammatory Effect of Kaempferol in Aged Kidney Tissues: The Involvement of Nuclear Factor-KB via Nuclear Factor-Inducing Kinase/IκB Kinase and Mitogen-Activated Protein Kinase Pathways. *J. Med. Food* **2009**, *12*, 351–358. [CrossRef]
210. Kelainy, E.G.; Ibrahim Laila, I.M.; Ibrahim, S.R. The Effect of Ferulic Acid against Lead-Induced Oxidative Stress and DNA Damage in Kidney and Testes of Rats. *Environ. Sci. Pollut. Res.* **2019**, *26*, 31675–31684. [CrossRef]
211. Karimi, M.Y.; Fatemi, I.; Kalantari, H.; Mombeini, M.A.; Mehrzadi, S.; Goudarzi, M. Ellagic Acid Prevents Oxidative Stress, Inflammation, and Histopathological Alterations in Acrylamide-Induced Hepatotoxicity in Wistar Rats. *J. Diet. Suppl.* **2020**, *17*, 651–662. [CrossRef]
212. Feng, H.; Cao, J.; Zhang, G.; Wang, Y. Kaempferol Attenuates Cardiac Hypertrophy via Regulation of ASK1/MAPK Signaling Pathway and Oxidative Stress. *Planta Med.* **2017**, *83*, 837–845. [CrossRef]
213. Aloud, A.A.; Veeramani, C.; Govindasamy, C.; Alsaif, M.A.; El Newehy, A.S.; Al-Numair, K.S. Galangin, a Dietary Flavonoid, Improves Antioxidant Status and Reduces Hyperglycemia-Mediated Oxidative Stress in Streptozotocin-Induced Diabetic Rats. *Redox Rep.* **2017**, *22*, 290–300. [CrossRef]
214. El Khashab, I.H.; Abdelsalam, R.M.; Elbrairy, A.I.; Attia, A.S. Chrysin Attenuates Global Cerebral Ischemic Reperfusion Injury via Suppression of Oxidative Stress, Inflammation and Apoptosis. *Biomed. Pharmacother.* **2019**, *112*, 108619. [CrossRef]
215. Al Olayan, E.M.; Aloufi, A.S.; AlAmri, O.D.; El-Habit, O.H.; Abdel Moneim, A.E. Protocatechuic Acid Mitigates Cadmium-Induced Neurotoxicity in Rats: Role of Oxidative Stress, Inflammation and Apoptosis. *Sci. Total Environ.* **2020**, *723*, 137969. [CrossRef]
216. Pan, X.; Shao, Y.; Wang, F.; Cai, Z.; Liu, S.; Xi, J.; He, R.; Zhao, Y.; Zhuang, R. Protective Effect of Apigenin Magnesium Complex on H2O2-Induced Oxidative Stress and Inflammatory Responses in Rat Hepatic Stellate Cells. *Pharm. Biol.* **2020**, *58*, 553–560. [CrossRef]
217. Han, X.-Y.; Du, W.-L.; Huang, Q.-C.; Xu, Z.-R.; Wang, Y.-Z. Changes in Small Intestinal Morphology and Digestive Enzyme Activity with Oral Administration of Copper-Loaded Chitosan Nanoparticles in Rats. *Biol. Trace. Elem. Res.* **2012**, *145*, 355–360. [CrossRef]

218. Chen, L.; Gnanaraj, C.; Arulselvan, P.; El-Seedi, H.; Teng, H. A Review on Advanced Microencapsulation Technology to Enhance Bioavailability of Phenolic Compounds: Based on Its Activity in the Treatment of Type 2 Diabetes. *Trends Food Sci. Technol.* **2019**, *85*, 149–162. [CrossRef]
219. Ramar, M.; Manikandan, B.; Raman, T.; Priyadarsini, A.; Palanisamy, S.; Velayudam, M.; Munusamy, A.; Marimuthu Prabhu, N.; Vaseeharan, B. Protective Effect of Ferulic Acid and Resveratrol against Alloxan-Induced Diabetes in Mice. *Eur. J. Pharmacol.* **2012**, *690*, 226–235. [CrossRef]
220. Fatima, N.; Hafizur, R.M.; Hameed, A.; Ahmed, S.; Nisar, M.; Kabir, N. Ellagic Acid in Emblica Officinalis Exerts Anti-Diabetic Activity through the Action on β-Cells of Pancreas. *Eur. J. Nutr.* **2017**, *56*, 591–601. [CrossRef]
221. Shi, G.-J.; Li, Y.; Cao, Q.-H.; Wu, H.-X.; Tang, X.-Y.; Gao, X.-H.; Yu, J.-Q.; Chen, Z.; Yang, Y. In Vitro and in Vivo Evidence That Quercetin Protects against Diabetes and Its Complications: A Systematic Review of the Literature. *Biomed. Pharmacother.* **2019**, *109*, 1085–1099. [CrossRef]
222. Sharma, D.; Kumar Tekade, R.; Kalia, K. Kaempferol in Ameliorating Diabetes-Induced Fibrosis and Renal Damage: An in Vitro and in Vivo Study in Diabetic Nephropathy Mice Model. *Phytomedicine* **2020**, *76*, 153235. [CrossRef]
223. Vinayagam, R.; Xu, B. Antidiabetic Properties of Dietary Flavonoids: A Cellular Mechanism Review. *Nutr. Metab.* **2015**, *12*, 60. [CrossRef]
224. El-Sonbaty, Y.A.; Suddek, G.M.; Megahed, N.; Gameil, N.M. Protocatechuic Acid Exhibits Hepatoprotective, Vasculoprotective, Antioxidant and Insulin-like Effects in Dexamethasone-Induced Insulin-Resistant Rats. *Biochimie* **2019**, *167*, 119–134. [CrossRef]
225. Hossain, C.M.; Ghosh, M.K.; Satapathy, B.S.; Dey, N.S.; Mukherjee, B. Apigenin Causes Biochemical Modulation, GLUT4 and Cd38 Alterations to Improve Diabetes and to Protect Damages of Some Vital Organs in Experimental Diabetes. *Am. J. Pharmacol. Toxicol.* **2014**, *9*, 39–52. [CrossRef]
226. Zhang, M.; He, L.; Liu, J.; Zhou, L. Luteolin Attenuates Diabetic Nephropathy through Suppressing Inflammatory Response and Oxidative Stress by Inhibiting STAT3 Pathway. *Exp. Clin. Endocrinol. Diabetes* **2021**, *129*, 729–739. [CrossRef]
227. Ren, B.; Qin, W.; Wu, F.; Wang, S.; Pan, C.; Wang, L.; Zeng, B.; Ma, S.; Liang, J. Apigenin and Naringenin Regulate Glucose and Lipid Metabolism, and Ameliorate Vascular Dysfunction in Type 2 Diabetic Rats. *Eur. J. Pharmacol.* **2016**, *773*, 13–23. [CrossRef]
228. Pei, B.; Sun, J. Pinocembrin Alleviates Cognition Deficits by Inhibiting Inflammation in Diabetic Mice. *J. Neuroimmunol.* **2018**, *314*, 42–49. [CrossRef]
229. Aref, A.; Tohamy, A.; E Abdel Moneim, A.; Sayed, R. Cinnamic Acid Attenuates Cisplatin-Induced Hepatotoxicity and Nephrotoxicity. *J. Basic Environ. Sci.* **2016**, 1–9.
230. Zhao, L.; Chen, F.; Zhang, Y.; Yue, L.; Guo, H.; Ye, G.; Shi, F.; Lv, C.; Jing, B.; Tang, H.; et al. Involvement of P450s and Nuclear Receptors in the Hepatoprotective Effect of Quercetin on Liver Injury by Bacterial Lipopolysaccharide. *Immunopharmacol. Immunotoxicol.* **2020**, *42*, 211–220. [CrossRef]
231. Wang, M.; Sun, J.; Jiang, Z.; Xie, W.; Zhang, X. Hepatoprotective Effect of Kaempferol Against Alcoholic Liver Injury in Mice. *Am. J. Chin. Med.* **2015**, *43*, 241–254. [CrossRef]
232. Aladaileh, S.H.; Abukhalil, M.H.; Saghir, S.A.M.; Hanieh, H.; Alfwuaires, M.A.; Almaiman, A.A.; Bin-Jumah, M.; Mahmoud, A.M. Galangin Activates Nrf2 Signaling and Attenuates Oxidative Damage, Inflammation, and Apoptosis in a Rat Model of Cyclophosphamide-Induced Hepatotoxicity. *Biomolecules* **2019**, *9*, 346. [CrossRef]
233. Pingili, R.B.; Pawar, A.K.; Challa, S.R.; Kodali, T.; Koppula, S.; Toleti, V. A Comprehensive Review on Hepatoprotective and Nephroprotective Activities of Chrysin against Various Drugs and Toxic Agents. *Chem. -Biol. Interact.* **2019**, *308*, 51–60. [CrossRef]
234. Rašković, A.; Gigov, S.; Čapo, I.; Kusturica, M.P.; Milijašević, B.; Kojić-Damjanov, S.; Martić, N. Antioxidative and Protective Actions of Apigenin in a Paracetamol-Induced Hepatotoxicity Rat Model. *Eur. J. Drug Metab. Pharmacokinet.* **2017**, *42*, 849–856. [CrossRef]
235. Yan, Y.; Jun, C.; Lu, Y.; Jiangmei, S. Combination of Metformin and Luteolin Synergistically Protects Carbon Tetrachloride-Induced Hepatotoxicity: Mechanism Involves Antioxidant, Anti-Inflammatory, Antiapoptotic, and Nrf2/HO-1 Signaling Pathway. *BioFactors* **2019**, *45*, 598–606. [CrossRef]
236. Reddy, M.K.; Reddy, A.G.; Kumar, B.K.; Madhuri, D.; Boobalan, G.; Reddy, M.A. Protective Effect of Rutin in Comparison to Silymarin against Induced Hepatotoxicity in Rats. *Vet World* **2017**, *10*, 74–80. [CrossRef]
237. Wang, Y.; Jiang, Y.; Fan, X.; Tan, H.; Zeng, H.; Wang, Y.; Chen, P.; Huang, M.; Bi, H. Hepato-Protective Effect of Resveratrol against Acetaminophen-Induced Liver Injury Is Associated with Inhibition of CYP-Mediated Bioactivation and Regulation of SIRT1–P53 Signaling Pathways. *Toxicol. Lett.* **2015**, *236*, 82–89. [CrossRef]
238. Olayinka, E.T.; Ola, O.S.; Ore, A.; Adeyemo, O.A. Ameliorative Effect of Caffeic Acid on Capecitabine-Induced Hepatic and Renal Dysfunction: Involvement of the Antioxidant Defence System. *Medicines* **2017**, *4*, 78. [CrossRef]
239. El-Sayed, E.-S.M.; Abd El-Raouf, O.M.; Fawzy, H.M.; Manie, M.F. Comparative Study of the Possible Protective Effects of Cinnamic Acid and Cinnamaldehyde on Cisplatin-Induced Nephrotoxicity in Rats. *J. Biochem. Mol. Toxicol.* **2013**, *27*, 508–514. [CrossRef]
240. Kocahan, S.; Dogan, Z.; Erdemli, M.E.; Taskin, E. Protective Effect of Quercetin Against Oxidative Stressinduced Toxicity Associated With Doxorubicin and Cyclophosphamide in Rat Kidney and Liver Tissue. *Iran. J. Kidney Dis.* **2017**, *11*, 124.
241. Vijayaprakash, S.; Langeswaran, K.; Gowtham Kumar, S.; Revathy, R.; Balasubramanian, M.P. Nephro-Protective Significance of Kaempferol on Mercuric Chloride Induced Toxicity in Wistar Albino Rats. *Biomed. Aging Pathol.* **2013**, *3*, 119–124. [CrossRef]

242. Farkhondeh, T.; Samarghandian, S.; Roshanravan, B. Impact of Chrysin on the Molecular Mechanisms Underlying Diabetic Complications. *J. Cell. Physiol.* **2019**, *234*, 17144–17158. [CrossRef]
243. Chen, L.; Tian, G.; Tang, W.; Luo, W.; Liu, P.; Ma, Z. Protective Effect of Luteolin on Streptozotocin-Induced Diabetic Renal Damage in Mice via the Regulation of RIP140/NF-KB Pathway and Insulin Signalling Pathway. *J. Funct. Foods* **2016**, *22*, 93–100. [CrossRef]
244. Wang, B.; Liu, D.; Zhu, Q.; Li, M.; Chen, H.; Guo, Y.; Fan, L.; Yue, L.; Li, L.; Zhao, M. Rutin Ameliorates Kidney Interstitial Fibrosis in Rats with Obstructive Nephropathy. *Int. Immunopharmacol.* **2016**, *35*, 77–84. [CrossRef]
245. Xue, H.-Y.; Yuan, L.; Cao, Y.-J.; Fan, Y.-P.; Chen, X.-L.; Huang, X.-Z. Resveratrol Ameliorates Renal Injury in Spontaneously Hypertensive Rats by Inhibiting Renal Micro-Inflammation. *Biosci. Rep.* **2016**, *36*, e00339. [CrossRef]
246. Mansouri, M.T.; Hemmati, A.A.; Naghizadeh, B.; Mard, S.A.; Rezaie, A.; Ghorbanzadeh, B. A Study of the Mechanisms Underlying the Anti-Inflammatory Effect of Ellagic Acid in Carrageenan-Induced Paw Edema in Rats. *Indian J. Pharm.* **2015**, *47*, 292–298. [CrossRef]
247. Lin, C.-Y.; Huang, C.-S.; Huang, C.-Y.; Yin, M.-C. Anticoagulatory, Antiinflammatory, and Antioxidative Effects of Protocatechuic Acid in Diabetic Mice. *J. Agric. Food Chem.* **2009**, *57*, 6661–6667. [CrossRef]
248. Darabi, P.; Khazali, H.; Mehrabani Natanzi, M. Therapeutic Potentials of the Natural Plant Flavonoid Apigenin in Polycystic Ovary Syndrome in Rat Model: Via Modulation of pro-Inflammatory Cytokines and Antioxidant Activity. *Gynecol. Endocrinol.* **2020**, *36*, 582–587. [CrossRef]
249. Xiong, J.; Wang, K.; Yuan, C.; Xing, R.; Ni, J.; Hu, G.; Chen, F.; Wang, X. Luteolin Protects Mice from Severe Acute Pancreatitis by Exerting HO-1-Mediated Anti-Inflammatory and Antioxidant Effects. *Int. J. Mol. Med.* **2017**, *39*, 113–125. [CrossRef] [PubMed]
250. Alhoshani, A.R.; Hafez, M.M.; Husain, S.; Al-sheikh, A.M.; Alotaibi, M.R.; Al Rejaie, S.S.; Alshammari, M.A.; Almutairi, M.M.; Al-Shabanah, O.A. Protective Effect of Rutin Supplementation against Cisplatin-Induced Nephrotoxicity in Rats. *BMC Nephrol.* **2017**, *18*, 194. [CrossRef]
251. Fan, R.; Pan, T.; Zhu, A.-L.; Zhang, M.-H. Anti-Inflammatory and Anti-Arthritic Properties of Naringenin via Attenuation of NF-KB and Activation of the Heme Oxygenase (HO)-1/Related Factor 2 Pathway. *Pharmacol. Rep.* **2017**, *69*, 1021–1029. [CrossRef] [PubMed]
252. Gu, X.; Zhang, Q.; Du, Q.; Shen, H.; Zhu, Z. Pinocembrin Attenuates Allergic Airway Inflammation via Inhibition of NF-KB Pathway in Mice. *Int. Immunopharmacol.* **2017**, *53*, 90–95. [CrossRef]
253. Fazayeli-Rad, A.R.; Afzali, N.; Farhangfar, H.; Asghari, M.R. Effect of Bee Pollen on Growth Performance, Intestinal Morphometry and Immune Status of Broiler Chicks. *Eur. Poult. Sci.* **2015**, *79*, 86. [CrossRef]
254. Münstedt, K.; Voss, B.; Kullmer, U.; Schneider, U.; Hübner, J. Bee Pollen and Honey for the Alleviation of Hot Flushes and Other Menopausal Symptoms in Breast Cancer Patients. *Mol. Clin. Oncol.* **2015**, *3*, 869–874. [CrossRef]
255. Yan, S.; Wang, K.; Wang, X.; Ou, A.; Wang, F.; Wu, L.; Xue, X. Effect of Fermented Bee Pollen on Metabolic Syndrome in High-Fat Diet-Induced Mice. *Food Sci. Hum. Wellness* **2021**, *10*, 345–355. [CrossRef]
256. Kurek-Górecka, A.; Górecki, M.; Rzepecka-Stojko, A.; Balwierz, R.; Stojko, J. Bee Products in Dermatology and Skin Care. *Molecules* **2020**, *25*, 556. [CrossRef]
257. Xi, X.; Li, J.; Guo, S.; Li, Y.; Xu, F.; Zheng, M.; Cao, H.; Cui, X.; Guo, H.; Han, C. The Potential of Using Bee Pollen in Cosmetics: A Review. *J. Oleo Sci.* **2018**, *67*, 1071–1082. [CrossRef]

antioxidants

Article

Polish and New Zealand Propolis as Sources of Antioxidant Compounds Inhibit Glioblastoma (T98G, LN-18) Cell Lines and Astrocytoma Cells Derived from Patient

Justyna Moskwa [1,*], Sylwia Katarzyna Naliwajko [1], Renata Markiewicz-Żukowska [1], Krystyna Joanna Gromkowska-Kępka [1], Jolanta Soroczyńska [1], Anna Puścion-Jakubik [1], Maria Halina Borawska [1], Valery Isidorov [2] and Katarzyna Socha [1]

1 Department of Bromatology, Faculty of Pharmacy with Division of Laboratory Medicine, Medical University of Białystok, Mickiewicza 2D Street, 15-222 Białystok, Poland; sylwia.naliwajko@umb.edu.pl (S.K.N.); renmar@poczta.onet.pl (R.M.-Ż.); krystyna.gromkowska-kepka@umb.edu.pl (K.J.G.-K.); jolanta.soroczynska@umb.edu.pl (J.S.); anna.puscion-jakubik@umb.edu.pl (A.P.-J.); borawska@umb.edu.pl (M.H.B.); katarzyna.socha@umb.edu.pl (K.S.)

2 Forest Faculty, Białystok University of Technology, Wiejska 45A Street, 15-351 Białystok, Poland; v.isidorov@pb.edu.pl

* Correspondence: justyna.moskwa@umb.edu.pl; Tel.: +48-8569745470

check for updates

Citation: Moskwa, J.; Naliwajko, S.K.; Markiewicz-Żukowska, R.; Gromkowska-Kępka, K.J.; Soroczyńska, J.; Puścion-Jakubik, A.; Borawska, M.H.; Isidorov, V.; Socha, K. Polish and New Zealand Propolis as Sources of Antioxidant Compounds Inhibit Glioblastoma (T98G, LN-18) Cell Lines and Astrocytoma Cells Derived from Patient. *Antioxidants* **2022**, *11*, 1305. https://doi.org/10.3390/antiox11071305

Academic Editor: Stanley Omaye

Received: 6 May 2022
Accepted: 27 June 2022
Published: 29 June 2022

Abstract: Gliomas, including glioblastoma multiforme and astrocytoma, are common brain cancers in adults. Propolis is a natural product containing many active ingredients. The aim of this study was to compare the chemical composition, total phenolic content and concentration of toxic elements as well as the anticancer potential of Polish (PPE) and New Zealand (Manuka—MPE) propolis extracts on diffuse astrocytoma derived from patient (DASC) and glioblastoma (T98G, LN-18) cell lines. The antioxidants such as flavonoids and chalcones (pinocembrin, pinobanksin, pinobanksin 3-acetate and chrysin) were the main components in both types of propolis. The content of arsenic (As) and lead (Pb) in MPE was higher than PPE. The anti-proliferative study showed strong activity of PPE and MPE propolis on DASC, T98G, and LN-18 cells by apoptosis induction, cell cycle arrest and attenuated migration. These findings suggest that despite their different geographic origins, Polish and New Zealand propolis are sources of antioxidant compounds and show similar activity and a promising anti-glioma potential in in vitro study. However, further in vivo studies are required in order to assess therapeutic potential of propolis.

Keywords: propolis; polyphenols content; glioma cells; cancer prevention and treatment

1. Introduction

Gliomas are tumors of neuroepithelial origin and represent approximately 40% of primary intracranial tumors. Diffuse astrocytomas belong to a category of diffuse gliomas which arise from glial cells. A glioma is a slow-growing brain tumor and tends to grow into and infiltrate neighboring, healthy tissue brain. Glioblastoma multiforme (GBM) is the most aggressive malignant tumor in the CNS and has a poor prognosis [1]. A characteristic feature of glioma is the diversity of histological features and cell composition. Currently, the main method of treatment this type of tumor is surgery, which offers rapid relief from the symptoms of high intracranial pressure and provides a chance to remove or reduce neurological defects. The next step is radio and chemotherapy [2]. However, patients with the GBM treated with radiotherapy combined with temozolomide (TZM) expect a median survival of only 15 months [3]. Therefore, natural compounds which could enhance currently available treatment modalities are sorely needed.

Propolis is a natural product composed of tree and plant resin, bee wax, pollen and gland secretions of bees. When compared to other natural products, propolis is unique, since it is of both plant and animal origin. It contains a wide range of active components, whose concentrations depends primarily on the geographical provenance, season of the year, and the breed of bees. There are several types of propolis: "Poplar" (European, Chinese, North and South American, including Manuka propolis from New Zealand), "Brazilian green" (containing artepillin-C), "Red" (from Cuba, Brazil, Mexico), "Birch" (from Russia), "Mediterranean" (Greece, Crete, Sicily, Malta), "Pacific" (from Okinawa, Taiwan, Indonesia) and "Clusia" (from Cuba and Venezuela) [4]. Hence, various biological activities of propolis have been reported by many authors. The most active compounds are flavonoids (e.g., chrysin, apigenin, pinocembrin, pinobanksin, kaempferol), aromatic acids (e.g., p-coumaric, ferulic), and esters (caffeic acid phenethyl ester—CAPE) [5,6]. A number of studies concerning the anti-tumor activity of propolis on various cancer cell lines such as human colorectal cancer (DLD-1) [4], human lung cancer (A549) [7], gastric cancer (HGC27) [8], and human prostate cancer (PC3) [9] have been published. The chemical composition and antiproliferative effect of propolis from Poland on the human glioblastoma multiforme cell line U87MG has been confirmed in our previous studies [5,10,11]. The research studies have focused on the potential utilization of propolis phenolic compounds in the development of new anti-cancer drugs [12,13]. The role of antioxidant action in cancer cells is complex and not completely understood. Scientific research shows that antioxidants are able to decrease the tumor formation risk by preventing ROS-induced oxidation of DNA and sub-sequent DNA damage [14], but on the other hand, Schafer et al. showed that antioxidant activity may promote the survival of preinitiated tumor cells in unnatural matrix environments and thus enhance malignancy [15].

It is well known that propolis has a very rich chemical composition, and its compounds show a multidirectional effect on the human body. The present study compare the antiproliferative activity of propolis from Poland and from New Zealand on different types of brain tumor—human diffuse astrocytoma cell line (DASC) derived from a patient with Grade II glioma and glioblastoma multiforme T98G and LN-18 cell lines.

2. Materials and Methods

2.1. Materials

DMEM/Ham's F12 with L-glutamine was purchased from PAA Laboratories GmbH (Pasching, Austria). Dulbecco's modified eagle medium (DMEM), fetal bovine serum (FBS), minimal essential medium eagle (MEM) with L-glutamine, trypsin-EDTA, penicillin, streptomycin were purchased from Gibco (Thermo Fisher Scientific, Waltham, MA, USA). Calcium-free phosphate-buffered saline (PBS) was received from Biomed (Lublin, Poland). Bis(trimethylsilyl)trifluoroacetamide (BSTFA) with an addition of 1% trimethylchlorosilane, C10–C40 n-alkane standard solution, methylthiazolyl diphenyl-tetrazolium bromide (MTT), dimethyl sulfoxide (DMSO), pyridine, trichloroacetic acid, and trizma base were obtained from Sigma-Aldrich (St. Louis, MO, USA). Ethanol at 95% was obtained (AWW Group, Poland). The scintillation cocktail was purchased from PerkinElmer (Boston, MA, USA) and methyl-3H thymidine from MP Biomedicals, Inc. (Irvine, CA, USA).

2.2. Sample Preparations

Propolis of *Apis mellifera* was collected in the Podlasie region (northeastern Poland). To prepare the ethanolic extract of Polish propolis (PPE), 20 g of crushed propolis was extracted on a shaker with 80 g of 70% ethanol for 12 h in a darkened place. The extract was centrifuged at 2500 rpm for 10 min at 20 °C, evaporated (40 °C) in a rotary evaporator (Rotavapor R-3, Buchi, Switzerland) and lyophilized. The dry Polish propolis extract (PPE) was protected from light and kept frozen at −20 °C. The yield of the prepared extracts (% *w*/*w*) in terms of the starting material was 47.6.

Propolis Manuka Health New Zealand (Bio 30) ethanolic tincture was purchased from the manufacturer. The tincture was evaporated (40 °C) in a rotary evaporator (Rotavapor R-3, Buchi, Switzerland) and lyophilized. The dry Manuka propolis extract (MPE) was protected from light and kept frozen at −20 °C.

The extracts were dissolved in DMSO and prepared as 1 mg/mL stock solution (calculated as dry extracts) in the culture medium.

2.3. Gas Chromatography–Mass Spectrometry (GC-MS) Analysis

At this stage, 5 mg of PPE and MPE were diluted with 220 μL of pyridine and 80 μL of BSTFA with an addition of 1% trimethylchlorosilane. The reaction mixture was sealed and heated for 0.5 h at 60 °C to form trimethylsilyl (TMS) derivatives.

GC-MS analyses of PPE and MPE were performed using GC-MS on an HP 6890 gas chromatograph with a mass selective detector MSD 5973 (Agilent Technologies, Santa Clara, CA, USA) equipped with a ZB-5MSi fused silica column (30 m, 0.25 mm i.d., 0.25 μm film thickness), with electronic pressure control and a split/splitless injector. Helium flow rate through the column was 1 mL/min in a constant flow mode. The injector worked at 250 °C in the split (1:50) mode. The initial column temperature was 50 °C, rising to 310 °C at 5 °C/min and the higher temperature was maintained for 15 min. MSD detector acquisition parameters were as follows: transfer line temperature 280 °C, MS Source temperature 230 °C and MS Quad temperature 150 °C. The EIMS spectra were obtained at the ionization energy of 70 eV. The MSD was set to scan 41–600 a.m.u. Following the integration, the fraction of each component in the total ion current was calculated. Hexane solutions of C_{10}–C_{40} *n*-alkanes were separated under the above conditions. Gas chromatographic linear programmed retention indices ($\boldsymbol{I}_T$) were calculated on the basis of the retention times of the *n*-alkanes hexane solution and separated components of the extract samples.

To identify the separated components, two independent analytical parameters were used: mass spectra and calculated retention indices. The mass spectrometric identification of non-derivatized components was performed with an automatic system for GC-MS data processing supplied by the NIST 14 library (NIST/EPA/NIH Library of Electron Ionization Mass Spectra). The mass spectra and retention indices of the components registered in the form of TMS derivatives were compared with those presented in a recently published database [16] and a private mass spectra library. Identification was considered reliable if the results of the computer search of the mass spectra library were confirmed by experimental *RI* values, i.e., if their deviation from the published database values did not exceed ± 10 u.i. (the average quantity of inter-laboratory deviation for non-polar stationary phases).

2.4. Total Phenolic Content Analysis

Total phenolic content (TPC) was measured using the Folin–Ciocalteu colorimetric method (FC). Absorbance versus a prepared blank was read at 760 nm using Cintra 3030 (GBC Scientific Equipment, Melbourne, Australia). The results were expressed as milligrams of gallic acid equivalent (GAE) per gram of a dry extract. The concentration of samples equaled 2 mg/mL (extract dissolved in 70% ethanol). Data were expressed as mean ± SD.

2.5. Toxic Elements Analysis (Arsenic, Cadmium, and Lead)

Coupled plasma mass spectrometry (ICP-MS, NexION 300D, PerkinElmer, USA) was applied to determine toxic element. Before analysis, propolis samples were mineralized according to a procedure proposed by Bielecka et al. [17]. A kinetic energy discrimination (KED) chamber was used in the case of As and the standard mode in the case of Cd and Pb. In order to correct for polyatomic interference in this configuration, kinetic energy discriminations and collisions were applied. The results were obtained in counts per second (cps) and based on calibration curves, were converted into concentrations. To determine the limit of detection (LOD), 10 independent blank determinations were made. A three-fold standard deviation (SD) from the mean value determined in concentration units was taken as the LOD. The LOD values were 0.018 μg/kg for As, 0.017 μg/kg for Cd, and 0.16 μg/kg for Pb. ICP-MS conditions for As, Cd, and Pb determination were described in our previous publication [17]. Quality control was performed by analyzing certified reference material (corn flour INCT-CF-3, Institute of Nuclear Chemistry and Technology, Warsaw, Poland) prior to the start of the analysis. The results of the quality control are summarized in Table 1.

Table 1. Results obtained in the quality control process.

Element	Precision (%)	Recovery (%)	Declared Concentration in CRM (μg/kg)
As	3.3	99.0	10
Cd	2.5	99.1	7
Pb	2.4	99.5	52

CRM—certified reference material.

2.6. Cell Culture

The study was performed using Diffuse Astrocytoma Stem-like Cells (DASC) and glioblastoma multiforme cell lines (T98G and LN-18). The DASC cell line was derived from a 43-year-old patient with diffuse astrocytoma (Grade II), as described in our previous research [18]. The study was approved by the local Ethics Committee [18]. T98G and LN-18 had been obtained from the American Type Culture Collection (ATCC, Rockville, MD, USA). The cells were cultured in a humidified incubator at 37 °C and 5% CO_2 atmosphere, in MEM (DASC and T98G) or DMEM (LN-18) supplemented with 10% heat inactivated FBS; 100 U/mL penicillin and 0.1 mg/mL streptomycin. Subconfluent cells were detached with a trypsin-EDTA solution in PBS and counted in a Neubauer hemocytometer. Assays were performed in triplicate.

2.7. Cell Viability Assay

Cell viability was measured using an MTT assay, as previously described for glioma cells [19]. The effects of PPE and MPE extracts on DASC, T98G and LN-18 cell lines were studied after 24 h, 48 h and 72 h of the treatment. The cells were cultured as follows: in a humidified incubator at 37 °C and 5% CO_2 atmosphere; in MEM or DMEM supplemented with 10% heat inactivated FBS; with 100 U/mL penicillin and 0.1 mg/mL streptomycin. Doses of propolis (10, 20, 30, 50, 100 μg/mL) were selected in our previous experiments [11]. Cells at a density of 1 × 105 cells/mL were seeded onto 96-well plates at a volume of 200 μL per well and grown for 22 h at 37 °C in a humidified 5% CO_2 incubator. The data were expressed as a percentage of the control (0.1% DMSO).

2.8. DNA Synthesis Assay

At this stage, [^{3}H]-thymidine assays were performed to study DNA synthesis in the cells after the treatment. The cells were seeded (1.5×10^5 cell/well) on 24-well plates in MEM or DMEM supplemented with 10% heat inactivated FBS, 100 U/mL penicillin and 0.1 mg/mL streptomycin, and exposed to the treatment medium containing DMSO (0.1%-control), PPE and MPE (30 μg/mL). The cells were cultured for 20, 44 and 68 h prior to adding 0.5 μCi of [^{3}H]-thymidine per well. After 4 h of incubation with [^{3}H]-thymidine, the medium was removed and the cells were washed twice with cold 0.05 M Tris-HCl and 5% trichloroacetic acid, then scraped and transferred to a scintillation cocktail. The level of [^{3}H]-thymidine incorporated in the newly synthesized DNA strand was assessed by a scintillation counter in relation to the DNA synthesis in the control cells. Amount of incorporated [^{3}H]-thymidine indicates the ability of cells proliferation.

2.9. Migration Assay (Scratch Assay)

For the scratch test, the DASC, T98G and LN-18 cells were cultured (0.5×10^6 cell/well) on 6-well plates, at 37 °C in a humidified atmosphere of 5% CO_2. After reaching 80–90% confluence, the cells in the well plates were scratched with a sterile 20 μL micropipette tip to the same length and width. After each well had been washed with PBS to remove debris, the cells were treated with PPE and MPE (30 μg/mL) and medium containing DMSO (0.1%, control), and then incubated for 42 h. The images of each treatment well were captured at

100× magnification, using an Olympus CKX 41 microscope and KcJunior program at each time point (0, 18, 42 h) and combined into one figure. The images acquired for each sample at different times were quantitatively analyzed using ImageJ 1.52v analysis software, a free image-processing and analysis program. The cell migration was calculated as a percentage of scratch area.

2.10. Cell Cycle Assay

The effect of PPE and MPE on the cell cycle was analyzed by the Advanced Image Cytometer NucleoCounter NC-3000 (ChemoMetec, Lillerød, Denmark), as described in our previously published study [19]. The DASC, T98G, and LN-18 cells were seeded into 6-well plates at a density of 1×10^6 cells per well. After 24 h of incubation, the cells were treated with PPE and MPE (30 μg/mL) or a medium containing DMSO (0.1%, control). After 24 h of cell treatment, the test was performed using 1–2 × 10^6 cells, according to the 2-step cell cycle assay protocol of the manufacturer (ChemoMetec, Lillerød, Denmark). The results are presented as the percentages of the cells in different cell cycle phases: subG1, G1/G0, S, and G2/M.

2.11. Annexin V Assay

Using image analysis, the NucleoCounter® NC-3000™(ChemoMetec, Lillerød, Denmark), we indicated a quantification of early apoptotic cells based on Annexin V binding and PI exclusion. Cells (2−4 × 10^5) were stained with Annexin V-CF488A conjugate along with Hoechst 33342. Just before analysis, cells were mixed with PI to stain nonviable cells. The DASC, T98G, and LN-18 cells were seeded into 6-well plates at a density of 1×10^5 cells per well, and after 24 h of incubation, they were treated with PPE and MPE (30 μg/mL). After 48 h of incubation with the studied agents, the assay was performed following the manufacturer's protocol for the Annexin V assay (ChemoMetec, Lillerød, Denmark).

2.12. Statistical Analysis

All data were analyzed using Statistica software, version 13.3. The results were expressed as mean ± SD and statistically compared to the control. Values were tested for a normal distribution using the Shapiro–Wilk test. Differences between two groups were analyzed using Student's *t*-test or Mann–Whitney U test. $p < 0.05$ was considered to be statistically significant.

3. Results

3.1. Chemical Composition and Total Phenolic Content of PPE and MPE

In this study, more than 100 individual compounds in PPE and more than 150 compounds in MPE were identified by GC-MS analysis (Supplementary material Table S1). A list of the main constituents is presented in Table 2, where both propolis extracts contained a lot of antioxidants compounds. Flavonoids and chalcones were the main components of both examined types of propolis (PPE, 49.4%; MPE, 52.1%) (Table 3). The main representatives of this group of compounds in PPE and MPE were pinocembrin (8.16% and 14.64%), pinobanksin (4.25% and 4.70%), pinobanksin 3-acetate (11.27% and 9.21%), chrysin (5.33% and 5.73%), galangin (8.95% and 9.60%), respectively, and their derivatives (Table 2). Cinnamic acid derivatives such as esters 3-methyl-2-bytenyl (E)-caffeate, benzyl (E)-caffeate, benzyl (E)-p-coumarate, 2-phenylethyl p-coumarate, benzyl (E)-ferulate, CAPE, cinnamyl (E)-p-coumarate and others were the second significant group of compounds in PPE and MPE (19.8% and 14.5%, respectively) (Table 2). Considerable quantities of aromatic acids were present in both studied propolis extracts, although propolis from Poland (PPE–18.3%) contained twice as great a quantity of aromatic acids as propolis from New Zealand (MPE–7.8%) (Table 3). The main representatives of this group were p-coumaric acid, (E)-ferulic acid and (E)-caffeic acid. PPE contained high levels of p-coumaric acid (9.80%) (Table 2). TPC determination confirmed that PPE and MPE are rich sources of polyphenolic antioxidants—243.7 ± 9.0 in PPE and 245.6 ± 5.9 mg GAE/g in MPE (Table 4).

Table 2. The main chemical compounds of the ethanolic extracts of propolis from Poland (PPE) and propolis from New Zealand (MPE), determined by GC-MS.

Components, TMS Derivative	I_T^{Exp}	I_T^{Lit}	PPE [%]	MPE [%]
Benzoic acid	1244	1247	1.80	0.33
Cinnamic acid	1542	1546	0.20	1.82
p-Coumaric acid	1944	1947	9.77	0.87
3,4-Dimethoxycinnamic acid	2030	2034	-	1.51
(E)-Ferulic acid	2101	2101	3.22	0.15
(E)-Caffeic acid	2155	2155	2.10	1.53
3-Methyl-3-butenyl (E)-caffeate	2371	2367	1.18	3.39
3-Methyl-1-butenyl (E)-caffeate	2374	2375	0.50	0.44
3-Methyl-2-bytenyl (E)-caffeate	2425	2421	1.65	2.36
Pinocembrin, mono-TMS	2460	2461	1.14	0.46
Benzyl (E)-p-coumarate	2516	2515	3.78	0.37
Pinocembrin	2551	2552	6.93	14.10
2-Phenylethyl p-coumarate	2603	2603	1.02	0.11
Pinobanksin	2613	2611	4.25	4.73
Pinobanksin 3-acetate, mono-TMS	2634	2632	1.26	0.21
Chrysin, mono-TMS	2655	2648	1.95	0.42
5,7-Dihydroxy-3-methoxyflavanone	2675	2673	2.02	2.04
Benzyl (E)-ferulate	2680	2680	1.64	0.45
Pinobanksin 3-acetate, di-TMS	2694	2693	10.01	9.00
Benzyl (E)-caffeate	2723	2722	3.79	2.70
Chrysin, di-TMS	2746	2745	5.33	5.73
Galangin, tri-TMS	2767	2769	8.95	9.60
CAPE	2805	2805	1.29	1.15
Cinnamyl (E)-p-coumarate	2836	2833	1.91	0.23
Sakuranetin	2877	2880	0.55	0.05
Quercetine	3218	3213	0.11	-

Table 3. Group composition of ethanolic extracts from Poland (PPE) and New Zealand (MPE) propolis.

Group of Compounds	PPE [%]	MPE [%]
Flavonoids and chalcones	49.4	52.1
Aromatic acids	18.3	7.8
Cinnamic acid esters	19.8	14.5
Phenylpropenoid glicerydes	1.3	0.0
Aliphatic and aromatic alcohol	0.2	0.8
Aliphatic acids	0.8	0.2
Carbohydrates	6.2	18.7
Sesquiterpenoids	0.0	0.2
Other compounds	4.0	5.7
Total	100.0	100.0

Table 4. Total phenolic content and toxic elements concentration of ethanolic extract from Poland (PPE) and New Zealand (MPE) propolis.

Extracts	TPC [mg GAE/g] Mean ± SD	Toxic Elements [mg/kg]		
		As	Cd	Pb
PPE	243.7 ± 9.0	0.00	0.01	0.16
MPE	245.6 ± 5.9	0.88	0.01	3.74

3.2. *Toxic Elements Content*

In this study, we determined the arsenic, cadmium, and lead content in Polish and Manuka propolis by ICP-MS method. The results are presented in Table 4.

3.3. *Cell Viability*

In this study, the impact on the viability was determined using different types of brain cancer cells—astrocytoma cell line derived from a patient (DASC) and two glioblastoma T98G and LN-18 cell lines from the ATCC. Dose- and time-dependent decreases in the viability (by MTT) of DASC were observed after 24, 48 and 72 h of incubation with both PPE and MPE (compared to the control) (Figure 1), and were comparable for both kinds of propolis. For the DASC cell line, we observed a significant reduction in cell numbers ($p < 0.05$) in all concentrations after 24, 48, and 72 h; for the dose of 30 µg/mL, it was 77.9 ± 4.3% and 81.3 ± 4.0% after 24 h, 58.6 ± 0.3% and 63.4 ± 7.8% after 48 h, and 47.0 ± 3.2% and 51.6 ± 8.1% after 72 h for PPE and MPE, respectively (Figure 1A–C). A significant (although lower than 10%) difference ($p < 0.05$) in the reduction in DASC cells treated with PPE in comparison to those treated with MPE was observed for the 100 µg/mL concentration after 48 h (approximately 7%) (Figure 1B) and for 20, 50, and 100 µg/mL concentrations after 72 h (8.4%, 6.9%, and 3.0%, respectively) (Figure 1C). For the T98G cell line, we observed a stronger, more significant reduction in cell numbers ($p < 0.05$) in all concentrations after 24, 48 and 72 h than for the DASC cell line; for the dose of 30 µg/mL, it was 78.4 ± 3.0% and 75.2 ± 2.3% after 24 h, 62.8 ± 1.3% and 50.8 ± 7.2% after 48 h, and 30.7 ± 7.7% and 22.0 ± 8.3% after 72 h for PPE and MPE, respectively (Figure 1D–F). Interestingly, dose-dependent decreases in the viability of T98G cells were observed after 24, 48 and 72 h, but only for the 10–50 µg/mL dose range. After the treatment of the 100 µg/mL dose, the decrease in viability was smaller than for the 50 µg/mL dose. This effect can be connected with the impact of the phytochemicals from propolis on activity of succinate dehydrogenase; however, some studies suggest that natural antioxidants may have a direct reductive potential and can interfere with MTT [20–22]. Therefore, for further study, a lower dose (30 ug/mL) of PPE and MPE has been used. A significant difference ($p < 0.05$) in the reduction in T98G cells treated with PPE in comparison to those treated with MPE was observed for the 50 µg/mL concentration after 24 h (Figure 1D), for 10, 20, 30, and 50 µg/mL concentrations after 48 h (Figure 1E), as well as and 20, 50, and 100 µg/mL concentrations after 72 h (Figure 1F). A significant ($p < 0.05$) reduction in cell number was observed for LN-18 in all concentrations of PPE and MPE after 24, 48 and 72 h. For the dose of 30 µg/mL, it was 81.6 ± 3.3% and 83.2 ±0.9% after 24 h, 49.1 ± 7.8% and 65.7 ± 8.0% after 48 h, 40.8 ± 2.5% and 41.1 ± 2.9% after 72 h for PPE and MPE, respectively. A significant difference ($p < 0.05$) in the reduction in LN-18 cells treated with PPE, as compared with those treated with MPE was observed for the 10, 30, 50, and 100 µg/mL concentrations after 48 h (Figure 1H), as well as for 50, and 100 µg/mL concentrations after 48 h (Figure 1I). Interestingly, PPE decreases the viability of LN-18 cells significantly more strongly than MPE.

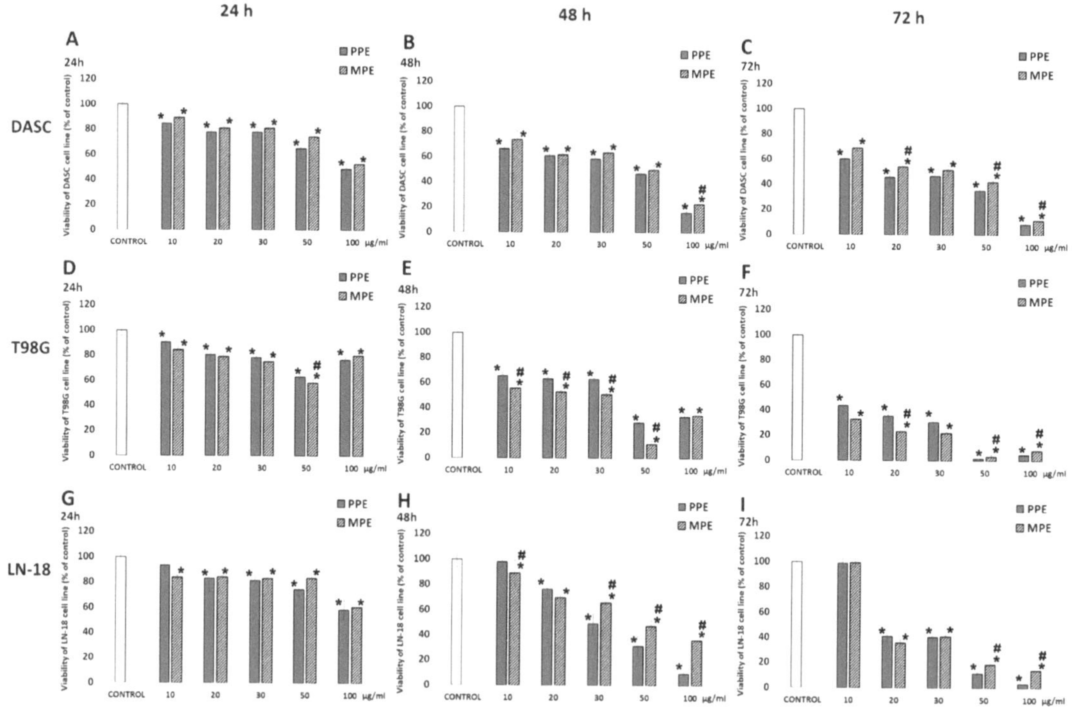

Figure 1. The viability of DASC (**A–C**), T98G (**D–F**) and LN-18 (**G–I**) cells after treatment with PPE and MPE (in concentrations 10, 20, 30, 50, 100 µg/mL) after 24, 48 and 72 h of incubation. The results are presented as a percentage of control. All statistical analyses were performed using Student's *t*-tests or Mann–Whitney U tests (significant changes: * $p < 0.05$ vs. control, # PPE vs. MPE).

3.4. DNA Biosynthesis

The impact of PPE and MPE on DNA biosynthesis in the [^{3}H]-thymidine incorporation assay was examined in order to confirm if the inhibition of cell viability was caused by a reduction in proliferation capacity. As regards the DASC cell line, we found that both PPE and MPE significantly inhibited proliferation—by approximately 10.2% and 13.2% after 48 h and by approximately 23.1% and 18.6% after 72 h, respectively (Figure 2A–C). For the T98G cell line, we observed a significant reduction in proliferation capacity ($p < 0.05$) only in the case of MPE: 18.4% after 24 h, 18.6% after 48 h and 39.6% after 72 h (Figure 2D–F). For the LN-18 cell line, we found a significant reduction in proliferation capacity ($p < 0.05$) with both PPE and MPE after 24, 48, and 72 h: approximately 40.6% and 44.5% after 24 h, 39.4% and 43.3% after 48 h, and 67.6% and 75.6% after 72 h, respectively (Figure 2G–I).

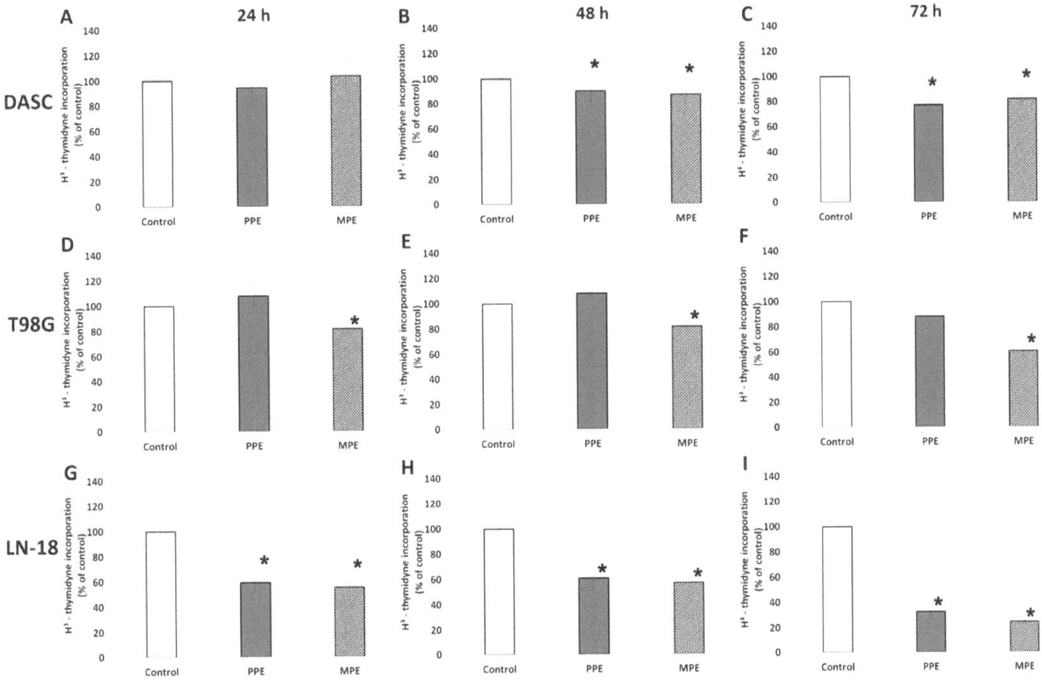

Figure 2. The [^{3}H]-thymidine incorporation into DASC, T98G and LN-18 cells after treatment with PPE and MPE. Legend: [^{3}H]-thymidine incorporation into DASC (**A–C**) and T98G (**D–F**) and LN-18 (**G–I**) cells after 24, 48, 72 h of incubation with PPE and MPE (in concentrations 30 µg/mL). The results are presented as a percentage of control. All statistical analyses were performed using Student's *t*-test (significant changes: * $p < 0.05$ vs. control).

3.5. Cells Migration

PPE and MPE impact on DASC, T98G and LN-18 cells migration was assessed using an in vitro scratch wound assay. Images of scratch areas from the time points 0, 12 and 42 and the percentage of the open wound area are illustrated in Figure 3. Our data show that MPE inhibited cell migration more strongly than PPE in the DASC (to 33.6% after 42 h) and LN-18 (to 26.0% after 42 h) cell lines. Regarding the T98G cell lines, both MPE and PPE inhibited cell migration to a similar extent (to 27.0%).

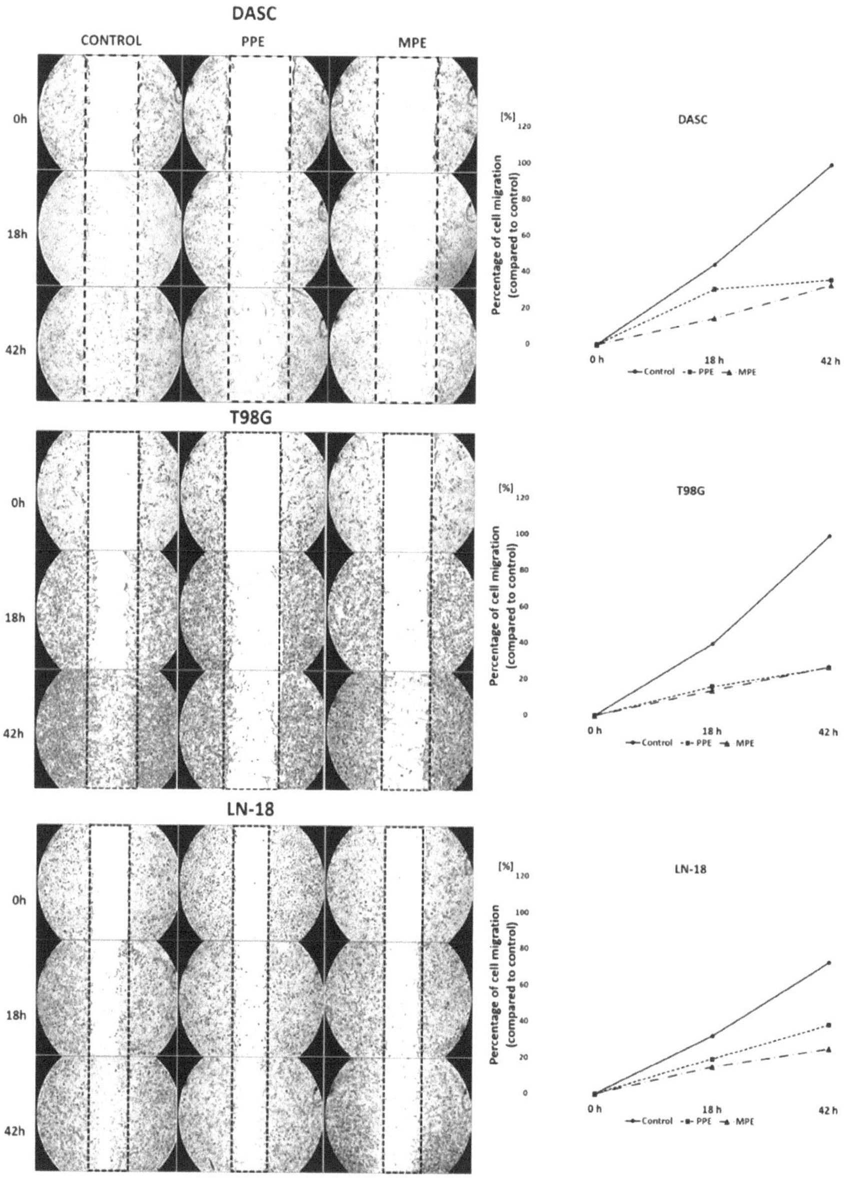

Figure 3. The effects of EEP (6.25 µg/mL) and MPE (30 µg/mL) on DASC, T98G and LN-18 on cells migration after 0, 18, 42 h of incubation.

3.6. Cell Cycle

The effects of PPE (30 µg/mL) and MPE (30 µg/mL) on the cycle of DASC, LN-18, and T98G cells after 48 h are illustrated in Figure 4. Our data demonstrate that PPE induced cell cycle arrest in the subG1/G1 phase in T98G (increased to 24.0% ± 2.6) and LN-18 (increased to 11.7% ± 0.6) cells compared to control ($p < 0.05$), but not in DASC cells. MPE induced cell cycle arrest in the subG1/G1 phase only in the T98G cell line (increased to 15.8% ± 0.5). The changes in DASC cell cycles were not observed.

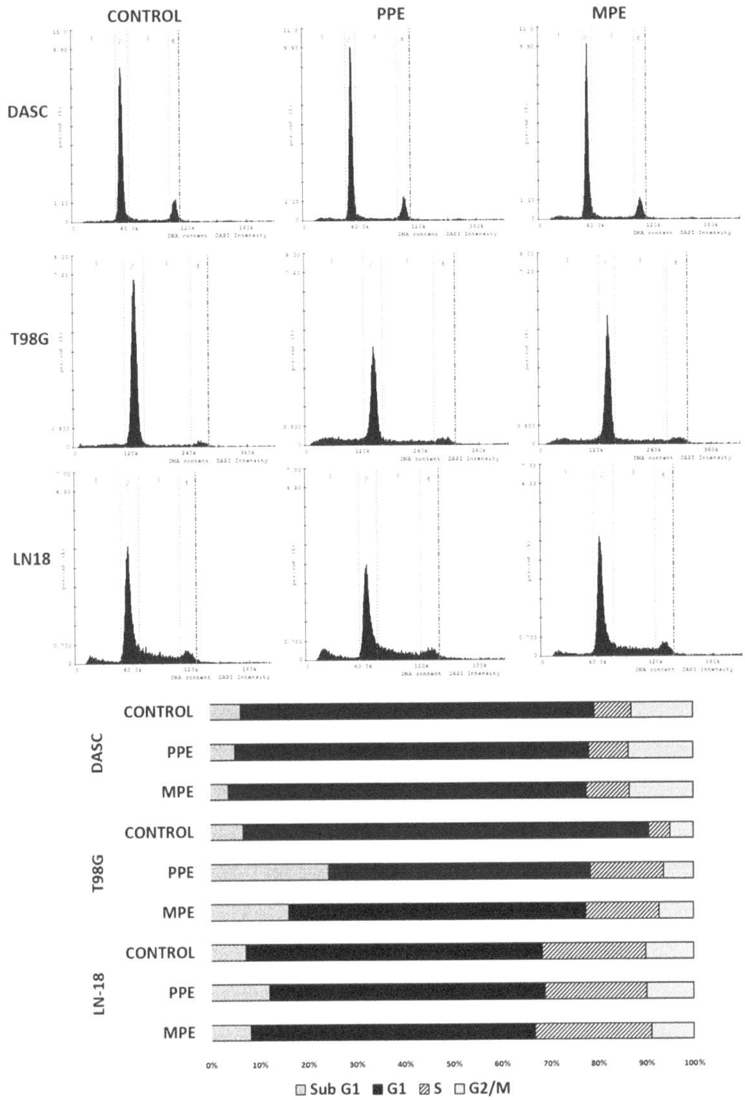

Figure 4. The effect of PPE and MPE on cell cycle analysis. DASC, T98G and LN-18 cells were incubated for 48 h with PPE (30 μg/mL) and MPE (30 μg/mL). Both the histogram and the bars present distributions of cells in subG1, G1, S and G2/M phases of the cell cycle.

3.7. Cell Apoptosis

In our study, we examined the impact of PPE and MPE treatment on glioma cell apoptosis (DASC, LN-18, T98G) by annexin V and PI staining. The results (Figure 5) showed that PPE caused increased early apoptosis (lower right quadrat) in DASC, T98G, and LN-18 cells (by approximately (75%, 38% and 77%, respectively, compared to control) and late apoptosis/necrosis (upper quadrat) in DASC and T98G cells (by 25% and 43%, respectively, compared to control). Treatment with MPE led to early apoptosis in DASC and LN-18 cells (by 60% and 74%, respectively, compared to control) and late apoptosis/necrosis in DASC and T98G cells (by 38% and 58%, respectively, compared to control).

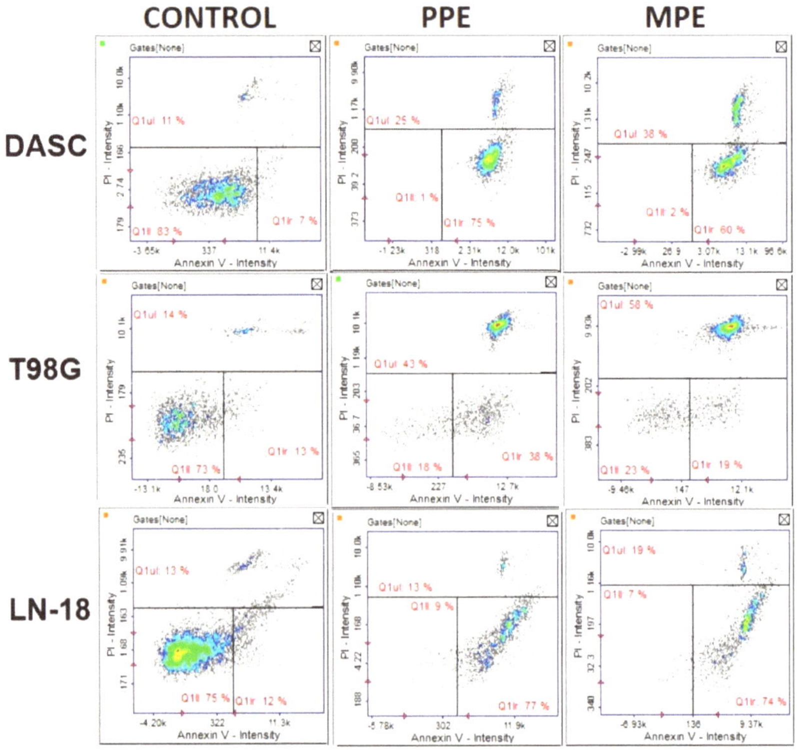

Figure 5. The quantitative assessment of DASC, T98G and LN–18 cells apoptosis induced by PPE and MPE (30 µg/mL) using Annexin V/PI staining.

4. Discussion

Propolis owes its complex chemical composition to the quality of the resinous materials gathered by honey bees from different floral sources available around the hive. The quality of the resins has an impact on the quality and bioactivity of propolis. The chemical composition of the tested propolis was characterized by a similar amount of the identified active components and the total content of phenols, which is consistent with the classification of propolis from New Zealand as the "Poplar" type. Kumazawa et al. [23] conducted a comparison of the antioxidant activity and composition, as well as total phenol and flavonoid content, of individual samples of ethanolic extracted propolis from 14 countries and showed that New Zealand-sourced propolis was similar in composition to propolis from Bulgaria, Uzbekistan, Hungary, and three South American countries: Chile, Uruguay and Argentina. In our analysis we found high content of compounds such as pinobanksin, pinobanksin 3-acetate pinocembrin, chrysin or galangin. These compounds are characteristic of propolis originating from bud exudates of *Populus nigra* [6,24]. The analysis also confirmed research results published by other authors who have demonstrated that New Zealand propolis has very high levels of pinocembrin and pinobanksin-3-O-acetate [4].

The TPC value in our study was on high level > 240 mg GAE/g in both propolis. Other authors detected varying amounts of TPC in propolis. The values ranged from 14.6 to 150.8 mg GAE/g in Polish propolis [25] and from 99 ± 4.0 to 775 ± 8.5 mg GAE/g in Manuka propolis [26]. The TPC value often depends on the extraction method utilized.

Diffusion of heavy metals in the environment, occurring as a result of various human activities, results in penetration of these elements into food and direct human exposure to their toxic effects. Toxic elements such as, Cd, and Pb, even in trace amounts, present a risk to human health, causing non-communicable diseases with long-term effects. In our study, the level of As and Pb was higher in MPE than in PPE (Table 4). Comparing the obtained results with the Commission Regulation (EC) No 629/2008 [27] standards for supplements (Pb, 3.0 mg/kg; Cd, 1.0 mg/kg), we found that the level of Pb (3.74 mg/kg) in MPE was exceeded, but the level of elements was assessed in the lyophilizates. The obtained Pb content in the lyophilized extract was recalculated to Pb content in the liquid extract (0.935 mg/kg) and did not exceed the standards. Polish propolis was also analyzed by Matuszewska et al. [28]. The concentrations of As, Cd and Pb (As, 0.07 mg/kg; Cd, 0.04 mg/kg; Pb, 0.64 mg/kg) were higher than in PPE but not MPE. High concentration of Pb (5.74 mg/kg), Cd (0.194 mg/kg), and As (0.657 mg/kg) in Polish propolis was also indicate Roman et al. [29]. It should be noted that our PPE was obtained from green areas, free from pollution (Podlasie region), while propolis analyzed by Roman et al. [29] from the urban regions. Scientific research confirms that high amounts of toxic elements in propolis may result from the level of urbanization of a region. Therefore, the content of these elements in propolis should be constantly monitored. Studies by Ahamed et al. [30] confirm that Pb can influence of viability, cell cycle, lipid peroxidation, and caspase activation in human lung epithelial (A549) cells. However, it should be noted that propolis is a product that contains a number of compounds with antioxidant potential. In a study by Mu et al. [31], the cell viability assay results indicated that three phenolic acids—chicoric acid, isochlorogenic acid C, and caffeic acid—alleviated the cytotoxicity induced by Pb2+.

Due to the presence of a large number of antioxidant substances, propolis exhibits powerful anticancer activity, which has been confirmed in many studies [12–14,32]. Our previous study has revealed that Polish propolis decreases viability and has an antiproliferative activity and additionally, synergistically cooperates with temozolomide (TMZ), enhancing its growth-inhibiting activity against U87MG glioblastoma cell line through the reduction in NF-κB activity [11]. Catchpole et al. [4] have demonstrated that propolis from New Zealand has a strong antiproliferative effect against gastro-intestinal cancer cells DLD-1, HCT-116, KYSE-30, and NCI-N87, due to the high level of phenolic compounds (pinocembrin, pinobanksin-3-O-acetate and others). Propolis from Brazil has been demonstrated to exert a strong inhibitory effect on cell growth in glioblastoma (U251 and U343) and fibroblast cell lines (MRC5), although not on apoptosis, demonstrating a cytostatic action [33]. In this study, comparing the effect of both propolis (PPE and MPE) extracts on different glioma cell lines, we found strong, decreasing viability and antiproliferative effects on DASC, T98G and LN-18 cells (Figures 1 and 2). Moreover, in scratch assay, it has been showed that PPE and MPE inhibit cell migration (Figure 3). Another study also confirmed the anti-migratory potential of propolis in cancer cells. Chang et al. [34] showed that treatment with different concentrations of Chinese propolis (25, 50 and, 100 μg/mL) and CAPE (25 μg/mL) significantly inhibited the proliferation and migration of the LPS-stimulated MDA-MB-231 breast cell line. Begnini et al. [35] reported that Brazilian Red Propolis (25 and 50 μg/mL) strongly inhibited migration in human bladder cancer 5637 cells.

Propolis shows anti-cancer activity through a various mechanisms. In our study, we examined the influence of MPE and PPE on cell cycle and apoptosis in the DASC, LN-18, and T98G cell lines. Both propolis extracts induced cell cycle arrest in T98G in the subG1/G1 phase, but PPE only in LN-18 cell line. Lack of cell cycle changes in DASC can be associated with a low proliferation capacity of that cells (Figure 4). What is more, PPE and MPE may induce cell necrosis, especially in T98G and DASC cells (Figure 5). Frión-Herrera et al. [36] showed that Cuban red propolis induced mitochondrial dysfunction and LDH release in breast cancer cell line (MDA MB-231), which indicated cell necrosis associated with reactive oxygen species production and decreased cell migration. The accumulation of cell population in the Sub-G1 phase may suggest that propolis did induce apoptosis. Interestingly, we also observed that PPE and MPE treatment induced cell cycle arrest in the S phase in T98G cells ($p < 0.05$). Other authors have also observed this effect. Jiang et al. [37] reported that Special Chinese propolis sourced from the Changbai Mountains showed anti-proliferation activity in SGC-7901 human gastric cancer cells by inducing both death receptor-induced apoptosis and mitochondria-mediated apoptosis, as well as cell cycle arrest in the S-phase. In this study, it has been observed that both PPE and MPE induced apoptosis in each glioma cell line (DASC, LN-18, T98G) (Figure 5). Zeynep et al. [38] also confirmed the apoptotic activity of propolis in C6 glioma cells. They showed that an ethanolic extract of propolis induced apoptosis in C6 glioma cells by activating the caspase cascade pathway, increasing caspase-8, -9, and -3 expression levels. The study by Noureddine et al. [39] showed that Lebanese propolis induced an increase in SubG0 fraction in Jurkat, glioblastoma (U251) and breast cancer (MDA-MB-231) cells. This increase in SubG0 was further investigated in Jurkat cells by annexinV/PI and showed an increase in the percentage of cells in early and late apoptosis as well as necrosis.

Many publications have explored significant anti-cancer properties of individual components of propolis. Szliszka and Krol [40] suggested that polyphenols from propolis sensitized tumor cells to TRAIL-induced apoptosis. The compounds, in combination with TRAIL, exhibit a strong cytotoxic effect on cancer cells [41,42]. Caffeic acid phenethyl ester (CAPE) inhibits NF-kB and enhanced the extrinsic pathway of apoptosis in cancer cells induced by TRAIL and Fas receptor stimulation [43]. The most recent research has demonstrated that CAPE displays significant cytotoxicity towards two glioma cell lines: Hs683 and LN319 [44]. Other authors have also confirmed that CAPE exhibits powerful antitumor effects on the following cancer cells: fibroblasts from oral submucous fibrosis (OSF), neck metastasis of Gingiva carcinoma (GNM) and tongue squamous cell carcinoma (TSCCa) [45]. Chrysin shows antiproliferative activity against human colorectal cancer cell line HCT-116, liver cancer cell line HepG2 and nasopharyngeal line CNE-1 to TNF-α-induced apoptosis [46]. Chrysin induces apoptosis in cancer cells by the activation of caspases and suppression of anti-apoptotic proteins such as IAP, c-FLIP, PI3K/Akt signal pathway, inhibition of IKK and NF-kB activity [47].

In this study, the potential activity of propolis extract against glioma cells was demonstrated, and the quality and safety of propolis were considered. Different glioma lines and astrocytoma cell line were used to compare whether the direction of action of propolis may be similar in different types of glioma. These are preliminary studies conducted in vitro.

Key factors in assessing a propolis extract, as well as other natural products in glioma treatment, are its bioavailability, metabolism, active compounds, and blood–brain barrier (BBB) permeability [48]. Bioavailability of some propolis compounds, such as flavonoids (chrysin and galangin), is low, and they are rapidly metabolized, which may limit the therapeutic potential of propolis extracts [49,50]. Despite that, Curti et al. [50] showed that oral uptake of brown propolis is followed by rapid metabolism and by cellular adaptation through the modulation of the concentration of first line antioxidant enzymes (SOD-1). Moreover, it cannot be excluded that the activity of propolis depends on synergisms between polyphenols and other active compounds [50]. An effective antiglioma agent must cross the BBB. BBB is permeable to some phenolic acids, such as caffeic acid (presented in studied extracts) [51] and flavonoids such as naringin, quercetin, genistein, epigallocatechin or its metabolites, but is neglected by some others, such as resveratrol and curcumin [52,53]. Future investigations including in vivo studies with cyclic administration of propolis, examination bioavailability and BBB permeability of propolis compounds are necessary for the evaluation of the therapeutic potential of propolis extracts.

5. Conclusions

In summary, the above results show that propolis from Poland and propolis from New Zealand (Manuka) have antiproliferative and pro-apoptotic activity on the human diffuse astrocytoma cell line (DASC) (Grade II glioma) derived from a patient and glioblastoma multiforme T98G and LN-18 cell lines from the ATCC. The anticancer potential was confirmed through induction of apoptosis, cell cycle arrest on subG1/G1 and S phase and attenuate migration. The PPE and MPE activity may be associated with the high content of antioxidant compounds in both types of propolis. The chemical composition of both propolis was comparable, with marginal differences in the amount of some compounds. The content of As and Pb in MPE was higher than in PPE.

In conclusion, Polish and New Zealand propolis extracts showed anti-glioma activity in in vitro study. However, further in vivo studies are required to confirm the therapeutic potential of propolis.

Supplementary Materials: The following supporting information can be downloaded at: https://www.mdpi.com/article/10.3390/antiox11071305/s1.

Author Contributions: J.M. and S.K.N. were responsible for conception, study design, obtaining funds, laboratory analysis, statistical analysis and writing the manuscript. V.I. was responsible for the GC-MS laboratory analysis and writing the manuscript. J.S. and K.S. performed ICP-MS analysis. R.M.-Ż., K.J.G.-K. and A.P.-J. were responsible for performing laboratory cell analysis. M.H.B. was responsible for management of the study and was responsible for revising the manuscript critically for important intellectual content. The final manuscript was revised by all co-authors. All authors have read and agreed to the published version of the manuscript.

Funding: This research was funded by Medical University of Bialystok in project numbered SUB/2/DN/21/005/2216. The scientific team was responsible for all stages of the study (design, methodology, analysis, interpretation of the data and writing the manuscript).

Institutional Review Board Statement: Not applicable.

Informed Consent Statement: Not applicable.

Data Availability Statement: Data is contained within the article and Supplementary Materials.

Conflicts of Interest: The authors declare no conflict of interest.

References

1. Ostrom, Q.T.; Cioffi, G.; Gittleman, H.; Patil, N.; Waite, K.; Kruchko, C.; Barnholtz-Sloan, J.S. CBTRUS Statistical Report: Primary brain and other central nervous system tumors diagnosed in the United States in 2012–2016. *Neuro-Oncology* **2019**, *21*, v1–v100. [CrossRef]
2. Thakkar, J.P.; Dolecek, T.A.; Horbinski, C. Epidemiologic and molecular prognostic review of glioblastoma. *Cancer Epidemiol. Biomark. Prev.* **2014**, *23*, 1985–1996. [CrossRef]
3. Barker, C.A.; Chang, M.; Beal, K.; Chan, T.A. Survival of patients treated with radiation therapy for anaplastic astrocytoma. *Radiat. Oncol.* **2014**, *48*, 381–386. [CrossRef]
4. Catchpole, O.; Mitchell, K.; Bloor, S.; Davis, P.; Suddes, A. Antiproliferative activity of New Zealand propolis and phenolic compounds vs human colorectal adenocarcinoma cells. *Fitoterapia* **2015**, *106*, 167–174. [CrossRef]
5. Markiewicz-Żukowska, R.; Car, H.; Naliwajko, S.K.; Sawicka, D.; Szynaka, B.; Chyczewski, L.; Isidorov, V.; Borawska, M.H. Ethanolic extract of propolis, chrysin, CAPE inhibit human astroglia cells. *Adv. Med. Sci.* **2012**, *57*, 208–216. [CrossRef]
6. Isidorov, V.A.; Szczepaniak, L.; Bakier, S. Rapid GC/MS determination of botanical precursors of Eurasian propolis. *Food Chem.* **2014**, *142*, 101–106. [CrossRef]
7. Demir, S.; Aliyazicioglu, Y.; Turan, I.; Misir, S.; Mentese, A.; Yaman, S.O.; Akbulut, K.; Kilinc, K.; Deger, O. Antiproliferative and proapoptotic activity of Turkish propolis on human lung cancer cell line. *Nutr. Cancer* **2016**, *68*, 165–172. [CrossRef]
8. Aru, B.; Güzelmeric, E.; Akgül, A.; Demirel, G.Y.; Kırmızıbekmez, H. Antiproliferative activity of chemically characterized propolis from Turkey and its mechanisms of action. *Chem. Biodivers.* **2019**, *16*, e1900189. [CrossRef]
9. Salim, E.I.; Abd El-Magid, A.D.; Farara, K.M.; Maria, D.S. Antitumoral and antioxidant potential of Egyptian propolis against the PC3 prostate cancer cell line. *Asian Pac. J. Cancer Prev.* **2015**, *16*, 7641–7651. [CrossRef]
10. Borawska, M.H.; Naliwajko, S.K.; Moskwa, J.; Markiewicz-Żukowska, R.; Puścion-Jakubik, A.; Soroczyńska, J. Anti-proliferative and anti-migration effects of Polish propolis combined with *Hypericum perforatum* L. on glioblastoma multiforme cell line U87MG. *BMC Complement. Altern. Med.* **2016**, *16*, 367. [CrossRef] [PubMed]
11. Markiewicz-Żukowska, R.; Borawska, M.H.; Fiedorowicz, A.; Naliwajko, S.K.; Sawicka, D.; Car, H. Propolis changes the anticancer activity of temozolomide in U87MG human glioblastoma cell line. *BMC Complement. Altern. Med.* **2013**, *13*, 50. [CrossRef] [PubMed]
12. Sforcin, J.M. Biological properties and therapeutic applications of propolis. *Phytother. Res.* **2016**, *905*, 894–905. [CrossRef] [PubMed]
13. Patel, S. Emerging adjuvant therapy for cancer: Propolis and its constituents. *J. Diet Suppl.* **2015**, *13*, 245–268. [CrossRef]
14. Harris, I.S.; DeNicola, G.M. The complex interplay between antioxidants and ROS in cancer. *Trends Cell Biol.* **2020**, *30*, 440–451. [CrossRef]

15. Schafer, Z.T.; Grassian, A.R.; Song, L.; Jiang, Z.; Gerhart-Hines, Z.; Irie, H.Y.; Gao, S.; Puigserver, P.; Brugge, J.S. Antioxidant and oncogene rescue of metabolic defects caused by loss of matrix attachment. *Nature* **2009**, *461*, 109–113. [CrossRef] [PubMed]
16. Isidorov, V.A. *Identification of Biologically and Environmentally Significant Organic Compounds: Mass Spectra and Retention Indices Library of Trimethylsilyl Derivatives*, 1st ed.; PWN: Warszawa, Poland, 2015.
17. Bielecka, J.; Markiewicz-Żukowska, R.; Nowakowski, P.; Grabia, M.; Puścion-Jakubik, A.; Mielcarek, K.; Gromkowska-Kępka, K.J.; Soroczyńska, J.; Socha, K. Content of toxic elements in 12 groups of rice products available on polish market: Human health risk assessment. *Foods* **2020**, *9*, 1906. [CrossRef] [PubMed]
18. Borawska, M.H.; Markiewicz-Żukowska, R.; Naliwajko, S.K.; Moskwa, J.; Bartosiuk, E.; Socha, K.; Surażyński, A.; Kochanowicz, J.; Mariak, Z. The interaction of bee products with temozolomide in human diffuse astrocytoma, glioblastoma multiforme and astroglia cell lines. *Nutr. Cancer* **2014**, *66*, 1247–1256. [CrossRef] [PubMed]
19. Moskwa, J.; Naliwajko, S.K.; Markiewicz-Żukowska, R.; Gromkowska-Kępka, K.J.; Nowakowski, P.; Strawa, J.W.; Borawska, M.H.; Tomczyk, M.; Socha, K. Chemical composition of Polish propolis and its antiproliferative effect in combination with *Bacopa monnieri* on glioblastoma cell lines. *Sci. Rep.* **2020**, *10*, 21127. [CrossRef]
20. Bruggisser, R.; von Daeniken, K.; Jundt, G.; Schaffner, W.; Tullberg-Reinert, H. Interference of plant extracts, phytoestrogens and antioxidants with the MTT tetrazolium assay. *Planta Med.* **2002**, *68*, 445–448. [CrossRef]
21. Wang, P.; Henning, S.M.; Heber, D. Limitations of MTT and MTS-based assays for measurement of antiproliferative activity of green tea polyphenols. *PLoS ONE* **2010**, *5*, e10202. [CrossRef]
22. Villota, H.; Moreno-Ceballos, M.; Santa-González, G.A.; Uribe, D.; Castañeda, I.C.H.; Preciado, L.M.; Pedroza-Díaz, J. Biological impact of phenolic compounds from coffee on colorectal cancer. *Pharmaceuticals* **2021**, *14*, 761. [CrossRef] [PubMed]
23. Kumazawa, S.; Hamasaka, T.; Nakayama, T. Antioxidant activity of propolis of various geographic origins. *Food Chem.* **2004**, *84*, 329–339. [CrossRef]
24. Popova, M.; Giannopoulou, E.; Skalicka-Woźniak, K.; Graikou, K.; Widelski, J.; Bankova, V.; Kalofonos, H.; Sivolapenko, G.; Gaweł-Bęben, K.; Antosiewicz, B.; et al. Characterization and biological evaluation of propolis from Poland. *Molecules* **2017**, *22*, 1159. [CrossRef] [PubMed]
25. Wieczynska, A.; Wezgowiec, J.; Wieckiewicz, W.; Czarny, A.; Kulbacka, J.; Nowakowska, D.; Gancarz, R.; Wilk, K.A. Antimicrobial activity, cytotoxicity and total phenolic content of different extracts of propolis from the west pomeranian region in Poland. *Acta Pol. Pharm.* **2017**, *74*, 715–722.
26. Gemiarto, A.T.; Ninyio, N.N.; Lee, S.W.; Logis, J.; Fatima, A.; Chan, E.W.; Lim, C.S. Isoprenyl caffeate, a major compound in manuka propolis, is a quorum-sensing inhibitor in *Chromobacterium violaceum*. *Antonie Leeuwenhoek* **2015**, *108*, 491–504. [CrossRef]
27. Commission Regulation (EC) No 629/2008 of 2 July 2008 Amending Regulation (EC) No 1881/2006 Setting Maximum Levels for Certain Contaminants in Foodstuffs [L 173/6]. Available online: https://eur-lex.europa.eu/legal-content/EN/TXT/PDF/?uri=CELEX:32008R0629&from=ES (accessed on 14 April 2022).
28. Matuszewska, E.; Klupczynska, A.; Maciołek, K.; Kokot, Z.J.; Matysiak, J. Multielemental Analysis of bee pollen, propolis, and royal jelly collected in west-central Poland. *Molecules* **2021**, *26*, 2415. [CrossRef]
29. Roman, A.; Madras-Majewska, B.; Popiela-Pleban, E. Comparative study of selected toxic elements in propolis and honey. *J. Apic. Sci.* **2011**, *55*, 97–106.
30. Ahamed, M.; Akhtar, M.J.; Majeed Khan, M.A. Single-walled carbon nanotubes attenuate cytotoxic and oxidative stress response of pb in human lung epithelial (A549) Cells. *Int. J. Environ. Res. Public Health* **2020**, *17*, 8221. [CrossRef]
31. Mu, Y.; Yu, J.; Ji, W.; Chen, L.; Wang, X.; Yan, B. Alleviation of Pb^{2+} pollution-induced oxidative stress and toxicity in microglial cells and zebrafish larvae by chicoric acid. *Ecotoxicol. Environ. Saf.* **2019**, *180*, 396–402. [CrossRef]
32. Watanabe, M.A.E.; Amarante, M.K.; Conti, B.J.; Sforcin, J.M. Cytotoxic constituents of propolis inducing anticancer effects: A review. *J. Pharm. Pharmacol.* **2011**, *63*, 1378–1386. [CrossRef]
33. Borges, K.S.; Brassesco, M.S.; Scrideli, C.A.; Soares, A.E.; Tone, L.G. Antiproliferative effects of Tubi-bee propolis in glioblastoma cell lines. *Genet. Mol. Biol.* **2011**, *34*, 310–314. [CrossRef] [PubMed]
34. Chang, H.; Wang, Y.; Yin, X.; Liu, X.; Xuan, H. Ethanol extract of propolis and its constituent caffeic acid phenethyl ester inhibit breast cancer cells proliferation in inflammatory microenvironment by inhibiting TLR4 signal pathway and inducing apoptosis and autophagy. *BMC Complement. Altern. Med.* **2017**, *17*, 471. [CrossRef] [PubMed]
35. Begnini, K.R.; Moura de Leon, P.M.; Thurow, H.; Schultze, E.; Campos, V.F.; Martins Rodrigues, F.; Borsuk, S.; Dellagostin, O.A.; Savegnago, L.; Roesch-Ely, M.; et al. Brazilian red propolis induces apoptosis-like cell death and decreases migration potential in bladder cancer cells. *Evid. Based Complement. Alternat. Med.* **2014**, *2014*, 639856. [CrossRef]
36. Frión-Herrera, Y.; Díaz-García, A.; Ruiz-Fuentes, J.; Rodríguez-Sánchez, H.; Sforcin, J.M. The cytotoxic effects of propolis on breast cancer cells involve PI3K/Akt and ERK1/2 pathways, mitochondrial membrane potential, and reactive oxygen species generation. *Inflammopharmacology* **2019**, *27*, 1081–1089. [CrossRef] [PubMed]
37. Jiang, X.S.; Xie, H.Q.; Li, C.G.; You, M.M.; Zheng, Y.F.; Li, G.Q.; Chen, X.; Zhang, C.P.; Hu, F.L. Chinese propolis inhibits the proliferation of human gastric cancer cells by inducing apoptosis and cell cycle arrest. *Evid. Based Complement. Alternat. Med.* **2020**, *2020*, 2743058. [CrossRef]
38. Zeynep, M.C.; Melike, E.; Melike, G.; Aytek, O.; Aynur, A. Cytotoxic and apoptotic effects of ethanolic propolis extract on C6 glioma cells. *Environ. Toxicol.* **2020**, *35*, 768–773.

39. Noureddine, H.; Hage-Sleiman, R.; Wehbi, B.; Fayyad-Kazan, H.; Hayar, S.; Traboulssi, M.; Alyamani, O.A.; Faour, W.H.; El Makhour, Y. Chemical characterization and cytotoxic activity evaluation of Lebanese propolis. *Biomed. Pharmacother.* **2017**, *95*, 298–307. [CrossRef]
40. Szliszka, E.; Krol, W. Polyphenols isolated from propolis augment TRAIL-induced apoptosis in cancer cells. *Evid. Based Complement. Alternat. Med.* **2013**, *2013*, 731940. [CrossRef]
41. Bernardi, S.; Secchiero, P.; Zauli, G. State of art and recent developments of anti-cancer strategies based on TRAIL. *Recent Pat. Anticancer Drug Discov.* **2012**, *7*, 207–217. [CrossRef]
42. Li, X.; Huang, Q.; Ong, C.N.; Yang, X.F.; Shen, H.M. Chrysin sensitizes tumor necrosis factor-α-induced apoptosis in human tumor cells via suppression of nuclear factor-κB. *Cancer Lett.* **2010**, *293*, 109–116. [CrossRef]
43. Szliszka, E.; Czuba, Z.P.; Domino, M.; Mazur, B.; Zydowicz, G.; Krol, W. Ethanolic extract of propolis (EEP) enhances the apoptosis-inducing potential of TRAIL in cancer cells. *Molecules* **2009**, *14*, 738–754. [CrossRef] [PubMed]
44. Morin, P.; St-Coeur, P.D.; Doiron, J.A.; Cormier, M.; Poitras, J.J.; Surette, M.E.; Touaibia, M. Substituted caffeic and ferulic acid phenethyl esters: Synthesis, leukotrienes biosynthesis inhibition, and cytotoxic activity. *Molecules* **2017**, *22*, 1124. [CrossRef] [PubMed]
45. Lee, Y.J.; Liao, P.H.; Chen, W.K.; Yang, C.C. Preferential cytotoxicity of caffeic acid phenethyl ester analogues on oral cancer cells. *Cancer Lett.* **2000**, *153*, 51–56. [CrossRef]
46. Li, X.; Wang, J.N.; Huang, J.M.; Xiong, X.K.; Chen, M.F.; Ong, C.N.; Shen, H.M.; Yang, X.F. Chrysin promotes tumor necrosis factor (TNF)-related apoptosis-inducing ligand (TRAIL) induced apoptosis in human cancer cell lines. *Toxicol. In Vitro* **2011**, *25*, 630–635. [CrossRef]
47. Sawicka, D.; Car, H.; Borawska, M.H.; Nikliński, J. The anticancer activity of propolis. *Folia Histochem. Cytobiol.* **2012**, *50*, 25–37. [CrossRef]
48. Zhai, K.; Siddiqui, M.; Abdellatif, B.; Liskova, A.; Kubatka, P.; Büsselberg, D. Natural compounds in glioblastoma therapy: Preclinical insights, mechanistic pathways, and outlook. *Cancers* **2021**, *13*, 2317. [CrossRef]
49. Mani, R.; Natesan, V. Chrysin: Sources, beneficial pharmacological activities, and molecular mechanism of action. *Phytochemistry* **2018**, *145*, 187–196. [CrossRef]
50. Curti, V.; Zaccaria, V.; Tsetegho Sokeng, A.J.; Dacrema, M.; Masiello, I.; Mascaro, A.; D'Antona, G.; Daglia, M. Bioavailability and in vivo antioxidant activity of a standardized polyphenol mixture extracted from brown propolis. *Int. J. Mol. Sci.* **2019**, *20*, 1250. [CrossRef]
51. Grabska-Kobylecka, I.; Kaczmarek-Bak, J.; Figlus, M.; Prymont-Przyminska, A.; Zwolinska, A.; Sarniak, A.; Wlodarczyk, A.; Glabinski, A.; Nowak, D. The presence of caffeic acid in cerebrospinal fluid: Evidence that dietary polyphenols can cross the blood-brain barrier in humans. *Nutrients* **2020**, *12*, 1531. [CrossRef]
52. Shimazu, R.; Anada, M.; Miyaguchi, A.; Nomi, Y.; Matsumoto, H. Evaluation of blood-brain barrier permeability of polyphenols, anthocyanins, and their metabolites. *J. Agric. Food Chem.* **2021**, *69*, 11676–11686. [CrossRef]
53. Figueira, I.; Garcia, G.; Pimpão, R.C.; Terrasso, A.P.; Costa, I.; Almeida, A.F.; Tavares, L.; Pais, T.F.; Pinto, P.; Ventura, M.R.; et al. Polyphenols journey through blood-brain barrier towards neuronal protection. *Sci. Rep.* **2017**, *7*, 11456. [CrossRef] [PubMed]

antioxidants

MDPI

Article

Antioxidant Effects of Korean Propolis in HaCaT Keratinocytes Exposed to Particulate Matter 10

In Ah Bae [1,2], Jae Won Ha [1,2], Joon Yong Choi [1,2] and Yong Chool Boo [1,2,3,4,*]

1 Department of Biomedical Science, The Graduate School, Kyungpook National University, 680 Gukchaebosang-ro, Jung-gu, Daegu 41944, Korea; sksnadlv@naver.com (I.A.B.); jaewon1226@knu.ac.kr (J.W.H.); halo134679@naver.com (J.Y.C.)
2 BK21 Plus KNU Biomedical Convergence Program, Kyungpook National University, 680 Gukchaebosang-ro, Jung-gu, Daegu 41944, Korea
3 Department of Molecular Medicine, School of Medicine, Kyungpook National University, 680 Gukchaebosang-ro, Jung-gu, Daegu 41944, Korea
4 Cell and Matrix Research Institute, Kyungpook National University, 680 Gukchaebosang-ro, Jung-gu, Daegu 41944, Korea
* Correspondence: ycboo@knu.ac.kr; Tel.: +82-53-420-4946

Abstract: Air pollution causes oxidative stress that leads to inflammatory diseases and premature aging of the skin. The purpose of this study was to examine the antioxidant effect of Korean propolis on oxidative stress in human epidermal HaCaT keratinocytes exposed to particulate matter with a diameter of less than 10 μm (PM_{10}). The total ethanol extract of propolis was solvent-fractionated with water and methylene chloride to divide into a hydrophilic fraction and a lipophilic fraction. The lipophilic fraction of propolis was slightly more cytotoxic, and the hydrophilic fraction was much less cytotoxic than the total extract. The hydrophilic fraction did not affect the viability of cells exposed to PM_{10}, but the total propolis extract and the lipophilic fraction aggravated the toxicity of PM_{10}. The total extract and hydrophilic fraction inhibited PM_{10}-induced ROS production and lipid peroxidation in a concentration-dependent manner, whereas the lipophilic fraction did not show such effects. High-performance liquid chromatography with photodiode array detection (HPLC-DAD) analysis showed that the hydrophilic fraction contained phenylpropanoids, such as caffeic acid, *p*-coumaric acid, and ferulic acid, whereas the lipophilic faction contained caffeic acid phenethyl ester (CAPE). The former three compounds inhibited PM_{10}-induced ROS production, lipid peroxidation, and/or glutathione oxidation, and ferulic acid was the most effective among them, but CAPE exhibited cytotoxicity and aggravated the toxicity of PM_{10}. This study suggests that Korean propolis, when properly purified, has the potential to be used as a cosmetic material that helps to alleviate the skin toxicity of air pollutants.

Keywords: Korean propolis; particulate matter; oxidative stress; keratinocytes; ferulic acid; caffeic acid; *p*-coumaric acid

Citation: Bae, I.A.; Ha, J.W.; Choi, J.Y.; Boo, Y.C. Antioxidant Effects of Korean Propolis in HaCaT Keratinocytes Exposed to Particulate Matter 10. *Antioxidants* **2022**, *11*, 781. https://doi.org/10.3390/antiox11040781

Academic Editors: Josipa Vlainić and Ivana Tlak Gajger

Received: 20 March 2022
Accepted: 12 April 2022
Published: 14 April 2022

1. Introduction

Industrial development and increased human activity are causing environmental pollution problems. In particular, air pollutants from natural and artificial sources cause fatal diseases, such as respiratory, cardiovascular, and brain-neurological diseases, and are an important cause of death for modern humans [1,2]. Air pollution has a detrimental effect on the health of the skin, the outermost organ of our body, and causes various inflammatory diseases, such as atopy, psoriasis, and acne, as well as premature skin aging [2,3]. Therefore, a dermatological or cosmetic defense strategy against air pollution should be devised to maintain skin health.

Air pollutants include gas components, such as ozone (O_3), nitrogen dioxide (NO_2), and sulfur dioxide (SO_2), and suspended particulate matter of various compositions [4].

The suspended particulate matter with a size of less than 10 μm is called PM_{10}, and it is a mixture of various organic compounds, such as aryl hydrocarbons, various heavy metals, such as cobalt, lead, and cadmium, and biological constituents [5,6]. PM_{10} can enter the body through various routes, such as the mouth, nose, eyes, and ears, and can also penetrate the skin through pores or the sites where the skin barrier is weak [7–10]. The components of PM_{10} generate reactive oxygen species (ROS) through chemical reactions or biological metabolism inside and outside cells [11–14], causing oxidative damage and inflammatory responses [15,16]. On the other hand, it is suggested that various types of antioxidants may help protect skin health by alleviating the oxidative stress and inflammatory response induced by PM_{10} [17].

Propolis is a natural product made by bees by mixing their discharges with the sap and pollen they collected from the plant. It is a green, yellow, or red-toned high-viscosity substance mainly used for building and repairing their hives. For thousands of years, propolis has been used in most civilized societies for various medicinal purposes [18,19]. The composition of industrial propolis from honey bees and stingless bees varies depending on the geographical locations in which bees and their vegetation are distributed; furthermore, its composition also varies depending on the climates and collection season of propolis [19]. Among the components of propolis, phenolic metabolites of plants are known to possess various biological activities including antioxidant activity [20]. Thus propolis rich in phenolic antioxidants has great potential to find utility in food, cosmetics, and medicines [20,21].

Korean propolis from various areas had high total phenolic content and strong antioxidant activity; the propolis from Cheongju had high contents of caffeic acid and caffeic acid phenethyl ester (CAPE) [22]. An ethanolic extract of Korean propolis provided ten phenylpropanoic acid esters, such as CAPE, caffeic acid benzyl ester, caffeic acid ethyl ester, ferulic acid benzyl ester, ferulic acid 3′,3′-dimethylallyl ester, 3,4-dimethoxycaffeic acid cinnamyl ester, *p*-coumaric acid cinnamyl ester, *p*-coumaric acid benzyl ester, cinnamic acid phenethyl ester, and cinnamic acid cinnamyl ester [23]. The components of Korean propolis, such as CAPE and quercetin, displayed potent antioxidant activities in vitro assays, and they inhibited tube formation and growth of human umbilical vein endothelial cells, supporting their potential anti-angiogenic activities [24]. Oral administration of Korean propolis attenuated oxidative stresses and neuronal degenerations induced by kainic acid in Sprague–Dawley rats, involving adenosine A1 receptor modulation [25].

The purpose of this study was to evaluate the antioxidant activity of Korean propolis in human epidermal keratinocytes exposed to airborne PM_{10}. The ethanol extract of Korean propolis was divided into a hydrophilic fraction and a lipophilic fraction, and their effects on cell viability, ROS production, lipid peroxidation, and glutathione levels in human HaCaT keratinocytes were compared in the presence or absence of PM_{10}. For the hydrophilic fraction, which was found to have relatively low toxicity and high antioxidant activity, component analysis and evaluation of the biological activity of the component were additionally performed. The results of this study suggested that Korean propolis, when properly purified, has the potential to be developed as a cosmetic material that helps to safely and effectively alleviate the skin toxicity of atmospheric particulate matter.

2. Materials and Methods

2.1. Reagents

Standardized fine dust (PM_{10}-like, European standard ERM-CZ120), CAPE, caffeic acid, *p*-coumaric acid, and ferulic acid were purchased from Sigma-Aldrich (St. Louis, MO, USA).

2.2. Preparation of the Total Extract and Fractions of Propolis

Propolis was purchased in Andong, Gyeongsangbuk-do, Korea. Propolis raw material (60 g) was extracted with ethanol (600 mL) at room temperature for 4 days. After filtering, the filtrate was concentrated using a rotary evaporator under reduced pressure to obtain the total extract (23 g). The total extract was solvent-fractionated using equal volumes of water and methylene chloride and each fraction was evaporated under reduced pressure to obtain the hydrophilic fraction (0.6 g), the lipophilic fraction (18.5 g), and insoluble material (2.6 g).

2.3. High-Performance Liquid Chromatography with Photodiode Array Detection (HPLC-DAD)

HPLC-DAD analysis of the total extract of propolis and its fractions was performed using Waters Alliance HPLC system (Waters, Milford, MA, USA) consisting of e2695 separation module and 2996 photodiode array detector. A Hector-M C_{18} column (4.6 mm × 250 mm, 5 μm) (RS Tech Co., Daejeon, Korea) was used as the stationary phase. A mixture of 0.1% phosphoric acid (solvent A) and acetonitrile (solvent B) was used as the mobile phase with the changing composition: 0–30 min, a linear gradient from 0 to 100% B; 30–40 min, 100% B; 40–45 min, a linear gradient from 100 to 0% B. The flow rate of the mobile phase was set at 1.0 mL min^{-1}, and the sample injection volume was 10 μL.

2.4. Cell Culture and Treatments

An immortalized human keratinocyte HaCaT cell line (CLS Cell Lines Service GmbH, Eppelheim, Germany) established by Dr. Norbert E. Fusenig [26] was cultured in a closed incubator at 37 °C in humidified air containing 5% CO_2. The growth medium was DMEM/F-12 medium (GIBCO-BRL, Grand Island, NY, USA) supplemented with 10% fetal bovine serum, 100 U mL^{-1} penicillin, 100 μg mL^{-1} streptomycin, 0.25 μg mL^{-1} amphotericin B, and 10 μg mL^{-1} hydrocortisone. For each experiment, cells were seeded on 96-well, 12-well, or 6-well culture plates (SPL Life Sciences, Pocheon, Korea) and cultured for 24 h prior to various treatments. The total extract, its fractions, and individual compounds were treated alone or in combination with PM_{10} (200 μg mL^{-1}) for 48 h to determine cell viability and lipid peroxidation, or for 60 min to measure ROS production.

2.5. Cell Viability Assay

HaCaT cells were seeded in 96-well culture plates at 4 × 10^3 cells/well and maintained in a 200 μL culture medium for 24 h. After various treatments for 48 h, the cells were washed with phosphate-buffered saline (PBS) to remove residual extract, compound, and PM_{10}, and their viability was measured using 3-(4,5-dimethylthiazol-2-yl)-2,5-diphenyl tetrazolium bromide (MTT) [27]. MTT (Sigma-Aldrich) was dissolved in PBS (5 mg mL^{-1}) and diluted 5 times with a culture medium to the final concentration of 1 mg mL^{-1}. The medium was dispensed by 100 μL per well in a 96-well plate and incubated at 37 °C for 2 h. After discarding the medium, cells were washed with PBS. The dye accumulated inside cells were extracted using 100 μL of dimethyl sulfoxide per well and the absorbance of the extracts was measured at 570 nm using a Spectrostar Nano microplate reader (BMG Labtech GmbH, Ortenberg, Germany).

2.6. Cellular ROS Production Assay

Cellular ROS production was assessed by using 2′,7′-dichlorodihydrofluorescein diacetate (DCFH-DA) [28]. The cells were plated onto 12-well culture plates at 1.4 × 10^5 cells/well for 24 h. Cells were pre-labeled with 10 μM DCFH-DA (Sigma-Aldrich) for 30 min. After various treatments for 60 min, cells were washed twice with PBS and the fluorescence images of cells were obtained with a LEICA DMI3000 B microscope (Leica Microsystems GmbH, Wetzlar, Germany). For quantitative analysis, cells were lysed with 150 μL of the lysis buffer (1% sodium dodecyl sulfate (SDS), 20 mM Tris-Cl, 2.5 mM EDTA, pH 7.5), and the cell lysates were centrifuged with an Eppendorf centrifuge 5418R (Eppendorf, Barkhausenweg, Hamburg, Germany) at 14,500× *g* for 15 min to obtain the supernatant. The fluorescence intensity (excitation at 485 nm and emission at 538 nm) of the supernatants was measured with a Gemini EM fluorescence microplate reader (Molecular Devices, Sunnyvale, CA, USA).

Flow cytometry was additionally used to analyze intracellular ROS production. After various treatments, the adherent cells were detached from the culture plates using a trypsin-

EDTA solution. Cells were centrifuged down with a Combi 408 centrifuge (Hanil, Daejeon, Korea at 316× *g* for 3 min, washed with PBS, and suspended in PBS. Flow cytometry for the cell suspension was conducted using BD FACSCalibur and data were analyzed using BD CellQuest (BD Biosciences, San Jose, CA, USA). Data are presented by the ratio (%) of cells with high DCFH-DA fluorescence due to intracellular ROS production to the total gated cells.

2.7. Lipid Peroxidation Assay

Cellular lipid peroxidation was assessed using 2-thiobarbituric acid (TBA) [29]. Cells were seeded at 2×10^5 cells per well in a 6-well plate and cultured for 24 h. After various treatments with a test material in combination with PM_{10} (200 μg mL^{-1}) for 48 h, cells were washed twice with PBS and lysed with 150 μL of the lysis buffer (1% SDS, 20 mM Tris-Cl, 2.5 mM EDTA, pH 7.5). The cell lysates were centrifuged with an Eppendorf centrifuge 5418R at 14,500× *g* for 15 min to remove cell debris and PM_{10}. The mixture of 100 μL cell lysate (200 μg protein), 50 μL 1.0% *m*-phosphoric acid, and 350 μL 0.9% TBA (Sigma-Aldrich) was heated at 95 °C in a water bath for 45 min. The reaction was also run with 100 to 400 nM 1,1,3,3-tetramethoxypropane (Sigma-Aldrich) as a donor of malondialdehyde (MDA) to construct a standard curve. The limit of detection for the fluorometric assay has been determined to be 5 nM. After cooling to room temperature, 500 μL n-butyl alcohol was added to the mixture, vortex-mixed, and then the mixture was centrifuged to separate into two layers. The fluorescence intensity of the n-butyl alcohol layer (excitation at 544 nm and emission at 590 nm) was measured by using a Gemini EM fluorescence microplate reader. Data are presented as MDA levels corrected for protein contents.

In vitro experiments for lipid peroxidation were performed using the lysates of control HaCaT cells without any treatments. The HaCaT cell lysate (200 μg protein) was diluted with the lysis buffer and reacted with PM_{10} (200 μg mL^{-1}) in the absence and presence of test material in a total volume of 200 μL in 1.5 mL Eppendorf tubes at 37 °C for 24 h. The reaction mixture was centrifuged to remove PM_{10}. The supernatant (100 μL) was used in the assay of MDA levels as above.

2.8. Glutathione Assay

Glutathione contents were measured by a recycling assay [30]. After culturing and treatments in 6-well plates as above, cells were extracted using 5% meta-phosphoric acid (150 μL per well), followed by centrifuging with an Eppendorf centrifuge 5418R at 14,500× *g* for 15 min. The supernatant was used for the glutathione assay using a GSH/GSSG assay kit (product number GT40) from Oxford Biomedical Research (Oxford, UK). The total content of reduced glutathione (GSH) plus oxidized glutathione (GSSG) was measured using the extract as it is, and the GSSG content was quantified after pre-scavenging GSH in the extract with a pyridine derivative. Absorbance change due to reduction of 5,5′-dithio-bis-2-nitrobenzoic acid by GSH was measured at 412 nm, and a calibration curve prepared using a GSSG standard was used for the determination of glutathione contents. The GSH content was calculated by subtracting the GSSG content from the total content of GSH plus GSSG.

2.9. Assay for Free Radical Scavenging Activities

Spectroscopic methods were used to measure the scavenging activities of the sample against 2,2′-azinobis-(3-ethylbenzothiazoline-6-sulfonic acid) cation radical ($ABTS^{\bullet+}$) and 2,2-diphenyl-1-picrylhydrazyl radial ($DPPH^{\bullet}$) [31–33]. The $ABTS^{\bullet+}$ solution was prepared by mixing 0.54 mM ABTS solution (Sigma-Aldrich) and 0.27 mM potassium persulfate solution (Sigma-Aldrich) in equal volumes and allowing them to react for 24 h at room temperature (25 °C) in the dark. Each serial dilution of a plant-derived material or compound in ethanol (100 μL) was reacted with 0.27 mM $ABTS^{\bullet+}$ in water (100 μL) of at 25 °C for 3 min, followed by measurement of the absorbance at 734 nm with a BioRad Model

680 microplate reader (Bio-Rad Laboratories, Inc., Hercules, CA, USA). For the DPPH• scavenging activity assay, a serially diluted sample in ethanol (100 µL) was mixed with 0.2 mM DPPH• (Alfa Aesar, Ward Hill, MA, USA) in ethanol and reacted at 25 °C for 30 min. The absorbance was measured at 517 nm using a microplate reader.

2.10. Statistical Analysis

SigmaStat v.3.11 software (Systat Software Inc., San Jose, CA, USA) was used for the statistical analysis of the experimental data. Data are expressed as mean ± standard deviation (SD) of three or more independent experiments. The presence of significantly different group means among all groups was determined using a one-way analysis of variance (ANOVA) at the $p < 0.05$ level. Then, Duncan's multiple range test was used to compare all groups to each other.

3. Results

3.1. Antioxidant Effects of Total Propolis Extract and Its Solvent Fractions in Cells

The total extract of Korean propolis and its hydrophilic and lipophilic fractions were prepared as depicted in Figure 1A. The yield of the total extract obtained by immersing the propolis raw material in ethanol was about 38.3%. The ratio of the hydrophilic fraction and the lipophilic fraction obtained by solvent fractionation of the total extract with water and methylene chloride was 1:31, and most of the extraction components were included in the lipophilic fraction.

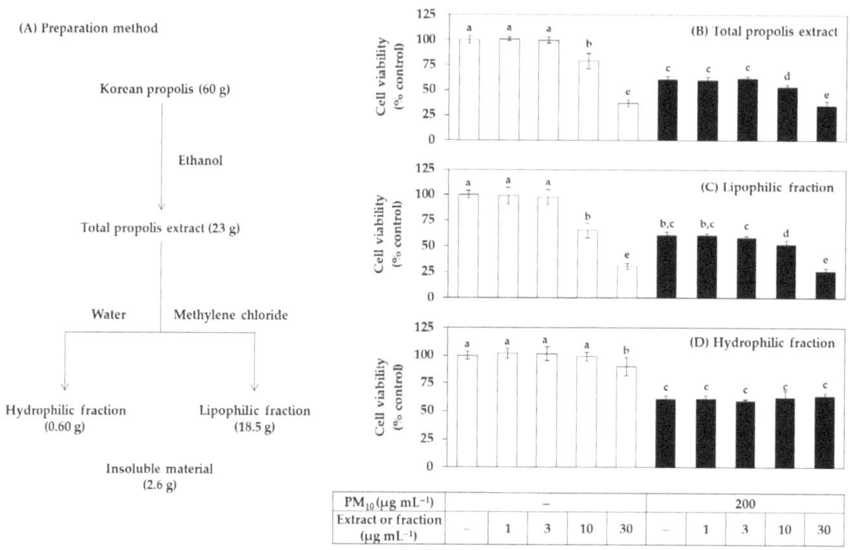

Figure 1. Effects of total propolis extract and its solvent fractions on the viability of human HaCaT keratinocytes exposed to PM_{10}. In (**A**), the total ethanolic extract of Korean propolis was divided into a hydrophilic and a lipophilic fraction by solvent partition between water and methylene chloride. Cells were treated with the total extract (**B**), a lipophilic fraction (**C**), or a hydrophilic fraction (**D**) at the specified concentration alone or in combination with PM_{10} (200 µg mL^{-1}) for 48 h. Cell viability was determined by the 3-(4,5-dimethylthiazol-2-yl)-2,5-diphenyl tetrazolium bromide (MTT) assay. Data are presented as mean ± SD (n = 5). (**D**) Duncan's multiple range test was performed to compare all group means to each other. Groups that share the same letters (a–e) do not have significantly different means at the $p < 0.05$ level.

In the first cell experiment, the effect of the total extract and its fractions on the viability of HaCaT cells in the presence or absence of PM_{10} exposure was investigated. The treatment concentration of PM_{10} was 200 μg mL^{-1}, which was selected in the previous study [34]. Cells were treated with extracts or fractions alone or in combination with PM_{10}, and cell viability was measured after 48 h. As shown in Figure 1B–D, the total propolis extract significantly reduced the cell viability at 10 μg mL^{-1} or higher, and the lipophilic fraction showed slightly stronger cytotoxicity than the total extract. However, the water fraction only reduced the cell viability by 10% at 30 μg mL^{-1}, but not at the lower test concentrations. That is, the cytotoxicity of the hydrophilic fraction was relatively weak compared to the total extract and the fat-soluble fraction. As expected, PM_{10} exposure reduced cell viability by 40%, but neither the total extract nor the two fractions had any mitigating effect. The total extract and the lipophilic fraction further reduced the viability of PM_{10}-exposed cells above 10 μg mL^{-1} because of their toxicity. On the other hand, the hydrophilic fraction did not change the viability of PM_{10}-exposed cells due to its weak toxicity.

Since both the total extract of propolis and its hydrophilic and lipophilic fractions had no cytotoxicity at a concentration of 3 μg mL^{-1} or less, it was evaluated whether they could reduce PM_{10}-induced oxidative stress in cells in a low concentration range. HaCaT cells were treated with the total extract or fraction at a concentration of 1 μg mL^{-1} or 3 μg mL^{-1} alone or in combination with PM_{10}. ROS production was measured after 60 min of PM_{10} exposure and lipid peroxidation was measured after 48 h. As shown in Figure 2, the total extract significantly reduced ROS production at 3 μg mL^{-1}. Among the two fractions, the hydrophilic fraction significantly reduced ROS production at 1–3 μg mL^{-1}, and the lipophilic fraction did not show such effects. The total extract and the hydrophilic fraction also significantly inhibited lipid peroxidation at 3 μg mL^{-1}, and the lipophilic fraction had no such effects.

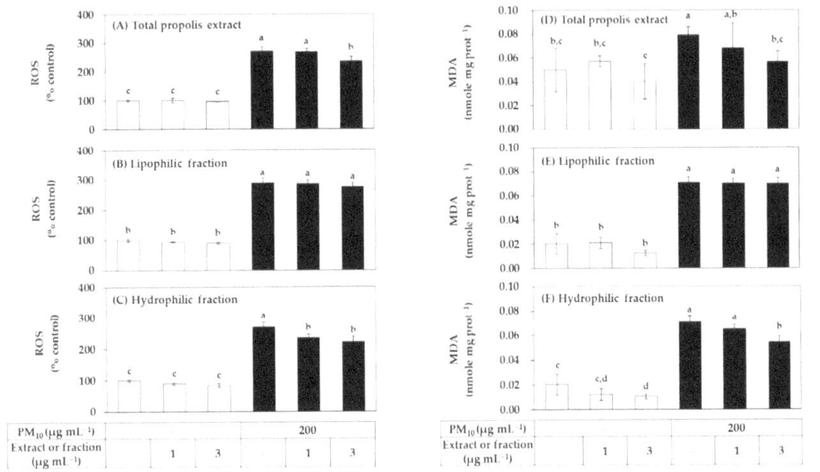

Figure 2. Effects of total propolis extract and its fractions on the reactive oxygen species (ROS) production and lipid peroxidation in HaCaT keratinocytes exposed to PM_{10}. Cells were treated with the total extract (**A**,**D**), a lipophilic fraction (**B**,**E**), or a hydrophilic fraction (**C**,**F**) at the specified concentration alone or in combination with PM_{10} (200 μg mL^{-1}) for 60 min for the determination of ROS production, or 48 h for the determination of lipid peroxidation. In (**A**–**C**), cells were pre-labeled with 10 μM 2′-7′dichlorofluorescin diacetate (DCFH-DA) for 30 min, and fluorescence of oxidized probe due to cellular ROS production was determined after treatments with the extracts and/or PM_{10}. In (**D**–**F**), lipid peroxidation levels of cell lysates were determined by the thiobarbituric acid (TBA) assay. Data are presented as malondialdehyde (MDA) levels corrected for protein contents. Data are presented as mean ± SD (n = 4 for (**A**–**C**); n = 3 for (**D**–**F**)). Duncan's multiple range test was performed to compare all group means to each other. Groups that share the same letters (a–d) do not have significantly different means at the $p < 0.05$ level.

Since the relatively low cytotoxicity of the hydrophilic fraction was seen in Figure 1, an additional experiment was conducted by extending the treatment concentration range of this fraction. As shown in Figure 3, the hydrophilic fraction inhibited PM_{10}-induced ROS generation and lipid peroxidation in a concentration-dependent manner in the range of 1–30 μg mL^{-1}. Combining the above results, it was suggested that the hydrophilic component of the propolis extract can relieve oxidative stress in cells exposed to PM_{10} more safely and effectively than the lipophilic component.

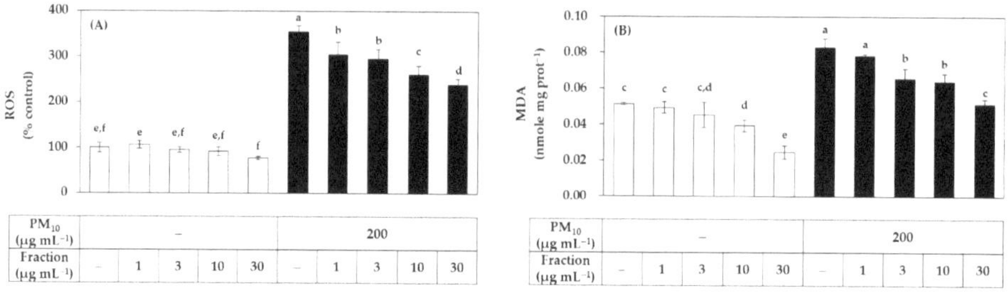

Figure 3. Effects of a hydrophilic fraction of propolis on the ROS production and lipid peroxidation in HaCaT keratinocytes exposed to PM_{10}. Cells were treated with a hydrophilic fraction at the specified concentration alone or in combination with PM_{10} (200 μg mL^{-1}) for 60 min for the determination of ROS production, or 48 h for the determination of lipid peroxidation. In (**A**), cells were pre-labeled with 10 μM (DCFH-DA) for 30 min and fluorescence of the oxidized probe due to cellular ROS production was determined after treatments with the extracts and/or PM_{10}. In (**B**), lipid peroxidation levels of cell lysates were determined by TBA assay. Data are presented as MDA levels corrected for protein contents. Data are presented as mean ± SD (n = 4). Duncan's multiple range test was performed to compare all group means to each other. Groups that share the same letters (a–f) do not have significantly different means at the $p < 0.05$ level.

3.2. *Analysis of Total Propolis Extract and Its Solvent Fractions*

HPLC-DAD analysis of the total extract of propolis and its lipophilic and hydrophilic fractions was performed. As shown in Figure 4, the total propolis extract and the two fractions show different phytochemical profiles. It was confirmed that one of the main components of the lipophilic fraction was CAPE. However, this study focused on the hydrophilic fraction based on the observed safety and effectiveness. The main peaks of the hydrophilic fraction were identified as caffeic acid, *p*-coumaric acid, and ferulic acid by comparing the retention times and absorption spectra of the standards. Among them, *p*-coumaric acid and ferulic acid are also partially included in the lipophilic fraction. The contents of caffeic acid, *p*-coumaric acid, and ferulic acid in the total extract were found to be similar to each other.

3.3. *Antioxidant Effects of Phenylpropanoid Compounds of Propolis in Cells*

Additional experiments were conducted to compare the biological activities of these three phenylpropanoid compounds. The effect of these compounds on the viability of HaCaT cells in the presence or absence of PM_{10} was investigated. As shown in Figure 5A, caffeic acid slightly reduced the cell viability at 100 μM, but all three phenylpropanoid compounds were found to be non-toxic at most concentrations tested. These three compounds did not affect the cell viability under PM_{10} exposure conditions. As shown in Figure 5B, CAPE exhibited cytotoxicity that reduced the cell viability by 50% at a concentration of 10 μM and aggravated the toxicity of PM_{10}, so it was excluded from subsequent experiments. The chemical structures of caffeic acid, *p*-coumaric acid, ferulic acid, and CAPE are shown in Figure 5C.

The effects of these three compounds on ROS production and lipid peroxidation in HaCaT cells exposed to PM_{10} were further compared. As shown in Figure 6, each compound reduced ROS production in a concentration-dependent manner, and the effect was in the order of ferulic acid > *p*-coumaric acid > caffeic acid, especially in the low concentration range.

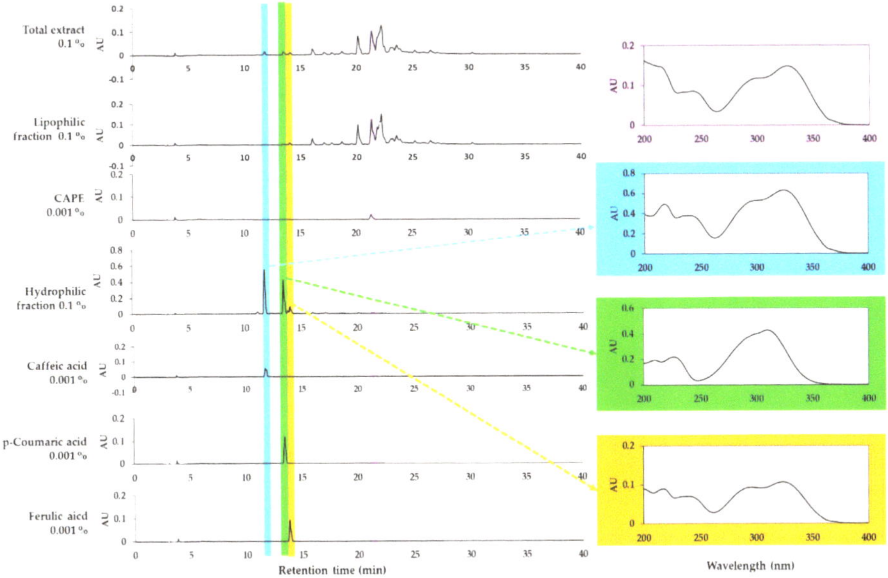

Figure 4. High-performance liquid chromatography-photodiode array detection (HPLC-DAD) analysis of the total extract of propolis and its solvent fractions. Authentic caffeic acid phenethyl ester (CAPE), caffeic acid, *p*-coumaric acid, and ferulic acid were used to identify the major peaks by comparing retention times and absorption spectra. Chromatograms detected at 310 nm and the absorption spectra of the designated peaks are shown.

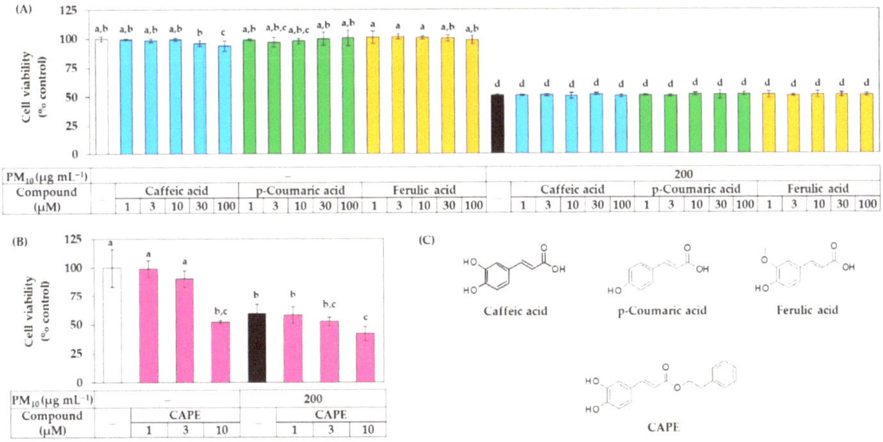

Figure 5. Effects of caffeic acid, *p*-coumaric acid, ferulic acid, and CAPE on viability in HaCaT keratinocytes exposed to PM_{10}. In (**A**,**B**), cells were exposed to PM_{10} (200 μg mL^{-1}) for 48 h in the absence and presence of each compound at the indicated concentrations. Cell viability was determined by the MTT assay. Data are presented as mean ± SD (n = 4 for (**A**); n = 5 for (**B**)). Duncan's multiple range test was performed to compare all group means to each other. Groups that share the same letters (a–d) do not have significantly different means at the $p < 0.05$ level. In (**C**), the chemical structure of caffeic acid, *p*-coumaric acid, ferulic acid, and CAPE are shown.

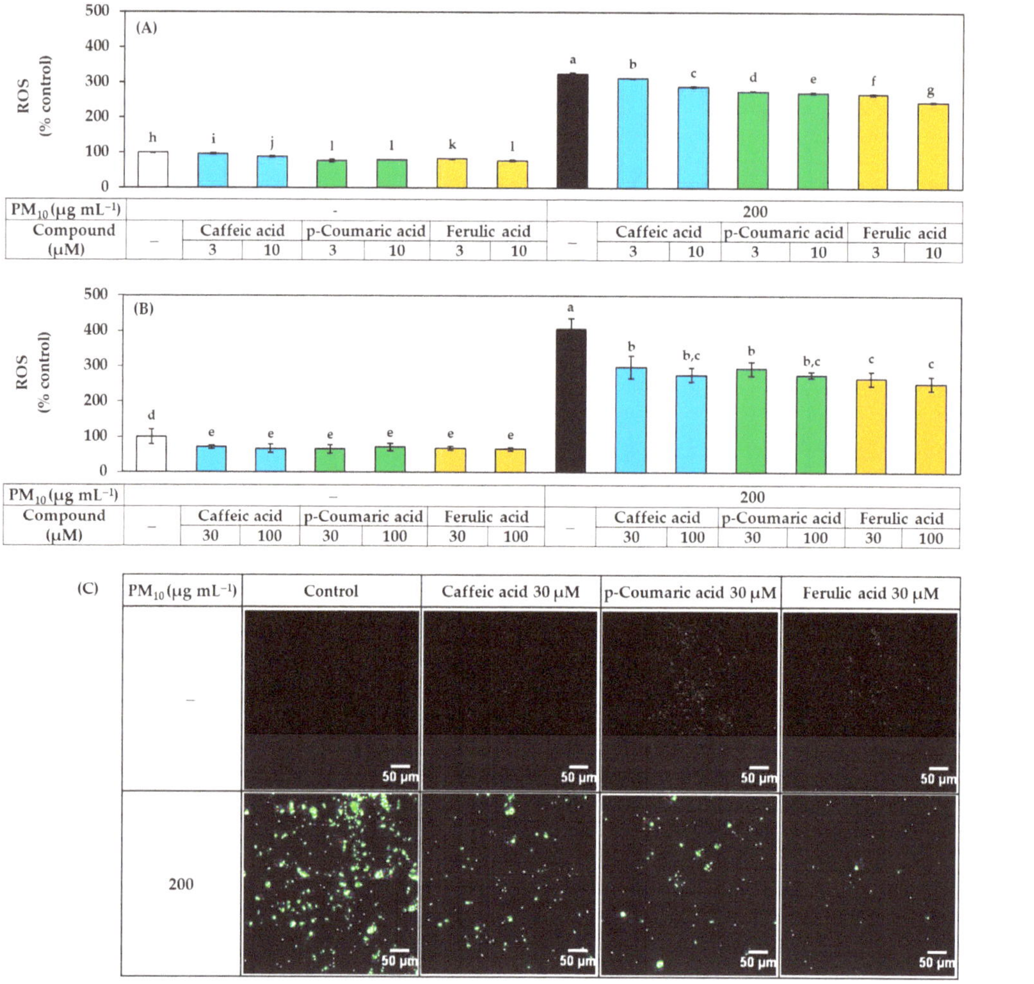

Figure 6. Effects of caffeic acid, *p*-coumaric acid, and ferulic acid on the ROS production in HaCaT keratinocytes exposed to PM_{10}. Cells were labeled with DCFH-DA, treated with each compound at the indicated concentrations, and exposed to PM_{10} (200 μg mL^{-1}) for 60 min or not. In (**A**,**B**), the fluorescence of the cell extracts was measured to quantitatively determine ROS levels. Data are presented as mean ± SD (n = 4). Duncan's multiple range test was performed to compare all group means to each other. Groups that share the same letters (a–l) do not have significantly different means at the $p < 0.05$ level. Typical images of cells fluorescing due to ROS production are shown in (**C**).

Flow cytometry was additionally used to analyze intracellular ROS production without cell disruption. Despite washing cells with PBS, PM_{10} resides on cells and forms aggregates with cells. As shown in Figure 7A, PM_{10} treatment increased the counts of particles or cells with low forward scattering and high side scattering. Thus, the gate was set to exclude the particles and cell aggregates. Figure 7B shows the plots of the cell counts versus fluorescence intensity. Figure 7C shows typical plots for the cells with different treatments. In Figure 7D, the ratios (%) of cells with high fluorescence to the total gated cells were compared between cells treated with PM_{10} in the absence and presence of caffeic acid, *p*-coumaric acid, ferulic acids, and ascorbic acid (a positive control antioxidant) at 30 μM. The results indicated that PM_{10} increased ROS production in cells and the change was significantly inhibited by ferulic acid and ascorbic acid in the order. Caffeic acid and *p*-coumaric acid had no significant effects. Thus, the antioxidant effect of ferulic acid inhibiting intracellular ROS production was evaluated to be comparable to that of ascorbic acid by flow cytometry.

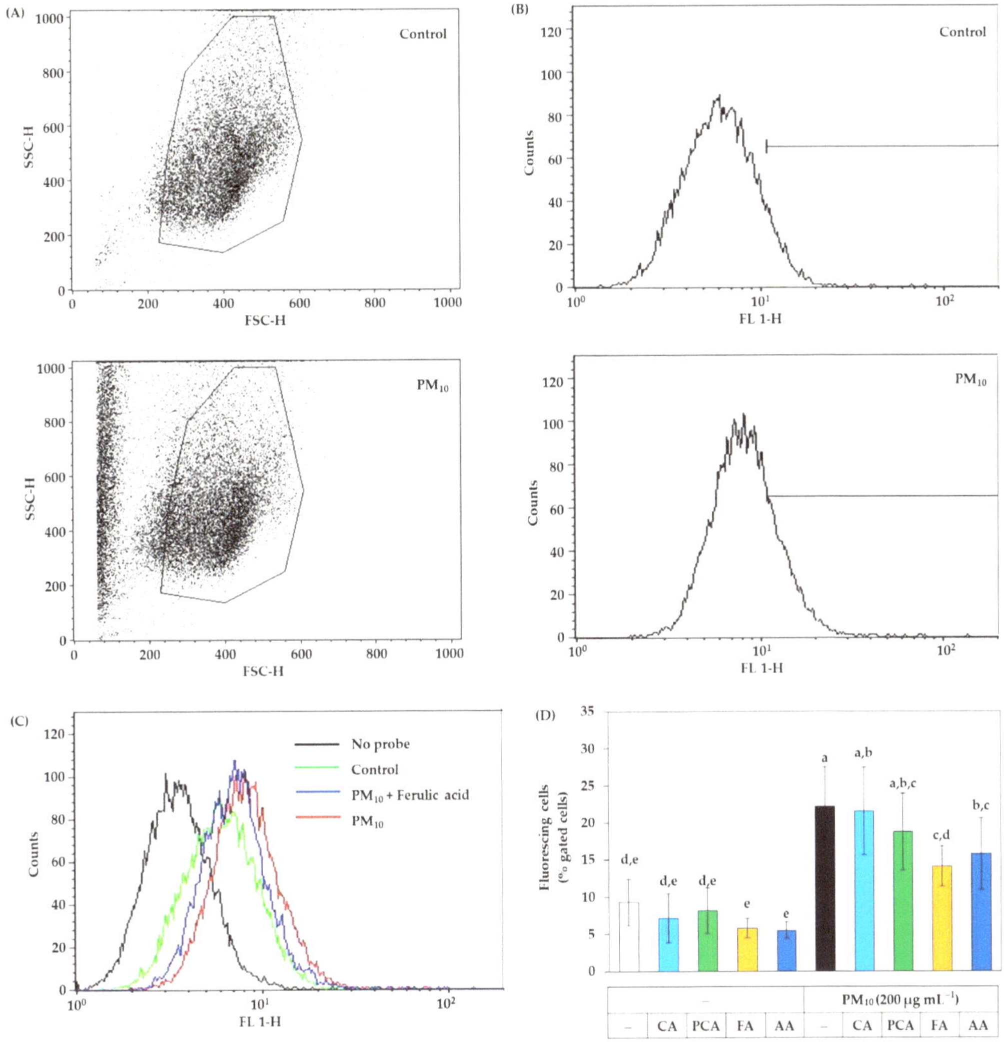

Figure 7. Flow cytometry for the ROS production in HaCaT keratinocytes exposed to PM_{10} in the absence and presence of caffeic acid (CA), *p*-coumaric acid (PCA), ferulic acid (FA), and ascorbic acid (AA). The adherent cells were labeled with DCFH-DA, treated with vehicle or each compound at 30 µM, and exposed to PM_{10} (200 µg mL^{-1}) for 60 min or not. Cells were washed, detached, centrifuged down, and suspended in PBS for flow cytometry. (**A**) The gate was set to exclude the PM_{10} particles and cell aggregates. (**B**) The plots of the cell counts versus fluorescence intensity are shown with a mark to define fluorescing cells. (**C**) Typical effects of PM_{10} in the absence and presence of FA on the distribution of cells with different fluorescence levels. (**D**) The ratios (%) of fluorescing cells to the total gated cells are presented. Data represent mean ± SD (n = 3). Duncan's multiple range test was performed to compare all group means to each other. Groups that share the same letters (a–e) do not have significantly different means at the $p < 0.05$ level.

As shown in Figure 8, all three compounds at 30–100 μM significantly inhibited lipid peroxidation in PM_{10}-exposed HaCaT cells, but at 10 μM, only ferulic acid showed a significant inhibitory effect, while the other two compounds had no significant effect. These results suggest that, although all three compounds have antioxidant activity that can relieve oxidative stress in cells, ferulic acid has relatively advantageous properties.

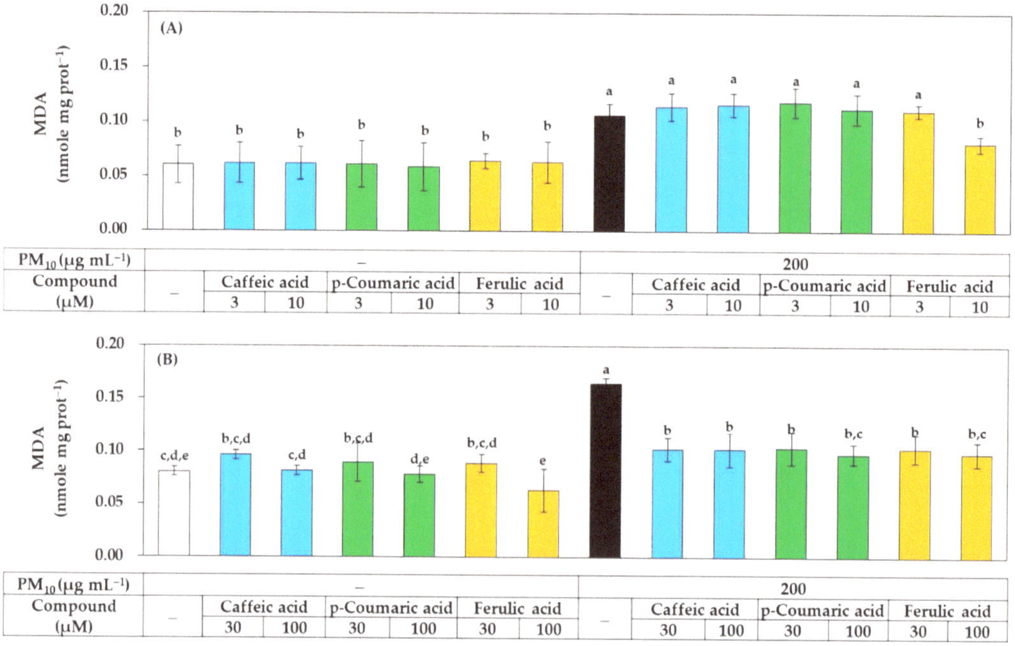

Figure 8. Effects of caffeic acid, *p*-coumaric acid, and ferulic acid on the lipid peroxidation in HaCaT keratinocytes exposed to PM_{10}. Cells were treated with each compound at 3–10 μM (**A**) or 30–100 μM (**B**) alone or in combination with PM_{10} (200 μg mL^{-1}) for 48 h. Lipid peroxidation levels of cell lysates were determined by TBA assay and data are presented as MDA levels corrected for protein contents. Data are presented as mean ± SD (n = 4). Duncan's multiple range test was performed to compare all group means to each other. Groups that share the same letters (a–e) do not have significantly different means at the $p < 0.05$ level.

The effects of caffeic acid, *p*-coumaric acid, and ferulic acid on the redox balance of cells were examined by quantifying cell glutathione in the presence or absence of PM_{10}. As shown in Figure 9A, in the absence of PM_{10}, ferulic acid, *p*-coumaric acid, and caffeic acid increased the total glutathione content in the order. However, the increases in total glutathione caused by ferulic acid and *p*-coumaric acid were significantly inhibited by PM_{10}. PM_{10} itself also significantly increased total glutathione, but this increase was inhibited by ferulic acid and *p*-coumaric acid. As shown in Figure 9B, in the absence of PM_{10}, ferulic acid and *p*-coumaric acid slightly increased the content of oxidized glutathione (GSSG). More notably, PM_{10} increased GSSG content by more than 10-fold, and the increase was strongly inhibited by ferulic acid, *p*-coumaric acid, and caffeic acid in the order. As shown in Figure 9C, in the absence of PM_{10}, ferulic acid, *p*-coumaric acid, and caffeic acid significantly increased the content of reduced glutathione (GSH) in this order. However, PM_{10} significantly inhibited the increase in GSH content by ferulic acid and *p*-coumaric acid. PM_{10} itself also slightly increased GSH content. As shown in Figure 9D, PM_{10} markedly increased the ratio of the GSSG content to the total glutathione content, and this change was significantly inhibited by ferulic acid and *p*-coumaric acid in the order. This complex phenomenon requires further study for interpretation but suggests that PM_{10} and phenylpropanoids, such as ferulic acid, may have diverse effects on the redox balance of cells. Nonetheless, these results demonstrate that the oxidative stress due to PM_{10} can be alleviated by phenylpropanoids, such as ferulic acid and *p*-coumaric acid.

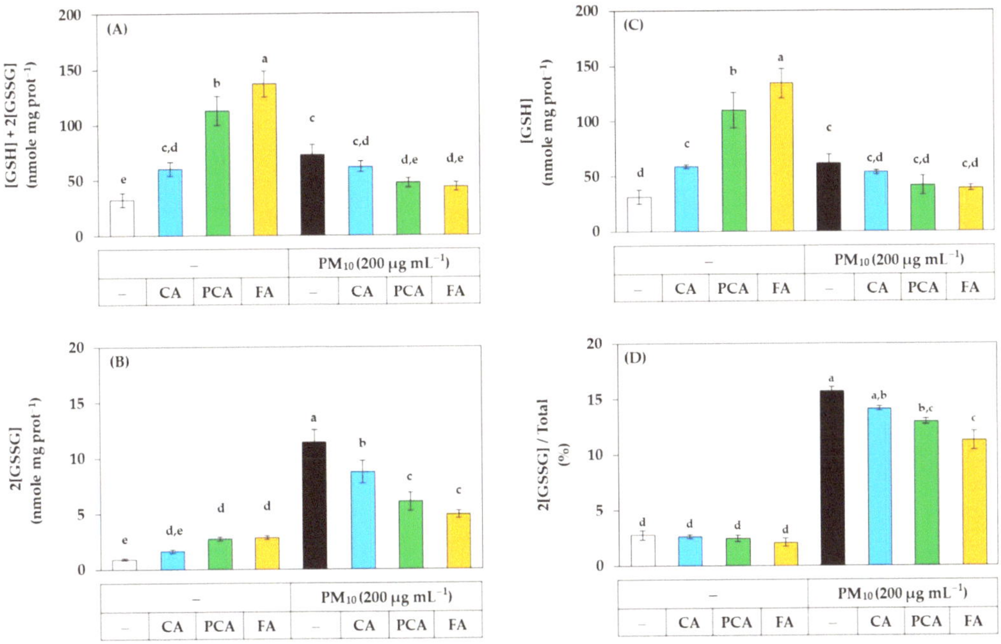

Figure 9. Effects of caffeic acid (CA), *p*-coumaric acid (PCA), and ferulic acid (FA) on the contents and ratios of glutathione (GSH) and glutathione disulfide (GSSG) in HaCaT keratinocytes exposed to PM_{10}. Cells were treated with each compound at 30 μM and cultured in the absence or presence of PM_{10} (200 μg mL^{-1}) for 24 h. The total contents of GSH plus GSSG (**A**) were subtracted by the GSSG contents (**B**) to calculate the GSH contents (**C**). The ratios of GSSG contents to the total contents were presented in (**D**). Data are presented as mean ± SD (n = 3). Duncan's multiple range test was performed to compare all group means to each other. Groups that share the same letters (a–e) do not have significantly different means at the $p < 0.05$ level.

3.4. Antioxidant Effects of Phenylpropanoid Compounds of Propolis In Vitro

Among the above three phenylpropanoids, ferulic acid had the strongest antioxidant effect in preventing ROS generation, lipid peroxidation, and GSH oxidation in cells, followed by *p*-coumaric acid and caffeic acid. What is the mechanism? Possibly, these compounds might directly and chemically inhibit the oxidation reactions catalyzed by PM_{10}. To examine this possibility, an in vitro experiment using the cell lysate was additionally performed. As shown in Figure 10A, when HaCaT cell lysate was exposed to PM_{10} in vitro, lipid peroxidation was induced, and strong inhibitory action was shown in the order of ferulic acid, *p*-coumaric acid, and caffeic acid. This trend matches the results obtained in cell experiments. Therefore, it is suggested that various phenylpropanoids, such as ferulic acid, can directly and chemically inhibit the oxidation reaction of cellular components by PM_{10}.

In many studies, scavenging activity for free radicals, such as $DPPH^{\bullet}$ and $ABTS^{\bullet+}$ is measured to search for general antioxidants or to evaluate their antioxidant activity. The activities of caffeic acid, *p*-coumaric acid, and ferulic acid that scavenge two types of free radicals in vitro were compared. As shown in Figure 10B, among the three compounds, caffeic acid scavenged $DPPH^{\bullet}$ most strongly, followed by ferulic acid and *p*-coumaric acid. As shown in Figure 10C, the $ABTS^{\bullet+}$ scavenging activity of ferulic acid was slightly stronger than that of caffeic acid, and *p*-coumaric acid was much weaker than the other two compounds. No special correlation was found between their reactivity to DPPH$^\bullet$ or ABTS$^{\bullet+}$ and their inhibitory effect on the PM_{10}-induced lipid peroxidation in vitro and in cells.

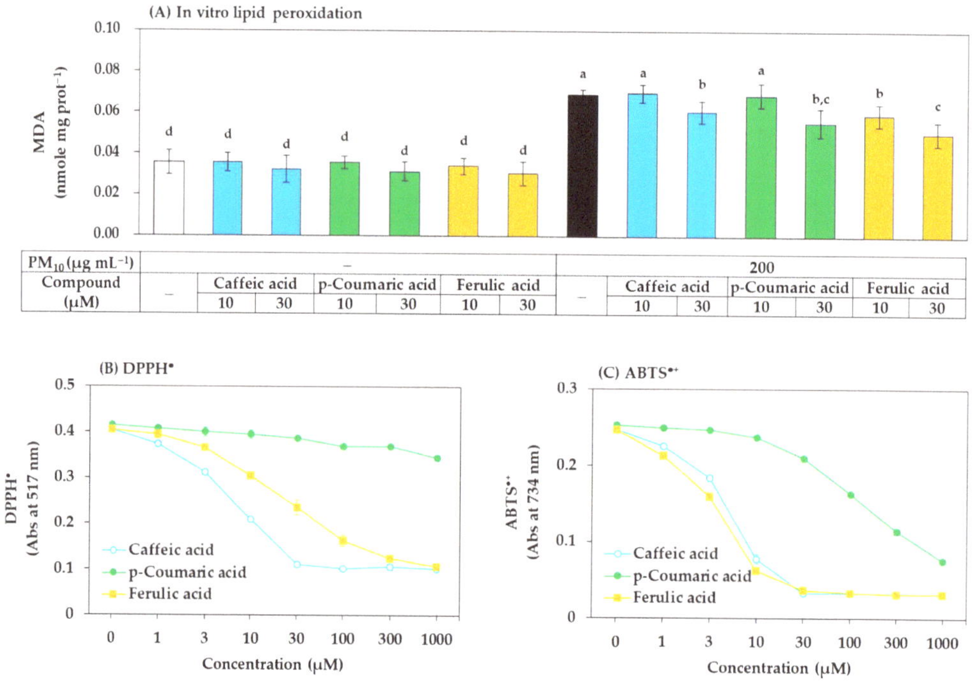

Figure 10. Effects of caffeic acid, *p*-coumaric acid, and ferulic acid on the lipid peroxidation of HaCaT cell lysate treated with PM10 in vitro, and their free radical scavenging activities against 2,2-diphenyl-1-picrylhydrazyl radical (DPPH$^\bullet$) and 2,2′-azinobis-(3-ethylbenzothiazoline-6-sulfonic acid) cation radical (ABTS$^{\bullet+}$) in vitro. (**A**) HaCaT cell lysate was treated with PM_{10} (200 µg mL^{-1}) for 24 h in the absence or presence of a compound at the specified concentration. DPPH$^\bullet$ (**B**) and ABTS$^{\bullet+}$ (**C**) were reacted with each compound at different concentrations, and their remaining levels were measured by absorbance at 517 nm and 734 nm respectively. Data are presented as mean ± SD (n = 3). Duncan's multiple range test was performed to compare all group means to each other. Groups that share the same letters (a–d) do not have significantly different means at the $p < 0.05$ level.

4. Discussion

This study showed the positive and negative effects of Korean propolis components on human epidermal keratinocytes exposed to PM_{10}. The total extract of propolis and its lipophilic fraction were cytotoxic, which significantly reduced the viability of keratinocytes, whereas no such cytotoxicity was observed for its hydrophilic fraction. The hydrophilic fraction of the propolis extract showed antioxidant activity that inhibited cellular ROS production and lipid peroxidation induced by exposure to PM_{10}, but the lipophilic fraction did not show such effects. Therefore, to use the propolis extract as a material for skin protection, it would be better to use it after removing harmful ingredients and enriching the active ingredients through a purification process rather than using it as it is.

Antioxidants that can directly or indirectly alleviate oxidative stress in cells are expected to be useful in reducing PM_{10}-induced skin inflammation and premature aging [17]. They can inhibit the production of ROS, scavenge ROS already produced, or enhance cellular antioxidant capacity by stimulating the gene expression of antioxidant enzymes mediated by nuclear factor erythroid 2-related factor 2 [35]. We have shown that various phenolic compounds contained in terrestrial and marine plants, such as punicalagin, (−)-epigallocatechin-3-gallate, chlorogenic acid, and dieckol reduce ROS production, lipid peroxidation, and inflammatory responses in HaCaT cells exposed to PM_{10} [16,34,36]. In this study, it was additionally reported that caffeic acid, *p*-coumaric acid, and ferulic acid, as phenolic compounds contained in propolis, have antioxidant actions to reduce PM_{10}-induced oxidative stress.

Plants are a source of various phytochemicals with diverse biological activities that are potentially useful to improve skin health and beauty [37–40]. Some phytochemicals act as either antioxidants or pro-oxidants and that show either cytotoxic or cytoprotective effects depending on their chemical nature and treatment conditions [41–44]. Therefore, it is important to select a phytochemical suitable for use and to optimize its biological activity by treating it at an optimal concentration for an optimal time. Since propolis contains various phenolic components derived from plants, various biological activities can be expected [18–20].

As observed in this study, the total extract of propolis has relatively strong cytotoxicity, and several previous studies reported the anti-proliferative and pro-apoptotic effects of propolis in various cancer cells. The extracts of the propolis from Chile, Brazil, Thailand, and Egypt have been shown to exert anti-proliferative and pro-apoptotic effects in various human cancer cell lines, such as mouth epidermoid carcinoma (KB), colon adenocarcinoma (Caco-2), androgen-insensitive prostate cancer (DU-145), laryngeal epidermoid carcinoma (Hep-2), cervical adenocarcinoma (HeLa), pulmonary adenocarcinoma (A549), and prostate cancer (PC3) cell lines [45–48]. A component of propolis, CAPE, was shown to induce apoptosis through activation of caspase-3, down-regulation of Bcl-2, and up-regulation of Bax in human leukemic HL-60 cells [49]. Caffeic acid and CAPE reduced glutathione levels and induced apoptosis of HeLa cells but not of Chinese hamster lung V79 fibroblast cells, suggesting that these compounds preferentially induce apoptosis of malignant cells through modulation of cellular redox state [50]. In the current study, CAPE showed strong toxicity to keratinocytes, and caffeic acid was also relatively more toxic than *p*-coumaric acid and ferulic acid, which matched well with previous studies.

On the other hand, the protective action of propolis extract in various cells has been reported. Uruguayan propolis induced the expression of endothelial nitric oxide synthase (NOS) while inhibiting endothelial NADPH oxidase, and thus it was suggested that the propolis can provide a cardiovascular protective benefit by increasing nitric oxide (NO) bioavailability in the endothelium [51]. Water extract of Brazilian green propolis and its constituents, 3,4-di-*O*-caffeoylquinic acid, 3,5-di-*O*-caffeoylquinic acid, chlorogenic acid, and *p*-coumaric acid exerted protective effects against the oxidative damage induced by glutathione depletion using L-buthionine-(S,R)-sulfoximine in cultured retinal ganglion cells, supporting its potential neuroprotective effects [52]. Ethyl acetate extracts of propolis from Algerian regions effectively scavenged free radicals, prevented lipid peroxidation, and inhibited myeloperoxidase activity, whereas its petroleum ether and chloroform extracts inhibited anticholinesterase activity [53,54]. Italian propolis with high polyphenolic components effectively inhibited lipid peroxidation of linoleic acid in SDS micelles and showed appropriate ultraviolet (UV) absorptivity to be used as broad-spectrum UVB and UVA photoprotection sunscreens [55]. However, there have been few studies focusing on the effect of propolis extracts on oxidative stress induced by atmospheric pollution.

In the current study, the hydrophilic fraction of the propolis extract was shown to have relatively weak cytotoxicity than the lipophilic fraction and have antioxidant activity to inhibit ROS generation and lipid peroxidation, suggesting that the hydrophilic fraction is more useful for protecting the skin from air pollution. This study also showed that the cytotoxicity of caffeic acid, *p*-coumaric acid, and ferulic acid contained in the hydrophilic fraction was very low compared to CAPE, which is one of the main components of the lipophilic fraction, and that these three compounds can mitigate the oxidative stress induced by PM_{10} in keratinocytes. The total content of caffeic acid, *p*-coumaric acid, and ferulic acid in the hydrophilic fraction of the propolis extract was estimated to be 14.2% (caffeic acid, 10.2%; *p*-coumaric acid, 3.63%; ferulic acid, 0.38%) by HPLC-DAD analysis, and 30 $\mu g\ mL^{-1}$ of this fraction corresponds to 4.3 $\mu g\ mL^{-1}$ (24.2 μM) of the compounds; caffeic acid, 3.1 $\mu g\ mL^{-1}$ (17.0 μM); *p*-coumaric acid, 1.1 $\mu g\ mL^{-1}$ (6.6 μM); ferulic acid, 0.1 $\mu g\ mL^{-1}$ (0.6 μM). The results of this study suggest that these three compounds in combination are partially responsible for the antioxidant activity of the hydrophilic fraction.

Despite the structural similarity of caffeic acid, *p*-coumaric acid, and ferulic acid, their cytotoxicity, reactivity to different ROS, antioxidant activity, and other biological activities are very different [56,57]. In our current study, the $DPPH^{\bullet}$ scavenging activity was caffeic acid > ferulic acid > *p*-coumaric acid, $ABTS^{\bullet+}$ scavenging activity was ferulic acid > caffeic acid > *p*-coumaric acid, and inhibitory activity against PM_{10}-induced lipid peroxidation was ferulic acid > *p*-coumaric acid > caffeic acid. Maurya et al. reported that ferulic acid showed weaker $DPPH^{\bullet}$ scavenging activity and stronger $ABTS^{\bullet+}$ scavenging activity than caffeic acid, and that ferulic acid inhibited 2,2′-azobis (2-methylpropionamidine) dihydrochloride (AAPH)-induced lipid peroxidation more effectively [58]. The results of both studies agree well with each other. Although caffeic acid exhibits stronger ROS scavenging activity than many other phenylpropanoids [56], it can act as a pro-oxidant rather than as an antioxidant under certain conditions [58]. It is presumed that ferulic acid has a higher probability to act as an efficient antioxidant rather than as a pro-oxidant in general cellular contexts, compared to caffeic acid and *p*-coumaric acid.

There are several methods that can measure PM_{10}-stimulated ROS in cellular models, but each has its own advantages and disadvantages. Direct measurement of the fluorescence of adherent cells can minimize changes that may occur during the extraction process, but unwashed black PM_{10} may affect the fluorescence measurement. In contrast, the method of extracting fluorescent probes from cells can more effectively remove PM_{10} and aggregated cell debris by centrifugation, but cannot completely rule out changes in the extraction process. Therefore, we used the two methods to complement each other. In flow cytometry, PM_{10} can be mistaken for small cells or can form cell aggregates, which can alter light scattering by cells. Gate settings that exclude cells highly affected by PM_{10} may distort the cell population to be analyzed, reducing the reliability of experimental data. There is also a high risk of PM_{10}-generated cell aggregates blocking the flow cell and causing mechanical failure. Thus, a special caution is required when using flow cytometry for the analysis of PM_{10}-treated cells.

5. Conclusions

This study investigated the effects of extracts of Korean propolis, its hydrophilic and lipophilic fractions, and several major components on the viability and oxidative stress of keratinocytes exposed to PM_{10}. In particular, the hydrophilic fraction and phenylpropanoid compounds, such as ferulic acid, contained in this fraction showed antioxidant action to inhibit PM_{10}-induced ROS generation, lipid peroxidation, and glutathione oxidation, suggesting their potential to be used as cosmetic and dermatological active ingredients. Since propolis contains both cytoprotective and cytotoxic components, a purification process to improve its safety and efficacy is required for use in skin protection. Additional in vivo experiments and clinical studies are needed to apply the results of this study to cosmetic or dermatology.

Author Contributions: Investigation, I.A.B., J.W.H. and J.Y.C.; writing, I.A.B. and Y.C.B.; conceptualization, supervision, and funding acquisition, Y.C.B. All authors have read and agreed to the published version of the manuscript.

Funding: This research was supported by a grant from the Korea Health Technology R&D Project through the Korea Health Industry Development Institute (KHIDI), funded by the Ministry of Health and Welfare, Republic of Korea (Grant No: HP20C0004).

Institutional Review Board Statement: Not applicable.

Informed Consent Statement: Not applicable.

Data Availability Statement: The data presented in this study are available in the article.

Conflicts of Interest: The authors declare no conflict of interest.

References

1. Mukherjee, A.; Agrawal, M. World air particulate matter: Sources, distribution and health effects. *Environ. Chem. Lett.* **2017**, *15*, 283–309. [CrossRef]
2. Ngoc, L.T.N.; Park, D.; Lee, Y.; Lee, Y.C. Systematic Review and Meta-Analysis of Human Skin Diseases Due to Particulate Matter. *Int. J. Environ. Res. Public Health* **2017**, *14*, 1458. [CrossRef] [PubMed]
3. Kim, K.E.; Cho, D.; Park, H.J. Air pollution and skin diseases: Adverse effects of airborne particulate matter on various skin diseases. *Life Sci.* **2016**, *152*, 126–134. [CrossRef] [PubMed]
4. Krzyzanowski, M. WHO Air Quality Guidelines for Europe. *J. Toxicol. Environ. Health A* **2008**, *71*, 47–50. [CrossRef]
5. Fuzzi, S.; Baltensperger, U.; Carslaw, K.; Decesari, S.; van Der Gon, H.D.; Facchini, M.C.; Fowler, D.; Koren, I.; Langford, B.; Lohmann, U.; et al. Particulate matter, air quality and climate: Lessons learned and future needs. *Atmos. Chem. Phys.* **2015**, *15*, 8217–8299. [CrossRef]
6. Zhu, X.; Qiu, H.; Wang, L.; Duan, Z.; Yu, H.; Deng, R.; Zhang, Y.; Zhou, L. Risks of hospital admissions from a spectrum of causes associated with particulate matter pollution. *Sci. Total Environ.* **2019**, *656*, 90–100. [CrossRef]
7. Lademann, J.; Schaefer, H.; Otberg, N.; Teichmann, A.; Blume-Peytavi, U.; Sterry, W. Penetration of microparticles into human skin. *Hautarzt* **2004**, *55*, 1117–1119. [CrossRef]
8. Song, S.; Lee, K.; Lee, Y.M.; Lee, J.H.; Il Lee, S.; Yu, S.D.; Paek, D. Acute health effects of urban fine and ultrafine particles on children with atopic dermatitis. *Environ. Res.* **2011**, *111*, 394–399. [CrossRef]
9. Ahn, K. The role of air pollutants in atopic dermatitis. *J. Allergy Clin. Immunol.* **2014**, *134*, 993–999. [CrossRef]
10. Jin, S.P.; Li, Z.; Choi, E.K.; Lee, S.; Kim, Y.K.; Seo, E.Y.; Chung, J.H.; Cho, S. Urban particulate matter in air pollution penetrates into the barrier-disrupted skin and produces ROS-dependent cutaneous inflammatory response in vivo. *J. Derm. Sci.* **2018**, *91*, 175–183. [CrossRef]
11. Verma, V.; Shafer, M.M.; Schauer, J.J.; Sioutas, C. Contribution of transition metals in the reactive oxygen species activity of PM emissions from retrofitted heavy-duty vehicles. *Atmos. Environ.* **2010**, *44*, 5165–5173. [CrossRef]
12. Tsuji, G.; Takahara, M.; Uchi, H.; Takeuchi, S.; Mitoma, C.; Moroi, Y.; Furue, M. An environmental contaminant, benzo(a)pyrene, induces oxidative stress-mediated interleukin-8 production in human keratinocytes via the aryl hydrocarbon receptor signaling pathway. *J. Dermatol. Sci.* **2011**, *62*, 42–49. [CrossRef] [PubMed]
13. Lee, D.U.; Ji, M.J.; Kang, J.Y.; Kyung, S.Y.; Hong, J.H. Dust particles-induced intracellular Ca2+ signaling and reactive Oxygen species in lung fibroblast cell line MRC5. *Korean J. Physiol. Pharmacol.* **2017**, *21*, 327–334. [CrossRef] [PubMed]
14. Ryu, Y.S.; Kang, K.A.; Piao, M.J.; Ahn, M.J.; Yi, J.M.; Hyun, Y.M.; Kim, S.H.; Ko, M.K.; Park, C.O.; Hyun, J.W. Particulate matter induces inflammatory cytokine production via activation of NFkappaB by TLR5-NOX4-ROS signaling in human skin keratinocyte and mouse skin. *Redox Biol.* **2019**, *21*, 101080. [CrossRef] [PubMed]
15. Lee, C.W.; Lin, Z.C.; Hsu, L.F.; Fang, J.Y.; Chiang, Y.C.; Tsai, M.H.; Lee, M.H.; Li, S.Y.; Hu, S.C.; Lee, I.T.; et al. Eupafolin ameliorates COX-2 expression and PGE2 production in particulate pollutants-exposed human keratinocytes through ROS/MAPKs pathways. *J. Ethnopharmacol.* **2016**, *189*, 300–309. [CrossRef] [PubMed]
16. Ha, J.W.; Song, H.; Hong, S.S.; Boo, Y.C. Marine Alga Ecklonia cava Extract and Dieckol Attenuate Prostaglandin E2 Production in HaCaT Keratinocytes Exposed to Airborne Particulate Matter. *Antioxidants* **2019**, *8*, 190. [CrossRef]
17. Boo, Y.C. Can Plant Phenolic Compounds Protect the Skin from Airborne Particulate Matter? *Antioxidants* **2019**, *8*, 379. [CrossRef]
18. Silva-Carvalho, R.; Baltazar, F.; Almeida-Aguiar, C. Propolis: A Complex Natural Product with a Plethora of Biological Activities That Can Be Explored for Drug Development. *Evid. Based Complement. Altern. Med.* **2015**, *2015*, 206439. [CrossRef]
19. Kocot, J.; Kielczykowska, M.; Luchowska-Kocot, D.; Kurzepa, J.; Musik, I. Antioxidant Potential of Propolis, Bee Pollen, and Royal Jelly: Possible Medical Application. *Oxid. Med. Cell. Longev.* **2018**, *2018*, 7074209. [CrossRef]
20. Kurek-Gorecka, A.; Rzepecka-Stojko, A.; Gorecki, M.; Stojko, J.; Sosada, M.; Swierczek-Zieba, G. Structure and antioxidant activity of polyphenols derived from propolis. *Molecules* **2013**, *19*, 78–101. [CrossRef]
21. Daleprane, J.B.; Abdalla, D.S. Emerging Roles of Propolis: Antioxidant, Cardioprotective, and Antiangiogenic Actions. *Evid.-Based Complementary Altern. Med.* **2013**, *2013*, 175135. [CrossRef] [PubMed]
22. Ahn, M.R.; Kumazawa, S.; Hamasaka, T.; Bang, K.S.; Nakayama, T. Antioxidant activity and constituents of propolis collected in various areas of Korea. *J. Agric. Food Chem.* **2004**, *52*, 7286–7292. [CrossRef] [PubMed]
23. Lee, I.K.; Han, M.S.; Kim, D.W.; Yun, B.S. Phenylpropanoid acid esters from Korean propolis and their antioxidant activities. *Bioorganic Med. Chem. Lett.* **2014**, *24*, 3503–3505. [CrossRef] [PubMed]
24. Ahn, M.R.; Kunimasa, K.; Kumazawa, S.; Nakayama, T.; Kaji, K.; Uto, Y.; Hori, H.; Nagasawa, H.; Ohta, T. Correlation between antiangiogenic activity and antioxidant activity of various components from propolis. *Mol. Nutr. Food Res.* **2009**, *53*, 643–651. [CrossRef] [PubMed]

25. Kwon, Y.S.; Park, D.H.; Shin, E.J.; Kwon, M.S.; Ko, K.H.; Kim, W.K.; Jhoo, J.H.; Jhoo, W.K.; Wie, M.B.; Jung, B.D.; et al. Antioxidant propolis attenuates kainate-induced neurotoxicity via adenosine A1 receptor modulation in the rat. *Neurosci. Lett.* **2004**, *355*, 231–235. [CrossRef]
26. Boukamp, P.; Petrussevska, R.T.; Breitkreutz, D.; Hornung, J.; Markham, A.; Fusenig, N.E. Normal keratinization in a spontaneously immortalized aneuploid human keratinocyte cell line. *J. Cell Biol.* **1988**, *106*, 761–771. [CrossRef]
27. Denizot, F.; Lang, R. Rapid colorimetric assay for cell growth and survival. Modifications to the tetrazolium dye procedure giving improved sensitivity and reliability. *J. Immunol. Methods* **1986**, *89*, 271–277. [CrossRef]
28. Eruslanov, E.; Kusmartsev, S. Identification of ROS using oxidized DCFDA and flow-cytometry. *Methods Mol. Biol.* **2010**, *594*, 57–72.
29. Lee, J.W.; Seok, J.K.; Boo, Y.C. Ecklonia cava Extract and Dieckol Attenuate Cellular Lipid Peroxidation in Keratinocytes Exposed to PM10. *Evid.-Based Complemen. Altern. Med.* **2018**, *2018*, 8248323. [CrossRef]
30. Tietze, F. Enzymic method for quantitative determination of nanogram amounts of total and oxidized glutathione: Applications to mammalian blood and other tissues. *Anal. Biochem.* **1969**, *27*, 502–522. [CrossRef]
31. Re, R.; Pellegrini, N.; Proteggente, A.; Pannala, A.; Yang, M.; Rice-Evans, C. Antioxidant activity applying an improved ABTS radical cation decolorization assay. *Free Radic. Biol. Med.* **1999**, *26*, 1231–1237. [CrossRef]
32. Xie, J.; Schaich, K.M. Re-evaluation of the 2,2-diphenyl-1-picrylhydrazyl free radical (DPPH) assay for antioxidant activity. *J. Agric. Food Chem.* **2014**, *62*, 4251–4260. [CrossRef] [PubMed]
33. Kilic, I.; Yesiloglu, Y. Spectroscopic studies on the antioxidant activity of p-coumaric acid. *Spectrochim. Acta A Mol. Biomol. Spectrosc.* **2013**, *115*, 719–724. [CrossRef] [PubMed]
34. Ha, J.W.; Boo, Y.C. Siegesbeckiae Herba Extract and Chlorogenic Acid Ameliorate the Death of HaCaT Keratinocytes Exposed to Airborne Particulate Matter by Mitigating Oxidative Stress. *Antioxidants* **2021**, *10*, 1762. [CrossRef] [PubMed]
35. Boo, Y.C. Natural Nrf2 Modulators for Skin Protection. *Antioxidants* **2020**, *9*, 812. [CrossRef] [PubMed]
36. Seok, J.K.; Lee, J.W.; Kim, Y.M.; Boo, Y.C. Punicalagin and (-)-Epigallocatechin-3-Gallate Rescue Cell Viability and Attenuate Inflammatory Responses of Human Epidermal Keratinocytes Exposed to Airborne Particulate Matter PM10. *Ski. Pharm. Physiol* **2018**, *31*, 134–143. [CrossRef]
37. Boo, Y.C. p-Coumaric Acid as An Active Ingredient in Cosmetics: A Review Focusing on its Antimelanogenic Effects. *Antioxidants* **2019**, *8*, 275. [CrossRef]
38. Boo, Y.C. Human Skin Lightening Efficacy of Resveratrol and Its Analogs: From in Vitro Studies to Cosmetic Applications. *Antioxidants* **2019**, *8*, 332. [CrossRef]
39. Boo, Y.C. Emerging Strategies to Protect the Skin from Ultraviolet Rays Using Plant-Derived Materials. *Antioxidants* **2020**, *9*, 637. [CrossRef]
40. Boo, Y.C. Arbutin as a Skin Depigmenting Agent with Antimelanogenic and Antioxidant Properties. *Antioxidants* **2021**, *10*, 1129. [CrossRef]
41. Pasciu, V.; Posadino, A.M.; Cossu, A.; Sanna, B.; Tadolini, B.; Gaspa, L.; Marchisio, A.; Dessole, S.; Capobianco, G.; Pintus, G. Akt downregulation by flavin oxidase-induced ROS generation mediates dose-dependent endothelial cell damage elicited by natural antioxidants. *Toxicol. Sci.* **2010**, *114*, 101–112. [CrossRef] [PubMed]
42. Posadino, A.M.; Cossu, A.; Giordo, R.; Zinellu, A.; Sotgia, S.; Vardeu, A.; Hoa, P.T.; Deiana, L.; Carru, C.; Pintus, G. Coumaric acid induces mitochondrial damage and oxidative-mediated cell death of human endothelial cells. *Cardiovasc. Toxicol.* **2013**, *13*, 301–306. [CrossRef] [PubMed]
43. Giordo, R.; Cossu, A.; Pasciu, V.; Hoa, P.T.; Posadino, A.M.; Pintus, G. Different redox response elicited by naturally occurring antioxidants in human endothelial cells. *Open Biochem. J.* **2013**, *7*, 44–53. [CrossRef] [PubMed]
44. Bouayed, J.; Bohn, T. Exogenous antioxidants–Double-edged swords in cellular redox state: Health beneficial effects at physiologic doses versus deleterious effects at high doses. *Oxid. Med. Cell Longev.* **2010**, *3*, 228–237. [CrossRef] [PubMed]
45. Russo, A.; Cardile, V.; Sanchez, F.; Troncoso, N.; Vanella, A.; Garbarino, J.A. Chilean propolis: Antioxidant activity and antiproliferative action in human tumor cell lines. *Life Sci.* **2004**, *76*, 545–558. [CrossRef] [PubMed]
46. Frozza, C.O.D.; Garcia, C.S.C.; Gambato, G.; de Souza, M.D.O.; Salvador, M.; Moura, S.; Padilha, F.F.; Seixas, F.K.; Collares, T.; Borsuk, S.; et al. Chemical characterization, antioxidant and cytotoxic activities of Brazilian red propolis. *Food Chem. Toxicol.* **2013**, *52*, 137–142. [CrossRef]
47. Khacha-Ananda, S.; Tragoolpua, K.; Chantawannakul, P.; Tragoolpua, Y. Antioxidant and Anti-cancer Cell Proliferation Activity of Propolis Extracts from Two Extraction Methods. *Asian Pac. J. Cancer Prev.* **2013**, *14*, 6991–6995. [CrossRef]
48. Salim, E.I.; Abd El-Magid, A.D.; Farara, K.M.; Maria, D.S. Antitumoral and Antioxidant Potential of Egyptian Propolis Against the PC3 Prostate Cancer Cell Line. *Asian Pac. J. Cancer Prev.* **2015**, *16*, 7641–7651. [CrossRef]
49. Chen, Y.J.; Shiao, M.S.; Hsu, M.L.; Tsai, H.T.; Wang, S.Y. Effect of caffeic acid phenethyl ester, an antioxidant from propolis, on inducing apoptosis in human leukemic HL-60 cells. *J. Agric. Food Chem.* **2001**, *49*, 5615–5619. [CrossRef]
50. Orsolic, N.; Bendelja, K.; Brbot-Saranovic, A.; Basic, I. Effects of caffeic acid and caffeic acid phenethyl ester, an antioxidants from propolis, on inducing apoptosis in HeLa human cervical carcinoma and Chinese hamster lung V79 fibroblast cells. *Period. Biol.* **2004**, *106*, 367–372.

51. Silva, V.; Genta, G.; Moller, M.N.; Masner, M.; Thomson, L.; Romero, N.; Radi, R.; Fernandes, D.C.; Laurindo, F.R.M.; Heinzen, H.; et al. Antioxidant Activity of Uruguayan Propolis. In Vitro and Cellular Assays. *J. Agric. Food Chem.* **2011**, *59*, 6430–6437. [CrossRef] [PubMed]
52. Nakajima, Y.; Shimazawa, M.; Mishima, S.; Hara, H. Water extract of propolis and its main constituents, caffeoylquinic acid derivatives, exert neuroprotective effects via antioxidant actions. *Life Sci.* **2007**, *80*, 370–377. [CrossRef] [PubMed]
53. Boufadi, Y.M.; Soubhye, J.; Riazi, A.; Rousseau, A.; Vanhaeverbeek, M.; Neve, J.; Boudjeltia, K.Z.; Van Antwerpen, P. Characterization and Antioxidant Properties of Six Algerian Propolis Extracts: Ethyl Acetate Extracts Inhibit Myeloperoxidase Activity. *Int. J. Mol. Sci.* **2014**, *15*, 2327–2345. [CrossRef] [PubMed]
54. Bouaroura, A.; Segueni, N.; Diaz, J.G.; Bensouici, C.; Akkal, S.; Rhouati, S. Preliminary analysis of the chemical composition, antioxidant and anticholinesterase activities of Algerian propolis. *Nat. Prod. Res.* **2020**, *34*, 3257–3261. [CrossRef]
55. Gregoris, E.; Fabris, S.; Bertelle, M.; Grassato, L.; Stevanato, R. Propolis as potential cosmeceutical sunscreen agent for its combined photoprotective and antioxidant properties. *Int. J. Pharm.* **2011**, *405*, 97–101. [CrossRef]
56. Kadoma, Y.; Fujisawa, S. A Comparative Study of the Radical-scavenging Activity of the Phenolcarboxylic Acids Caffeic Acid, p-Coumaric Acid, Chlorogenic Acid and Ferulic Acid, With or Without 2-Mercaptoethanol, a Thiol, Using the Induction Period Method. *Molecules* **2008**, *13*, 2488–2499. [CrossRef]
57. Kilani-Jaziri, S.; Mokdad-Bzeouich, I.; Krifa, M.; Nasr, N.; Ghedira, K.; Chekir-Ghedira, L. Immunomodulatory and cellular antioxidant activities of caffeic, ferulic, and p-coumaric phenolic acids: A structure-activity relationship study. *Drug Chem. Toxicol.* **2017**, *40*, 416–424. [CrossRef]
58. Maurya, D.K.; Devasagayam, T.P.A. Antioxidant and prooxidant nature of hydroxycinnamic acid derivatives ferulic and caffeic acids. *Food Chem. Toxicol.* **2010**, *48*, 3369–3373. [CrossRef]

antioxidants

MDPI

Article

Comparative Study of Antiviral, Cytotoxic, Antioxidant Activities, Total Phenolic Profile and Chemical Content of Propolis Samples in Different Colors from Turkiye

Nazli Boke Sarikahya [1,*,†], Ekin Varol [2,†], Gaye Sumer Okkali [1], Banu Yucel [2], Rodica Margaoan [3,*] and Ayse Nalbantsoy [4]

1 Department of Chemistry, Faculty of Science, Ege University, Bornova, 35100 Izmir, Türkiye
2 Department of Animal Science, Faculty of Agriculture, Ege University, Bornova, 35100 Izmir, Türkiye
3 Advanced Horticultural Research Institute of Transylvania, University of Agricultural Sciences and Veterinary Medicine, 400372 Cluj-Napoca, Romania
4 Department of Bioengineering, Faculty of Engineering, Ege University, Bornova, 35100 Izmir, Türkiye
* Correspondence: nazli.sarikahya@ege.edu.tr (N.B.S.); rodica.margaoan@usamvcluj.ro (R.M.)
† These authors contributed equally to this work.

Abstract: Propolis is a valuable natural substance obtained by honey bees after being collected from the bark, resin of trees, plant leaves and mixed with their saliva, and has been widely used for various biological activities. The properties of propolis can vary widely by botanical origin, location of the hives and colony population. It is thought that the color of propolis is one of the main factors determining its acceptability and originates from the flower markers, pollen and nectar of some plants and is directly related to its chemical content. It is important to compare and standardize the colors, chemical content and biological activities of propolis in our country, which has a rich endemic plant diversity. Thus, in this study, the color indexes of 39 propolis samples from different locations in Turkiye were determined by Lovibond Tintometer, for the first time. The color index, total phenolic content, cytotoxic and antioxidant activities relationship of propolis and two commercial propolis samples were also investigated by HCA and PCA. Turkish propolis, which is defined by its color indices, chemical contents and many different activity potentials, such as antioxidant, antiviral and cytotoxic activity, will find use in many fields from medicine to cosmetics with this study.

Keywords: propolis; antioxidant activity; color measuring; Lovibond Tintometer; antiviral; cytotoxicity; total phenolic content; HCA; PCA

check for updates

Citation: Boke Sarikahya, N.; Varol, E.; Sumer Okkali, G.; Yucel, B.; Margaoan, R.; Nalbantsoy, A. Comparative Study of Antiviral, Cytotoxic, Antioxidant Activities, Total Phenolic Profile and Chemical Content of Propolis Samples in Different Colors from Turkiye. *Antioxidants* **2022**, *11*, 2075. https://doi.org/10.3390/antiox11102075

Academic Editors: Josipa Vlainić and Ivana Tlak Gajger

Received: 30 September 2022
Accepted: 14 October 2022
Published: 21 October 2022

1. Introduction

Propolis is a valuable natural substance that is collected and produced by honey bees (*Apis mellifera*) and it has been widely used for its antioxidant, antibacterial, antifungal and anti-inflammatory properties since the early ages of humanity [1]. The word 'propolis' is originated from Greek 'pro' which means 'in front' and 'polis' which means 'city', describing this natural product that has a function to guard the entrance to the beehives [2,3]. Propolis is resinous material collected by the honeybees from the bark and resin of trees and various plant sources which is obtained after mixing with their saliva. It is collected, transformed and used by bees to seal holes in their honeycombs, smooth out the internal walls and protect the entrance against intruders. It has a strong odor and can be found in hard, frozen, flexible, sticky and liquid forms depending on the temperature. However, it is found in solid form when first collected from the hive and commonly used in liquid forms by dissolving it in solvents such as ethanol, ether and methanol [4]. In general, it is composed of 50% resin and vegetable balsam, 30% wax, 10% essential oils, 5% pollen and 5% various other substances. More than 800 compounds have been identified in this 5% residue [1,5]. These compounds can be listed as flavonoids [6], terpenes [7], phenolics [8], aldehydes [9], steroids [10], carbohydrates [11], aminoacids [3], aliphatic and aromatic acids and esters [12]. It has wide-spectrum biological effects due to its rich chemical and mineral content. Studies have shown that it has antibacterial, antioxidant [13], antiviral [10], antitumor [14], immunomodulatory [15], anti-inflammatory [16], hepatoprotective [17], cardioprotective [18], neuroprotective [19], antidiabetic [20], regenerative [21], anesthetic [22], antiallergic [23] and biological effects. Parallel to these activities, its effects on inflammatory diseases such as gingivitis, osteoarthritis, mastitis, rhinitis and asthma were also investigated [24]. Propolis is widely used in traditional medicine in many countries due to all these features.

Propolis can be classified depending on its physicochemical properties like color, texture and chemical composition. These properties of propolis can vary widely by botanical origin, location of the hives and colony population. It has a wide color range from brown-yellow, brown-green or brown-red to dark-red. The color of propolis is considered one of the main factors determining its acceptability in accordance with previous reports that have revealed most of its floral markers to be its flavonoids/phenolic compounds which come from the nectar or pollen of specific plants [25]. For instance, Birch and Brazilian Baccharis propolis have a greenish color, while the red propolis from the tropics is reddish [26]. Brazilian propolis is famous all over the world as green propolis characterized by higher levels of phenolic compounds, while the dark and black ones are characterized by mostly triterpenoids. It is produced, predominantly in the southeast of the country, in areas of Cerrado. It is obtained from the apical buds and young leaves of *Baccharis dracunculifolia* (Asteraceae) and has a green color as it contains chlorophyll propolis. The main constituents of Brazilian green propolis are prenylated phenylpropanoids and chlorogenic acid. Flavonoids are also constituents of green propolis, as well as condensed tannins [26,27]. Briefly, the chemical composition of propolis varies depending on the plant, region, season, colony and techniques of collecting propolis; the color, smell, medicinal character and chemical composition of each propolis show differences [28].

According to this scientific backing, in this study we focused on the determination of color, which is one of the determinants of floral origin, chemical content and therapeutic properties of propolis. The color of propolis should be defined in future standardization and a criterion in determining the method of use in apitherapeutic applications. It is important to compare and standardize the colors, chemical content and biological activities of propolis in Turkiye, which has a rich endemic plant diversity. Thus, in this study, the color indexes of 39 propolis samples from different locations in Turkiye were determined by Lovibond Tintometer, for the first time. The color index, chemical, total phenolic contents, antiviral, cytotoxic, and antioxidant activity relationships of these propolis samples and two well-known commercial propolis samples were also investigated by HCA and PCA analyses.

2. Materials and Methods

2.1. Materials

Propolis samples were supplied by beekeepers from 39 different geographical regions of Turkiye. The exact collection points and locations can be seen in Table S1 and Figure 1.

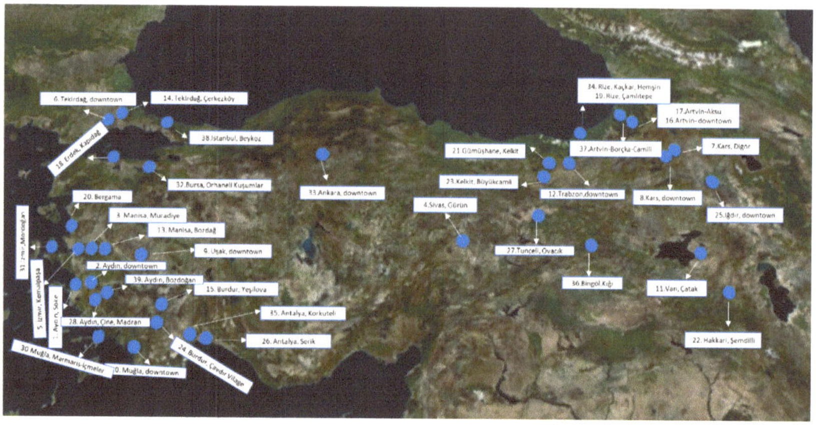

Figure 1. The exact locations of propolis samples collected from Turkiye [29].

2.2. Color Determination

Color determination analysis was done by Lovibond Tintometer (PFX880). This instrument incorporates calibrated color standards for the particular scale of interest and is operated as a stand-alone instrument. The dried 50–100 mg of the propolis samples were dissolved in ethanol in a 5 mL volumetric flask. The flask was kept in an ultrasonic bath mixed and warmly heated until a clear solution was obtained. Then, the solution was filtered through a 0.45 μm Millipore Millex-HV filter and was placed in the tube of the instrument. The samples were kept at 4 °C until the analysis [30,31].

2.3. Chemical Content

The liquid chromatography and high-resolution mass spectrometry (LC-HRMS) methods were developed to analyze the chemical composition of propolis samples from 39 different locations in Turkiye. The identification was performed through the comparison of chromatographic retention times and MS spectra with commercially available standard compounds and the literature findings according to Sarikahya et. al. (2021) [32]. The preparation of samples, chromatographic and optimization conditions for LC-HRMS analysis can be seen in Supplementary Materials.

2.4. Antioxidant Capacity and Total Phenolic Content

The antioxidant activities of propolis extracts were determined using 2,2-diphenyl-1-picrylhydrazyl (DPPH) (Sigma-Aldrich, St. Louis, MS, USA) as a free radical, the CUPRAC total antioxidant capacity (TAC) and the ferric-reducing ability (FRAP) of propolis extracts were analyzed by our previous study. Total phenolic contents were also determined according to the Folin–Ciocalteu colorimetric method [29]. All experiments were done in triplicates and all data were shown as mean ± SD.

2.5. Cytotoxicity Assay

The cytotoxicity of propolis samples was determined by MTT [3-(4,5-dimethyl-2-thiazolyl)-2,5-diphenyl-2H-tetrazolium bromide)] assay [33]. The test is based on the principle of cleavage of MTT that forms formazan crystals by cellular succinate dehydrogenases in viable cells and doxorubicin was used as a positive control [29]. PC-3 (human prostate adenocarcinoma), MDA-MB-231 (human breast adenocarcinoma), HeLa (human epitheloid cervix adenocarcinoma), A-549 (human alveolar adenocarcinoma) cancer cell lines and normal cell line HEK-293 (human embryonic kidney) were used for assessing cytotoxicity of the propolis samples.

2.6. *In Ovo Antiviral Activity*

Antiviral activities of the 39 different propolis samples were measured as virucidal activity against IBV by *in ovo* [34]. Specific pathogen-free embryo chicken eggs (SPF-ECEs) were purchased from Izmir Bornova Veterinary Control Institute, Turkiye. Favipiravir, used as a broad-spectrum antiviral agent, was purchased from a local pharmacy. While evaluating the antiviral activities of propolis samples, the selection was made according to the content of caffeic acid and some flavonoids such as isosakuranetin, naringenin, rhamnocitrin, diosmetin, and chrysin. 5% DMSO was used as vehicle control and favipiravir was used as positive antiviral control [32]. The protocol for the antiviral test was approved by the Ege University, Local Ethical Committee of Animal Experiment (No: 2020-051).

2.7. *Statisitics*

Hierarchical cluster analysis (HCA) and principal component analysis (PCA) were performed using the Paleontological Statistics (PAST) software (version 4.11, Oslo, Norway) [35]. HCA was performed on a Bray–Curtis similarity with complete linkage. Heatmap and dendrograms were generated using the Euclidean distance based on Ward's algorithm for clustering [36].

3. Results and Discussion

Propolis is one of the most important bee products consumed daily as an immune system supporter and antioxidant agent [35]. It is produced in a wide range of different formulations in the world market. Also, there is a variety of propolis types classified depending on color in each country such as Brazilian green propolis, Portugal red propolis, Egyptian red propolis, etc. [36]. In recent studies, it is determined that the study of the correlations between the parameters examined revealed a significant correlation between the phenolic composition, antioxidant activity and color. The chemical content of the commercially available propolis, such as European poplar propolis and Brazilian green and red propolis, has been studied and standardized [37]. Although Turkiye has a rich flora and significant endemic plant diversity, which can be a good source for propolis, color, antioxidant activity potential and chemical content comparisons have not been studied so far. Therefore, this study is dedicated to further providing information about the color index, chemical composition and total antioxidant capacity, antiviral, cytotoxic activities and phenolic content of 39 propolis samples from different locations in Turkiye (Figure 1, Table S1) and two well-known commercial propolis samples.

One of the first physicochemical properties used to commercially describe propolis is its color. As we have seen in many studies in the literature, the color of propolis is an important indicator of biological activity and phenolic content. However, there is no official method for propolis color identification and to the best of our knowledge, there is limited literature references concerning the comparative study of color indices, chemical content and biological activity on propolis. In the study conducted by Coelho et al. (2017), colorimetric analyzes of Southeast Brazilian propolis were performed with a Minolta colorimeter CR-400 device, using the CIELAB color system. In the study, it was shown that the Brazilian green propolis has richer phenolic and poorer flavonoid content than the different colored propolis samples collected from the same regions. They also stated that green propolis, which is rich in phenolics, has a higher capacity to capture free radicals and therefore has a higher antioxidant activity [38]. In another color determination study, the colors were determined by the CIELAB system in physicochemical studies on Portuguese propolis by Falcao et al. (2013). It has been shown that even though the hues of the colors are important, the dark green propolis sample has less phenolic content than the light green. Since antioxidant activity is related to phenolic compounds, it is stated that the antioxidant activity capacity of these dark green compounds is low [39]. On the other hand, Machado et al. (2016) performed a comparative study on four different colors of propolis (yellow, red, brown, red) especially focusing on yellow propolis [40].

In this context, this work aims to study the physicochemical parameter color index of 39 propolis samples with various colors (Figure 2) from different locations in Turkiye by Lovibond Tintometer, for the first time. The tintometer is a subtractive colorimeter, which used red, blue and yellow glass standards. Almost everywhere in the world, the Lovibond color scale is nowadays considered an acceptable means of assigning precise color values to edible oils, waxes and fats [31]. In our study red, blue and yellow glass standards in Lovibond Tintometer were used for color detection of 39 propolis samples and two positive controls. The units of red varied from 0.5 to 42.0 units indicating a color change from pale red to dark brown. The red color index of propolis number **31** (Mordogan district of Izmir province) is detected as 42.0 which is the highest red index among forty-one propolis samples. In our previous study, diosmetin is the most abundant chemical substance in propolis **31**. This propolis also showed significant cytotoxicity against the A549 cell line (IC_{50} value of 3.32 ± 0.21 μg/mL). This variation in activity may result from the variability of the non-major compounds in the extract. Another reason for this red color index is that honey bees collect pitch from roads as an adhesive in propolis. If the color and texture are evaluated, it can be said that such a situation may occur in the Izmir propolis sample. The high red color index in propolis in Izmir-Karaburun-Mordogan also corroborates the possibility that it is due to the presence of common Red Pine (*Pinus brutia*) forests in this region. It is determined that increasing the red color index in propolis also increases the hardness of propolis [41]. Furthermore, propolis obtained from pine trees has a lower wax content and a higher resin content. It is seen that the highest yellow index value of 70 is predominant in propolis samples, and this value is found in eight propolis. The lowest yellow index was determined as 2.7 in Rize propolis number **19**. Rize, Çamlıtepe propolis sample contains the highest amount of ellagic acid (12.87 mg/g) as major phenolic acid, and it was determined that the Eastern Black Sea Region, including Rize, Çamlıtepe, is rich in citrus and pine trees. This propolis, which has the lowest yellow index, also has the lowest CUPRAC total antioxidant capacity (TAC) and total phenolic content according to the Folin–Ciocalteu colorimetric method, but the DPPH antioxidant capacity is high. It is clearly seen that the total phenolic content decreases as the yellow color index decreases. These results are also compatible with the literature [38]. The blue color index which is the highest one is detected as 3.4 for both propolis samples from Sivas-Gürün and Tekirdag (**4** and **6**) which is interesting and strengthens the prediction that it originates from *Juniperus excelsa* (*Brown juniper*), which is common in both regions. In addition, the antioxidant capacity of propolis samples **4** and **6** were determined as 83.05% and 83.13%, respectively, as one of the highest values.

Methylquercetin, an antioxidant flavonoid compound, was detected at high levels in propolis collected from Bergama, Izmir (**20**), Hakkari (**22**) and Igdir (**25**), and the yellow index of these propolis samples is varying between 4.2–24.4 in significant value. *Aesculus hippocastanum* L. (horse chestnut) and *Crataegus* L. (hawthorn) trees were widely spread in these regions. These propolis samples also exhibited considerable toxicity on HeLa cells [32]. Quercetin content, which is another antioxidant component, was the highest value with 54.52 mg/g in the propolis sample from Hakkari (**22**). This propolis sample (**22**) contains significantly more phenolic and flavonoid compounds such as chrysin, caffeic acid phenethyl ester, apigenin, acacetin, quercetin, naringenin, rhamnocitrin and diosmetin than any other sample. When it comes to the plant origin of this propolis sample **22**, *Juglans regia* L., *Quercus* spp. L., *Origanum vulgare* L., *Astragalus* L., *Elaeagnus angustifolia* L., *Cotoneaster* spp. Medik and *Morus alba* L plants spread in the region and are responsible for the chemical content. Epigallocatechin, which is the active ingredient of green tea, was determined in Bergama, Izmir (**20**), Muradiye, Manisa (**3**), Trabzon (**12**) and Gumushane (**21**) of which yellow index and antioxidant activity are considerably high. These results establish a direct relationship between yellow color and antioxidant, cytotoxic activity potentials [32] (Tables S2) (Figures 3 and 4).

Figure 2. The color variety of the raw propolis samples collected from different geographical locations in Turkiye.

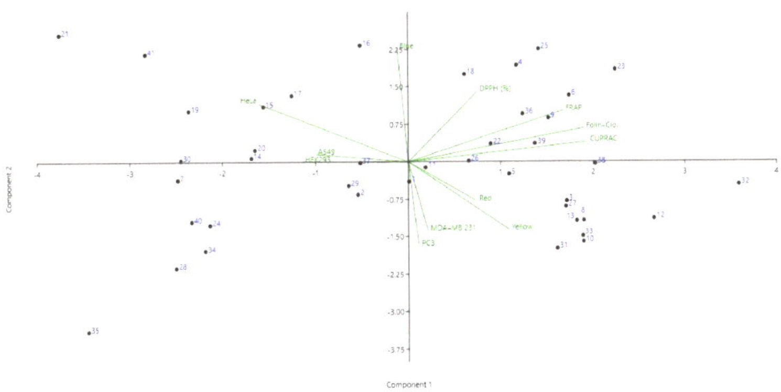

Figure 3. Principal component analysis (PCA) biplot obtained for the propolis samples, color index, antioxidant activities and cytotoxicity. The first two PC explained 65% of the data variance.

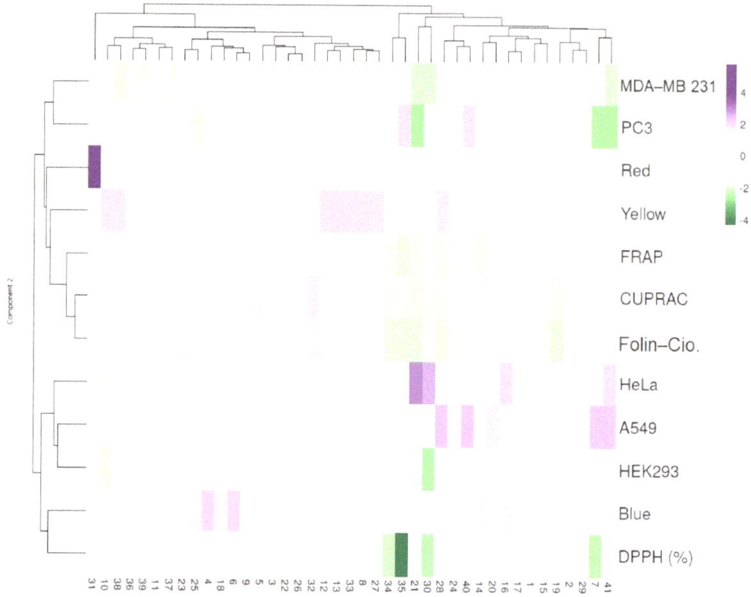

Figure 4. Hierarchical clustering and heat map visualization of the 41 propolis samples based on their color, antioxidant activity and cytotoxicity. Columns indicate the propolis samples and rows the color index, antioxidant activity and cytotoxic assays. Cells are colored based on the quantity in each propolis sample, where purple represents a strong positive correlation and green a strongly negative correlation. The row dendrogram resulted from the correlation between the color index, antioxidant activity and cytotoxic assays; the column dendrogram showed the correlation between propolis samples.

Recent studies proved that the formation of the most common oxidants in the body, including the superoxide (O_2^{-}•), hydroxyl (OH•), peroxyl (ROO•), alkoxyl (RO•) and hydroperoxyl (HO_2) radicals, which are collectively known as reactive oxygen species, has been implicated in the oxidative deterioration of food products, as well as in several human pathologies caused by oxidative stress processes. These free radicals are formed via a reduction reaction of molecular oxygen and generate unoccupied electrons, which cause oxidative stress when they are out of equilibrium [42]. Propolis, a rich source of phenolic and flavonoid compounds, can act as an antioxidant with high potentialities in scavenging free radicals associated with various biological activities. The total antioxidant and phenolic capacities in TR equivalents of the same 39 propolis samples were examined by DPPH, CUPRAC, FRAP, and Folin methods and also chemical contents of this propolis were determined in our previous study [29]. The DPPH free radical scavenging model system is a simple method to evaluate the antioxidant activity of compounds in which the purple chromogen radical 2,2-diphenyl-1-picrylhydrazil (DPPH) is reduced to the corresponding pale-yellow hydrazine by the antioxidant component [38]. According to the DPPH results of this study, the antioxidant activity of only seven samples (numbers: **5**, **7**, **30**, **31**, **33**, **34**, **35**) out of 39 samples was found to be below the tested two commercial propolis samples (Brazilian green propolis and Bio-Bee propolis). The highest antioxidant capacity was found in samples **20**, **17**, **37**, **36**, **16** and **2** varied between 84.77 ± 0.02% and 86.17 ± 0.16%. The red and yellow color index of these propolis samples varies between 1.4–3.0 and 5.5–51.0 Lovibond Tintometer, respectively. While the blue color indexes were found as 2.6 and 2.9 only in the propolis samples numbered **16** and **17**, respectively, this value was found to be 0.1 in the others. It is noteworthy that **16** and **17**, which have similar blue color indices, were collected from the same province. The highest benzaldehyde content was determined in the propolis from Artvin (**16**), and it might arise from the presence of *Brosimum alicastrum* Swartz and *Picea orientalis* (L.) trees. Comparing the commercially available propolis samples **40** (Bio-Bee propolis extract) and **41** (Brazilian green propolis), it is seen that the blue color index and antioxidant capacity increased relative to each other. However, it was not concluded that the increase in the blue color index in Turkish propolis directly increased the antioxidant activity. In addition, the variation of propolis color to green with the increase of blue index and yellow color intensity showed that the chemical content of propolis was rich in phenolics. The number of phenolic compounds caffeic acid, ellagic acid, chlorogenic acid, *trans*-4-hydroxycinnamic acid in these green propolis is ranging between 0.84–636.09, 0.01–12.87, 0.01–0.70 and 1.76–62.41 μg/g, respectively (Table 1). The results obtained in the present work are in agreement with the study conducted by Coelho et al. (2017) [38].

Table 1. Antioxidant capacity, total phenolic contents and color index of propolis samples collected from Turkiye.

No	COLOR DETECTION			Antiox. Act. [a] (%)	Total Phenolic Capacities [a] (mmol TR/g)		
	Red	Yellow	Blue	DPPH	CUPRAC	FRAP	Folin–Cio.
1	1.6	8.1	0.1	81.48 ± 0.01	4.42 ± 0.17	2.01 ± 0.08	7.39 ± 0.25
2	3.0	51.0	0.1	84.77 ± 0.02	1.78 ± 0.35	0.65 ± 0.02	4.49 ± 0.13
3	3.0	34.0	0.1	82.22 ± 0.01	6.17 ± 0.21	1.69 ± 0.06	10.03 ± 0.31
4	3.5	18.0	3.4	83.05 ± 0.05	6.49 ± 0.24	1.97 ± 0.08	8.56 ± 0.32
5	0.5	7.3	0.1	77.49 ± 0.02	8.07 ± 0.09	1.23 ± 0.06	10.49 ± 0.26
6	6.2	7.3	3.4	83.13 ± 0.03	7.07 ± 0.08	2.05 ± 0.09	12.33 ± 0.16
7	3.0	23.0	0.1	67.96 ± 0.01	3.47 ± 0.10	1.06 ± 0.01	4.39 ± 0.09
8	5.8	70.0	0.9	80.35 ± 0.02	7.13 ± 0.19	1.43 ± 0.02	11.42 ± 0.17
9	3.6	11.5	2.6	80.83 ± 0.01	7.59 ± 0.21	2.11 ± 0.09	10.00 ± 0.32
10	8.9	70.0	0.1	81.27 ± 0.02	4.24 ± 0.03	1.32 ± 0.06	8.56 ± 0.22
11	1.5	14.0	0.1	84.22 ± 0.00	3.92 ± 0.09	1.59 ± 0.07	6.86 ± 0.19
12	11.5	70.0	0.1	81.72 ± 0.03	7.34 ± 0.08	1.76 ± 0.08	12.42 ± 0.33
13	3.0	70.0	0.1	79.77 ± 0.15	6.08 ± 0.11	1.76 ± 0.04	9.39 ± 0.36
14	2.4	5.3	2.9	80.67 ± 0.01	3.14 ± 0.13	0.31 ± 0.01	5.56 ± 0.13
15	1.8	12.9	1.9	84.67 ± 0.02	1.47 ± 0.07	2.23 ± 0.09	2.09 ± 0.07
16	3.1	7.6	2.9	84.87 ± 0.03	5.31 ± 0.13	1.68 ± 0.08	12.28 ± 0.36
17	2.6	10.5	2.6	85.76 ± 0.01	3.61 ± 0.09	1.08 ± 0.02	9.14 ± 0.19
18	4.4	13.9	2.9	84.51 ± 0.34	5.03 ± 0.11	2.27 ± 0.09	10.65 ± 0.21
19	1.4	2.7	2.6	81.72 ± 0.43	0.71 ± 0.01	0.34 ± 0.01	0.96 ± 0.02
20	1.4	5.3	0.1	86.17 ± 0.16	2.09 ± 0.04	0.55 ± 0.01	7.57 ± 0.24
21	2.4	4.2	2.6	80.68 ± 0.22	0.75 ± 0.01	0.27 ± 0.01	1.18 ± 0.07
22	3.8	24.4	1.4	79.68 ± 0.01	6.69 ± 0.22	1.60 ± 0.08	9.85 ± 0.19
23	1.2	25.0	2.2	83.59 ± 0.00	7.26 ± 0.24	2.08 ± 0.07	13.53 ± 0.32
24	1.3	20.0	0.1	83.09 ± 0.01	1.70 ± 0.09	0.54 ± 0.01	2.91 ± 0.15
25	3.3	15.0	2.9	81.19 ± 0.02	6.71 ± 0.23	2.17 ± 0.09	9.30 ± 0.21
26	5.7	21.9	0.9	81.34 ± 0.01	5.07 ± 0.19	1.58 ± 0.07	9.10 ± 0.21
27	3.2	70.0	0.1	82.32 ± 0.00	6.20 ± 0.34	2.03 ± 0.04	10.38 ± 0.23
28	5.5	70.0	0.1	84.40 ± 0.02	0.96 ± 0.02	0.32 ± 0.17	1.65 ± 0.07
29	1.5	21.0	0.1	84.28 ± 0.02	2.55 ± 0.05	0.73 ± 0.02	5.28 ± 0.19
30	2.5	16.0	1.6	68.03 ± 0.02	1.17 ± 0.03	0.60 ± 0.04	3.12 ± 0.12
31	42.0	8.9	0.1	77.00 ± 0.04	6.74 ± 0.18	0.96 ± 0.07	8.37 ± 0.22
32	9.9	57.0	0.1	83.26 ± 0.02	8.24 ± 0.31	2.24 ± 0.06	13.43 ± 0.36
33	7.5	70.0	0.1	79.09 ± 0.01	6.04 ± 0.34	1.67 ± 0.03	10.16 ± 0.33
34	1.2	7.6	0.1	69.08 ± 0.00	0.78 ± 0.01	0.32 ± 0.01	1.16 ± 0.09
35	1.4	8.3	0.1	55.98 ± 0.02	1.11 ± 0.03	0.21 ± 0.01	1.24 ± 0.80
36	3.4	10.0	0.1	85.27 ± 0.02	5.00 ± 0.21	1.73 ± 0.07	10.01 ± 0.23
37	2.0	23.0	0.1	85.56 ± 0.07	2.46 ± 0.05	0.88 ± 0.02	6.55 ± 0.34

Table 1. *Cont.*

No	COLOR DETECTION			Antiox. Act. [a] (%)	Total Phenolic Capacities [a] (mmol TR/g)		
	Red	Yellow	Blue	DPPH	CUPRAC	FRAP	Folin–Cio.
38	6.1	70.0	0.9	83.87 ± 0.02	5.53 ± 0.17	1.74 ± 0.07	11.11 ± 0.53
39	2.7	27.0	0.1	80.99 ± 0.13	6.12 ± 0.20	1.71 ± 0.06	9.85 ± 0.39
40 [b]	2.0	4.9	0.1	79.75 ± 0.07	2.67 ± 0.07	1.02 ± 0.02	4.80 ± 0.07
41 [b]	1.7	10.6	0.9	82.07 ± 0.09	2.68 ± 0.11	0.92 ± 0.02	5.47 ± 0.19

[a] Sarikahya et al., 2021 [b] Propolis products available on the market as a positive control, data are expressed as mean ± SD.

Propolis samples numbered **32** demonstrated the highest CUPRAC antioxidant capacity (8.24 ± 0.31 mmol TR g^{-1}) and phenolics contents according to the Folin–Ciocalteu method (13.43 ± 0.36 mmol TR g^{-1}), whereas **19** showed the lowest total antioxidant capacity. This propolis sample was also found to have the lowest antioxidant capacity with 0.71 (mmol TR g^{-1}) CUPRAC and 0.96 (mmol TR g^{-1}) compared to Folin–Ciocalteu methods (Table 1). These results can be attributed to the high yellow index, which was 57.0 Lovibond Tintometer in propolis number **32**, compared to 2.7 Lovibond Tintometer in propolis sample number **19**. It is concluded that the lowest antioxidant capacity and yellow index value were determined in Rize-Çamlıtepe (**19**), which is the sample with the weaker chemical content. It is also anticipated that there is a close relationship between the color of the vegetation, which can be a source of propolis for this region and the antioxidant capacity. The Rize-Çamlıtepe region is covered with broad-leaved forests, and there is a propolis source from the Birch family (Betulaceae), the dominant species of Black Alder (*Alnus glutinosa* subsp. *barbata*). It is determined that as the yellow color index decreases, the hardness and antioxidant activity of propolis decreases [41].

The phenolic and flavonoid compounds are correlated with the cytotoxic activity of propolis. Additionally, other compounds identified in the propolis such as triterpenes and sterols are well-known to be responsible for a variety of infectious diseases such as Alzheimer's, diabetes, hypertension, obesity and cancer [43]. It has been proven in the literature that propolis has cytotoxic activity against Hep-2 (squamous cell carcinoma cell line), Caco-2 (human colon adenocarcinoma), HL60 (human promyelocytic leukemia), MG63 (human osteosarcoma), A549 (human lung adenocarcinoma cell line), MDA-MB-231 (breast cancer cell line), PANC-1 (human pancreatic cancer cell line), HeLa (epitheloid cervix carcinoma) and MCF7 (breast cancer cell line). However, its activity against cell lines HeLa, MCF7 and A549 stands out [43–45]. For this purpose, the cytotoxicity results of different color propolis extracts were discussed through a panel of cancerous and nontumor cells in this study. The results of our previous study exhibited that propolis samples numbered **10** (Mugla), **25** (Igdir), **31** (Izmir-Mordogan), **32** (Bursa) and **38** (Istanbul) had the highest cytotoxicity for HeLa, A549 and PC-3 cancer cell lines (Table 2) (Figure S1) [29]. Propolis extracts with high yellow and red color index were more cytotoxic to HeLa cells followed by A549 cells than other cells. In general, propolis samples with high yellow and red color indexes showed significant cytotoxicity, especially on HeLa cells. The first standout of these is propolis number **10**, which was most active in HeLa cells with an index of 8.9 red and 70.0 yellow. In the literature, we found similar results regarding the cytotoxic effect of propolis against HeLa. They proved that yellow propolis is rich in triterpene and has high cytotoxic activity against human ovarian cancer. However, spectrometric color analysis was not carried out in this study (Machado et al., 2016) [40]. Another study found that green propolis extract exhibited an antagonistic effect with doxorubicin in HeLa cells [43].

Table 2. Cytotoxic activity results and color index of propolis samples collected from Turkiye.

No	COLOR DETECTION			CYTOTOXIC ACTIVITY MTT IC_{50} (μg/mL) [a]				
	Red	Yellow	Blue	MDA-MB 231	HeLa	A549	PC3	HEK293
1	1.6	8.1	0.1	47.82 ± 3.58	16.70 ± 3.35	19.05 ± 3.36	35.26 ± 2.88	43.19 ± 5.14
2	3.0	51.0	0.1	28.33 ± 0.85	17.04 ± 0.57	16.18 ± 3.40	19.79 ± 1.02	13.25 ± 4.55
3	3.0	34.0	0.1	>50	11.63 ± 2.63	14.15 ± 1.97	32.99 ± 5.90	12.21 ± 4.69
4	3.5	18.0	3.4	32.86 ± 2.65	20.14 ± 4.27	15.45 ± 4.23	22.44 ± 1.84	11.35 ± 4.90
5	0.5	7.3	0.1	24.13 ± 4.90	12.44 ± 5.34	15.06 ± 1.33	35.02 ± 2.18	19.61 ± 5.36
6	6.2	7.3	3.4	>50	12.19 ± 3.83	24.38 ± 3.12	33.85 ± 1.19	22.65 ± 2.69
7	3.0	23.0	0.1	49.34 ± 0.50	29.67 ± 4.19	>50	ND	46.95 ± 0.46
8	5.8	70.0	0.9	46.39 ± 4.57	11.96 ± 0.63	16.32 ± 5.69	38.05 ± 5.81	29.72 ± 4.89
9	3.6	11.5	2.6	46.60 ± 5.88	15.75 ± 1.17	17.33 ± 0.24	35.50 ± 5.75	21.99 ± 3.59
10	8.9	70.0	0.1	18.08 ± 5.55	1.78 ± 0.01	7.79 ± 0.33	38.55 ± 3.58	8.61 ± 2.62
11	1.5	14.0	0.1	23.79 ± 3.43	9.68 ± 0.50	20.05 ± 1.02	32.12 ± 4.71	28.46 ± 2.84
12	11.5	70.0	0.1	25.87 ± 2.99	4.46 ± 0.74	17.55 ± 0.54	34.97 ± 1.89	29.22 ± 4.54
13	3.0	70.0	0.1	25.82 ± 4.55	5.88 ± 0.76	14.93 ± 1.65	35.78 ± 1.34	25.82 ± 3.73
14	2.4	5.3	2.9	38.81 ± 4.74	14.37 ± 1.29	36.68 ± 5.52	40.56 ± 5.10	36.17 ± 5.80
15	1.8	12.9	1.9	32.91 ± 2.03	29.14 ± 5.10	16.07 ± 2.45	28.71 ± 4.40	>50
16	3.1	7.6	2.9	28.62 ± 6.34	37.47 ± 2.45	34.58 ± 3.75	29.04 ± 5.58	48.04 ± 2.08
17	2.6	10.5	2.6	35.31 ± 5.83	32 ± 0.68	28.68 ± 0.87	34.55 ± 3.21	>50
18	4.4	13.9	2.9	35.45 ± 5.39	23.20 ± 1.10	28.91 ± 5.22	30.59 ± 2.22	33.09 ± 3.55
19	1.4	2.7	2.6	20.95 ± 1.77	17.42 ± 1.20	22.25 ± 4.06	15.39 ± 4.08	19.60 ± 3.58
20	1.4	5.3	0.1	25.04 ± 5.56	22.40 ± 1.77	42.18 ± 4.81	26.71 ± 5.82	31.16 ± 3.01
21	2.4	4.2	2.6	ND	48.04 ± 0.75	ND	ND	49.60 ± 0.62
22	3.8	24.4	1.4	43.90 ± 4.51	14.47 ± 3.37	21.95 ± 0.40	21.20 ± 2.21	30.68 ± 5.30
23	1.2	25.0	2.2	19.55 ± 5.67	7.67 ± 1.78	15.38 ± 0.52	15.73 ± 0.60	21.28 ± 1.51
24	1.3	20.0	0.1	>50	16.66 ± 3.08	31.84 ± 1.82	33.97 ± 5.80	49.43 ± 1.23
25	3.3	15.0	2.9	16.45 ± 3.85	8.59 ± 0.98	11.78 ± 0.50	8.12 ± 0.56	27.74 ± 3.17
26	5.7	21.9	0.9	35.32 ± 5.25	10.24 ±0.62	21.31 ± 0.13	25.36 ± 4.97	33.20 ± 0.56
27	3.2	70.0	0.1	>50	14.01 ± 0.74	24.82 ± 1.41	28.84 ± 4.79	31.33 ± 1.59
28	5.5	70.0	0.1	>50	25.72 ± 2.07	>50	41.83 ± 2.90	>50
29	1.5	21.0	0.1	41.20 ± 0.41	14.01 ± 3.17	6.88 ± 2.27	19.28 ± 5.58	29.40 ± 4.08
30	2.5	16.0	1.6	ND	44.20 ± 3.68	10.70 ± 3.59	39.03 ± 4.54	ND
31	42.0	8.9	0.1	>50	13.21 ± 1.46	3.32 ± 0.21	19.70 ± 0.18	27.39 ± 3.92
32	9.9	57.0	0.1	42.14 ± 0.85	6.79 ± 2.12	2.84 ± 0.60	22.84 ± 4.74	15.22 ± 3.44
33	7.5	70.0	0.1	>50	4.33 ± 2.09	9.04 ± 3.96	24.27 ± 5.18	39.24 ± 5.54
34	1.2	7.6	0.1	23.95 ± 0.34	14.09 ± 0.51	5.21 ± 0.09	35.57 ± 5.39	20.90 ± 2.09
35	1.4	8.3	0.1	>50	20.06 ± 5.50	16.72 ± 1.18	>50	>50
36	3.4	10.0	0.1	5.86 ± 2.37	11.98 ± 0.38	8.14 ± 0.56	20.29 ± 4.85	14.32 ± 1.13
37	2.0	23.0	0.1	4.83 ± 2.99	15.45 ± 2.50	7.29 ± 3.11	32.18 ± 4.96	31.33 ± 5.81
38	6.1	70.0	0.9	4.10 ± 1.82	13.98 ± 1.70	4.60 ± 0.44	36.47 ± 5.68	21.01 ± 5.59
39	2.7	27.0	0.1	5.38 ± 3.16	6.71 ± 2.16	2.88 ± 0.42	21.23 ± 0.94	32.74 ± 0.36
40 [b]	2.0	4.9	0.1	>50	27.20 ± 2.67	>50	>50	>50
41 [b]	1.7	10.6	0.9	ND	37.64 ± 2.08	>50	ND	>50
	Doxorubicin			13.14 ± 4.24	1.51 ± 0.38	14.09 ± 2.16	>20	1.10 ± 0.01

[a] Sarikahya et al., 2021. [b] Propolis products available on the market as a positive control, data are expressed as mean ± SD.

According to the PCA (Figure 3), from left to right the 1st quadrant highlights samples **21** and **41** with similar antioxidant capacity and cytotoxicity, along with samples **14–17**, **19–21** and **30** with blue index color and moderate cytotoxic activity, particularly in HeLa and A549. Furthermore, an increased antioxidant capacity with regard to DPPH (%) was noticed compared with the other propolis samples, whereas lower capacity was observed in CUPRAC. The 2nd quadrant emphasized samples **23** and **25** with similar cytotoxicity and antioxidant activities, along with samples **4**, **6**, **18**, **22**, **26**, **36** and **39** which exhibited moderate to low cytotoxic activity. In the 3rd quadrant, the propolis samples (**3**, **8**, **10**, **12**, **13**, **31–33**) had lower activity against HeLa and HEK293, whereas higher activities were noticed in MDA-MB 231 and PC3. Furthermore, these samples presented high red and yellow color indexes. The last quadrants presented the propolis samples with the lowest blue color indexes and relatively low CUPRAC and FRAP activities. Out of these, samples **7**, **28**, **35** and **40** demonstrated increased cytotoxicity against all tested lines (Figure 3).

Many studies in the literature demonstrated that most of the propolis samples taken from the temperate zone showed antiviral activity and it is known that flavonoids and esters of phenolic acids are responsible for this activity [46]. Similarly, all 39 propolis samples in our study showed remarkable inhibition of the virus at a concentration of 1 µg/g (Table 3) (Figure S2). The most effective HA titer inhibition was observed as 64 in sample **9** (Usak), which had a blue color index of 2.6, which was higher than the other samples. It also has the best inhibition of HA titer for 0.1 µg/g decreased the virus activity five-fold in comparison with virus control for the 0.1 µg/g concentration. Studies have shown that certain structures in propolis, such as flavonoids and phenolics, cause antiviral effects on the virus. In parallel with the literature, it has been determined that green propolis sample number **9** is rich in terms of flavonoids such as naringenin, rhamnocitrin and phenolic compounds such as caffeic acid. These molecules inhibit the virus by affecting replication mechanisms of viruses and viral envelopes [47].

Table 3. Mortality and HA titers of chosen propolis extracts.

Samples	Concentration (µg/g)	Egg Mortality	% Mortality	HA Titer	HA Titer (log2)
Untreated SPF-ECE control		0/4	0%	0	0
Only virus control		0/4	0%	2048	11
Vehicle control (Virus treated with %5 DMSO)		0/4	0%	2048	11
Favipiravir (Positive antiviral agent)	10	0/4	0%	512	9
	25	0/4	0%	256	8
Propolis Sample **9** (Usak)	0.1	0/4	0%	64	6
	1	1/4	25%	2	1
Propolis Sample **11** (Catak-Van)	0.1	0/4	0%	128	7
	1	1/4	25%	2	1
Propolis Sample **14** (Cerkezköy-Tekirdag)	0.1	0/4	0%	512	9
	1	1/4	25%	2	1
Propolis Sample **19** (Camlitepe-Rize)	0.1	1/4	25%	256	8
	1	0/4	0%	2	1
Propolis Sample **22** (Semdinli-Hakkari)	0.1	1/4	25%	1024	10
	1	0/4	0%	2	1
Propolis Sample **26** (Serik-Antalya)	0.1	0/4	0%	512	9
	1	0/4	0%	2	1
Propolis Sample **30** (İcmeler-Marmaris)	0.1	1/4	25%	256	8
	1	0/4	0%	2	1
Propolis Sample **37** (Borcka-Artvin)	0.1	0/4	0%	256	8
	1	1/4	25%	2	1

To better comprehend the similarities and differences between the propolis samples, a dendrogram of the hierarchical clustering and heatmap was constructed and is presented in Figure 4. The first cluster highlights the propolis samples with red (particularly in 31) and yellow color indexes which were correlated with the antioxidant activities as seen by the increased levels in FRAP, CUPRAC and Folin–Ciocalteu. Conversely, a negative correlation was noticed in the cytotoxicity assay. The following cluster highlights the samples with blue color index which exhibited lower antioxidant activities particularly in DPPH (**7**, **30** and **35**) as seen by the negative correlation. On the contrary, samples **7**, **16**, **21**, **28**, **40** and **41** presented increased cytotoxic activity mainly in HeLa and A549 (Figure 4).

As a consequence, the color index of propolis samples differs according to the plant source. A few studies have been done to determine the color index of propolis in literature [38–40]. The paler the color resulted in the lower the phenolic content and the antioxidant capacity [8]. The correlations between the phenolic composition and the color revealed that the darker propolis showed a higher total phenolic content ($p \leq 0.05$) [39]. It also can be observed that the yellow color was negatively correlated with the phenolic content and with the antioxidant activity ($p < 0.01$) for some propolis samples. Therefore, the yellower and paler the color, the lower the phenolic content and the antioxidant capacity, in accordance with that previously observed. Similarly, another study that determined different Spanish propolis samples showed that the lighter color of Spanish propolis could be due to the collection region that is further to the north than propolis becoming darker as one moves towards the south, due to the differences in the local flora. It also reported a significant correlation between the observed color and the antioxidant activity [8,39].

4. Conclusions

The propolis products, which are becoming increasingly important and are used as a dietary supplement these days attract a great deal of attention in the pharmacy, cosmetic, food industries and apitherapy. Various countries have focused on determining the chemical composition and different biological activities of propolis to establish their own standards for propolis. Turkiye is the second biggest honey producer in the world with its annual production of 81.115 tons and provides a convenient apicultural environment in terms of flowers [48]. Therefore, considering that propolis is similarly a bee product, Turkish propolis, which is defined by its color indices, chemical contents and potential for many different activities such as antioxidant, antiviral and cytotoxic activity, will find use in many fields from medicine to cosmetics [49]. These results also may be defined in future standardization and a criterion in determining the method of use in apitherapeutic applications for propolis samples. In addition, this study will be guided in the formation of many scientific and industrial studies.

Supplementary Materials: The following supporting information can be downloaded at: https://www.mdpi.com/article/10.3390/antiox11102075/s1, Table S1: The exact collection localities of propolis samples from Turkiye; Table S2: Compounds and their amounts (mg/g extract) in extracts of propolis from Turkiye; Figure S1: Cytotoxicity and anti-inflammatory effects of propolis samples; Figure S2: Physiological changes in the embryos after 48h incubation with propolis-virus mixture; Materials and Methods.

Author Contributions: N.B.S., E.V., G.S.O., B.Y., R.M. and A.N.: Conceptualization, Investigation, Writing—original draft. A.N.: Biological activity discussion. N.B.S., E.V., G.S.O., B.Y. and A.N.: Data Curation. N.B.S., E.V., G.S.O., B.Y., R.M. and A.N.: Resources. N.B.S., B.Y. and R.M.: Supervision, Writing—Review & Editing. All authors have read and agreed to the published version of the manuscript.

Funding: The publication was supported by funds from the National Research Development Projects to finance excellence (PFE)-14/2022-2024, granted by the Romanian Ministry of Research and Innovation.

Institutional Review Board Statement: Not applicable.

Informed Consent Statement: Not applicable.

Data Availability Statement: Not applicable.

Conflicts of Interest: The authors declare no conflict of interest.

References

1. Kasote, D.; Bankova, V.; Viljoen, A.M. Propolis: Chemical Diversity and Challenges in Quality Control. *Phytochem. Rev.* **2022**, *24*, 1–25. [CrossRef] [PubMed]
2. Wagh, V.D. Propolis: A Wonder Bees Product and Its Pharmacological Potentials. *Adv. Pharmacol. Sci.* **2013**, *2013*, 308249. [CrossRef]
3. Bankova, V.S.; de Castro, S.L.; Marcucci, M.C. Propolis: Recent Advances in Chemistry and Plant Origin. *Apidologie* **2000**, *31*, 3–15. [CrossRef]
4. Song, Y.S.; Park, E.-H.; Hur, G.M.; Ryu, Y.S.; Kim, Y.M.; Jin, C. Ethanol Extract of Propolis Inhibits Nitric Oxide Synthase Gene Expression and Enzyme Activity. *J. Ethnopharmacol.* **2002**, *80*, 155–161. [CrossRef]
5. Hossain, R.; Quispe, C.; Khan, R.A.; Saikat, A.S.M.; Ray, P.; Ongalbek, D.; Yeskaliyeva, B.; Jain, D.; Smeriglio, A.; Trombetta, D.; et al. Propolis: An Update on Its Chemistry and Pharmacological Applications. *Chin. Med.* **2022**, *17*, 100. [CrossRef]
6. Benkovic, V.; Knezevic, A.H.; Orsolic, N.; Basic, I.; Ramic, S.; Viculin, T.; Knezevic, F.; Kopjar, N. Evaluation of Radioprotective Effects of Propolis and Its Flavonoid Constituents: In Vitro Study on Human White Blood Cells. *Phytother. Res.* **2009**, *23*, 1159–1168. [CrossRef]
7. Popova, M.P.; Chinou, I.B.; Marekov, I.N.; Bankova, V.S. Terpenes with Antimicrobial Activity from Cretan Propolis. *Phytochemistry* **2009**, *70*, 1262–1271. [CrossRef] [PubMed]
8. Revilla, I.; Vivar-Quintana, A.M.; González-Martín, I.; Escuredo, O.; Seijo, C. The Potential of near Infrared Spectroscopy for Determining the Phenolic, Antioxidant, Color and Bactericide Characteristics of Raw Propolis. *Microchem. J.* **2017**, *134*, 211–217. [CrossRef]
9. Sokolov, I.V.; Torgov, I.V. Phenylpropanoid Aldehydes in Propolis and Its Sources. *Chem. Nat. Compd.* **1990**, *26*, 261–262. [CrossRef]
10. El Hady, F.K.A.; Hegazi, A.G. Egyptian Propolis: 2. Chemical Composition, Antiviral And Antimicrobial Activities Of East Nile Delta Propolis. *Z. Nat. C* **2002**, *57*, 386–394. [CrossRef] [PubMed]
11. Pasupuleti, V.R.; Sammugam, L.; Ramesh, N.; Gan, S.H. Honey, Propolis, and Royal Jelly: A Comprehensive Review of Their Biological Actions and Health Benefits. *Oxid. Med. Cell Longev.* **2017**, *2017*, 1259510. [CrossRef] [PubMed]
12. Said, S.A.; Khan, S.A.; Ahmad, I.; Ali, H.S. Chemical Composition of Egyptian and UAE Propolis. *Pak. J. Pharm. Sci.* **2006**, *19*, 58–61.
13. Mašek, T.; Perin, N.; Racané, L.; Cindrić, M.; Čipčić Paljetak, H.; Perić, M.; Matijašić, M.; Verbanac, D.; Radić, B.; Šuran, J.; et al. Chemical Composition, Antioxidant and Antibacterial Activity of Different Extracts of Poplar Type Propolis. *Croat. Chem. Acta.* **2018**, *91*, 81–89. [CrossRef]
14. Noureddine, H.; Hage-Sleiman, R.; Wehbi, B.; Fayyad-Kazan, H.; Hayar, S.; Traboulssi, M.; Alyamani, O.A.; Faour, W.H.; ElMakhour, Y. Chemical Characterization and Cytotoxic Activity Evaluation of Lebanese Propolis. *Biomed. Pharmacother.* **2017**, *95*, 298–307. [CrossRef]
15. Santos, L.A.; Rosalen, P.L.; Dias, N.A.; Grisolia, J.C.; Nascimento Gomes, B.J.; Blosfeld-Lopes, L.; Ikegaki, M.; de Alencar, S.M.; Burger, E. Brazilian Red Propolis Shows Antifungal and Immunomodulatory Activities against Paracoccidioides Brasiliensis. *J. Ethnopharmacol.* **2021**, *277*, 114181. [CrossRef] [PubMed]
16. Catchpole, O.; Mitchell, K.; Bloor, S.; Davis, P.; Suddes, A. Anti-Gastrointestinal Cancer Activity of Cyclodextrin-Encapsulated Propolis. *J. Funct. Foods* **2018**, *41*, 1–8. [CrossRef]
17. Shukla, S.; Bhadauria, M.; Jadon, A. Evaluation of Hepatoprotective Potential of Propolis Extract in Carbon Tetrachloride Induced Liver Injury in Rats. *Indian J. Biochem. Biophys.* **2005**, *42*, 321–325. [PubMed]
18. Neto, J.C.; Paulino, E.T.; Rodrigues, A.K.B.F.; da Silva, J.C.G.; Bernardino, A.C.; dos Santos Oliveira, J.M.; do Nascimento, T.G.; de Souza Oliveira, W.; Santos, J.C.C.; Smaniotto, S.; et al. Cardioprotective Effect of Hydroalcoholic Extract of Brazilian Red Propolis against Isoproterenol-Induced Myocardial Infarction in Rats. *Phytomed. Plus.* **2022**, *2*, 100190. [CrossRef]
19. Nanaware, S.; Shelar, M.; Sinnathambi, A.; Mahadik, K.R.; Lohidasan, S. Neuroprotective Effect of Indian Propolis in β-Amyloid Induced Memory Deficit: Impact on Behavioral and Biochemical Parameters in Rats. *Biomed. Pharmacother.* **2017**, *93*, 543–553. [CrossRef] [PubMed]
20. el Rabey, H.A.; Al-Seeni, M.N.; Bakhashwain, A.S. The Antidiabetic Activity of *Nigella Sativa* and Propolis on Streptozotocin-Induced Diabetes and Diabetic Nephropathy in Male Rats. *Evid. Based Complementary Altern. Med.* **2017**, *2017*, 5439645. [CrossRef]
21. Ahangari, Z.; Naseri, M.; Jalili, M.; Mansouri, Y.; Mashhadiabbas, F.; Torkaman, A. Effect of Propolis on Dentin Regeneration and the Potential Role of Dental Pulp Stem Cell in Guinea Pigs. *Cell J.* **2012**, *13*, 223–228. [PubMed]
22. da Rosa, C.; Bueno, I.L.; Quaresma, A.C.M.; Longato, G.B. Healing Potential of Propolis in Skin Wounds Evidenced by Clinical Studies. *Pharmaceuticals* **2022**, *15*, 1143. [CrossRef] [PubMed]
23. Liew, K.Y.; Kamise, N.I.; Ong, H.M.; Aw Yong, P.Y.; Islam, F.; Tan, J.W.; Tham, C.L. Anti-Allergic Properties of Propolis: Evidence From Preclinical and Clinical Studies. *Front Pharmacol.* **2021**, *12*, 785371. [CrossRef] [PubMed]
24. Peycheva, S.; Apostolova, E.; Gardjeva, P.; Peychev, Z.; Kokova, V.; Angelov, A.; Slavov, A.; Murdjeva, M. Effect of Bulgarian Propolis on the Oral Microflora in Adolescents with Plaque-Induced Gingivitis. *Rev. Bras. Farmacogn.* **2019**, *29*, 271–277. [CrossRef]
25. Alvarez-Suarez, J.M.; González- Paramás, A.M.; Santos-Buelga, C.; Battino, M. Antioxidant Characterization of Native Monofloral Cuban Honeys. *J. Agric. Food. Chem.* **2010**, *58*, 9817–9824. [CrossRef] [PubMed]

26. Salatino, A.; Teixeira, É.W.; Negri, G.; Message, D. Origin and Chemical Variation of Brazilian Propolis. *Evid. Based Complementary Altern. Med.* **2005**, *2*, 33–38. [CrossRef] [PubMed]
27. Righi, A.A.; Negri, G.; Salatino, A. Comparative Chemistry of Propolis from Eight Brazilian Localities. *Evid. Based Complementary Altern. Med.* **2013**, *2013*, 267878. [CrossRef] [PubMed]
28. Keskin, Ş.; Yatanaslan, L.; Karlidag, S. FARKLI İLLERDEN TOPLANAN PROPOLİS ÖRNEKLERİNİN KİMYASAL KARAKTERİZASYONU. *Uludağ Arıcılık Dergisi* **2020**, *20*, 81–88. [CrossRef]
29. Böke Sarıkahya, N.; Gören, A.C.; Sümer Okkalı, G.; Çöven, F.O.; Orman, B.; Kırcı, D.; Yücel, B.; Kışla, D.; Demirci, B.; Altun, M.; et al. Chemical Composition and Biological Activities of Propolis Samples from Different Geographical Regions of Turkey. *Phytochem. Lett.* **2021**, *44*, 129–136. [CrossRef]
30. Xin, J.H. *Total Colour Management in Textiles*; Elsevier: Amsterdam, The Netherlands, 2006; pp. 44–56.
31. Tan, Y.A.; Kuntom, A.; Lee, C.K.; Low, K.S. Comparative Evaluation of Palm Oil Color Measurement Using a Prototype Palm Oil Colorimeter. *J. Am. Oil. Chem. Soc.* **2004**, *81*, 733–736. [CrossRef]
32. Boke Sarikahya, N.; Goren, A.C.; Sumer Okkali, G.; Kırmızıgül, S. Quantification of Saponins from Twenty-Two Cephalaria Species. *Rec. Nat. Prod.* **2021**, *15*, 537–546. [CrossRef]
33. Mosmann, T. Rapid Colorimetric Assay for Cellular Growth and Survival: Application to Proliferation and Cytotoxicity Assays. *J. Immunol. Methods* **1983**, *65*, 55–63. [CrossRef]
34. Ghoke, S.S.; Sood, R.; Kumar, N.; Pateriya, A.K.; Bhatia, S.; Mishra, A.; Dixit, R.; Singh, V.K.; Desai, D.N.; Kulkarni, D.D.; et al. Evaluation of Antiviral Activity of Ocimum Sanctum and Acacia Arabica Leaves Extracts against H9N2 Virus Using Embryonated Chicken Egg Model. *BMC Complement Altern. Med.* **2018**, *18*, 174. [CrossRef]
35. Sforcin, J.M. Propolis and the Immune System: A Review. *J. Ethnopharmacol.* **2007**, *113*, 1–14. [CrossRef]
36. Wieczorek, P.P.; Hudz, N.; Yezerska, O.; Horčinová-Sedláčková, V.; Shanaida, M.; Korytniuk, O.; Jasicka-Misiak, I. Chemical Variability and Pharmacological Potential of Propolis as a Source for the Development of New Pharmaceutical Products. *Molecules* **2022**, *27*, 1600. [CrossRef] [PubMed]
37. Popova, M.; Silici, S.; Kaftanoglu, O.; Bankova, V. Antibacterial Activity of Turkish Propolis and Its Qualitative and Quantitative Chemical Composition. *Phytomedicine* **2005**, *12*, 221–228. [CrossRef] [PubMed]
38. Coelho, J.; Falcão, S.I.; Vale, N.; Almeida-Muradian, L.B.; Vilas-Boas, M. Phenolic Composition and Antioxidant Activity Assessment of Southeastern and South Brazilian Propolis. *J. Apic. Res.* **2017**, *56*, 21–31. [CrossRef]
39. Falcão, S.I.; Tomás, A.; Vale, N.; Gomes, P.; Freire, C.; Vilas-Boas, M. Phenolic Quantification and Botanical Origin of Portuguese Propolis. *Ind. Crops Prod.* **2013**, *49*, 805–812. [CrossRef]
40. Machado, C.S.; Mokochinski, J.B.; de Lira, T.O.; de Cassia Evangelista de Oliveira, F.; Cardoso, M.V.; Ferreira, R.G.; Frankland Sawaya, A.C.H.; Ferreira, A.G.; Pessoa, C.; Cuesta-Rubio, O.; et al. Comparative Study of Chemical Composition and Biological Activity of Yellow, Green, Brown, and Red Brazilian Propolis. *Evid. Based Complementary Altern. Med.* **2016**, *2016*, 6057650. [CrossRef] [PubMed]
41. Akcay, C.; Ayata, U.; Birinci, E.; Yalcin, M.; Kolayli, S. Some Physical, Biological, Hardness, and Color Properties of Wood Impregnated with Propolis. *Forestist* **2022**, *72*, 283–293. [CrossRef]
42. Hunyadi, A. The Mechanism(s) of Action of Antioxidants: From Scavenging Reactive Oxygen/Nitrogen Species to Redox Signaling and the Generation of Bioactive Secondary Metabolites. *Med. Res. Rev.* **2019**, *39*, 2505–2533. [CrossRef] [PubMed]
43. Surek, M.; Fachi, M.M.; de Fatima Cobre, A.; de Olivera, F.F.; Pontarolo, R.; Crisma, A.R.; de Souza, W.M.; Felipe, K.B. Chemical Composition, Cytotoxicity, and Antibacterial Activity of Propolis from Africanized Honeybees and Three Different Meliponini Species. *J. Ethnopharmacol.* **2021**, *269*, 113662. [CrossRef] [PubMed]
44. Forma, E.; Bryś, M. Anticancer Activity of Propolis and Its Compounds. *Nutrients* **2021**, *13*, 2594. [CrossRef] [PubMed]
45. El-aziz, E.A.E.A.; Elgayar, S.F.; Mady, F.M.; Abourehab, M.A.S.; Hasan, O.A.; Reda, L.M.; Alaaeldin, E. The Potential of Optimized Liposomes in Enhancement of Cytotoxicity and Apoptosis of Encapsulated Egyptian Propolis on Hep-2 Cell Line. *Pharmaceutics* **2021**, *13*, 2184. [CrossRef] [PubMed]
46. Magnavacca, A.; Sangiovanni, E.; Racagni, G.; Dell'Agli, M. The Antiviral and Immunomodulatory Activities of Propolis: An Update and Future Perspectives for Respiratory Diseases. *Med. Res. Rev.* **2022**, *42*, 897–945. [CrossRef] [PubMed]
47. Schnitzler, P.; Neuner, A.; Nolkemper, S.; Zundel, C.; Nowack, H.; Sensch, K.H.; Reichling, J. Antiviral Activity and Mode of Action of Propolis Extracts and Selected Compounds. *Phytother. Res.* **2010**, *24*, 20–28. [CrossRef] [PubMed]
48. Yılmaz, H.; Yavuz, Ö. Content of Some Trace Metals in Honey from South-Eastern Anatolia. *Food Chem.* **1999**, *65*, 475–476. [CrossRef]
49. Popova, M.P.; Bankova, V.S.; Bogdanov, S.; Tsvetkova, I.; Naydenski, C.; Marcazzan, G.L.; Sabatini, A.-G. Chemical Characteristics of Poplar Type Propolis of Different Geographic Origin. *Apidologie* **2007**, *38*, 306–311. [CrossRef]

antioxidants

MDPI

Article

Chemical Components and Antioxidant Activity of *Geotrigona* sp. and *Tetragonisca fiebrigi* Stingless Bee Cerumen Reduce Juglone-Induced Oxidative Stress in *Caenorhabditis elegans*

Isamara Carvalho Ferreira [1], Raíssa Cristina Darroz Côrrea [1], Sarah Lam Orué [1], Daniel Ferreira Leite [1], Paola dos Santos da Rocha [1], Claudia Andrea Lima Cardoso [2], Rosilda Mara Mussury [3], Patricia Vit [4], Kely de Picoli Souza [1], Edson Lucas dos Santos [1] and Jaqueline Ferreira Campos [1,*]

1 Research Group on Biotechnology and Bioprospecting Applied to Metabolism (GEBBAM), Federal University of Grande Dourados, Rodovia Dourados-Itahum, Km 12, Dourados 79804-970, MS, Brazil; isamara.ferreira137@academico.ufgd.edu.br (I.C.F.); raissa.correa416@academico.ufgd.edu.br (R.C.D.C.); sarah.orue005@academico.ufgd.edu.br (S.L.O.); daniel.leite059@academico.ufgd.edu.br (D.F.L.); paolasantosrocha@ufgd.edu.br (P.d.S.d.R.); kelypicoli@ufgd.edu.br (K.d.P.S.); edsonsantos@ufgd.edu.br (E.L.d.S.)

2 Course of Chemistry, State University of Mato Grosso do Sul, Rodovia Dourados-Itahum, Km 12, Dourados 79804-970, MS, Brazil; claudia@uems.br

3 Faculty of Biological and Environmental Sciences, Federal University of Grande Dourados, Rodovia Dourados-Itahum, Km 12, Dourados 79804-970, MS, Brazil; maramussury@ufgd.edu.br

4 Apitherapy and Bioactivity, Food Science Department, Faculty of Pharmacy and Bioanalysis, Universidad de Los Andes, Mérida 5101, Venezuela

* Correspondence: jaquelinefcampos@ufgd.edu.br

Citation: Ferreira, I.C.; Côrrea, R.C.D.; Orué, S.L.; Leite, D.F.; da Rocha, P.d.S.; Cardoso, C.A.L.; Mussury, R.M.; Vit, P.; de Picoli Souza, K.; dos Santos, E.L.; et al. Chemical Components and Antioxidant Activity of *Geotrigona* sp. and *Tetragonisca fiebrigi* Stingless Bee Cerumen Reduce Juglone-Induced Oxidative Stress in *Caenorhabditis elegans*. *Antioxidants* **2023**, *12*, 1276. https://doi.org/10.3390/antiox12061276

Academic Editors: Josipa Vlainić and Ivana Tlak Gajger

Received: 10 May 2023
Revised: 5 June 2023
Accepted: 7 June 2023
Published: 15 June 2023

Abstract: Cerumen is a bee product produced exclusively by stingless bees, resulting from a mixture of beeswax and plant resins. The antioxidant activity of bee products has been investigated since oxidative stress is associated with the onset and progression of several diseases that can lead to death. In this context, this study aimed to investigate the chemical composition and antioxidant activity of cerumen produced by the *Geotrigona* sp. and *Tetragonisca fiebrigi* stingless bees, in vitro and in vivo. The chemical characterization of cerumen extracts was performed by HPLC, GC, and ICP OES analyses. The in vitro antioxidant potential was evaluated by $DPPH^{\bullet}$ and $ABTS^{\bullet+}$ free radical scavenging methods, and in human erythrocytes subjected to oxidative stress with AAPH. In vivo, the antioxidant potential was evaluated in *Caenorhabditis elegans* nematodes subjected to oxidative stress with juglone. Both cerumen extracts presented phenolic compounds, fatty acids, and metallic minerals in their chemical constitution. The cerumen extracts showed antioxidant activity by capturing free radicals, reducing lipid peroxidation in human erythrocytes, and reducing oxidative stress in *C. elegans*, observed by the increase in viability. The results obtained indicate that cerumen extracts from *Geotrigona* sp. and *Tetragonisca fiebrigi* stingless bees may be promising against oxidative stress and associated diseases.

Keywords: bioprospecting; meliponines; oxidative stress

1. Introduction

Stingless bees are included in the subfamily Meliponinae (Hymenoptera, Apidae) consisting of 58 genera and more than 550 species [1]. These meliponines play a crucial role in plant pollination, being responsible for about 40–90% of the pollination of native species or those cultivated in the tropics [2]. Furthermore, bee products from stingless bees, which include honey, propolis, wax, and cerumen, are used in folk medicine [3] and have pharmacological activities reported in the literature, such as antioxidant action [4–7].

The antioxidant activity of bee products can be related to the presence of bioactive molecules [8], including phenolic compounds, aromatic acids, alcohols, and sugars [9], that are capable of neutralizing oxidative stress effects [10]. Oxidative stress results from the imbalance between the production of reactive species, such as oxygen, and the number of antioxidant agents to neutralize them [11]. This state of imbalance can be associated with the onset and/or progression of several diseases, such as diabetes, cardiovascular diseases, and cancer, which are among the leading causes of death in the world today [12,13].

From this perspective, we highlight the stingless bees *Geotrigona* sp. (Moure, 1943) and *Tetragonisca fiebrigi* (Schwarz, 1938), distributed in different countries of Latin America, as producers of cerumen [14,15]. Cerumen is a product elaborated exclusively by meliponines, consisting of a mixture of beeswax and plant resins [16]. Previous studies have described the biological activities of cerumen from other bee species, such as antioxidant activity [4]. However, no studies report the chemical composition and the biological potential of the cerumen produced by *Geotrigona* sp. and *Tetragonisca fiebrigi*.

To evaluate the pharmacological effects of different substances, especially on the modulation of oxidative stress, the *Caenorhabditis elegans* model has been widely used due to the homology of its genes with human genes [17]. In this context, this study aimed to investigate the chemical composition and evaluate the antioxidant activity of the cerumen produced by the stingless bees *Geotrigona* sp. and *T. fiebrigi* in vitro and nematodes *C. elegans*.

2. Materials and Methods

2.1. Preparation of Ethanolic Extracts of Cerumen from Geotrigona sp. (EEC-G) and Tetragonisca fiebrigi (EEC-T)

The cerumen of the stingless bee *Geotrigona* sp. was purchased in Quimís, Manabí province, Ecuador, and that of *Tetragonisca fiebrigi* was collected in Dourados, Mato Grosso do Sul, Brazil, after Dalmo Henrique Franco Silva specialist identification. The cerumen was stored at −20 °C until the preparation of the extract.

The cerumen samples were solubilized at a ratio of 4.5 mL with 80% ethanol for each gram of cerumen [18]. The solutions were kept in a water bath at 70 °C in a closed container and periodically homogenized until complete dissolution. Finally, the material was filtered to obtain ethanolic extracts of cerumen from *Geotrigona* sp. (EEC-G) and *T. fiebrigi* (EEC-T).

2.2. Chemical Composition

2.2.1. HPLC Analysis

The phenolic compounds were identified and quantified using ultra-high-efficiency liquid chromatography (UHPLC) coupled to a diode array detector (DAD) with wavelength monitoring between 200 and 800 nm. A Shimadzu Shim-pack XR-ODS column (Shimadzu, Kyoto, Japan) with a particle size of 2.2 μm and a particle size of 2 mm × 75 mm was used. The injection rate employed was 2 μL with a 0.4 mL/min flow rate. The gradient used consisted of (A) water with 0.1% acetic acid (*v/v*) and (B) acetonitrile with 0.1% acetic acid (*v/v*). Identification was performed using analytical standards (3,4-hydroxycinnamic acid, catechin, gallic acid, ferulic acid, coumaric acid, maleic acid, rutin, quercertin, apigenin, and vanillin) obtained from Merck (Darmstadt, Germany) A curve was developed with the analytical standards at concentrations between 1 and 1000 μg/mL for quantification.

2.2.2. GC Analysis

The fatty acid content was determined using the method described by Bligh and Dier [19]. For fatty acid extraction, 1 mL of EEC-G or EEC-T was mixed with 10 mL of chloroform, 20 mL of methanol, and 8 mL of distilled water (1:2:0.8) and kept in constant homogenization for 30 min. Subsequently, 10 mL of chloroform and 10 mL of the 1.5% sodium sulfate solution were added and kept in homogenization for two minutes. Then, the excess water and methanol were removed from the tubes.

The material remained in an oven at 100 °C until the evaporation of the solvent. After drying, the material was solubilized, and the sample was methylated with 0.25 mol/L sodium methylate in diethyl ether methanol (1:1). Gas chromatography (GC) equipped with a flame ionization detector was performed for the analyses. An SP2560 30 m × 0.25 mm × 0.20 m column was used with a temperature gradient: 80 °C, 0 min, 7 °C/min up to 240 °C; injector (1/30 split) at 220 °C and detector at 250 °C. Hydrogen was used as the carrier gas (2.0 mL/min) with an injection volume of 1 μL. Peak identification was performed by comparison with FAME C8–C22 methylated fatty acid standards. Quantification was performed by external standardization.

2.2.3. ICP-OES Analysis

The metal mineral content was determined by acid extraction using nitric acid (HNO_3) and hydrogen peroxide (H_2O_2) 30%. To determine the metal content, the samples were prepared with deionized water. The glassware and bottles used for storage and/or sample preparation were decontaminated with nitric acid (HNO_3) 10% (*v/v*) beforehand, incubated for 24 h, and then washed with deionized water. For the assay, EEC-G or EEC-T were subjected to acid extraction, performed in borosilicate glass tubes in a digester block with a timer (SL-25/40, Solab, Piracicaba, Brazil), and subjected to pre-digestion with 3.5 mL of HNO_3 for 24 h at room temperature (overnight). After this period, they were digested for 45 min at 150 °C. Then, 1.5 mL H_2O_2 was added and kept for another 45 min in the digester block under the same heating. The blank only contained the mixture of HNO_3 and H_2O_2.

The obtained solution was diluted to 10 mL with deionized water and kept at rest for cooling. The analyses were performed by inductively coupled plasma optical emission spectrometry (ICP-OES). The equipment was adjusted using solutions between 10 and 100 μg/L containing the elements Al, Ba, Ca, Cu, Fe, K, Mg, Mn, Na, P, S, and Zn and the following operational parameters: radiofrequency power 1500 W, radiofrequency generator 27.12 MHz, sample aspiration flow rate 1.5 mL/min, argon flow rate (L/min) 15.00 (plasma), and 1.00 (auxiliary) 0.45 (nebulizer). The multi-element determination of the sample was expressed in mg/L and the analyses were performed in triplicate.

2.3. *In Vitro Antioxidant Activity*

2.3.1. 2,2-Diphenyl-1-picrylhydrazyl (DPPH$^{\bullet}$) Free Radical Scavenging

For the in vitro evaluation of the antioxidant activity of EEC-G and EEC-T, the DPPH-free radical capture assay was performed [20]. For this, 200 μL of EEC-G or ECC-T (0.001–3 mg/mL) was added to 1800 μL of 0.11 mM DPPH$^{\bullet}$. The solution was kept in the dark at room temperature for 30 min. Then, the absorbance (Abs) was measured in a spectrophotometer at 517 nm. Ascorbic acid and butylated hydroxytoluene (BHT) were used as the standard. Three independent experiments were performed in duplicate. The percentage of DPPH$^{\bullet}$ capture was calculated according to the following equation:

$$\text{DPPH-capturing activity (\%)} = (1 - \text{Abs sample}/\text{Abs control}) \times 100, \quad (1)$$

and the concentration able to inhibit 50% of the free radical (IC_{50}) was calculated.

2.3.2. Elimination of the Free Radical 2,2′-Azino-bis(3-ethylbenzothiazoline-6-sulfonic acid) (ABTS$^{\bullet+}$)

The antioxidant activity of the cerumen extracts was also evaluated using a 2,2′-azinobis(3-ethylbenzothiazoline-6-sulfonic acid) radical capture assay (ABTS$^{\bullet+}$) [21]. The ABTS$^{\bullet+}$ radical was prepared by adding 88 μL of a potassium persulfate solution (140 mM) to 5 mL of the aqueous ABTS solution (7 mM). After 12 h, this solution was diluted in absolute ethanol until an absorbance of 0.700 ± 0.05 at 734 nm was obtained. Then, 20 μL of EEC-G or EEC-T (0.001–2 mg/mL) was added to 1980 μL of ABTS$^{\bullet+}$ radical solution and incubated for 6 min in the dark at room temperature. The absorbance was measured in a spectrophotometer at 734 nm.

Ascorbic acid and BHT were used as reference antioxidants. Three independent experiments were performed in duplicate. The percentage of $ABTS^{\bullet+}$ radical inhibition was calculated according to the following equation:

$$\text{ABTS radical inhibition (\%)} = ((\text{Abs control} - \text{Abs sample})/\text{Abs control}) \times 100, \quad (2)$$

and the concentration able to inhibit 50% of the free radical (IC_{50}) was calculated.

2.4. *Antioxidant Activity in Human Erythrocytes*

After approval by the Research Ethics Committee of the University Center of Grande Dourados (UNIGRAN), Brazil (CEP process number 123/12-UNIGRAN), 20 mL of peripheral blood was collected from a healthy donor in tubes containing sodium citrate. The material was centrifuged at 1500 rpm for 10 min. Subsequently, the blood plasma and leukocyte layers were discarded. The erythrocytes were washed three times with 0.9% sodium chloride (NaCl) solution. A 10% red cell suspension in 0.9% NaCl was prepared for the assays.

2.4.1. Hemolysis and Oxidative Hemolysis

The anti-hemolytic assay was performed on human erythrocytes to evaluate the ability of cerumen extracts to provide antioxidant protection in human cells [5]. For this, a 10% erythrocyte suspension was pre-incubated at 37 °C for 30 min in the presence of different concentrations of EEC-G or EEC-T (0.01–1 mg/mL). Then, to evaluate the effect of the extracts on hemolysis and oxidative hemolysis, 0.9% NaCl or the oxidizing agent 2,2′-Azobis(2-amidinopropane) dihydrochloride (AAPH) 50 mM, respectively, was added. The samples were kept at 37 °C for 2 h under constant stirring. After this period, the samples were centrifuged at 3000 rpm for 5 min, and 200 µL of the supernatant was added to 1800 µL of 0.9% NaCl. The absorbance was measured in a spectrophotometer at 540 nm. A 0.36% ethanol solution was used as a solvent control. A solution of ascorbic acid (0.01–1 mg/mL) was used as a positive control. Total erythrocyte hemolysis was induced with distilled water. Three independent experiments were performed in duplicate. The percentage of hemolysis was determined by the equation:

$$\text{Hemolysis (\%)} = (\text{Abs sample}/\text{Abs total hemolysis}) \times 100. \quad (3)$$

2.4.2. Malondialdehyde Dosage (MDA)

To evaluate the efficiency of cerumen extracts on lipid peroxidation, the MDA dosage was realized in human erythrocytes subjected to oxidative stress induced with AAPH [5]. For this, a 10% erythrocyte suspension was pre-incubated at 37 °C for 30 min with different concentrations of EEC-G or EEC-T (0.01–1 mg/mL). Subsequently, a 50 mM AAPH solution was added and the samples were kept at 37 °C for 2 h under constant stirring. After this period, the samples were centrifuged at 3000 rpm for 5 min. Then, 500 µL of the supernatant was collected and added to 1 mL of 10 nM thiobarbituric acid (TBA). All samples were incubated at 96 °C for 45 min and then cooled for 15 min. After this period, 4 mL of n-butyl alcohol was added to the samples, which were centrifuged at 3000 rpm for 5 min. The absorbance of 2 mL of the supernatant was measured in a spectrophotometer at 532 nm. A 0.36% ethanol solution was used as a solvent control. Ascorbic acid (0.01–1 mg/mL) was used as a positive control. MDA at 20 µM was used as a standard. Three independent experiments were performed in duplicate. MDA content was expressed in nmol/mL, obtained by the following equation:

$$\text{MDA (nmol/mL)} = \text{Abs sample} \times (20 \times 220.32/\text{Abs MDA standard}). \quad (4)$$

2.5. *Antioxidant Activity in C. elegans*

To perform the in vivo assays, the wild-type strain N2 of *C. elegans* nematodes, obtained from the *Caenorhabditis* Genetics Center (University of Minnesota, Minneapolis, MN,

USA), was used. The animals were kept in Petri dishes containing Nematode Growth Medium (NGM) agar and fed with *Escherichia coli* OP50 bacteria. For the assays, the nematodes were synchronized by treatment with hypochlorite (1%) and sodium hydroxide (NaOH 5 M) [22].

2.5.1. Acute Toxicity Assay

The acute toxicity assay was performed to evaluate the effect of EEC-G and EEC-T on the viability of *C. elegans* [23]. For this, 10 to 20 synchronized nematodes at the L4 stage were transferred to 96-well plates containing M9 medium. Then, different concentrations of EEC-G or EEC-T (0.005–1 mg/mL) were added. The treated nematodes were kept in BOD at 20 °C for 24 h. After this period, the viability of the nematodes was evaluated under a stereomicroscope. The individuals were considered alive when they moved after being touched with a platinum micro spatula. Three experiments were performed in triplicate.

2.5.2. Oxidative Stress Resistance Assay

To evaluate the effect of EEC-G and EEC-T on the oxidative stress induced by 5-hydroxynaphthalene-1,4-dione (juglone) in *C. elegans*, the oxidative stress resistance assay was performed [24]. For this, 10 to 20 nematodes at the L4 stage were transfected into 96-well plates containing liquid M9 medium. The nematodes were pretreated with different concentrations of EEC-G or EEC-T (0.01–1 mg/mL) for 1 h at 20 °C. Then, the animals were exposed to the oxidizing agent juglone (40 μM) and incubated at 20 °C for 24 h. After this period, the nematode viability was assessed using a stereomicroscope. The individuals were considered alive when they moved after being touched with a platinum micro spatula. Three experiments were performed in triplicate.

2.6. Statistical Analysis

The data obtained were expressed as the mean ± standard error of the mean (SEM). The 50% inhibitory concentration (IC_{50}) with 95% confidence limits was determined by non-linear regression. For analysis and comparison between the experimental groups, univariate analysis of variance (ANOVA) was used, followed by Dunnett's post hoc test. The results were considered significant when $p \leq 0.05$.

3. Results

3.1. Chemical Composition

The phenolic compounds, fatty acids, and metal minerals identified in EEC-G and EEC-T are presented in Table 1. HPLC analysis revealed the presence of different phenolic compounds for EEC-G and EEC-T. Among the phenolic compounds present in the cerumen extracts, catechin for EEC-G and rutin for EEC-T were the most noteworthy.

Table 1. Phenolic compounds, fatty acids, and metallic minerals were identified in the ethanolic extract of cerumen from *Geotrigona* sp. (EEC-G) and *Tetragonisca fiebrigi* (EEC-T).

Compounds	EEC-G (mg/L)	EEC-T (mg/L)
Phenolic compounds		
3,4-hydroxycinnamic acid	3.040 ± 0.032	-
Catechin	10.000 ± 0.044	-
Gallic acid	-	6.473 ± 0.020
Rutin	-	12.993 ± 0.022
Vanillin	4.020 ± 0.035	-
Dicarboxylic acids		
Maleic acid	1.970 ± 0.023	-
Fatty acids		
Caprylic acid	0.303 ± 0.003	0.347 ± 0.003
Capric acid	1.607 ± 0.007	1.503 ± 0.003
Lauric acid	2.590 ± 0.012	2.543 ± 0.009

Table 1. *Cont.*

Compounds	EEC-G (mg/L)	EEC-T (mg/L)
Myristic acid	9.973 ± 0.035	10.583 ± 0.015
Pentadecanoic acid	0.103 ± 0.003	0.103 ± 0.003
Palmitic acid	37.193 ± 0.143	35.953 ± 0.046
Margaric acid	0.103 ± 0.003	0.107 ± 0.003
Stearic acid	10.983 ± 0.020	11.447 ± 0.009
Arachidic acid	3.347 ± 0.020	3.417 ± 0.018
Behenic acid	3.067 ± 0.023	2.987 ± 0.015
Myristoleic acid	7.447 ± 0.035	7.107 ± 0.020
Palmitoleic acid	8.683 ± 0.035	9.173 ± 0.032
Oleic acid	12.083 ± 0.071	12.027 ± 0.032
Linoleic acid	13.710 ± 0.059	13.523 ± 0.020
Linolenic acid	14.237 ± 0.041	14.023 ± 0.024
Metallic minerals		
Aluminum (Al)	0.713 ± 0.012	0.593 ± 0.009
Barium (Ba)	0.263 ± 0.003	0.293 ± 0.009
Calcium (Ca)	1.510 ± 0.017	1.343 ± 0.231
Copper (Cu)	0.260 ± 0.006	0.290 ± 0.017
Iron (Fe)	0.713 ± 0.012	0.620 ± 0.015
Potassium (K)	1.623 ± 0.024	1.630 ± 0.035
Magnesium (Mg)	1.710 ± 0.015	1.630 ± 0.059
Manganese (Mn)	0.303 ± 0.007	0.303 ± 0.012
Sodium (Na)	0.310 ± 0.010	0.347 ± 0.015
Zinc (Zn)	0.227 ± 0.003	0.217 ± 0.003
Non-metallic minerals		
Phosphorus (P)	0.777 ± 0.015	0.720 ± 0.015
Sulfur (S)	0.253 ± 0.003	0.267 ± 0.009

The values are expressed as the mean ± SEM. Analyses were performed in triplicate.

The GC analysis showed no difference in the composition of fatty acids between the cerumen extracts. EEC-G and EEC-T presented 14 fatty acids, with an emphasis on palmitic acid. The ICP OES analysis did not reveal any difference in the composition of metallic minerals between the cerumen extracts. EEC-G and EEC-T showed 12 metallic minerals, including copper, magnesium, and zinc.

3.2. *In Vitro Antioxidant Activity*

EEC-G and EEC-T showed antioxidant activity by scavenging DPPH• and ABTS•+ free radicals (Table 2).

Table 2. In vitro antioxidant activity of the ethanolic extract of cerumen from *Geotrigona* sp. (EEC-G) and from *Tetragonisca fiebrigi* (EEC-T).

Sample	DPPH•	ABTS•+
	IC_{50} (mg/mL)	IC_{50} (mg/mL)
Ascorbic acid	0.004 ± 0.00029	0.003 ± 0.00006
BHT	0.031 ± 0.005	0.009 ± 0.0009
EEC-G	1.001 ± 0.062	0.496 ± 0.040
EEC-T	1.251 ± 0.068	0.254 ± 0.023

BHT: butylated hydroxytoluene; DPPH•: 2,2-diphenyl-1-picrylhydrazyl; ABTS•+: 2,2′-azino-bis(3-ethylbenz othiazoline-6-sulfonic acid); IC_{50}: concentration capable of inhibiting 50% of the free radical. The values are expressed as the mean ± SEM. Three independent experiments were performed in duplicate.

In DPPH• free radical scavenging assay, EEC-G showed lower IC_{50} when compared to EEC-T. In the ABTS•+ free radical scavenging assay, EEC-T showed two times lower IC_{50} than EEC-G.

3.3. Antioxidant Activity in Human Erythrocytes

EEC-G and EEC-T showed antioxidant activity in human erythrocytes (Figure 1). EEC-T did not promote hemolysis in human erythrocytes at the evaluated concentrations when compared to erythrocytes incubated with 0.9% NaCl only (control). However, EEC-G induced hemolysis in human erythrocytes at the concentrations of 0.5 and 1 mg/mL in a similar manner to ascorbic acid at the concentration of 1 mg/mL (Figure 1A).

When human erythrocytes were incubated with the oxidizing agent AAPH, EEC-T reduced hemolysis at concentrations of 0.1, 0.25, 0.5, and 1 mg/mL by 52.37%, 62.92%, 51.58%, and 58.20%, respectively, when compared to erythrocytes incubated with AAPH alone. EEC-G reduced hemolysis in human erythrocytes subjected to oxidative stress with the oxidizing agent AAPH at concentrations of 0.01, 0.1, and 0.25 mg/mL by 36.78%, 54.80%, and 32.02%, respectively, compared to erythrocytes incubated with AAPH only (Figure 1B).

EEC-T reduced the lipid peroxidation of human erythrocytes promoted by oxidative stress induced by the oxidizing agent AAPH, observed by lower MDA content generated at concentrations of 0.1, 0.25, 0.5, and 1 mg/mL by 52.81%, 72.01%, 66.66%, and 58.48%, respectively, compared to erythrocytes incubated with AAPH only. EEC-G reduced MDA content at the concentration of 0.1 mg/mL by 37.86% compared to erythrocytes incubated with AAPH alone (Figure 1C).

3.4. Acute Toxicity and Antioxidant Activity in C. elegans

The effect of EEC-G and EEC-T on *C. elegans* viability is shown in Figure 2A. EEC-T showed no toxicity at the evaluated concentrations. However, EEC-G reduced the viability of *C. elegans* at the concentration of 1 mg/mL by 22.43%.

EEC-G and EEC-T showed antioxidant activity in *C. elegans* exposed to the oxidizing agent juglone (Figure 2B). EEC-T reduced the oxidative stress damage promoted by juglone in *C. elegans*, observed by the increase in viability at concentrations of 0.1 and 1 mg/mL by 33.87% and 46.63%, respectively, when compared to nematodes treated with juglone only. EEC-G increased viability at 0.1 mg/mL by 38.04% compared to nematodes treated with juglone only.

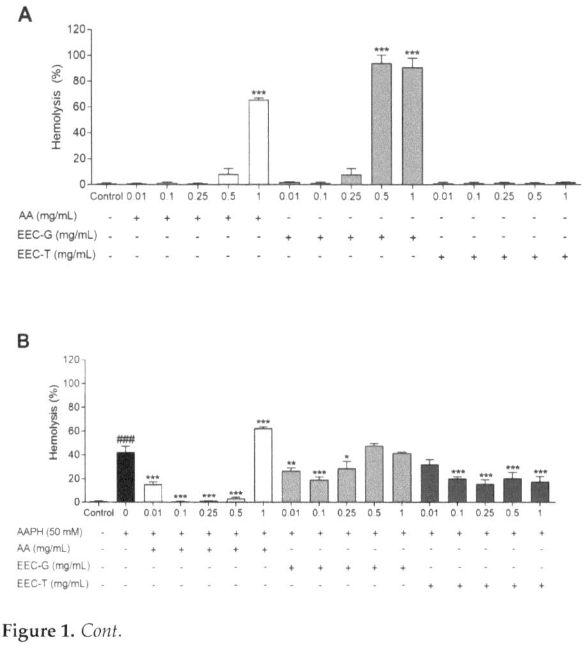

Figure 1. *Cont.*

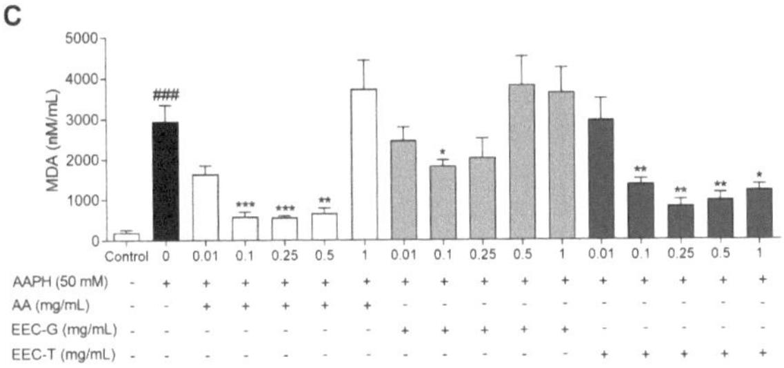

Figure 1. Antioxidant activity in human erythrocytes: (**A**) percentage of hemolysis of human erythrocytes treated with different concentrations of ascorbic acid, EEC-G, and EEC-T (0.01–1 mg/mL); (**B**) percentage of hemolysis AAPH and (**C**) MDA content of human erythrocytes treated with different concentrations of ascorbic acid, EEC-G, and EEC-T (0.01–1 mg/mL) and induced to oxidative stress with the oxidant agent AAPH. AA: ascorbic acid; EEC-G: ethanolic extract of *Geotrigona* sp. cerumen; EEC-T: ethanolic extract of *Tetragonisca fiebrigi* cerumen; AAPH: 2,2′-Azobis(2-amidinopropane) dihydrochloride; MDA: malondialdehyde. +: presence; −: absence. # versus control; * versus 0 (AAPH 50 mM); ### $p < 0.001$; * $p < 0.05$; ** $p < 0.01$; *** $p < 0.001$. Values are expressed as the mean ± SEM. Three independent experiments were performed in duplicate.

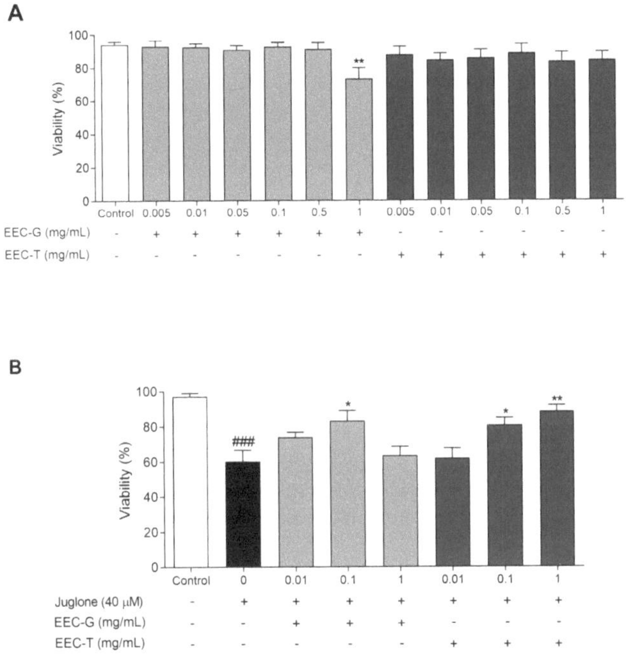

Figure 2. Acute toxicity and antioxidant activity in *C. elegans*: (**A**) percentage of viability of *C. elegans* treated with different concentrations of EEC-G and EEC-T (0.005–1 mg/mL); (**B**) percentage of viability of *C. elegans* treated with different concentrations of EEC-G and EEC-T (0.01–1 mg/mL) and induced oxidative stress with the oxidant agent juglone. EEC-G: ethanolic extract of *Geotrigona* sp. cerumen; EEC-T: ethanolic extract of *Tetragonisca fiebrigi* cerumen. +: presence; −: absence. # versus control; * versus 0 (juglone 40 μM); ### $p < 0.001$; * $p < 0.05$; ** $p < 0.01$. Values are expressed as the mean ± SEM. Three independent experiments were performed in triplicate.

4. Discussion

This study shows, for the first time, the chemical composition and the antioxidant activity of the cerumen of stingless bees *Geotrigona* sp. and *T. fiebrigi* in vitro, on human erythrocytes, and in vivo, on *C. elegans* nematodes.

The chemical constitution of *Geotrigona* sp. and *T. fiebrigi* showed phenolic compounds, fatty acids, and metallic minerals. The chemical profile of the cerumen differed mainly in their phenolic constituents. EEC-G presented a higher proportion of flavonoid catechin, while EEC-T presented a higher concentration of flavonoid rutin. Among the fatty acids, palmitic acid stood out, and among the metallic minerals, manganese, zinc, copper, and iron were present in both extracts.

Phenolic compounds are described by their antioxidant potential, attributed to their chemical structure, consisting of an aromatic ring and free hydroxyls. Thus, these hydroxyls can donate their electrons to stabilize radical molecules [25].

The EEC-G showed better antioxidant activity by capturing the $DPPH^{\bullet}$ radical, while the EEC-T showed better antioxidant activity by capturing the $ABTS^{\bullet+}$ radical. The difference in the response of cerumen extracts is probably related to its phenolic profile, since the rutin present in EEC-T is a glycoside that has the flavonolic aglycone quercetin (lipophilic) and the disaccharide rutinose (hydrophilic) in its chemical structure, which together have many free hydroxyls and amphiphilic characteristics, which makes the antioxidant result more evident using the $ABTS^{\bullet+}$ free radical scavenging method [26,27].

Besides stabilizing radical molecules, phenolic compounds can also modulate the activity of antioxidant enzymes, contributing to the response to oxidative stress [25]. EEC-T reduced oxidative stress at concentrations of 0.1 to 1 mg/mL in human erythrocytes, observed by the lower percentage of hemolysis and corroborated by the lower malondialdehyde content generated. Rutin, only present in EEC-T, besides eliminating free radicals directly, is also able to increase the activity of the antioxidant enzyme glutathione peroxidase (GPx) by increasing the production of glutathione reductase (GR) responsible for maintaining the substrate for GPx action. Rutin also inhibits xanthine oxidase, which is involved in the generation of free radicals [28].

Additionally, antioxidant enzymes such as superoxide dismutase and catalase depend on metallic minerals such as iron, copper, zinc, and manganese for their function, which were observed in the chemical spectra of both extracts. Therefore, the inhibition of free radicals directly, together with the increase in the activity of antioxidant enzymes, may contribute to the reduction in lipid peroxidation observed in human erythrocytes submitted to oxidative stress.

EEC-G showed an antioxidant effect at a 0.1 mg/mL concentration and pro-oxidant effect at higher concentrations (0.5 and 1 mg/mL). Some antioxidants, such as ascorbic acid, have been described to have pro-oxidant potential at high concentrations [29]. Catechin, one of the EEC-G constituents, has been described as having a pro-oxidant action by its ability to induce pore formation that increases mitochondrial permeability [30]. Moreover, its pro-oxidant activity can be attributed to the increase in catechin-derived oxidants after the initial elimination of the superoxide radical anion [31,32].

In *C. elegans*, EEC-T ameliorated juglone-induced oxidative stress at concentrations of 0.1 and 1 mg/mL, and EEC-G only showed activity at 0.1 mg/mL and reduced viability at the highest concentration (1 mg/mL), probably due to its pro-oxidant characteristic at high concentrations.

Juglone is an organic compound capable of generating large amounts of the radical superoxide anion, which induces oxidative stress and reduces nematode viability [33]. During the redox imbalance promoted by the oxidizing agent juglone, signaling pathways, such as DAF-16/FOXO, may have been upregulated by the chemical compounds present in the cerumen extracts. DAF-16 activation increases the gene expression of *sod-3*, which triggers the activity of the mitochondrial antioxidant system through the conversion of the superoxide radical anion by the SOD-3 enzyme to hydrogen peroxide, which is considered less deleterious and, subsequently, can be converted into water and oxygen by catalase (CAT) and GPx enzymes [34].

In previous studies, it was shown that rutin, also present in EEC-T, reduced the levels of reactive oxygen species, activated DAF-16 migration, and increased *sod-3* expression in *C. elegans* [35–37]. Furthermore, in other biological models, rutin reduced MDA levels and increased the activity of antioxidant enzymes [38,39].

Additionally, it has been reported that palmitic acid and oleic acid, also present in *Geotrigona* sp. and *T. fiebrigi*, extracts, can increase the resistance to oxidative stress in *C. elegans* by activating DAF-16 [40,41]. Moreover, the presence of other constituents in the extracts, such as manganese, may contribute to the efficiency in the activity of the antioxidant enzyme SOD-3 in *C. elegans*.

5. Conclusions

Together, these results show, for the first time, the chemical composition and the effect of cerumen from *Geotrigona* sp. and *T. fiebrigi* stingless bees on the reduction of oxidative stress in human erythrocytes and *C. elegans*. Our results provide new perspectives for the development of future studies investigating the mechanisms of the antioxidant action of these products, as well as for potential use in diseases associated with oxidative stress.

Author Contributions: Conceptualization, I.C.F., R.C.D.C., S.L.O., D.F.L., P.d.S.d.R., K.d.P.S., E.L.d.S. and J.F.C.; methodology, I.C.F., R.C.D.C., S.L.O., D.F.L., P.d.S.d.R. and C.A.L.C.; software, I.C.F., R.C.D.C., S.L.O., D.F.L., P.d.S.d.R., K.d.P.S., E.L.d.S. and J.F.C.; validation, I.C.F., P.d.S.d.R., K.d.P.S., E.L.d.S., J.F.C., C.A.L.C., R.M.M. and P.V.; formal analysis, I.C.F., P.d.S.d.R., K.d.P.S., E.L.d.S., J.F.C. and C.A.L.C.; investigation, I.C.F., R.C.D.C., S.L.O., D.F.L., P.d.S.d.R., K.d.P.S., E.L.d.S. and J.F.C.; resources, K.d.P.S., E.L.d.S. and J.F.C.; data curation, I.C.F., R.C.D.C., P.d.S.d.R., K.d.P.S., E.L.d.S., J.F.C., C.A.L.C., R.M.M. and P.V.; writing—original draft preparation, I.C.F., R.C.D.C., S.L.O., D.F.L., P.d.S.d.R., K.d.P.S., E.L.d.S. and J.F.C.; writing—review and editing, I.C.F., R.C.D.C., S.L.O., D.F.L., P.d.S.d.R., K.d.P.S., E.L.d.S., J.F.C., C.A.L.C., R.M.M. and P.V.; visualization, I.C.F., R.C.D.C., P.d.S.d.R., K.d.P.S., E.L.d.S., J.F.C., C.A.L.C., R.M.M. and P.V.; supervision, P.d.S.d.R., K.d.P.S., E.L.d.S. and J.F.C.; project administration, K.d.P.S., E.L.d.S. and J.F.C.; funding acquisition, K.d.P.S., E.L.d.S. and J.F.C. All authors have read and agreed to the published version of the manuscript.

Funding: This research was funded by Pró-Reitoria de Ensino de Pós-Graduação e Pesquisa da Universidade Federal da Grande Dourados (PROPP-UFGD); Fundação de Apoio ao Desenvolvimento do Ensino, Ciência e Tecnologia do Estado de Mato Grosso do Sul; Coordenação de Aperfeiçoamento de Pessoal de Nível Superior (CAPES), Conselho Nacional de Desenvolvimento Científico e Tecnológico (CNPq); and Financiadora de Estudos e Projetos (Finep).

Institutional Review Board Statement: The study was conducted in accordance with the Declaration of Helsinki, and approved by the Ethics Committee of UNIGRAN (protocol code 123/12 on 20 November 2012).

Informed Consent Statement: Informed consent was obtained from all subjects involved in the study.

Data Availability Statement: The data presented in this study are available on request from the corresponding author.

Conflicts of Interest: The authors declare no conflict of interest.

References

1. Grüter, C. Evolution and Diversity of Stingless Bees. In *Stingless Bees*; Fascinating Life Sciences; Springer: Cham, Switzerland, 2020; pp. 43–86. [CrossRef]
2. Meléndez Ramírez, V.; Ayala, R.; Delfín González, H. Crop Pollination by Stingless Bees. In *Pot-Pollen in Stingless Bee Melittology*; Vit, P., Pedro, S., Roubik, D., Eds.; Springer: Cham, Switzerland, 2018; pp. 139–153. [CrossRef]
3. Al-Hatamleh, M.A.I.; Boer, J.C.; Wilson, K.L.; Plebanski, M.; Mohamud, R.; Mustafa, M.Z. Antioxidant-Based Medicinal Properties of Stingless Bee Products: Recent Progress and Future Directions. *Biomolecules* **2020**, *10*, 923. [CrossRef]
4. Hamilton, K.D.; Brooks, P.R.; Ogbourne, S.M.; Russell, F.D. Natural products isolated from Tetragonula carbonaria cerumen modulate free radical-scavenging and 5-lipoxygenase activities in vitro. *BMC Complement. Altern. Med.* **2017**, *17*, 232. [CrossRef]

5. Campos, J.F.; dos Santos, U.P.; Rocha, P.D.S.D.; Damião, M.J.; Balestieri, J.B.P.; Cardoso, C.A.L.; Paredes-Gamero, E.J.; Estevinho, L.M.; Souza, K.D.P.; dos Santos, E.L. Antimicrobial, Antioxidant, Anti-Inflammatory, and Cytotoxic Activities of Propolis from the Stingless Bee *Tetragonisca fiebrigi* (Jataí). *Evid. -Based Complement. Altern. Med.* **2015**, *2015*, 296186. [CrossRef]
6. Lopes, A.J.O.; Vasconcelos, C.C.; Garcia, J.B.S.; Pinheiro, M.S.D.; Pereira, F.A.N.; Camelo, D.D.S.; de Morais, S.V.; Freitas, J.R.B.; da Rocha, C.Q.; Ribeiro, M.N.D.S.; et al. Anti-Inflammatory and Antioxidant Activity of Pollen Extract Collected by *Scaptotrigona affinis postica*: In silico, in vitro, and in vivo Studies. *Antioxidants* **2020**, *9*, 103. [CrossRef] [PubMed]
7. dos Santos, H.F.; Campos, J.F.; dos Santos, C.M.; Balestieri, J.B.P.; Silva, D.B.; Carollo, C.A.; Souza, K.D.P.; Estevinho, L.M.; Dos Santos, E.L. Chemical Profile and Antioxidant, Anti-Inflammatory, Antimutagenic and Antimicrobial Activities of Geopropolis from the Stingless Bee *Melipona orbignyi*. *Int. J. Mol. Sci.* **2017**, *18*, 953. [CrossRef]
8. Martinello, M.; Mutinelli, F. Antioxidant Activity in Bee Products: A Review. *Antioxidants* **2021**, *10*, 71. [CrossRef] [PubMed]
9. Kocot, J.; Kiełczykowska, M.; Luchowska-Kocot, D.; Kurzepa, J.; Musik, I. Antioxidant Potential of Propolis, Bee Pollen, and Royal Jelly: Possible Medical Application. *Oxid. Med. Cell. Longev.* **2018**, *2018*, 7074209. [CrossRef] [PubMed]
10. Durazzo, A.; Lucarini, M.; Plutino, M.; Pignatti, G.; Karabagias, I.K.; Martinelli, E.; Souto, E.B.; Santini, A.; Lucini, L. Antioxidant Properties of Bee Products Derived from Medicinal Plants as Beekeeping Sources. *Agriculture* **2021**, *11*, 1136. [CrossRef]
11. Sies, H. Oxidative Stress: Concept and some Practical Aspects. *Antioxidants* **2020**, *9*, 852. [CrossRef]
12. Pizzino, G.; Irrera, N.; Cucinotta, M.; Pallio, G.; Mannino, F.; Arcoraci, V.; Squadrito, F.; Altavilla, D.; Bitto, A. Oxidative Stress: Harms and Benefits for Human Health. *Oxid. Med. Cell. Longev.* **2017**, *2017*, 8416763. [CrossRef]
13. The Top 10 Causes of Death. Available online: https://www.who.int/news-room/fact-sheets/detail/the-top-10-causes-of-death (accessed on 10 February 2023).
14. Camargo, J.M.F.; Pedro, S.R.M.; Melo, G.A.R.; Meliponini Lepeletier. 1836. In Catalogue of Bees (Hymenoptera, Apoidea) in the Neotropical Region; Moure, J.S., Urban, D., Melo, G.A.R., (Orgs). 2013. Available online: http://www.moure.cria.org.br/catalogue (accessed on 10 February 2023).
15. Camargo, J.M.F.; Moure, J.S. Meliponini neotropicais: O gênero *Geotrigona* Moure, 1943 (Apinae, Apidae, Hymenoptera), com especial referência à filogenia e biogeografia. *Arq. Zool.* **1996**, *33*, 95–161. [CrossRef]
16. Roubik, D.W. Stingless bee nesting biology. *Apidologie* **2006**, *37*, 124–143. [CrossRef]
17. Ishii, N.; Ishii, T.; Hartman, P.S. Oxidative Stress and C. elegans Models. In *Aging Mechanisms*; Mori, N., Mook-Jung, I., Eds.; Springer: Tokyo, Japan, 2015; pp. 111–122. [CrossRef]
18. Alencar, S.; Oldoni, T.; Castro, M.; Cabral, I.; Costa-Neto, C.; Cury, J.; Rosalen, P.; Ikegaki, M. Chemical composition and biological activity of a new type of Brazilian propolis: Red propolis. *J. Ethnopharmacol.* **2007**, *113*, 278–283. [CrossRef] [PubMed]
19. Bligh, E.G.; Dyer, W.J. A rapid method of total lipid extraction and purification. *Can. J. Biochem. Physiol.* **1959**, *37*, 912–917. [CrossRef]
20. Gupta, D.; Gupta, R.K. Bioprotective properties of Dragon's blood resin: In vitro evaluation of antioxidant activity and antimicrobial activity. *BMC Complement. Altern. Med.* **2011**, *11*, 13. [CrossRef]
21. Re, R.; Pellegrini, N.; Proteggente, A.; Pannala, A.; Yang, M.; Rice-Evans, C. Antioxidant activity applying an improved ABTS radical cation decolorization assay. *Free Radic. Biol. Med.* **1999**, *26*, 1231–1237. [CrossRef]
22. Bonamigo, T.; Campos, J.F.; Alfredo, T.M.; Balestieri, J.B.P.; Cardoso, C.A.L.; Paredes-Gamero, E.J.; Souza, K.D.P.; dos Santos, E.L. Antioxidant, Cytotoxic, and Toxic Activities of Propolis from Two Native Bees in Brazil: *Scaptotrigona depilis* and *Melipona quadrifasciata anthidioides*. *Oxidative Med. Cell. Longev.* **2017**, *2017*, 1038153. [CrossRef]
23. Dengg, M.; van Meel, J.C. Caenorhabditis elegans as model system for rapid toxicity assessment of pharmaceutical compounds. *J. Pharmacol. Toxicol. Methods* **2004**, *50*, 209–214. [CrossRef] [PubMed]
24. Xiao, L.; Li, H.; Zhang, J.; Yang, F.; Huang, A.; Deng, J.; Liang, M.; Ma, F.; Hu, M.; Huang, Z. Salidroside Protects Caenorhabditis elegans Neurons from Polyglutamine-Mediated Toxicity by Reducing Oxidative Stress. *Molecules* **2014**, *19*, 7757–7769. [CrossRef] [PubMed]
25. Tsao, R. Chemistry and Biochemistry of Dietary Polyphenols. *Nutrients* **2010**, *2*, 1231–1246. [CrossRef]
26. Ganeshpurkar, A.; Saluja, A.K. The Pharmacological Potential of Rutin. *Saudi Pharm. J.* **2017**, *25*, 149–164. [CrossRef]
27. Gulcin, İ. Antioxidants and antioxidant methods: An updated overview. *Arch. Toxicol.* **2020**, *94*, 651–715. [CrossRef] [PubMed]
28. Enogieru, A.B.; Haylett, W.; Hiss, D.C.; Bardien, S.; Ekpo, O.E. Rutin as a Potent Antioxidant: Implications for Neurodegenerative Disorders. *Oxidative Med. Cell. Longev.* **2018**, *2018*, 6241017. [CrossRef] [PubMed]
29. Du, J.; Cullen, J.J.; Buettner, G.R. Ascorbic acid: Chemistry, biology and the treatment of cancer. *Biochim. Biophys. Acta Rev. Cancer* **2012**, *1826*, 443–457. [CrossRef] [PubMed]
30. Bernatoniene, J.; Kopustinskiene, D.M. The Role of Catechins in Cellular Responses to Oxidative Stress. *Molecules* **2018**, *23*, 965. [CrossRef]
31. Caro, A.A.; Davis, A.; Fobare, S.; Horan, N.; Ryan, C.; Schwab, C. Antioxidant and pro-oxidant mechanisms of (+) catechin in microsomal CYP2E1-dependent oxidative stress. *Toxicol. In Vitro* **2019**, *54*, 1–9. [CrossRef]
32. Lu, N.; Chen, P.; Yang, Q.; Peng, Y.-Y. Anti- and pro-oxidant effects of (+)-catechin on hemoglobin-induced protein oxidative damage. *Toxicol. In Vitro* **2011**, *25*, 833–838. [CrossRef]
33. Park, H.-E.H.; Jung, Y.; Lee, S.-J.V. Survival assays using Caenorhabditis elegans. *Mol. Cells* **2017**, *40*, 90–99. [CrossRef]

34. Moreno-Arriola, E.; Cárdenas-Rodríguez, N.; Coballase-Urrutia, E.; Pedraza-Chaverri, J.; Carmona-Aparicio, L.; Ortega-Cuellar, D. *Caenorhabditis elegans*: A Useful Model for Studying Metabolic Disorders in Which Oxidative Stress Is a Contributing Factor. *Oxidative Med. Cell. Longev.* **2014**, *2014*, 705253. [CrossRef]
35. Cordeiro, L.M.; Machado, M.L.; da Silva, A.F.; Baptista, F.B.O.; da Silveira, T.L.; Soares, F.A.A.; Arantes, L.P. Rutin protects Huntington's disease through the insulin/IGF1 (IIS) signaling pathway and autophagy activity: Study in Caenorhabditis elegans model. *Food Chem. Toxicol.* **2020**, *141*, 111323. [CrossRef]
36. Cordeiro, L.M.; Soares, M.V.; da Silva, A.F.; Machado, M.L.; Baptista, F.B.O.; da Silveira, T.L.; Arantes, L.P.; Soares, F.A.A. Neuroprotective effects of rutin on ASH neurons in *Caenorhabditis elegans* model of Huntington's disease. *Nutr. Neurosci.* **2022**, *25*, 2288–2301. [CrossRef] [PubMed]
37. Kampkötter, A.; Nkwonkam, C.G.; Zurawski, R.F.; Timpel, C.; Chovolou, Y.; Wätjen, W.; Kahl, R. Investigations of protective effects of the flavonoids quercetin and rutin on stress resistance in the model organism Caenorhabditis elegans. *Toxicology* **2007**, *234*, 113–123. [CrossRef] [PubMed]
38. Kandemir, F.M.; Ozkaraca, M.; Yildirim, B.A.; Hanedan, B.; Kirbas, A.; Kilic, K.; Aktas, E.; Benzer, F. Rutin attenuates gentamicin-induced renal damage by reducing oxidative stress, inflammation, apoptosis, and autophagy in rats. *Ren. Fail.* **2015**, *37*, 518–525. [CrossRef]
39. Magalingam, K.B.; Radhakrishnan, A.; Haleagrahara, N. Protective effects of quercetin glycosides, rutin, and isoquercetrin against 6-hydroxydopamine (6-OHDA)-induced neurotoxicity in rat pheochromocytoma (PC-12) cells. *Int. J. Immunopathol. Pharmacol.* **2015**, *29*, 30–39. [CrossRef] [PubMed]
40. Jattujan, P.; Srisirirung, S.; Watcharaporn, W.; Chumphoochai, K.; Kraokaew, P.; Sanguanphun, T.; Prasertsuksri, P.; Thongdechsri, S.; Sobhon, P.; Meemon, K. 2-Butoxytetrahydrofuran and Palmitic Acid from Holothuria scabra Enhance *C. elegans* Lifespan and Healthspan via DAF-16/FOXO and SKN-1/NRF2 Signaling Pathways. *Pharmaceuticals* **2022**, *15*, 1374. [CrossRef]
41. Wei, C.-C.; Yen, P.-L.; Chang, S.-T.; Cheng, P.-L.; Lo, Y.-C.; Liao, V.H.-C. Antioxidative Activities of Both Oleic Acid and Camellia tenuifolia Seed Oil Are Regulated by the Transcription Factor DAF-16/FOXO in Caenorhabditis elegans. *PLoS ONE* **2016**, *11*, e0157195. [CrossRef]

MDPI AG
Grosspeteranlage 5
4052 Basel
Switzerland
Tel.: +41 61 683 77 34

Antioxidants Editorial Office
E-mail: antioxidants@mdpi.com
www.mdpi.com/journal/antioxidants

www.ingramcontent.com/pod-product-compliance
Lightning Source LLC
Chambersburg PA
CBHW041424210926
51850CB00013B/3348
* 9 7 8 3 7 2 5 8 3 2 0 8 8 *